W

THE GEOGRAPHY OF URBAN TRANSPORTATION

The Geography of Urban Transportation

THIRD EDITION

Edited by

Susan Hanson
Genevieve Giuliano

THE GUILFORD PRESS
New York   London

© 2004 The Guilford Press
A Division of Guilford Publications, Inc.
72 Spring Street, New York, NY 10012
www.guilford.com

Printed in the United States of America

This book is printed on acid-free paper.

Last digit is print number: 9   8   7   6   5   4   3

Library of Congress Cataloging-in-Publication Data

The geography of urban transportation / edited by Susan Hanson, Genevieve Giuliano.—3rd ed.
  p. cm.
Includes bibliographical references and index.
ISBN 1-59385-055-7 (hardcover: alk. paper)
1. Urban transportation.   I. Hanson, Susan, 1943– .   II. Giuliano, Genevieve.
HE305.G46 2004
388.4—dc22

                                                                    2004004578

To our children

Kristin and Bill
Erik and Meg—Will and Luke

Dante and Tammy
Marcello and Christine

and the memory of
Vincent

Preface

Transportation is arguably the backbone of urban life; without it, activities in cities grind to a halt. Transportation is also the source of many seemingly intractable urban problems, namely, congestion, pollution, inequality, and reliance on fossil fuels. This third edition of *The Geography of Urban Transportation* sustains the fundamental line of argument that informed the book's previous incarnations: how citizens and policymakers conceptualize a problem informs how they go about studying and analyzing it, and analysis, in turn, informs policy formulation, decision making, and ultimately the shape of the urban transportation system itself.

For many years the urban transportation problem was equated with congestion, and the analytical structure devised to address the problem (the four-stage urban transportation model system) aimed to guide the building of capacity-increasing new infrastructures, most often highways. However, with growing concerns about air pollution and other environmental damage, mobility of those without access to a private vehicle, and the long-term consequences of an urban transport system almost entirely dependent upon the private vehicle, pressure

for policy change accumulated throughout the 1980s. The passage of the Intermodal Surface Transportation Efficiency Act in 1991 was the culmination of these forces and symbolized a fundamental change in perspective. The urban transportation problem was no longer conceived of simply as congestion; questions of environmental management, historic preservation, and citizen participation, among others, were placed firmly on the mainstream transportation agenda. In this third edition, we trace out how this shift in thinking has altered the nature of the urban transportation planning process, changed the policy context, and enlarged the scope for citizen input.

Like its predecessor editions, this one is aimed at advanced undergraduates and beginning graduate students. This edition retains the basic three-part structure of the previous two editions. The first part sets the scene by explaining core concepts, providing overviews of passenger and freight movements in the urban context, describing the history of transportation and urban form, and assessing the likely impact of information technology on travel patterns and urban form. The second part introduces students to the urban transportation planning process and contemporary trends in this process, places the planning process within the political (and often politicized) context it seeks to inform, and provides an overview of the use of GIS in transportation planning. In the final part, each chapter takes up a pressing policy issue—public transit, land use, energy, finance, equity, and environmental impacts. The last chapter draws together these policy concerns by thinking about ways that public policy might more effectively engage with managing the automobile.

Compared to the first and second editions, this edition has a sharper focus on policy. We have achieved this altered emphasis by refocusing the contents of each chapter and by adding several new chapters, specifically, one on intercity travel, one on transportation finance, and the final chapter on managing the auto.

The Geography of Urban Transportation encourages students to see the links among problem formulation, research design, analytical approach, and planning decisions. We hope that students can appreciate how the current geography of urban transportation can be understood in large part as the outcome of policy choices, themselves a result of how planners, citizens, business and labor interests, and elected officials have conceptualized problems, envisioned solutions, and taken action. And we hope that understanding will enable students to imagine—and actively work for—new transportation geographies.

Thanks are due to many. First and foremost we thank the contributors, whose research and ideas are the heart of this book. Chapter authors worked patiently with us as we sought to bring consistency and coherence to a book authored by 13 people from many different specialty fields. We are also grateful to Kristal Hawkins at The Guilford Press, who helped to shepherd this volume to completion. We thank Sarah Loy and Jaime Haber for their assistance in preparing the volume for publication. Finally, we would like to acknowledge the many pleasures of collaboration. Susan (who edited the first two rounds of *The Geography of Urban Transportation* solo) greatly enjoyed sharing the job with Gen this time around; as expected, it really did turn out to be more fun working collaboratively.

List of Abbreviations

AVR average vehicle ratio

BRT bus rapid transit

CAA Clean Air Act (first passed in 1971)

CAAA Clean Air Act Amendments (first passed in 1977)

CAFE Corporate Average Fuel Economy

CBD central business district

CO carbon monoxide

CO$_2$ carbon dioxide

DOT Department of Transportation

DRAM disaggregate residential allocation module

EMPAL employment allocation module

EPA Environmental Protection Agency

FAZ forecast analysis zone

FHA Federal Highway Administration

FTA Federal Transit Administration

GAO General Accounting Office

GHG greenhouse gases

GIS geographic information system

GIS-T geographic information system for transportation

GMA Growth Management Act (passed in the State of Washington in 1990)

GPS global positioning system

HOV high-occupancy vehicle (e.g., carpool)

IPCC Intergovernmental Panel on Climate Change

ISTEA Intermodal Surface Transportation Efficiency Act (passed in 1991; also referred to as the Surface Transportation Act of 1991)

ITLUP Integrated Transportation and Land Use Package

ITS Intelligent Transportation System

IVHS Intelligent Vehicle Highway System

LEV low-emission vehicle

LOS level of service

LPG liquefied petroleum gases

LRT light-rail transit

MPG miles per gallon

MPO metropolitan planning organization

MSA metropolitan statistical area (a city and its surrounding suburbs)

MTA Metropolitan Transportation Authority

MTP Metropolitan Transportation Plan

NAAQS National Ambient Air Quality Standards

NEPA National Environmental Policy Act (passed in 1969)

NHTS National Household Transportation Survey

NHTSA National Highway Traffic Safety Administration

NPTS Nationwide Personal Travel Survey

OECD Organisation for Economic Cooperation and Development

OPEC Organization for Petroleum Exporting Countries

PECAS Production, Exchange, and Consumption Allocation Model

PEM proton exchange membrane

PMT person miles traveled

PNGV Partnership for a New Generation of Vehicles

POV privately owned vehicle

PSRC Puget Sound Regional Council

RTI Road Transport Informatics

RTPO Regional Transportation Planning Organization

SACMET Sacramento Metropolitan Travel Model

SAZ small-area zone

SIP State Implementation Plan

SO$_2$ sulfur dioxide

SOV single-occupancy vehicle

STP Surface Transportation Program

STPP Surface Transportation Policy Program

SUV sports utility vehicle

SULEV super low-emission vehicle

TAM transportation adequacy measure

TAZ traffic (or travel) analysis zone

TCM transportation control measure

TEA-21 Transportation Equity Act for the 21st Century

TIP Transportation Improvement Plan

TOD transit-oriented development

TNR transportation needs report

TSM transportation systems management

TSP transportation strategic plan

USDOT United States Department of Transportation

USEPA United States Environmental Protection Agency

USGS United States Geological Survey

UTMS Urban Transportation Modeling System

UTPS Urban Transportation Planning System

V/C ratio volume-to-capacity ratio

VMT vehicle miles traveled

VOC volatile organic compound

ZEV zero-emission vehicle

Contents

III. POLICY ISSUES

PART I

SETTING THE SCENE

CHAPTER I

The Context of Urban Travel

Concepts and Recent Trends

SUSAN HANSON

Many trace the dawn of the modern civil rights movement in the United States to events that transpired on a city bus in Montgomery, Alabama, on December 1, 1955, when Rosa Parks refused an order from a municipal bus driver to give up her seat to a white man. Her arrest and the subsequent Montgomery bus boycott (1955–1959), in which blacks refused to patronize the segregated city bus system, proved the power of collective action and brought Martin Luther King, Jr., to prominence. That the civil rights movement should have been born on a city bus is just one measure of how urban transportation is woven into the fabric of U.S. life.

Can you imagine what life would be like without the ease of movement that we now take for granted? The blizzards that periodically envelop major cities give individuals a fleeting taste of what it is like to be held captive (quite literally) in one's own home (or some other place) for several days. With roads buried under 6 feet of packed snow, you cannot obtain food, earn a living, get medical care for a sick child, or visit friends. As recent earthquakes in California and floods in the Midwest have illustrated, the collapse of a single bridge can disrupt the daily lives of tens of thousands of people and hundreds of businesses. The blackout that enveloped much of the U.S. Northeast and Midwest for a few days in August 2003 brought life to a standstill.

Transportation is vital to U.S. urban life and to life in other places as well because it is an absolutely necessary means to an end: It allows people to carry out the diverse range of activities that make up daily life. Because cities consist of spatially separated, highly specialized land uses—food stores, laundromats, hardware stores, banks, drugstores, hospitals, libraries, schools, post offices, and so on—people must travel if they want to obtain necessary goods and services. Moreover, home and work are in the same location for only a few people (about 3.3% of the U.S. workforce in 2000), so that to earn an income as well as to spend it one must travel.

Although people do occasionally engage in travel for its own sake (as in taking a Sunday drive or a family bike ride), most urban travel occurs as a by-product of trying to accomplish some other (nontravel) activity such as work, shopping, or mailing a letter. Only about half of 1% of all trips in the United States are trips for pleasure driving (U.S. Department of Transportation, Federal Highway Administration, 1994, p. 4-72). In this sense, the demand for urban transportation is referred to as a *derived demand* because it is derived from the need or desire to do something else. A trade-off always exists between doing an activity at home (such as eating a meal, watching a video, or doing laundry) or paying the costs of movement to accomplish that activity or a similar one somewhere else (such as at a restaurant, a movie theater, or a laundromat).

All movement incurs a cost of some sort, which is usually measured in terms of time or money. Some kinds of travel, such as that by auto, bus, or train, incur both time and monetary costs; other trips, such as those made on foot, involve an outlay primarily of time. In deciding which mode(s) to use on a given trip (e.g., car or bus), travelers often trade off time versus money costs, as the more costly travel modes are usually the faster ones. A trade-off is also involved in the decision to make a trip: the traveler weighs the expected benefits to be gained at the destination against the expected costs of getting there. Each trip represents a triumph of such anticipated benefits over costs, although for the many trips that are made out of habit this intricate weighing of costs and benefits does not occur before each and every trip.

Although transportation studies have emphasized the costs of travel, recent research suggests that for many people daily mobility can also be a source of pleasure and is not simply an aggravation to be endured in order to accomplish a necessary activity, like going to work. Some people, for example, enjoy the time they spend alone in the car on the commute, saying it's the only time during the day they have to themselves. Contrary to most transportation theory, these people don't seek to minimize the time or distance traveled on the journey to work or other trips (Mokhtarian, Solomon,

& Redmond, 2001). In this case, the demand for travel is not purely "derived" from the demand to accomplish other activities, but something undertaken for its own good.

This chapter introduces some key concepts in urban transportation and sets the stage for the chapters that follow. In particular, I describe (1) the concepts of accessibility, mobility, and equity; (2) certain aspects of the urban context within which travel takes place; (3) recent trends in U.S. travel patterns; and (4) the policy context within which transportation analysis and planning in the United States are set. The overall goal of this book is help you understand the central role of transportation and transportation planning in shaping metropolitan areas.

CORE CONCEPTS

Accessibility and Mobility

Two concepts that are central to understanding transportation are accessibility and mobility. *Accessibility* refers to the number of opportunities, also called "activity sites," available within a certain distance or travel time. *Mobility* refers to the ability to move between different activity sites (e.g., from home to grocery store). As the distances between activity sites have become longer (because of lower density settlement patterns), accessibility has come to depend more and more on mobility, particularly in privately owned vehicles (POVs).

Accessibility and Land Use Patterns

Let me give an example from my own neighborhood in inner-city Worcester, Massachusetts. About 40 years ago, many different kinds of activities were located within three blocks of my house: a supermarket, a clothing store, a drugstore, a post office, several churches, a large park, three elementary schools, a bookstore, a bakery, a dry cleaning store, a laundromat, a barbershop, and several restaurants. In addition, several large manufacturing employers (a steel plant, a carpet-making firm, a textile machine manufacturer) were located on the other side of the residential neighborhood. Anyone who could walk had excellent accessibility to goods and services and even to employment. Access depended on pedestrian mobility rather than vehicular mobility. Although many of these places are still here, the manufacturing companies and the supermarket have closed; food stores in general have become significantly larger and simultaneously fewer in number and farther apart. Our neighborhood no longer has a grocery store within walking distance. Access to food now requires mobility by bus, car, or taxi. The successful creation of ever larger (and increasingly fewer) food stores *depends* on ever-escalating levels of mobility; we can now travel much farther by car, for example, in about the same amount of time it took us to get someplace on foot.

This example illustrates how the need for mobility can be seen as the *consequence* of the spatial separation between different types of land uses in the city, but enhanced mobility can also be seen as *contributing* to increased separation of land uses. Because improved transportation facilities enable people to travel farther in a given amount of time than they could previously, transportation improvements contribute to the growing spatial separation between activity sites (especially between home and work) in urban areas. As you will learn in the ensuing chapters, the goal of transportation planning has been to increase people's mobility, sometimes equating increased mobility with increased accessibility. Ausabel, for example, observes that transportation planners have engineered systems that seem "coded to seek low-cost speed to enable individuals to maximize range" (1992, p. 879). Planners and policymakers now recognize, however, that, through attention to land use planning, that is, by creating high-density urban neighborhoods much like my Worcester

neighborhood of 40 years ago, accessibility can be achieved without increasing mobility.

This symbiotic relationship between transportation and land use is one reason geographers are interested in urban transportation. One could never hope to understand the spatial structure of the metropolis or to grasp how it is changing without a knowledge of movement patterns. The accessibility of places has a major impact upon their land values (and hence the use to which the land is put), and the location of a place within the transportation network determines its accessibility. Thus, in the long run, the transportation system (and the travel on it) shapes the land use pattern. In Chapter 2 Thomas R. Leinbach shows how this principle works at the intercity scale, and in Chapter 3 Peter O. Muller provides numerous historical examples of this interaction between transportation innovation and urban land use patterns at the intrametropolitan scale. In the short run, however, the existing land use configuration helps to shape travel patterns. The intimate relationship between transportation and land use is explicitly acknowledged by the fact that at the heart of every city's long-term land use plan is a transportation plan. In Chapter 4 Donald G. Janelle explores the fascinating question of how information technologies, such as the Internet, cell phones, and video conferencing, are changing the relationship between distance and accessibility, and therefore the relationship between accessibility and land use.

Measuring Accessibility

We can talk about the accessibility of *places* (i.e., how easily certain places can be reached) or of *people* (i.e., how easily a person or a group of people can reach activity sites). As we saw in the example above, an individual's level of accessibility will depend largely on where activity sites are located vis-à-vis the person's home and the transportation network, but it will also be affected by when such sites are open and even by how much time someone can spare for making trips. Urban planners and scholars have long argued that the ease with which people can reach employment locations, retail and service outlets, and recreational opportunities should be considered in any assessment of the health of a city. They have implied that accessibility should be a central part of any measure of the quality of life (see, e.g., Chapin, 1974; Scott, 2000; Wachs & Kumagi, 1973). Measuring accessibility in a meaningful way can be difficult, however.

Personal accessibility is usually measured by counting the number of activity sites (also called "opportunities") available at a given distance from the person's home and "discounting" that number by the intervening distance. Often accessibility measures are calculated for specific types of opportunities, such as shops, employment places, or medical facilities. One measure of accessibility is presented in Equation 1,

$$A_i = \sum_j O_j d_{ij}^{-b} \tag{1}$$

where A_i is the accessibility of person i, O_j is the number of opportunities at distance j from person i's home, and d_{ij} is some measure of the separation between i and j (this could be travel time, travel costs, or simple distance). Such an accessibility index is a measure of the number of potential destinations available to a person and how easily they can be reached. Accessibility is usually assessed in relation to the person's home because that is the base from which most trips originate; personal accessibility indices could (and perhaps should) also be computed around other important bases, such as the workplace.

The accessibility of a place to other places in the city can be measured in a similar way, using Equation 2:

$$A_i = \sum_j O_j d_{ij}^{-b} \tag{2}$$

where A_i is the accessibility of zone i, O_j is the number of opportunities in zone j, and

d_{ij} is, as before, a measure of the separation between i and j.

Although Equations 1 and 2 are structurally alike, the difference between the two is important. The first measures the accessibility of individuals, and the second indicates the accessibility of places (or zones) within a city. The second measure treats all those living in zone i as if they have the same level of accessibility to activity sites in the city; it does not distinguish among different types of people within a zone, such as those with or without a car.

Both these measures of accessibility are highly simplified representations; neither really addresses mobility nor includes dimensions such as the ability to visit places at different times of day. A third measure—that of space–time autonomy—takes both accessibility and mobility into consideration; it is a more satisfying measure conceptually than measure (1) but far more difficult operationally. The concept of space–time autonomy has been developed in the context of time geography and focuses on the constraints that impinge on a person's freedom of movement (Hagerstrand, 1970). These constraints include:

- *Capability constraints*—the limited ability to perform certain tasks within a given transportation technology and the fact that we can be in only one place at a time; for example, if the only means of transport available to you are walking and biking, the number of activity sites you can visit in, say, half an hour is lower than it would be if you had access to a car.
- *Coupling constraints*—the need to undertake certain activities at certain places with other people; for instance, that lunch meeting with your boss can only be scheduled when you both can be in the same place at the same time.
- *Authority constraints*—the social, political, and legal restrictions on access—for example, you can only conduct

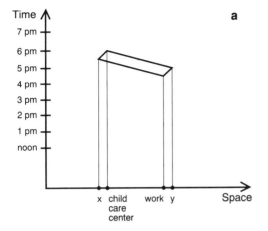

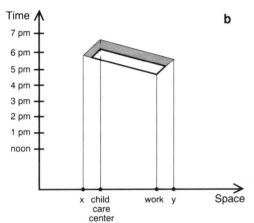

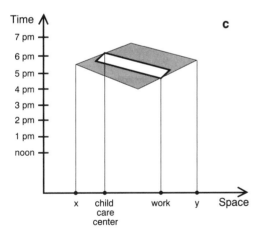

business at the bank or the post office during the hours they are open, and certain locations are off-limits to citizens without access permits.

Your access to places and activities is restricted by these constraints.

A measure of an individual's space–time autonomy is the *space–time prism*, a visual representation of the possibilities in space and time that are open to a person, given certain constraints (see Figure 1.1). The larger the prism, shown in each frame of Figure 1.1 as a parallelogram, the greater the individual's space–time autonomy in a specific situation.

Figure 1.1a, for example, shows the space–time autonomy for a person who is currently (at 5:00 P.M.) at work and who must arrive at the childcare center no later than 6:00 P.M. to pick up his daughter; the distance between these two locations is shown on the "space" axis. Somewhere in between he must stop at a food store to buy soup, bread, cheese, and a lottery ticket. In addition to these location and time constraints, the father in this example must conduct all travel either on foot or by bicycle. The slope of the lines in Figure 1.1 shows the maximum speed (in 1.1a, presumably by bicycle) that this person can travel. The prism outlines the envelope within which lies the set of all places that are accessible to him given these constraints. If no food store exists between *x* and *y* (shown on the "space" axis), then he lacks accessibility in this instance.

The concept of a space–time prism can also illustrate how changes in constraints can affect accessibility. If, in this example,

FIGURE 1.1. One measure of space–time autonomy is the space–time prism. (a) The prism defines the set of possibilities that are open to this father who must travel on foot or by bike from his place of work where he is at 5:00 P.M. to the childcare center, which closes at 6:00 P.M. (b) Effects of extended hours; the shading shows the increases in space–time autonomy if the childcare center were to extend its hours from 6:00 P.M. to 6:30 P.M. (c) Effects of car availability; the shading shows the increase in space–time autonomy if a car is available, thereby permitting higher speed travel.

the childcare center were to extend its hours until 6:30 P.M., the prism defining the set of possibilities would be enlarged (see shaded area in Figure 1.1b), and this man's space–time autonomy would be increased. Or suppose he traveled by car: he could then travel farther in the same amount of time, and the prism would therefore be larger. Notice that this greater speed is shown by the slope of the lines in Figure 1.1c, which is not as steep as in 1.1a and 1.1b where he is assumed to be traveling by bike. The shading in Figure 1.1c indicates the increase in space–time autonomy that would result from the availability of a car. Notice that the outer spatial limits of possibilities, shown in each case by x and y on the space axis, shift outward as constraints are eased. In general, the prisms show the relationship between time and space, and you can see that as the time constraints facing this father are reduced, the greater the space within which he can move.

Many factors can, then, affect space–time autonomy. For example, flextime work schedules, longer store hours, and purchasing a second car all enhance space–time autonomy by adding margins to the space–time prism. Lower speed limits, rigid school hours, and traffic congestion all constrain choice. Large families impose coupling constraints, which often affect women more than men. Babysitters, daycare centers, and children's growing up all reintroduce issues of space–time autonomy. You can see, however, that measuring space–time autonomy by including all of these relevant factors would be complicated.

Increasing people's space–time autonomy seems desirable in that it implies a greater accessibility to places and more discretion for spending one's time. We might question, however, the need for ever-increasing space–time autonomy and ever-increasing personal mobility. One question that transportation geographers and many others have begun to ponder is whether or not there is such a thing as too much mobility!

Equity

As we can see from the concept of space–time autonomy, someone's ability to reach places depends only in part on the relative location of those places; it also depends on *mobility*, the ability to move to activity sites, which in the United States usually requires an automobile. We have seen how the spatial organization of contemporary society demands—indeed assumes—mobility; yet not all urban residents enjoy the high level of mobility that the contemporary city requires for the conduct of daily life. Assessing the equity of a transportation system or a transportation policy requires that we consider who gains accessibility and who loses it as a result of how that system or policy is designed; it requires that we consider to what degree people's travel patterns are the outcomes of choice or constraints. How are the costs and benefits of transportation systems shared among different groups of people?

At the time of the Montgomery, Alabama, bus boycott in the late 1950s, as now, a disproportionate share of people with fewer economic resources relied on buses for transportation. At the start of the boycott in 1955, blacks comprised 45% of Montgomery's population but fully 75% of the bus ridership, and the majority of bus riders were women (Powledge, 1992; for an excellent treatment of the Montgomery bus boycott, see Garrow, 1988). People without access to cars are especially likely to lack the mobility necessary to reach job locations or other activity sites. In fact, lower-income people travel significantly less (they "consume" less transportation) than do higher-income people. In 2000, households with incomes under $25,000 were 22.5% of all U.S. households, but accounted for only 15.2% of all vehicle miles traveled.[1] Equity issues are so important in transportation that we devote a chapter to this topic (see Deka, Chapter 12, this volume).

This lack of mobility among certain groups of people is one part of the urban transportation problem. Because transportation is so essential to the very fabric of urban life, transportation issues and problems are inextricably bound up with other societal issues and problems such as economic well-being, social inequities, health, and environmental pollution. Before we consider these other aspects of the urban transportation problem, the next section of the chapter provides some background on the context within which travel takes place in the North American city and considers how this U.S. context differs from that in other countries.

THE CHANGING URBAN CONTEXT

How have U.S. cities been changing in recent decades? In particular, how have residential and employment patterns been changing? In addition to looking at patterns for U.S. cities as a whole, we focus on one city in particular—Worcester, Massachusetts—because the trends revealed here are similar to those in other U.S. cities and allow us to examine intraurban patterns and trends.

Residential Patterns

Table 1.1 presents data on some important demographic trends from 1970 to 2000 for

TABLE 1.1. Demographic Trends, Metropolitan Areas in the United States, 1970–2000

	1970[b]	2000
Population of MSAs[a]	139,418,811	225,981,711
Number of households in MSAs	43,862,993	84,351,108
Percentage of households in MSAs that are single-person	18.1	25.9
Percentage of MSA population living in central city	45.8	37.8
Percentage of households with no vehicle		
MSA	18.6	11.0
City	28.4	18.4
Suburbs	9.2	6.2
Percentage of households with more than one vehicle		
MSA	35.6	54.2
City	26.2	41.9
Suburbs	44.7	62.0
Percentage of population over 65 years of age		
MSA	9.3	11.9
City	10.8	11.5
Suburbs	8.0	12.1
Percentage of families below the poverty level		
MSA	8.5	8.7
City	11.0	13.6
Suburbs	6.3	6.0
Percentage of families headed by women		
MSA	11.5	18.0
City	15.5	25.0
Suburbs	8.3	14.2

[a]MSAs, Metropolitan Statistical Areas, which includes a central city/cities and the surrounding suburbs.
[b]For 1970, figures refer to SMSAs (Standard Metropolitan Statistical Areas) as defined at that time.
Source: Adapted from the Censuses of Population and Housing (U.S. Bureau of the Census, 1970, 2000).

U.S. metropolitan areas as a whole, and Table 1.2 contains data for the Worcester, Massachusetts, metropolitan area for the same variables from 1960 to 2000. The census figures in these two tables disclose a number of trends that hold important implications for travel patterns and for access, mobility, and urban transportation planning. Although the following discussion focuses on Worcester, you should put Worcester in the context of other U.S. metro areas by referring frequently to Table 1.1.

First, while the population of the Worcester metropolitan area (Metropolitan Statistical Area, or MSA) as a whole has grown somewhat (by 55% between 1960 and 2000), the number of households and the number of single-person households have increased dramatically, by 101% and 299%, respectively). The proportion of single-person households increased from only 13.6% of all MSA households in 1960 to nearly 27% of all households in 2000. The greater increase in households relative to population has implications for trip making because the number of trips made per person per day generally declines as household size increases. The trend to more households and more single-person households contributes significantly, then, to an overall growth in travel.

A second trend is that despite the increase in population and in households, the proportion of the population residing in the

TABLE 1.2. Demographic Trends, Worcester, Massachusetts, 1960–2000

	1960	1970	1980	1990	2000
Population of MSA[a]	323,306	344,320	372,940	436,905	502,511
Number of households in MSA	94,680	104,694	130,785	161,350	191,011
Percentage of households in MSA that are single-person	13.6	17.6	23.2	24.5	26.9
Percentage of MSA population living in central city	57.7	51.3	43.4	38.9	34.4
Percentage of households with no vehicle					
MSA	22.0	17.7	14.5	11.9	10.5
City	29.3	26.2	23.0	20.5	18.1
Suburbs	No data	7.6	7.5	6.3	6.3
Percentage of households with more than one vehicle					
MSA	15.2	28.6	43.2	52.4	53.1
City	11.1	19.4	29.1	37.6	36.9
Suburbs	No data	39.3	53.0	62.1	61.9
Percentage of population over 65 years of age					
MSA	11.9	12.0	13.4	14.2	13.3
City	13.6	14.7	16.3	16.0	14.1
Suburbs	9.5	9.2	11.3	13.0	12.8
Percentage of families below the poverty level					
MSA	No data	5.4	7.5	6.4	7.1
City		7.1	11.2	12.2	14.1
Suburbs		3.7	4.7	3.2	4.0
Percentage of families headed by women					
MSA	No data	11.3	15.1	16.0	11.6
City		15.2	21.1	24.2	15.6
Suburbs		7.2	10.9	11.5	9.4

[a]MSA, Metropolitan Statistical Area, which includes the City of Worcester and the surrounding suburbs.
Source: Adapted from the Censuses of Population and Housing (U.S. Bureau of the Census, 1960, 1970, 1980, 1990, 2000).

central city has steadily declined. A larger proportion of the metropolitan population (66% in 2000 vs. only 42% in 1960) now lives in the lower-density suburbs, which are more difficult to serve efficiently with public transportation. It's worthwhile looking at the data in Table 1.2 for the decades intervening between 1960 and 2000 because you can see how persistent the move to suburbanization has been, with an ever-diminishing proportion of the metropolitan population living in the city. As later chapters in this book make clear, urban and suburban areas have different transportation needs and priorities.

Third, although the proportion of households having no vehicle has dropped since 1960 both for the MSA and for the City of Worcester, the percentage of households without a car has remained noticeably higher in the city than in the suburbs. This latter point is to be expected, given the higher incidence of elderly and low-income households in the central city and given the greater availability of public transportation there. Nevertheless, despite the fact that a smaller proportion of households is now carless than was the case in 1960 ("only" 10.5% in 2000 vs. 22% in 1960), many people must still rely for mobility upon the bus, taxis, a bicycle, their own feet, or rides from other people. Fourth, although the proportion of carless households has declined, the proportion of households with more than one vehicle has grown in both city and suburbs; by 2000, more than half of Worcester-area households had more than one car.

A fifth trend that is evident from examining Table 1.2 is that since 1960 the number (and proportion) of households that are likely to have special transportation needs—the elderly, the poor, and female-headed households—has risen somewhat, and certainly has not declined. Throughout this period, a higher proportion of the central-city population than of the suburban population has been elderly, though the central-city–suburban gap has been closing. In fact, the proportion of the MSA's elderly who live in the central city has been declining over the past 40 years—from 66% of the elderly in 1960 to 36% in 2000. This means that the number of elderly people living in the suburbs has increased markedly. Because some older people do not drive, their presence in the suburbs—where bus service is often infrequent or nonexistent—raises questions about how the mobility needs of this group can be met. The travel problems of single-parent households, headed mostly by women, stem from the difficulty of running a household single-handedly; earning an income, shopping, obtaining medical care and childcare all must be done by the one adult in the household, sometimes without the aid of an automobile.

Employment Patterns

Since the 1960s, jobs have been decentralizing from the central city to the suburbs. Traditionally, especially from the standpoint of transportation planning, the suburbs were viewed as bedrooms for the central-city workforce. Radial transportation systems, focused on the urban core, were organized in large part around moving workers from the suburbs to the central city in the morning and back to the suburbs again in the evening. But this simple pattern now describes only a small portion of current reality. In Worcester in 1960, for example, 42% of suburban workers had jobs in the central city; by 2000 only 22% of employed people living in the suburbs worked in the central city. Similarly, the proportion of the metropolitan labor force that works in the City of Worcester as opposed to surrounding suburbs has declined from more than two-thirds in 1960 to only about one-third in 2000.[2]

Hughes (1991) has documented the extent to which employment has moved from central-city Newark, New Jersey, and into the surrounding region. Although the Newark region as a whole experienced considerable job growth in the 30 years after 1960, the spatial distribution of employment shifted dramatically within the region, from the central city to the suburbs. Central-city job loss coupled with suburban job growth makes access to employment extremely difficult for people who live in the central city but do not have a car. In his study Hughes documents how relatively few suburban jobs in the Newark region can be reached by people living in central Newark without a car; they would have to take a commuter train, and even then, how do they reach the employment site from a suburban train station?

Hughes links this decentralization of employment over the past few decades to the increase in poverty in inner-city Newark. Clearly, as we saw in the case of Worcester, large numbers of residences as well as jobs have been moving to the suburbs in the past four decades. But because of the unequal access of different groups of people to suburban housing, not all social groups have been able to decentralize to the same degree. In particular, low incomes as well as racial discrimination in the housing market have prevented many people, especially those from minority groups, from moving to the suburbs. Hughes's analysis, as well as work by other scholars (e.g., Wilson, 1987), underlines how the reality of residential segregation in U.S. cities, together with changes in job location, has important implications for people's access—or lack of access—to employment opportunities. The term *spatial mismatch* refers to this "mismatch" between inner-city residential location and suburban job location, without the automobility needed to "connect the dots" (see Holzer, 1991, and Mouw, 2000, for reviews of the spatial mismatch literature).

In a detailed study of the Boston metropolitan area, Shen (2001) extends and deepens our understanding of the spatial access of low-skilled job seekers to employment. In particular, Shen argues that analysts should focus on the location of job *openings* rather than on the location of *employment* as Hughes (1991) did, and he shows that preexisting employment, con-

centrated in the central city, is the main source of job openings. Shen's analysis also demonstrates that residential location (e.g., city vs. suburb) is not as important as transportation mode is in accounting for differences in job seekers' access to jobs. That is, job seekers who travel by car will have higher than average accessibility to job openings from just about any residential location, whereas job seekers who depend on public transit will have substantially lower than average accessibility from most residential locations (Shen, 2001, p. 65). Devajyoti Deka (Chapter 12, this volume) takes up these issues of equity in access in greater detail.

The Issue of Scale

Our discussions of residential and employment location patterns provide a useful snapshot of some important urban processes that have transportation implications: the decentralization of population and employment and the concentrations of low-income, carless, and female-headed households in the central city. But the spatial resolution of the information discussed thus far is quite generalized; in tracing out demographic trends in Worcester and employment trends in Newark, we made no finer distinction than that between the central city and the suburbs. For understanding many problems, data at these scales are sufficient, but if transportation policies and facilities are to be tailored to the specific needs of different kinds of people such as the elderly or the carless, then it is important to know as precisely as possible where, *within* the suburbs and *within* the central city, members of these target groups live.

Maps at the level of the census tract (an area comprising 4,000–5,000 people on average) or the census block group (an area within a census tract, encompassing about 1,000 people) reveal the degree to which people and households with certain characteristics are clustered in certain areas within the city or within the suburbs. Census tract maps for the City of Worcester, Massachusetts, provide examples. Compare Figure 1.2, which shows the distribution of the population over age 65 in 1960, with Figure 1.3, which shows the distribution of elderly people in 2000. In these four decades not only has the overall proportion of the population that is over age 65 increased (notice that the bounds on the quartiles are much higher in the map for 2000 [Figure 1.3] than they were in 1960 [Figure 1.2]), the spatial distribution of elderly people within the city has become more diffuse. There are now far more tracts with at least 16% of their populations above the age of 65.

By contrast, the spatial pattern of families below the poverty line has changed relatively little between 1970 (Figure 1.4) and 2000 (Figure 1.5), although the overall percentage of the city's households falling below the poverty line has doubled (from 7.1% in 1970 to 14.1% in 2000) and the proportion of poverty households in high-poverty tracts is dramatically higher in 2000 than it was in 1970 (compare the bounds on the fourth quartile for 1960 with that for 2000). The northwest quadrant of the city has remained a high-income, low-poverty area.

Maps at the scale of the census tract can also highlight demographic characteristics with similar spatial patterns. Compare the 2000 census tract maps of poverty (Figure 1.5), female-headed households (Figure 1.6), and households without a vehicle (Figure 1.7). These maps show a spatial coincidence of female-headed households, the carless, and the poor within the City of Worcester; that is, the same areas tend to have a high percentage of female-headed households, households in poverty, and households without an automobile. These spatial correlations are not as strong for suburban tracts (cf. Figures 1.8 and 1.9). While the overall proportions of female-headed households and households without vehi-

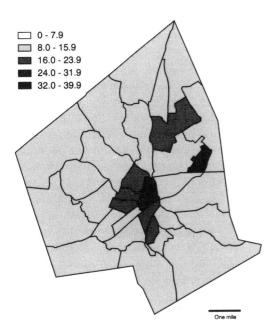

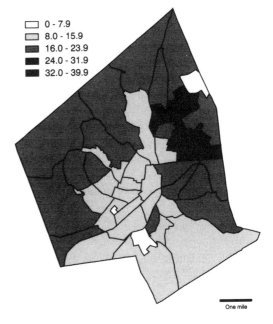

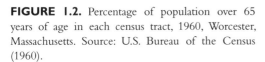

FIGURE 1.2. Percentage of population over 65 years of age in each census tract, 1960, Worcester, Massachusetts. Source: U.S. Bureau of the Census (1960).

FIGURE 1.3. Percentage of population over 65 years of age in each census tract, 2000, in Worcester, Massachusetts. Source: U.S. Bureau of the Census (2000).

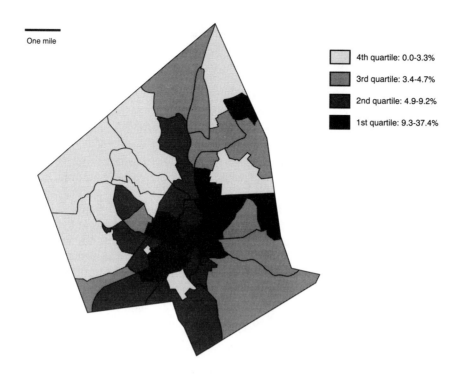

One mile

4th quartile: 0.0-3.3%

3rd quartile: 3.4-4.7%

2nd quartile: 4.9-9.2%

1st quartile: 9.3-37.4%

FIGURE 1.4. Percentage of households in poverty by census tract, 1970, Worcester, Massachusetts. Source: U.S. Bureau of the Census (1970).

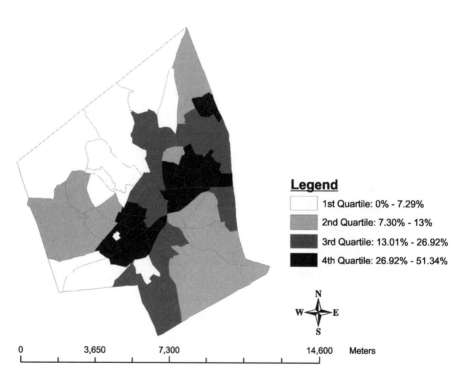

Legend

1st Quartile: 0% - 7.29%

2nd Quartile: 7.30% - 13%

3rd Quartile: 13.01% - 26.92%

4th Quartile: 26.92% - 51.34%

N W E S

0   3,650   7,300   14,600   Meters

FIGURE 1.5. Percentage of households in poverty by census tract, 2000, in Worcester, Massachusetts. Source: U.S. Bureau of the Census (2000).

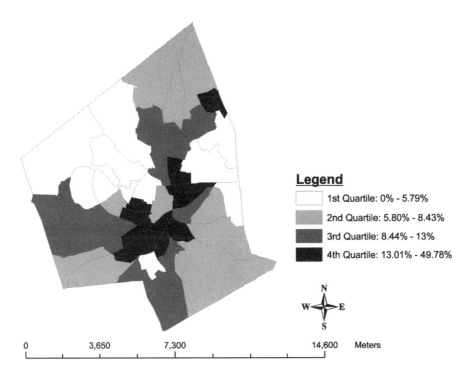

FIGURE 1.6. Percentage of households headed by women in each census tract, 2000, in Worcester, Massachusetts. Source: U.S. Bureau of the Census (2000).

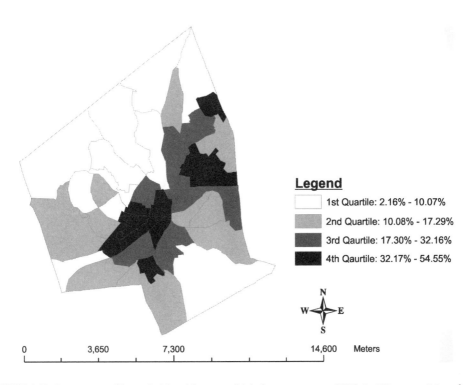

FIGURE 1.7. Percentage of households without a vehicle by census tract, 2000, in Worcester, Massachusetts. Source: U.S. Bureau of the Census (2000).

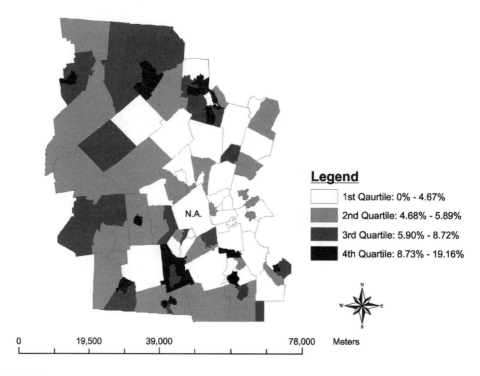

FIGURE 1.8. Percentage of households headed by women in each census tract, 2000, in Worcester, Massachusetts, suburbs. Source: U.S. Bureau of the Census (2000).

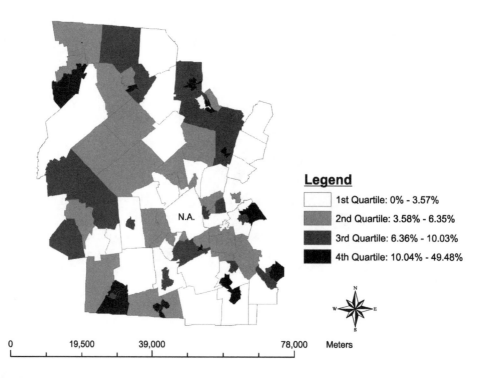

FIGURE 1.9. Percentage of households without vehicle in each census tract, 2000, Worcester, Massachusetts, suburbs. Source: U.S. Bureau of the Census (2000).

cles are lower for suburban tracts than for city census tracts (cf. the legends of Figures 1.6 and 1.8 and of Figures 1.7 and 1.9), what is striking about the suburban maps is the high level of variability among suburban tracts in these two measures (percentage of households headed by women and percentage of households without a vehicle). These maps forcefully demonstrate the folly of generalizing about "the suburbs" as if they were a homogeneous region, even within one MSA.

Policies aimed at providing mobility for low-income carless people might effectively be focused on the census tracts that have the largest percentages of households with these characteristics. You can see that such policies would be far easier to implement in the city, where target tracts are clustered together, than in the suburbs, where they are widely dispersed. Even when an area-targeted policy can be implemented, it provides services to many households who live in the targeted tracts but *do* have a car or are not in poverty, and it would miss the many carless households that do *not* live in the target census tracts. Also, there are numerous *individuals* (rather than households) who are carless for much of the day—people, for example, who remain at home while someone else takes the household's one car to work. The census tract maps are little help in locating these people.

What these maps show is the familiar pattern of people with similar characteristics clustering together in space. What they do not show is the extent to which different types of people live within each census tract or the extent to which certain variables that covary at the area (tract) level also covary at the individual level. For instance, what percentage of female-headed households within a tract are below the poverty line or do not have access to a car?

Consider the three hypothetical census tracts in Figure 1.10. All have an average household income of $30,000, and in this purely fictitious example we have information not only for the tract but also for households within the tract. In the tract in Figure 1.10a, every household's income is identical, exactly $30,000, so the zonal average income is an accurate measure of the individual household incomes within the zone. In Figure 1.10b, however, the zonal average masks two distinct subareas within the zone. In one part, every household's income is $35,000, and in the other, every household's income is $22,000. In Figure 1.10c the $35,000 households are interspersed with the $22,000 households.

The point is that the complete zonal homogeneity depicted in Figure 1.10a simply does not occur in the real world; data for areas (or zones) smooth out whatever internal heterogeneity exists. People in $35,000 households are likely to have quite differ-

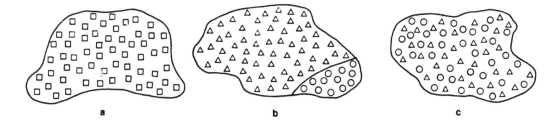

a   b   c

△ **Household with $35,000 income**

□ **Household with $30,000 income**

○ **Household with $22,000 income**

FIGURE 1.10. Hypothetical distributions of households with different income levels.

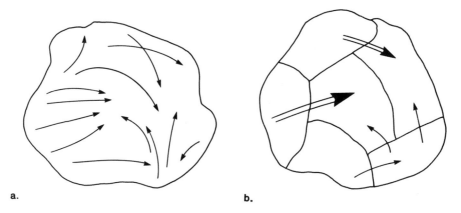

a.   b.

FIGURE 1.11. (a) Individual trips, showing points of origin and destination. (b) Individual trips aggregated by origin and destination zones. Thickness of arrow indicates volume of flow between zones.

ent travel patterns from people in $22,000 households, but the zonal data will portray only an "average" behavior for the people of the zone. The more homogeneous an area is, the closer the zonal data will come to approximating the characteristics of the individuals living within that zone. Census tract boundaries or the boundaries of traffic zones (areal units often used in transportation studies) sometimes split relatively homogeneous areas, adding heterogeneity to the resulting zones. In general, the larger a tract or zone, the less likely it is that all the households living there will share similar characteristics.

The kind of areal (or zonal) data shown in the Worcester maps are useful for providing an overview of population distributions and employment locations within the MSA, for showing where certain population characteristics coincide in space, and for suggesting where certain transportation policies might best be deployed. They are not particularly useful for indicating what characteristics occur together at the household or individual level or for investigating how and why people make travel decisions or how they might respond to a particular transportation policy such as increased headways on a bus route (i.e., longer times between buses) or the installation of a bicycle lane on a certain route. Such questions re-

quire data for individuals rather than for areas.

Transportation analysts use both area (*aggregate*) and individual-level (*disaggregate*) data in studying movement patterns in cities. Studies taking an aggregate approach use data for areal units called "traffic zones" and group separate trips together according to their zone of origin and their zone of destination (see Figure 1.11). The focus is on the flows between zones: how many trips does a particular zone "produce" (in other words, how many trips leave zone i) or "attract" (how many of those trips end in zone j)? What are the characteristics of zone i and zone j that might account for the volume of flows leaving i and arriving at j? Can you explain the size of the flow from i to j in terms of the attractiveness of j and its distance from i? These are some of the questions an aggregate approach to transportation analysis seeks to answer.

In recent years, transportation analysts have begun using individual- and household-level data in addition to areal data. These disaggregate data often use more finely grained spatial codes as well, such as street addresses instead of zones. The conceptual base of the disaggregate approach is the person's daily travel activity pattern, rather than flows between zones. Figure 1.12 shows a schematic, bird's-eye-view of a hypothetical daily travel pattern; you could try mapping your own travel behavior like this over the course of several days. The questions posed in a disaggregate context are aimed at individuals or households rather than at zones: What sociodemographic characteristics are related to a household's level of trip making? What factors affect why a person selects one destination or mode rather than another? What proportion of those who live in the suburbs and work in the central city will shift from commuting by drive-alone auto to a car pool or van pool if a high-speed, high-occupancy vehicle (HOV) lane is installed on their journey-to-work route?

The scale distinction—between aggregate and disaggregate approaches—threads throughout many of the chapters in this book, particularly those in Part II, which focuses on the ways planners analyze movement patterns in order to design and implement changes to the urban transportation system. It is important to understand at the outset the close interdependencies among the scale at which you collect data, the types of models you can build (i.e., how you can simplify and make more comprehensible some of the overwhelming complexities that characterize flow patterns), and the kinds of policy analysis you can carry out. Always ask, At what scale is this transportation issue or problem being conceptualized.

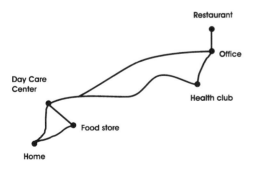

FIGURE 1.12. One person's hypothetical daily travel pattern.

TRENDS IN U.S. TRAVEL PATTERNS

Americans have more mobility, particularly the kind that is provided by motorized vehicles, than people anywhere else on earth. In 1997, Americans logged 4.6 trillion passenger miles of travel by all motorized modes, and 92% of those miles were by automobile; this total amounts to more than 14,000 miles of travel per person per year (U.S. Department of Transportation, Bureau of Transportation Statistics, 1999, pp. 35–37). Moreover, American mobility has been increasing in recent years, despite threats of fuel shortages and concerns about environmental pollution. Between 1975 and 1997, person miles of travel (a person mile of travel, or PMT, is one person traveling 1 mile) in the United States grew by 77%. In part, this increase simply reflects population growth, but even more important are increases in the number of trips made per person and in average trip length (Pisarski, 1992; U.S. Department of Transportation, Bureau of Transportation Statistics, 2001, pp. 85–86).

These trends are tied to two other ones: the increasing number of vehicles on the road and the growing proportion of the U.S. population that is licensed to drive. In 1960, there was only about one car for every 3.8 people in the country; by 2000 there was one car for every 1.3 of the 281 million people in the United States (United Nations, 1961, 2001). The growth in rates of auto ownership has been accompanied by increasing proportions of licensed drivers in the population. Whereas in 1960 only 71.7% of people over age 16 had a driver's license, by 2001 93% of men and 87% of women (90% of the population over age 16) were licensed to drive (U.S. Bureau of the Census, 1962; U.S. Department of Transportation 2003b). In fact, in recent years women have disproportionately accounted for increased levels of travel in the United States. Although women's daily rate of trip making used to be lower than men's, by 1995 it was the same as men's at 4.3 local trips per day, while men's average trip length of 10 miles is still greater than women's with 8 miles (U.S. Department of Transportation, Federal Highway Administration, 2001, pp. 4-9, 4-12).

These increases in licensed drivers and in the number of private vehicles have fueled an increase in vehicle miles traveled (VMT; 1 VMT is 1 mile traveled by a vehicle; if a vehicle has four passengers, then 1 VMT would equal 4 PMTs [passenger miles traveled]). Between 1989 and 1999, VMT increased by almost 30%, or 2.5% annually. At the same time, vehicle miles per capita grew by 16% (U.S. Department of Transportation, Bureau of Transportation Statistics, 2001, p. 87). These increases reflect shifts from walking and transit to autos as well as lower levels of vehicle occupancy. Average vehicle occupancy for all trips fell from 1.9 in 1977 to 1.6 in 1990 and has stayed at that level since (Pisarski, 1992, p. 12, U.S. De-

partment of Transportation, 2003a, p. 11). Fewer people are now passengers and more are drivers, so that, increasingly, more cars are being used to serve the same number of riders. The combined impact of these trends is evident in the steady increase in VMT (see Figure 1.13).

Urban transportation planning has for decades focused largely on the work trip. This overarching concern with the journey to work reflects several factors. First, of all the purposes for which people travel (including work, socializing, recreation, shopping, and personal business), work used to account for the largest proportion of trips. Second, work trips are associated with the morning and evening "peaking problem"; because most people have to be at work between 7:00 A.M. and 9:00 A.M. and leave 8 hours later, work trips have been concentrated in time. The peak load associated with the work trip has placed the greatest demands on the transportation system. As

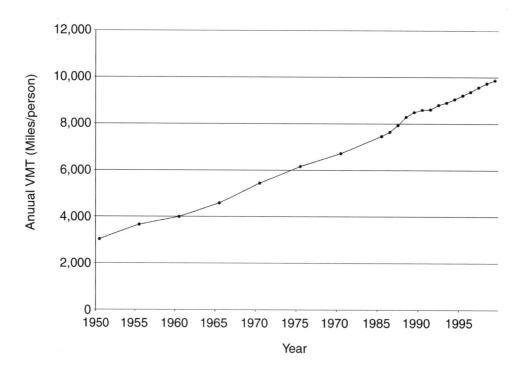

FIGURE 1.13. Annual vehicle miles traveled per person, 1950–1999. Source: U.S. Department of Energy (2001).

you will see in the chapters describing the urban transportation planning process (Chapters 5, 6, and 7, this volume), urban transportation planners have traditionally aimed to provide a transportation system with enough capacity to handle the work trip, under the assumption that such a system can then easily accommodate travel for other purposes. (Obviously, this assumption will be erroneous if nonwork trips have a markedly different spatial configuration from work trips.) A final reason for transportation planners to focus on the work trip is that people tend to travel longer distances for work than for other purposes. The average work trip length in 2001 (13.4 miles) was double the average distance traveled for shopping (6.7 miles), for example (calculated from National Household Travel Survey [NHTS]).

The first two of these patterns have been changing in recent years, however. First, the proportion of travel for nonwork purposes (e.g., socializing, recreation, personal business) has increased significantly. Whereas in 1969, work and work-related travel accounted for more than 41% of all local trips, by 2001, it accounted for only about 15% (U.S. Department of Transportation, 2003a, p. 10). Although travel for all purposes has grown substantially, nonwork travel has increased at a faster rate than work travel has. This increase in nonwork travel can be traced to increases in the number of affluent households and two-earner households, which spur more trips to childcare centers, restaurants, shops, fitness centers, and the like. Another reason is the decline in household size (and therefore a greater number of households for a given population), because "it is the care and upkeep of households, almost independent of the number of persons in the household that frequently governs trip making" (U.S. Department of Transportation, Bureau of Transportation Statistics, 1994, p. 54). Commuting costs are now a smaller proportion of the average household's total transportation cost than in the past.

A second change in work travel is that the morning and evening peaks have been declining, so that even though overall VMT and congestion have grown, travel is becoming more spread out over the day and less concentrated in the peak commuting hours (U.S. Department of Transportation, Bureau of Transportation Statistics, 1999, p. 57; U.S. Department of Transportation, 2003a, p. 11). This change is in part related to the replacement of jobs in manufacturing with service-sector jobs, which tend to be more geographically dispersed and to have more staggered hours of employment. It is also due to the increase in part-time and contingent work, which shifts many work trips to off-peak times.[3] These two trends (the increase in nonwork travel and the decline in rush-hour peaks) suggest that the blame for traffic congestion can no longer be placed solely on the work trip.

Not only do people travel longer distances to work than they do for other purposes, work travel *distances* have been increasing, whereas *travel time* to work has been holding steady. In 1975, average travel distance to work was about 9 miles and the average travel time was approximately 20 minutes (U.S. Bureau of the Census, 1979). In 1995, the average work trip covered 11.6 miles and took 21 minutes (U.S. Department of Transportation, Federal Highway Administration, 2001, p. 6-11). These national averages mask a great deal of variability, of course, among different groups of people, defined, for example, by place of residence (e.g., central city, suburb, small town), age, gender, and travel mode. The relatively constant travel time despite the increase in mileage (and the resulting increase in average speed) reflects in part the improvements in the transportation system (especially highways) that make it easy to travel farther within a given amount of time. But the increase in average work trip speed also reflects shifts in travel mode away from carpooling, mass transit, and walking and into single-occupant vehicles (Pisarski, 1992).

It is important to understand a number of other important characteristics of work travel, including the size and composition of the workforce, the structure of commute trips, and the modes of transportation used. Since the late 1960s, the creation of jobs (and therefore the number of workers) has outpaced population growth. Between 1980 and 1990, for example, the U.S. population grew by less than 10%, but the number of workers increased by more than 19% (U.S. Department of Transportation, Bureau of Transportation Statistics, 1994, p. 52).[4] The entry of large numbers of women into the paid labor force has been a major contributor to this trend; whereas in 1960 36% of women age 16 to 64 were in the paid labor force, by 2000 that figure had risen to 58%, and more than three-fifths (63%) of mothers of children under age 6 were in the labor force (U.S. Bureau of the Census, 1960, 2000). Women now comprise more than 45% of those making work trips.

The spatial pattern of commuting flows has become more complex; the traditional suburb-to-central-city commute has not been the dominant work trip type since at least as long ago as 1970 (Plane, 1981). In 2000, if we exclude work trips made within and between nonmetropolitan areas and look only at trips made within metropolitan areas, the national pattern of commuting flows looks quite intricate (see Table 1.3). The within-suburb or suburb-to-suburb commute is clearly dominant, accounting for about two-fifths of all metropolitan work trips. The "traditional" commute (suburb to central city) accounts for less than one-fifth (only 17.4%) of all journeys to work, and the reverse commute (central city to suburb) is 7.6% of commuting flows.

Given the complexity of the flow patterns depicted in Table 1.3, it is perhaps not surprising that the proportion of work trips made by auto has consistently increased while the proportion made on public transit (bus, commuter rail, subway) has continued to decline (Figure 1.14). In 2000, less than 5% of work trips in the United States were made on transit, a figure that masks a great deal of place-to-place variability (U.S. Bureau of the Census, 2000). (For a thorough discussion of public transit, see Chapter 8.) By 2000, the proportion of people driving alone to work had increased (to about 80%), while the proportion carpooling had decreased from roughly 20% in 1980 to 12%. Those commuting by private vehicle in 2000 accounted for 88% of all work trips (U.S. Department of Transportation, Bureau of Transportation Statistics, 2001, p. 98).

Taken together, these trends—more vehicles on the road, increasing VMT, longer trips in terms of distance—add up not only to more mobility, which many Americans clearly value, but also to many of the problems associated with transportation and primarily with the car, including congestion; air, water, and noise pollution; energy consumption; urban sprawl; traffic accidents; and health problems. A study undertaken by the Natural Resources Defense Council (NRDC) (Miller & Moffet, 1993) finds that a substantial portion of the cost of automo-

TABLE 1.3. Commuting Flows in U.S. Metropolitan Areas

Suburbs to central city	18,175,489	17.4%
Within suburbs	40,745,878	39.0%
From suburbs to outside home MSA	7,650,705	7.3%
Central city to suburbs	7,984,014	7.6%
Within central city	27,425,079	26.3%
From central city to outside home MSA	2,402,466	2.3%

Note. Base (workers over 16 living in MSAs): 104,383,631. Source: U.S. Bureau of the Census (2000).

Percentage of Workers

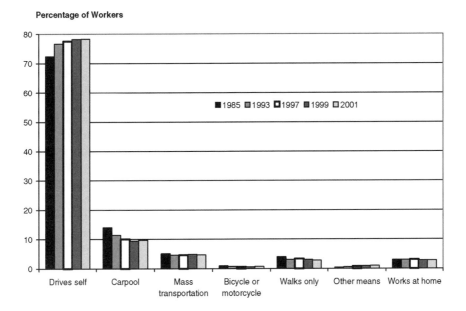

FIGURE 1.14. How people get to work: 1985–2001. Source: U.S. Department of Housing and Urban Development, American Housing Survey, various years.

bile travel is borne not by the user, but by government and by society, including future generations. The NRDC study argues that whereas people perceive transit as being more heavily subsidized than the auto (in that a large portion of the costs of transit are not paid directly by the user), "the subsidies that transit receives are easily scrutinized [because they are generally direct government payments] while the subsidies that automobiles receive are hidden, not easily quantified, and widely dispersed" (1993, p. 67). In fact, the study estimates that about 85% of the subsidies that autos receive are external costs (i.e., not paid for by the user), incurred, for example, by congestion, parking, accidents, and air, noise, and water pollution. The actual estimated cost of travel by auto per person mile traveled (PMT) was between 38 cents and 52 cents, with between 33.5 cents and 42.4 cents of that cost being borne by society rather than by the traveler (Miller & Moffet, 1993, p. 66).

Although increases in automobile ownership and in VMT by car are evident in most countries, the patterns described in this sec-

tion for the United States are not replicated everywhere; in fact, careful study of travel patterns in other countries (see, e.g., Pucher & Lefevre, 1996) shows that economic efficiency and high levels of personal mobility are possible without the extreme automobile dependency that characterizes the U.S. transportation system. Despite European trends that mimic those in the United States (increases in car usage, reductions in walk, bike, and transit trips) Pucher and Lefevre (1996) report, for example, that public transport is still far more widely used in Western Europe than in the United States, accounting for between 10 and 20% of total urban travel in Europe, but only 3% in the United States; similarly, Europeans are more likely to get places via walking or biking (they make more than one-third of their trips by these modes) than are Americans, who make only 10% of their trips via walk or bicycle. As many of the chapters in this book make clear, public policy plays a vital role in shaping international differences in land use and transportation patterns.

In the United States, public policies have

helped to create a society in which we use the time saved via technological improvements in transportation to consume more distance (keeping the total amount of time spent traveling relatively constant) rather than reallocating that time from transportation to other activities. Yet one of the reasons we value the convenience, flexibility, and speed of the auto is that we are a society in which people always seem to be pressed for time. As you will see in Chapters 5 and 6, much of transportation planning has been aimed at saving travel time; this emphasis, operationalized by placing a monetary value on the time "saved" by traveling at faster speeds, has pushed the transportation system toward building ever more, and ever wider, roads (Whitelegg, 1993, pp. 94–96), high-speed railways, and airports. How do we break out of this cycle? Should we even try to? Recent policy shifts within the United States suggest a move toward a more comprehensive approach to thinking about transportation issues.

THE POLICY CONTEXT

In the early 1990s the policy context for transportation planning in the United States changed dramatically with the passage of two key pieces of federal legislation: the Clean Air Act Amendments (CAAA, passed in 1990) and the Intermodel Surface Transportation Efficiency Act (ISTEA, passed in 1991). The Clean Air Act of 1970 identified the automobile as a major contributor to the nation's air pollution problems and explicitly enlisted transportation planners in the effort to meet air quality goals. The 1990 CAAA required that the transportation planning process be broadened to integrate clean air planning and transportation planning at the regional level. Specifically, the CAAA set out goals for cleaner vehicles, for cleaner fuels, and for transportation programs to meet air quality standards.

ISTEA allocated funding support and set out institutional processes to meet these goals. As Howe put it, ISTEA embodied "a whole new attitude toward transportation planning" (1994, p. 11). ISTEA stated, "It is the policy of the United States to develop a National Intermodel Transportation System that is economically efficient and environmentally sound, provides the foundation for the Nation to compete in the global economy, and will move people and goods in an energy-efficient manner." As you can see from this policy statement, ISTEA construed the transportation problem far more broadly than had previous policies to include energy consumption, air pollution, and economic competitiveness as goals in addition to increasing mobility. In 1998 Congress passed the Transportation Equity Act for the 21st Century (TEA-21), legislation that continues to support the transportation planning and funding philosophy embodied in ISTEA.

ISTEA and TEA-21 have increased the flexibility of the regional agencies responsible for transportation planning, known as Metropolitan Planning Organizations (MPOs), in their approaches to solving transportation problems. Funds that earlier had been reserved for highway projects can now be used for all surface modes of transportation, including walking, bicycling, and public transit, which the planning process had neglected in the past. Significantly, ISTEA encouraged the building of bicycle and pedestrian facilities and gave priority to managing the existing transportation system more efficiently rather than increasing supply (i.e., building more roads). Under ISTEA and TEA-21, regional planning agencies have enhanced power in the transportation planning arena, and public participation (the involvement of the users of the transportations system) is an integral part of the planning process. Other goals of ISTEA and TEA-21 include preserving the integrity of communities and providing increased mobility for the elderly, the disabled, and the economically disadvantaged.

All this is a far cry from the days, not so long ago, when transportation planning meant highway building. Throughout the remaining chapters in this book you will see how, together, the CAAA, the ISTEA, and its successor act the TEA-21 have had a significant impact on the way planners conceptualize and try to solve urban transportation planning problems.

What are the issues, the problems, the questions that transportation analysts seek to understand and to remedy? Some are evident from the above discussions of recent trends in travel and the contemporary urban context within which travel and transportation planning take place. The increasing separation between home and work and between activity sites in general—together with the growth in population, in households, in the civilian labor force, and in consumption—mean not only that more travel is required for each individual to carry out his or her round of daily activities but also that more and more people are traveling more and more miles. Congestion has long been viewed as the main urban transportation problem to be "solved," mainly by constructing more and more highways with ever greater capacity. Since the 1950s, however, we have learned the ironic lesson that increased highway capacity generally cannot keep pace with the increased travel demand that is attracted by faster movement and lower-cost travel; as a result, even with more highway capacity roads remain congested.

The CAAA, ISTEA, and TEA-21 articulate a range of transportation-related policy concerns—other than traffic congestion—and a number of these are addressed in Part III of this book. Not every major transportation-related problem is accorded a separate chapter in Part III. One example is health. The growing distance between activity sites along with the overwhelming automobile orientation of U.S. society makes travel on foot or by bicycle difficult and often dangerous. In 1995 pedestrian and bicycle travel accounted for only 6.3 % and less than 0.05%, respectively, of all person miles traveled but fully 14% of all traf-

fic fatalities (U.S. Department of Transportation, Federal Highway Administration, 2001). One might argue, therefore, that part of the urban transportation problem is the threat to health and safety posed by the monopoly that motorized vehicles seem to have in urban travel. Air pollution, water pollution, and traffic accidents (some 40,000 traffic deaths per year in the United States) are all health problems that can be related to the current configuration of urban transportation. There is also the question as to whether the current U.S. transportation system discourages physical activity and encourages a sedentary lifestyle; how would you go about investigating that question?

The policy concerns that *are* addressed in Part III reflect the range of questions that transportation geographers and planners are grappling with: transit, land use change, energy consumption, transportation finance, equity issues, and environmental impacts. As several chapter authors make clear, politics surrounds decision making in all of these policy arenas. Careful analysis by transportation planners may conclude that a particular plan or policy would best serve the transportation needs of a community. Whether or not that plan or policy is implemented, however, is the result of a political process. Because every transportation-related decision will benefit some people more than others—and because who the "winners" and "losers" will be is often defined by *where they live*—the politics of urban transportation often has a distinct geographic dimension.

One major current transportation issue is the appropriate role for public transport in U.S. cities. In the 1960s and the early 1970s planners (and the public) looked to transit to reduce air pollution, energy consumption, and congestion, as well as to revitalize downtown areas and to promote mobility for the carless. It is now clear that, although public transportation is not a panacea for all these urban problems, transit does fill an important niche in many, if not all, U.S. cities. What are the reasons behind the precarious finances of transit companies in U.S. cities? What is an appropriate role for public transportation in a country as devoted to the private automobile as is the United States?

The intimate relationship between transportation and land use was acknowledged at the outset of this chapter, but what are the policy implications of this close relationship? To what extent are transportation projects responsible for increasing urban land values and for generating urban development? Can urban sprawl be attributed to large-scale transportation improvements? Are certain transportation investments, such as light-rail rapid transit lines, an effective means of changing urban land use patterns (e.g., intensifying urban land use or revitalizing certain parts of the city)?

Transportation is a major consumer of energy, especially energy from petroleum. In 1999, transportation of all types consumed more than one-quarter of the energy used in the United States but about two-thirds of the petroleum consumed (U.S. Department of Transportation, Bureau of Transportation Statistics, 2001, pp. 171, 173). Although the United States has less than 5% of the world's population, it consumes 42% of the world's gasoline (United Nations, 1994). In the 1970s the price of energy rose substantially, and the reality of petroleum shortages forced its way into the American consciousness. What impact have these earlier changes in energy price and availability had upon American energy consumption? How effective can transportation policies be in reducing the consumption of fossil fuels?

Transportation investments involve huge amounts of money; more than 11% of the U.S. economy is tied to transportation (U.S. Department of Transportation, Bureau of Transportation Statistics, 2003). What is the economic rationale for investing public funds in transportation systems? How should public monies for transportation be raised and how should they be allocated? What factors determine how and where that public money gets spent? How can we assess whether or not transportation funds are being allocated equitably across geographic areas and various social groups? How does the history of transportation finance shed light on contemporary transportation systems, especially highways and public transit?

Because social status in the U.S. city is closely related to location, as is illustrated in this chapter in the maps of Worcester, Massachusetts, the placement of different transportation projects will affect various social groups differently. One dimension of the urban transportation problem is, then, who pays for and who benefits from any given transportation investment. Are public transportation costs and benefits, in particular, distributed evenly among transit users? How can transportation services be provided in an equitable manner? Similarly, are various social groups equally or differentially exposed to the environmental costs associated with urban mobility (e.g., noise, air pollution, traffic accidents)?

Because most travel in the United States is conducted in motor vehicles, another dimension of the urban transportation problem is the set of environmental impacts stemming from facility construction and from the use of motor vehicles. Although the amount of air pollution generated per automobile has declined significantly in the past 20 years, increases in VMT mean that transportation sources remain a primary contributor to air quality problems. For example, transportation accounts for 77% of carbon monoxide, 47% of volatile organic compounds, which contribute to ground-level ozone formation, and 56% of nitrogen dioxide released into the air (U.S. Environmental Protection Agency, 2001). Transportation analysts are now federally mandated to play a key role in maintaining air quality standards. How can transportation investments be made so as also to minimize other adverse environmental impacts such as noise and water pollution and wildlife habitat fragmentation?

Each of the chapters in Part III takes up an issue that is closely linked to questions of sustainability. Because transportation is so completely intertwined with all aspects of urban life, sustainable transportation has to be at the core of any effort to promote sustainable development. While difficult to define, *sustainable development* involves meeting current needs in ways that improve economic, environmental, and social conditions while not jeopardizing the ability of future generations to meet their own needs (Brundtland Commission, 1987). Strategies for sustainable transportation include those that reduce overall vehicle trip frequencies and trip lengths and those that facilitate walking, bicycling, and using public transportation (Deakin, 2002). As Deakin points out, "What makes sustainable transportation planning different from past practice is that social and environmental objectives are an integral part of sustainable transportation planning, rather than constraints or the focus of mitigation efforts" (2002, p. 9). With the U.S. transportation sector estimated to be the "the single largest source of greenhouse gas emissions in the world" (U.S. Department of Transportation, 1998, cited in Deakin, 2002, p. 3), current transportation practices in the United States are far from sustainable. Will transportation become more sustainable through reduced travel or through further technology improvement, or through some of each? We invite you to think carefully about how citizens and transportation professionals might improve the sustainability of urban transportation.

Each of the policy chapters examines the evidence that bears upon an issue related to sustainability. An interesting theme that emerges from these chapters is that careful empirical analysis often yields results that challenge long-held ideas. Some of these established, accepted notions emerged from microeconomic theory; others came from earlier, less carefully controlled empirical work. But the message that comes through again and again in Part III is that we cannot assume that an assertion is true simply be-cause it has been accepted and unquestioned for a long time. So, we invite you to read critically and to think about how *you* would go about improving transportation in cities.

NOTES

1. Calculated from the 2000 National Household Travel Survey.
2. In 1960, 67.2% of Worcester's MSA labor force worked in the City of Worcester, and in 2000 the percentage was 32.2. The MSA boundaries changed over this period as well; in 1960, the MSA included 20 towns, and by 2000 it included 35 towns. Although the number of workers in the city increased slightly in these four decades (from about 81,500 to 82,800), suburban employment grew at a far greater rate.
3. Temporary work is growing rapidly in the United States While aggregate nonfarm employment grew at an annual rate of 2% between 1972 and 1995, employment in temporary services grew 11.8% per annum during the same period (Segal & Sullivan, 1997). One-fifth of all new jobs creation since 1984 was through temporary agencies (Capelli et al., 1997).
4. Between 1990 and 2000, population and workforce both grew at 11.6%.

ACKNOWLEDGMENT

The author thanks Guido Schwarz and Mang Lung Cheuk (both of Clark University) and Sung Ho Ryo (University of Southern California) for their research assistance.

REFERENCES

Ausabel, J. (1992). Industrial ecology: Reflections on a colloquium. *Proceedings of the National Academy of Sciences USA, 89,* 879–884.

Bruntland Commission (World Commission on Environment and Development). (1987). *Our common future.* New York: Oxford University Press.

Cappelli, P., Bassi, L., Katz, H., Knoke, D., Osterman, P., & Useem, M. (1997). *Change at work: How American industry and workers are coping with corporate restructuring and what work-*

ers must do to take charge of their own careers. Oxford, UK: Oxford University Press.

Chapin, F. S., Jr. (1974). *Human activity patterns in the city: Things people do in time and in space.* New York: Wiley.

Deakin, E. (2002). Sustainable transportation: U.S. dilemmas and European experiences. *Transportation Research Record, No. 1792,* 1–11.

Garrow, D. (1988). *Bearing the cross: Martin Luther King Jr. and the Southern Christian Leadership Conference.* New York: Vintage Books.

Hagerstrand, T. (1970). What about people in regional science? *Papers, Regional Science Association, 24,* 7–21.

Holzer, H. (1991). The spatial mismatch hypothesis: What has the evidence shown? *Urban Studies, 28,* 105–122.

Howe, L. (1994). Winging it with ISTEA. *Planning, 60*(1), 11–14.

Hughes, M. (1991). Employment decentralization and accessibility: A strategy for stimulating regional mobility. *Journal of the American Planning Association, 57,* 288–298.

Miller, P., & Moffet, J. (1993). *The price of mobility: Uncovering the hidden costs of transportation.* Washington, DC: Natural Resources Defense Council.

Mokhtarian, P., Solomon, I., & Redmond, L. (2001). Understanding the demand for travel: It's not purely "derived." *Innovation, 14*(4), 355–380.

Mouw, T. (2000). Are black workers missing the connection?: The effect of spatial distance and employee referrals on interfirm racial segregation. *Demography, 39,* 507–528.

Pisarski, A. E. (1992). *Travel behavior issues in the 90s.* Washington, DC: U.S. Department of Transportation.

Plane, D. A. (1981). The geography of urban commuting fields: Some empirical evidence from New England. *Professional Geographer, 33,* 182–188.

Powledge, F. (1992). *Free at last?: The civil rights movement and the people who made it.* New York: Harper Perennial.

Pucker, J., & Lefevre, C. (1996). *Urban transport crisis in Europe and North America.* London: MacMillan Press.

Scott, L. M. (2000). Evaluating intra-metropolitan accessibility in the information age: Operational issues, objectives, and implementation. In D. G. Janelle & D. C. Hodge

(Eds.), *Information, place, and cyberspace: Issues in accessibility* (pp. 21–45). Heidelberg, Germany: Springer.

Segal, L. M., & Sullivan, D. G. (1997). The temporary labor force. *Economic Perspectives, Review from the Federal Reserve Bank of Chicago, 19*(2), 2–19.

Shen, Q. (2001). A spatial analysis of job openings and access in a U.S. metropolitan area. *Journal of the American Planning Association, 67,* 53–68.

United Nations. (1961). *Statistical yearbook.* New York: Author.

United Nations. (1994). *1992 energy statistics yearbook.* New York: Author.

United Nations. (2001). *Statistical yearbook (46th ed.).* New York: Author.

U.S. Bureau of the Census. (1960). *1960 Census of population and housing.* Washington, DC: U.S. Department of Commerce.

U.S. Bureau of the Census. (1962). *Statistical abstract of the United States 1962.* Washington, DC: U.S. Government Printing Office.

U.S. Bureau of the Census. (1970). *1970 Census of population and housing.* Washington, DC: U.S. Department of Commerce.

U.S. Bureau of the Census. (1979). *The journey to work in the United States: 1975.* Washington, DC: U.S. Government Printing Office.

U.S. Bureau of the Census. (1980). *1980 census of the population. General social and economic characteristics: United States summary.* Washington, DC: U.S. Government Printing Office.

U.S. Bureau of the Census. (1990). *1990 census of the population and housing.* Washington, DC: U.S. Department of Commerce.

U.S. Bureau of the Census. (2000). *2000 Census of the population and housing.* Washington, DC: U.S. Department of Commerce.

U.S. Department of Energy. (2001). *Transportation energy databook.* Washington, DC: U.S. Government Printing Office.

U.S. Department of Housing and Urban Development. (1985, 1993, 1997, 1999, 2001). *American Housing Survey.* Washington, DC: U.S. Government Printing Office.

U.S. Department of Transportation. (1998). *Transportation and global climate change—A review and analysis of the literature.* Washington, DC: U.S. Department of Transportation, Research and Special Programs Administration.

U.S. Department of Transportation. (2003a). *NHTS 2001 highlights report (BTS03-05).*

Washington, DC: U.S. Government Printing Office

U.S. Department of Transportation. (2003b). *Summary statistics on demographic characteristics and total travel, 2001 National Household Travel Survey.* Available online at http://nhts.ornl.gov/2001/html_files/trends_ver6.shtml

U.S. Department of Transportation, Bureau of Transportation Statistics. (1994). *Transportation statistics annual report 1994.* Washington, DC: U.S. Government Printing Office.

U.S. Department of Transportation, Bureau of Transportation Statistics. (1999). *Transportation statistics annual report 1999.* Washington, DC: U.S. Government Printing Office.

U.S. Department of Transportation, Bureau of Transportation Statistics. (2001). *Transportation statistics annual report 2000.* Washington, DC: U.S. Government Printing Office.

U.S. Department of Transportation, Bureau of Transportation Statistics. (2003). *Pocket Guide to transportation 2003* (BTS03-01). Washington, DC: Author. Also available online at http://www.bts.gov/publications/pocket_guide_to_transportation/2003/index.html

U.S. Department of Transportation, Federal Highway Administration. (1994). *1990 NPTS Databook* (Vol. 1). Wahington, DC: U.S. Government Printing Office.

U.S. Department of Transportation, Federal Highway Administration. (2001). *1995 NPTS databook.* Wahington, DC: U.S. Government Printing Office.

U.S. Environmental Protection Agency. (2001). *National air quality and emissions trend report, 1999.* Washington, DC: Author.

Wachs, M., & Kumagi, T. G. (1973). Physical accessibility as a social indicator. *Socioeconomic Planning Sciences, 7,* 437–456.

Whitelegg, J. (1993). *Transport for a sustainable future: The case for Europe.* New York: Belhaven Press.

Wilson, W. J. (1987). *The truly disadvantaged: The inner city, the underclass, and public policy.* Chicago: University of Chicago Press.

CHAPTER 2

City Interactions

The Dynamics of Passenger and Freight Flows

THOMAS R. LEINBACH

In contrast to the previous chapter, the current one takes a broader perspective on the topic of urban transportation and examines both passenger and freight movements at the *interurban* scale, that is, movements *between* cities. Spatial interactions between cities reflect the connectivity of these nodes in the urban system. These interurban movements are important because they affect the spatial patterns, structure, and development of cities. For example, while both ports and airports are growth generators, they may also be congestion and constraint points. Indeed, collection and distribution from these nodes is a major urban transportation problem. In addition, node-related traffic contributes to local noise, congestion, and air pollution. In this light, the objectives of this chapter are, first, to provide an overview of the nature of intercity passenger and freight flows; second, to explain the reasons for the patterns and usage in recent decades; and third, to discuss the impacts and problems associated with increased flows between cities.

Focusing on the intercity scale necessarily brings all the basic modes of transport into the discussion. Given the importance of road transport, it is especially important that we examine passenger movements by auto and bus and freight movements by truck. Of special interest here is the importance of significant infrastructure, in particular the Interstate Highway System, but rail and air are also important and have been topics of intense debate within the United States over the last decade. We discuss the efficiency and viability of rail passenger service in the United States. In addition, the events of September 11, 2001, and the overall structure and viability of the airline industry are important themes. In examining these interaction patterns it is important to remember that rail is the only mode that is "fully private," that is, completely responsible for rolling stock, terminals, and rights-of-way. Highways and airport infrastructure are financed in part via more general user fees and are publicly operated and maintained. These differences help to explain the different patterns of usage across sectors. In addition, rising per capita incomes, declining real costs of transportation associated with fuel price, deregulation-driven shipping prices and airfares, and productivity gains (bigger vehicles and more seamless modal integration) are important. Finally, we need to consider the substantial economic restructuring that has taken place within the United States as we move to a "new" economy driven by technology, lighter manufacturing, and services.

While the emphasis throughout the chapter is on the United States, globalization trends are touching every facet of society. Furthermore, the contrasting situations of passenger and freight transport in other countries offer interesting evidence and lessons in mobility. Access to the Internet for online ticketing has greatly facilitated access to the airlines and has expanded international tourism. Multinational firms and the new international division of labor have created diverse production and distribution linkages, which require material and product movements over a wide range of distances. For these reasons alone, it is critical that an international perspective be an integral part of this chapter.

After a general discussion of domestic passenger and freight activity, the chapter is organized by mode. It begins with a discussion of highway transportation for both passenger travel and freight movements. This is followed by a discussion of rail interactions similarly in both passenger and freight dimensions. It includes the status of Amtrak and the topic of high-speed rail. Subsequently, air transportation between, and the impact upon, cities are discussed. The final section deals with maritime interactions and the role of ports in the domestic transportation system.

THE NATURE OF PASSENGER AND FREIGHT FLOWS

Patterns of passenger travel in the United States are in part related to the degree to which different modes of transportation are accessible to communities. A map of intercity accessibility, of course, reflects the underlying transport grid of highways and railways as well as the pattern of airports (see Figure 2.1). It is obvious that more heavily populated areas are more accessible and have access to multiple modes of transport. Perhaps as striking, however, are the "empty" spaces on the map where modal access is limited. In part these empty spaces reflect population voids, but they also indicate the degree to which rural populations are automobile-dependent.

Data for two recent time periods confirms the extent to which personal vehicles, and especially the automobile, have come to dominate personal travel in the United States (see Figure 2.2). While journeys to work and intraurban travel are included, nearly 86% of all passenger-kilometers are accounted for by personal vehicles (bus accounts for only 3%), the vast majority of which are automobiles. A variety of statements have documented and discussed the U.S. fetish with the automobile and point up the impacts on society both in general and particular (see, e.g., Kay, 1997; Zuckermann, 1991). Despite the conspicuous presence of air travel within the United States, it accounts for only about 11% of the passenger-kilometers traveled, while both rail and transit together account for less than 1%. It is clear that rail as a form of passenger travel is not important; we discuss the reasons for this below.

The data depicted in Figure 2.2 support the increasing mobility and trend toward travel in the United States. From 1990 to 1996, passenger-kilometers increased by 20%; meanwhile, growth in personal vehicle kilometers increased by nearly 20% also. But the greatest gain was in passenger travel by air, at 24%. Bus travel is only a minor contributor to passenger travel. But of even smaller consequence is intercity rail travel, which has declined by 20% over the same period.

Still another perspective on mobility can be gained from the National Household Travel Surveys (NHTS; previously called the Nationwide Personal Travel Survey [NPTS]), an inventory of daily and long-distance travel (Hu & Young, 1999; U.S. Department of Transportation, Na-

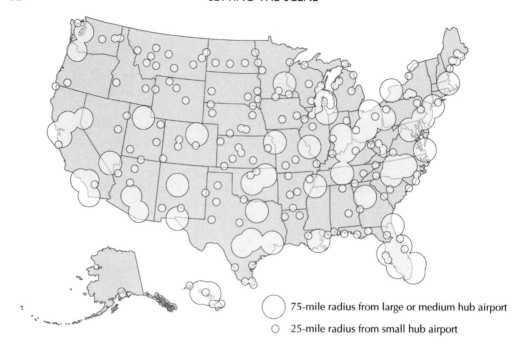

75-mile radius from large or medium hub airport

25-mile radius from small hub airport

○ 25-mile radius from Amtrak station

FIGURE 2.1. Access to transportation services, 1999. Source: Spear and Weil (1999). Available online at *http://www.bts.gov/publications/tsar/2000/chapter4/access_to_intercity_public_transportation_services_map.html.*

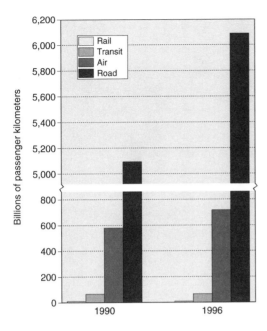

FIGURE 2.2. U.S. domestic passenger travel by mode, 1990, 1996 (billions [thousand millions] of passenger-kilometers). Source: U.S. Department of Transportation, Bureau of Transportation Statistics (2000b, p. 121).

tional Household Travel Survey, 2001). The survey includes demographic characteristics of households, people, and vehicles, as well as detailed information on daily and longer distance travel for all purposes by all modes. NHTS survey data are collected from a sample of U.S. households and expanded to provide national estimates of trips and miles by travel mode, trip purpose, and a host of household attributes. The NHTS data can be used to investigate topics in transportation safety, congestion, the mobility of various population groups, and the relationship of personal travel to economic productivity, the impact of travel on the human and natural environment, and other important subjects (e.g., Mehndiratta & Hansen, 1998). These data provide planners and decision makers with up-to-date information to assist them in effectively improving the mobility, safety, and security of our nation's transportation systems. Trends and changes

in travel behavior can be identified and understood by analyzing the data series, which for daily travel now encompass more than 30 years.

A National Personal Transportation Survey (NPTS) conducted in 1995 consists of a sample of more than 42,000 households in both urban and rural settings. The data were collected for all personal trips over all household-based vehicles. Results of the survey revealed that while the average household size remained relatively stable (about 2.6 persons) from 1990 to 1995, a typical household traveled about 4,000 more miles in 1995, or on average about 34,459 miles. This increase took the form of more, but shorter, trips for most trip purposes. But both work and shopping trips increased in length, on average to 11.6 and 6.08 miles in 1995, respectively. Interestingly, trips in the category of "work-related business" became shorter over the period from 1990 to 1995, declining from 28.2 to 20.3 miles (Hu & Young, 1999, p. 12). This finding raises the interesting question of the possible influence of an electronic substitute for physical travel (see Janelle, Chapter 4, this volume; also see Cairncross, 1997; Janelle & Hodge, 2000). Over the same 5-year period, the average number of commuting trips per household increased (perhaps as a result of more workers per household), as did the average length of such trips. In 1995 commuting overtook social and recreational trips as the longest distance trip type, 11.8 miles on average.

The 2001 NHTS collected travel data from a national sample of the civilian, noninstitutionalized population of the United States. People living in college dormitories, nursing homes, other medical institutions, prisons, and military bases were excluded from the sample. The 2001 NHTS data set has approximately 66,000 households— 26,000 households in the national sample, and 40,000 households from nine add-on areas.

Some preliminary analyses of the preliminary data released in January 2003 provide initial insights regarding intercity (trips in excess of 50 miles) passenger travel. First, a relatively small proportion of the sample (only 2.5%) made trips of 50 miles or more. Of those making trips of this distance, the purposes for the longest trips were, in rank order, those associated with medical/dental services, school or religious activity, work-related, travel as a student, and goods purchasing. Income size was only weakly correlated with trip lengths in excess of 50 miles. Apart from air, recreational vehicles are associated with the longest trip distances. A fuller set of data will ultimately be available for public use.

Modal Freight Movements, 1990–1996

The movement of freight within the United States is critical to the economy. For the same periods of 1990 and 1996, the modal distribution of freight presents a striking contrast to that of passengers (see Figure 2.3). Whereas rail passenger traffic is insignificant in the United States, rail accounted for more than one-third of the metric ton-kilometers of freight shipped in 1996. Rail is especially important for moving heavy freight, such as the traffic in coal from Wyoming and West Virginia to midwestern demand points and East Coast ports, respectively. Freight moving by truck on the Interstate Highway System accounts for nearly 25% of the total ton-kilometers. The segments accounting for the heaviest traffic occur from Indiana to Illinois, Pennsylvania to New Jersey, Michigan to Ohio, and New Jersey to New York. The production of manufactured goods requiring input materials and the demand for finished goods are high in these heavily populated states. Water and pipeline traffic each account for around 20% of the total freight movements. In the former, the most conspicuous movements are along the Missouri, Mississippi, and Ohio River systems as well as on the Gulf Coast between Louisiana and Texas. Airfreight *within* the United States is

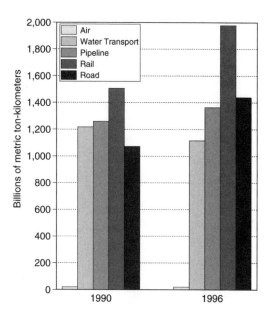

FIGURE 2.3. U.S. domestic freight activity by mode, 1990, 1996 (billions [thousand millions] of passenger-kilometers). Source: U.S. Department of Transportation, Bureau of Transportation Statistics (2000b, p. 59).

quite small as a proportion of the total, less than 1%. Obviously, the dominant explanation here is that only goods of relatively high value per unit weight can bear the high costs of airfreight.

Overall the growth of freight over the 6-year period was around 17%. Noteworthy is the fact that airfreight showed the greatest gain, 47%. This in part reflects economic restructuring within the United States as a "new economy" derives growth from technology applications in light manufacturing operations (Leinbach, 2001). It also reflects somewhat the surge of courier business and the rise of the integrators, such as FedEx and the United Parcel Service (UPS), which provide door-to-door, time-definite deliveries. Both of these carriers, which have their own aircraft, are involved in freight movements intranationally as well as internationally. To some extent, the growth in domestic airfreight movements does reflect the forward and backward staging points for materials and finished goods to and from international locations. Road and rail also strengthened, at 34 and 31%, respectively. Pipeline movements of crude oil and natural gas increased only modestly, while total water movements actually declined. Although traffic on the inland waterways grew slightly, it was the decline especially in coastal, and to a lesser extent Great Lakes, shipping that accounted for the lack of growth. It is well known that railway and highway competition have gradually eroded coastal shipping's share of freight. Competition from the Interstate Highway System and rail has likewise affected freight movements on the Great Lakes.

Domestic and International Freight Projections

While these data on recent trends is important, so too is the projection of freight flows in both volume and value. The U.S. transportation system carried over 15 billion tons of freight, valued at over $9 trillion,

in 1998. Domestic freight movements accounted for nearly $8 trillion of the total value of shipments. By 2020, the U.S. transportation system is expected to handle cargo valued at nearly $30 trillion (see Table 2.1). The nation's highway system, and its huge truck fleet, moved 71% of the total tonnage and 80% of the total value of U.S. shipments in 1998. While trucks made the vast majority of local deliveries, they also carried large volumes of freight between regional and national markets. Water and rail also moved significant shares of total tonnage, but they accounted for much smaller shares when measured on a value basis. As expected, airfreight moved less than 1% of total tonnage, but carried 12% of the total value of shipments in 1998. Again, this is a reflection of economic restructuring within the United States (U.S. Department of Transportation, Federal Highway Administration, 2002).

Domestic freight volumes are expected to grow by more than 65%, increasing from 13.5 billion tons in 1998 to 22.5 billion

TABLE 2.1. U.S. Freight Shipments by Tons and Value

Mode	Tons (millions)			Value (billions)		
	1998	2010	2020	1998	2010	2020
Domestic						
Air	9	18	26	545	1,308	2,246
Highway	10,439	14,930	18,130	6,656	12,746	20,241
Rail	1,954	2,528	2,894	530	848	1,230
Water	1,082	1,345	1,487	146	250	358
Total, domestic	13,484	18,820	22,537	7,876	15,152	24,075
International						
Air	9	16	24	538	1,198	2,284
Highway	419	733	1,069	772	1,724	3,131
Rail	358	518	699	116	248	432
Water	136	199	260	17	34	57
Other[a]	864	1,090	1,259	NA	NA	NA
Total, international	1,787	2,556	3,311	1,444	3,203	5,904
Total	15,271	21,376	25,848	9,320	18,355	29,980

[a]Includes international shipments that moved via pipeline or by unspecified mode.
Note. NA, not available. Source: Federal Highway Administration (2002). Available online at *http:// ops.fhwa.dot.gov/freight/publications/faf.html.*

tons in 2020. The forecast shows that the air and truck modes will experience the fastest growth. Domestic air cargo tonnage is projected to nearly triple over this period, although its share of total tonnage is expected to remain small. Trucks are expected to move over 75% more tons in 2020, capturing a somewhat larger share of total tonnage. Both of these developments will place more pressure on air hubs and major highways. While volumes moved by rail and water modes are also projected to increase over the forecast period, they will not grow as dramatically, primarily because of anticipated slower growth in demand for many of the key commodities carried by these modes.

International trade accounted for 12% of total U.S. freight tonnage in 1998 and is forecast to grow faster than domestic trade. International trade is projected to increase by 2.8% annually between 1998 and 2020, nearly doubling in volume. This growth in international trade is likely to present challenges to U.S. ports and border gateways (U.S. Department of Transportation, Federal Highway Administration, 2002).

International Perspectives

While the pace and extent of change varies, common trends are underway worldwide. Passenger and freight transportation continue to gain increasing shares of motor vehicle movements. At the same time, many countries are experiencing rapid growth in air passengers and airfreight, but these reflect a smaller modal share. In general, the changes are occurring faster in countries with rapidly developing economies compared to those already industrialized. A comparison of modal shares of passenger-kilometer travel for the United States, the European Union, and Japan clearly depict the dominance of automobile travel (see Figure 2.4). Motor vehicles, especially private passenger cars, are dominant in most Organization for Economic Cooperation and Development (OECD) countries.

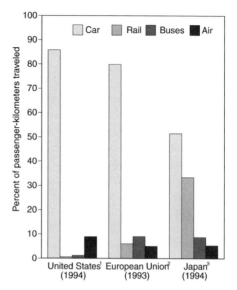

[1] United States. Car total includes data for passenger cars, taxis, and light trucks. Rail total includes light rail, heavy rail, and commuter and intercity rail. Bus total includes data for transit motor buses and intercity buses. Air total includes data for certificated air carriers, domestic carriers, and general aviation.

[2] European Union. Car total includes data for private cars. Rail total includes data for intercity and commuter rail, and light rail (where available). Bus total includes local buses and intercity coaches. Air total includes domestic carriers.

[3] Japan. Car total includes data for private and commercial passenger cars. Rail total includes Japan railways and private rail. Bus total includes buses for private and commercial use. Air total includes general aviation and domestic carriers.

FIGURE 2.4. United States, European Union, and Japan: Modal shares of passenger-kilometers, 1994. Source: U.S. Department of Transportation, Bureau of Transportation Statistics (1997, p. 230).

The auto in the mid–1990s accounted for 86% of passenger-kilometers in the United States and 80% in Western Europe. In Japan this figure is only 52%, while in China passenger road transport, especially by bus, accounts for 46% of passenger-kilometers. In the early 1990s road transport accounted for 85% of passenger travel in India. On the other hand, the importance of rail passenger travel in Japan is reflected in the graph in Figure 2.4, as is the higher proportion of rail travel in Europe in contrast with that of the United States. Well-developed rail systems, high population densities, and the longer tradition of rail travel in Europe and Japan are relevant explanations for the differences.

As with passenger travel, freight activity is increasing worldwide. Among the OECD countries, annual growth in freight traffic has ranged from 1% (France, United Kingdom, and The Netherlands) to 4% (Italy, Japan, and Spain) over the past 25 years. Much higher rates of growth took place in some developing countries—for example, China grew at a rate of 7.5%. The U.S. comparison figure for this period is roughly 2% per annum.

In comparing Europe and North America, several generalizations are possible. First, the greater economic integration of the European nations brought about by the European Union has generally led to substantial increases in both freight and international passenger traffic in these countries. Second, regulation, investment policies, and shorter average travel distances have led to a much greater share of passenger rail and a smaller share of passenger air transport in Europe compared to North America. Under the recent moves toward deregulation new low-cost air carriers are emerging in Europe and the United States that are providing competition and encouraging air travel. Given the shorter distances in Europe, the increases in freight are mostly truck-related; this has led to congestion problems and the demand for new infrastructure.

The nature of freight activity is also changing across the globe. The use of road transport accelerated more than any other mode in most countries. The flexibility and dominant infrastructure associated with roads has gradually eroded the rail share of freight transport. In the developing world the shift toward road transport has been most pronounced. Despite this dominant trend, the increasing demand for efficient and secure delivery of high-value goods in the global production system has influenced the rapid growth of airfreight. Worldwide forecasts of domestic and international airfreight (ton-kilometers) are predicted to increase at an annual rate of 6% between 1999 and 2010 (International Civil Aviation Organization, 2001).

The factors accounting for the growth of passenger and freight traffic interact in complex ways (Bowen & Leinbach, 1995; Debbage, 1994). In much of the developed world, population and workplaces have shifted away from urban centers to more dispersed suburban locations, as described by Muller (Chapter 3, this volume). This forces higher usage of the automobile for work- and social-related purposes. Mass transit is less economical given this shift to less densely populated areas. In the developing world, by contrast, rural populations are migrating to the city or to periurban areas for employment, placing severe pressure on urban areas. Moreover, because circular migration, where individuals move between rural and urban areas on a temporary or short-term basis, is common, interurban transportation facilities are also stressed.

HIGHWAY INTERACTIONS

As the economy has grown rapidly over the past 10 years, passenger travel in the United States has risen more steeply than in earlier decades. The U.S. highway network consists of 4 million miles of roads and streets. Although the Interstate Highway System (IHS) accounts for only 1% of all highway mileage, it carries 25% of the total vehicle miles of travel (Gifford, 1994; Lewis, 1997; Moon, 1994; Rose, 1990). Enabling legislation for the largest public works project in history came in the form of the Federal Highway Act of 1956 and provided not only the authorization to construct but also appropriated the necessary funds for implementation (see Taylor, Chapter 11, this volume). Much of the IHS was constructed between 1960 and 1980, but particular linkages came even later (Moon, 1994).

The impact of the system on the intercity transportation of goods and people has been nothing short of phenomenal (Lewis, 1997), and may even be the single most important influence on the U.S. metropolis in the last 50 years. The IHS transformed cities in ways its planners never anticipated. The system was supposed to save central cities by rescuing them from automobile congestion and to provide high-speed long-distance travel from city to city. But the massive new urban highways, intended to move traffic rapidly in and out of downtown, quickly became snarled in ever-growing congestion, and their construction devastated many urban neighborhoods. Meanwhile, the new peripheral "beltways," originally designed to enable long-distance travelers to bypass crowded central cities, turned into the Main Streets of postwar suburbia. Cheap rural land along the beltways became the favored sites for new suburban housing, shopping malls, industrial parks, and office parks that drew people and businesses out of the central cities. Finally, the IHS was financed by a highway trust fund supported by the abundant revenue from federal gasoline taxes. Those funds were available only for highways, and the federal government paid 90% of the cost of the new highways. By contrast, localities paid a much higher percentage for investment in mass transit. This was a powerful incentive to neglect mass transit and focus a region's transportation investments only on roads. More than any other measure, the 1956 Interstate Highway Act created the decentralized, automobile-dependent metropolis we know today (Fishman, 1999).

Many studies have examined the ways in which the IHS has disrupted urban neighborhoods. The increased accessibility of the urban freeway stimulated decentralization, so that industry was no longer tied to the central city but could retain access to a diverse and abundant labor force. The IHS divided neighborhoods. Critics claimed a racial bias in the location of routes that caused the disintegration of local communities. The "kink" in Nashville's I-40 is one such example (see Bae, Chapter 13, this volume; Rose, 1990).

The IHS has had a major role in economic restructuring. One of the most conspicuous cases has been the I-75 corridor that runs from Sault Ste. Marie, Ontario,

to Miami, transecting Michigan, Ohio, Kentucky, Tennessee, Georgia, and Florida (Moon, 1994, pp. 79–80). A huge amount of capital investment has been located within the corridor to take advantage of the accessibility. Perhaps as much as 15% of all U.S. capital spending has been invested here, including 25% of Japanese foreign direct investment in the United States. Especially conspicuous is the concentration of auto assembly plants and suppliers and defense contracting. Explanations for these developments range from branch plant development and industrial decentralization to just-in-time manufacturing (Rubenstein, 1992), low land costs, and heavy subsidies from individual states. Coupled with this form of economic impact has been that associated with the effect of extended commuting ranges (e.g., Mitchelson & Fisher, 1987) and the land use change around IHS interchanges (e.g., Norris, 1987).

National estimates of interstate traffic volumes are available from 1992 Federal Highway Administration data (*www.fhwa.dot.gov/ohim/tvtw*) where vehicle miles of travel are depicted for rural and urban segments on an aggregate monthly basis. For 2001 these figures for rural and urban are 273.8 and 398.5 billion vehicle miles of travel, respectively. Between 1992 and 2001, the rural and urban IHS vehicle miles of travel grew by 32 and 31%, respectively. Traffic volumes for individual segments of the IHS are available from most state departments of highways. This data reveals what can be easily observed as one drives the system: congestion is prevalent in many areas and maintenance and upgrading are badly needed given the system's heavy traffic and age. Six years ago only 40% of the IHS was considered to be in good condition, while 40% was in poor or mediocre condition. Estimated costs for the repairs at that time were $315 billion (Field, 1996).

Evidence suggests that the fascination with the automobile is only loosely tied to economic conditions and that mobility will continue to expand. It is not clear to what extent information technology will displace actual physical travel (see Janelle, Chapter 4, this volume). The number of users of cellular phones and personal digital assistants capable of sending and receiving faxes, email, and accessing the Internet has jumped from 77 million in 1998 to well over 300 million in 2001. Although telework is practiced by only a small fraction of the labor force, it has become more common and will continue to increase. Despite the availability of tools necessary to advance telecommuting, a variety of factors prevent greater market penetration (e.g., Gillespie & Richardson, 2000). The necessity of team development and the complexity of systems mean that face-to-face contacts are critically important in commerce.

RAIL INTERACTIONS

The Origins and Evolution of Amtrak

After more than a century of intercity passenger operations on private freight railroads in the United States, the Rail Passenger Service Act of 1970 established the National Railroad Passenger Corporation commonly known as Amtrak (*www.amtrak.com*). Congress created the corporation to relieve freight railways of the burden of unprofitable rail passenger service and to preserve rail passenger service over a national system of congressionally designated routes. The viability of passenger rail service continues to be threatened by the existence of relatively inexpensive air travel and also by the widespread access to private automobiles. As noted above, intercity rail passenger travel accounted for less than 1% of total domestic passenger kilometers in 1996.

After its creation Amtrak launched a concerted effort to rebuild rail equipment and infrastructure. In October 2000, the Amtrak network consisted of a basic national grid (see Figure 2.5). Higher frequency train service exists in the Northeast Corridor from Boston south to Richmond, Virginia, and

on a number of segments focused on Chicago and the hubs of Los Angeles and San Francisco. Transcontinental service has been available through four separate corridors on a daily basis. In 1975 annual ridership levels stood at just over 17 million riders; by 1999 ridership had reached nearly 22 million passengers (see Figure 2.6). The growth in ridership over the 24-year period represents only a bit over 1% on an annual basis.

In 1997, after 26 years in which it granted some $22 billion in federal operating subsidies and capital investment that produced very little gains in ridership, Congress began to debate the viability of continuing to fund Amtrak. From these debates emerged the Amtrak Reform and Accountability Act, which provided that Amtrak would no longer be a government corporation or hold a passenger monopoly, would be allowed to add new routes and close routes that were not profitable, would receive $2.2 billion in tax relief aid, and would have to achieve operational self-sufficiency 5 years after enactment of the legislation. Just as important was the creation of the Amtrak Reform Council, whose mandate was to assist the Amtrak Corporation in achieving self-sufficiency and to regularly provide Congress with information on its progress toward profitability.

High-Speed Rail

Since this legislation, several noteworthy developments have occurred. One is the further recognition of the importance of high-speed rail (HSR). HSR systems link metropolitan areas that are 100–500 miles apart with operational speeds in excess of 200 kilometers per hour. Although initial infrastructure costs are high, performance capabilities point increasingly to the use of magnetic levitation (maglev) systems that employ magnetic forces to lift, propel, and guide the train's cars. The world's first HSR system, Japan's Bullet Train or "Shinkansen," started running between Tokyo and Osaka in 1964. This service was followed

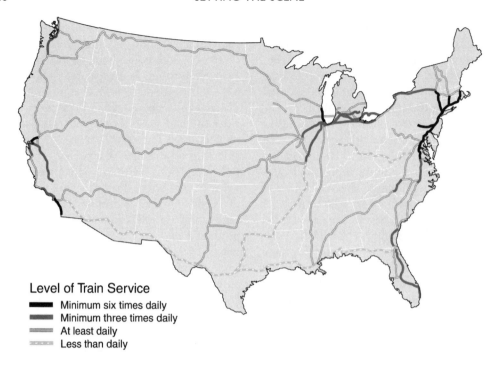

Level of Train Service

- Minimum six times daily
- Minimum three times daily
- At least daily
- Less than daily

FIGURE 2.6. Amtrak ridership, 1975–1999. Source: U.S. Department of Transportation, Bureau of Transportation Statistics (2000a, p. 2-16).

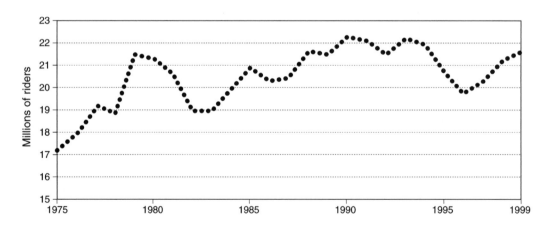

FIGURE 2.5. Amtrak passenger rail system, 2000. Source: U.S. Department of Transportation, Bureau of Transportation Statistics (2000a, p. 2-16).

by France's TGV (Train à Grande Vitesse) service between Paris and Lyon in 1983. Germany and Spain initiated HSR service in 1991 and 1992, respectively. Systems are under development or operating in a variety of Asian locations including Taiwan and South Korea. In 1994, France and the United Kingdom initiated London–Paris HSR service using the privately constructed and operated Channel Tunnel or "Chunnel." Amtrak's Metroliner started HSR service in 1986 between New York and Washington, DC.

In recognition of the importance of developing HSR service, the High Speed Rail Investment Act of 2001 provides for the issuance of $12 billion in bonds, by Amtrak, over 10 years for qualified HSR projects chosen by the secretary of transportation. Bond proceeds can be used for the acquisition, financing, or refinancing of HSR rights-of-way, track structure, rolling stock, and the development of intermodal facilities within the 12 designated HSR corri-

dors (see Figure 2.7). Projects are selected on the basis of state matching funds, potential contribution to highway and airport congestion relief, improvement of commuter rail operations, and potential economic performance, including the development of "in-city terminals" as economic development nodes in central cities that would counter the sprawl tendencies of out-of-town airports. Projects are also intended to promote regional balance in infrastructure investment. States will provide at least a 20% match for a particular project.

In this light, there has been an important effort to restructure rail service in the Northeast Corridor with the addition of the new high-speed Acela train. Further, the plan has been to "proceed aggressively" on HSR corridor development outside of the Northeast. In January 2002, Amtrak announced that trains were running at 90 miles per hour over a 45-mile stretch of track in southeast Michigan. The increased speeds are the first fruits of a 5-year, multi-

FIGURE 2.7. Designated high-speed rail corridors. Source: U.S. Department of Transportation, Federal Railway Administration (2000).

million-dollar effort to bring HSR to the Chicago–Detroit rail corridor. The funds helped to improve grade crossing safety and install a state-of-the-art train signaling system. As evaluation of the new technology continues, the 45-mile stretch will extend to 65 miles of track and speeds will increase from 90 to 110 miles per hour. To date, 77 miles of the 279-mile corridor have been upgraded at a cost of $1 million per mile, which is 10–20 times less expensive than highway construction.

The expansion of HSR corridors is important for a variety of reasons. Among the most important is that such projects may "free the airports." HSR networks can bleed the short-haul capacity out of the system, freeing airports to do what they do best: handle long-distance trips (Sharkey, 2002). Obviously, such a development would be controversial, given the likely impact on air services to smaller communities. In part because of this increased attention to HSR, Amtrak revenues have increased by 38% since 1997.

Despite signs of progress, in mid-2002 Amtrak was again plagued by the lack of a sufficient cash flow to keep services fully functional. While Amtrak cut spending by $285 million in 2001, it incurred a loss of $1.1 billion, its largest loss since 1971. The restructuring plan of the Amtrak Reform Council, proposed in February 2002, suggested that a small federal agency be responsible for administering the nation's passenger rail program and be linked to two subsidiary companies: one to conduct train operations and the other to own, operate, maintain, and improve Amtrak's Northeast Corridor system. Amtrak essentially rejected this proposal and suggested that there must be a stronger federal commitment to define, develop, and invest in the passenger rail system.

We are again reminded after still another funding crisis that "the railroads built America but modern America seems not to be built for the railroads." A major problem in deriving a sensible plan for reform once more is the political element, which like the rail network itself is a patchwork of local interests. Whether politicians are for or against Amtrak is truly related to whether the train stops in their district. Needed is a middle ground between the Bush administration's break-up-and-privatize solution versus the basic Amtrak model with heavier federal subsidies. In truth, perhaps the government should at long last take the long view of rail as one transport option alongside highways and air ("Amtrak derailed," 2002). In this perspective there is a need to examine technological and organizational breakthroughs and apply policy innovations that have been successful in renewing passenger rail in Europe and Asia (Perl, 2002).

Rail Freight

The experience of freight railroads presents a striking contrast. Foundering under Interstate Commerce Commission regulations, freight railroads rebounded after passage of the Railroad Revitalization and Regulatory Act of 1976, which partially deregulated rail rates and expedited merger processing. A reflection of this is that Conrail, a publicly owned railroad company established by the federal government under the Regional Rail Reorganization Act of 1973, replaced seven bankrupt northeastern rail lines and commenced operations on April 1, 1976. Four years later, the Staggers Act gave railroads the freedom to set rates and also granted them the ability to abandon service on unprofitable lines. During the following 8 years (1980–1988) freight rates on a ton basis declined by 38% but achieved a 7.5% net rate of return on investment while producing stronger freight flows (see Figure 2.8).

The railroad industry has greatly consolidated, moving from 42 large carriers in 1980 to four megacarriers in the late 1990s. It has also incorporated much higher labor productivity (Burns, 1998). Noteworthy among the numerous mergers was that between the Burlington Northern Corporation and the Santa Fe Pacific Corporation,

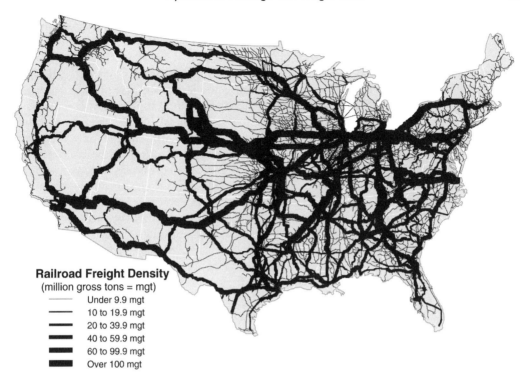

Railroad Freight Density
(million gross tons = mgt)

——— Under 9.9 mgt
——— 10 to 19.9 mgt
▬▬▬ 20 to 39.9 mgt
▬▬▬ 40 to 59.9 mgt
▬▬▬ 60 to 99.9 mgt
▬▬▬ Over 100 mgt

FIGURE 2.8. Railway network and volume of freight, 1998. Source: U.S. Department of Transportation, Bureau of Transportation Statistics (2000a, p. 2-44).

which created the nation's largest railway in 1994. The combination is a 33,000-mile western railway system strong in both commodity shipments and the haulage of truck trailers and containers by rail. The impact of these railroad mergers is still unfolding as evidence is accumulated (e.g., Park, Babcock, & Lemke, 1999). In addition, under deregulation more than 90,000 miles of unprofitable track have been abandoned or sold. Perhaps most important, decontrolling the rail industry accelerated the movement toward intermodalism. Using freight containers that can easily be switched from ship to rail to truck, railroads in partnership with ocean carriers have used double-stack freight trains—which carry containers set one atop another—to move shipments from seaports deep into the interior of the United States.

Compared to the transformation that has occurred in the United States, an even more striking example of liberalization has taken place in Canada (Fossey, 1996). In November 1995, the Canadian government privatized the Canadian National Railway (CNR), which now fends for itself in the free market. The transition changed the mindset of employees and has allowed CNR a much greater degree of flexibility in its operations and cost-cutting.

It is important here to stress that the rail freight systems of the United States and Canada have important impacts on rural areas. This is particularly true as innovations are developed in an effort to make the rail freight system more cost-efficient. One such recent development is the displacement of traditional grain elevators by high-speed grain terminals (Machalaba, 2001). The high-speed terminals have the capacity to fill a mile-long train of 110 cars with grain in about 15 hours, compared to the traditional grain elevator, which can only accommodate half that number of cars in a much longer time. These "shuttle train loaders" are rapidly transforming the grain transport industry. During the past 6 years 73 loaders have been built and an additional 18 are planned along the Burlington Northern Santa Fe's tracks. In addition, 73 loaders are in place along the railways of the Union Pacific. Of course, change rarely affects places evenly. The worry among traditional grain elevator operators is that farmers will begin to bypass their facilities in favor of the shuttle train loaders, with major consequences for the viability of small towns that have already experienced a huge population exodus. Yet the savings and the positive impact is clear. The Alton Grain terminal in Hillsboro, North Dakota, cost $9.6 million to build. The operator gets a rate of $2,100 per car to ship wheat to the Pacific Northwest. This charge is about $400 lower than that charged by a traditional facility, and the savings translate to 12 cents more per bushel for farmers who deliver wheat to Alton.

Smaller railroads are an important part of the U.S. railroad system, especially to agricultural and rural shippers. By offering an alternative to the abandonment of low-traffic routes, rural branch lines have preserved rail service to small communities. To survive and prosper, local and regional railroads cannot just maintain but must increase the market for railroad transportation. The scarcity of traffic leads to an exit process, which is preceded by service degradation. But through knowledge of local conditions and an orientation to customer service, small railroads can succeed. Especially important are innovative service offerings perhaps involving value-added functions such as warehousing, transloading, and just-in-time delivery that firmly integrates the small railroad into the logistics system. The small railroad also faces the challenge related to regional and national economic shifts. The extractive and manufacturing industries that generated high volumes of traffic in the past have given way to a service-oriented economy focused on the global marketplace. Many companies in the service sector are not traditional rail freight customers.

Most important is the fact that the 21st century will see a renewed focus on intermodal freight transportation (Leinbach & Bowe, in press). With the development of containerization in the mid-1900s, the reorientation toward deregulation near the end of the century, and a new focus on logistics and global supply chain requirements, the stage is set for continued intermodal growth. *Intermodal shipping*, defined as single-bill shipments using multiple modes, provides a flexible response to changing supply chain management requirements in global markets. Although trailer and container traffic is most commonly thought of, many other commodities can be intermodal. For example, all grain movements move off farm by truck or water before being connected by rail. Coal, fertilizers, and building products move intermodally. In fact, all air express movements are inherently intermodal, with truck links connecting to airline haul at origin and destination. A major point in favor of intermodal shipping argues that increasing rail traffic reduces truck traffic on the nation's highways, which would enhance highway safety. In 1999 and 2000, the freight rail industry invested more than $14 billion, representing one-fifth of its revenues for that period, in intermodal infrastructure. Information systems were upgraded, terminals were added, and state-of-the-art locomotives and intermodal cars were purchased.

The future of the freight railroads is certain to be dynamic. Rail freight is expected to grow at a rate of over 2% per year through 2025. But it is likely that merger activity will continue and that further consolidation may take place. There is also the possibility that nonrailroad firms could acquire railways in order to diversify or complement their current operations—as UPS has done. Finally the issue of access to rail lines of competing railways may be contentious. If access is mandated to ensure competitiveness, such actions may produce negative financial impacts. The pricing of access is critical in order not to discourage the owning railway to invest in roadway. As always with increased financial pressure, it will be interesting to see how railways are able to improve services and reduce costs (U.S. Department of Transportation, 2000a, p. 2-45).

AIR INTERACTIONS

As noted above, air travel is the fastest growing mode of transportation in the United States. Passenger traffic has nearly tripled since 1975 (see Figure 2.9) and is expected to reach one billion enplanements within the next decade (U.S. Department of Transportation, Federal Aviation Administration, 2000). The ratio of enplaned passengers per airline employee rose by 25% over the same period. This figure reflects rising per capita incomes, the use of faster and larger aircraft, changes in work rules and practices, and the adoption of various marketing strategies. The pattern of enplanements in major markets is equally impressive (see Figure 2.10). Especially conspicuous over this period is the passenger growth in Sunbelt cities in connection with the economic transformation of the "New South." Noteworthy are the increased enplanements in Atlanta, as well as in Miami, Orlando, Houston, Dallas–Ft. Worth, Phoenix, Las Vegas, and Los Angeles. A significant part of this growth,

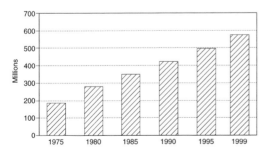

FIGURE 2.9. Domestic enplanements on U.S. commercial carriers: 1975–1999. Source: U.S. Department of Transportation, Bureau of Transportation Statistics (2000a, p. 2-19).

too, is associated with the world's largest industry: travel and tourism. Dominant origin–destination pairs in air traffic basically reflect the urban hierarchy where Boston–New York–Washington, D.C., Los Angeles–San Francisco, New York–Miami, and New York–Chicago are prime interaction corridors.

What explains this high growth of air passenger traffic? In truth, a variety of factors. Among these are the not-inconsequential "lure" of air travel, the growth of discretionary income that allows people to travel more frequently and to farther destinations, the growth of business travel, and the strong price competition among carriers. The growth of "no-frills" airlines, such as Southwest and more recently Jet Blue and Air Trans, which operate in selective niche markets, also contributes (Leinbach & Bowen, 2004). With the introduction of the Internet in 1995 and the growth of e-commerce, airlines have altered their marketing strategies. E- or "paperless" tickets are now the standard, such that airlines levy an additional charge for paper tickets. But more important have been the promotion of travel bargains, "mileage programs," and the emergence of a host of virtual discount travel agents. Expedia, Orbitz, Cheaptickets, and others are now common in the average household's travel lexicon.

As we reflect on the emergence of new carriers, we are reminded of the wave of liberalization in the U.S. transport sector. Indeed, the airline industry has been a major part of this deregulation. In 1975 the Civil Aeronautics Board (CAB) controlled market entry by new airlines, regulated cargo rates and passenger fares, provided government subsidies to the airlines, and monitored airline relations. The Airline Deregulation Act of 1978 created conditions for competition, removed restrictions on the provision of domestic services, allowed market-based fares, recognized the need to continue to service small communities, and abolished the CAB. Deregulation dramatically changed the airline industry in a num-

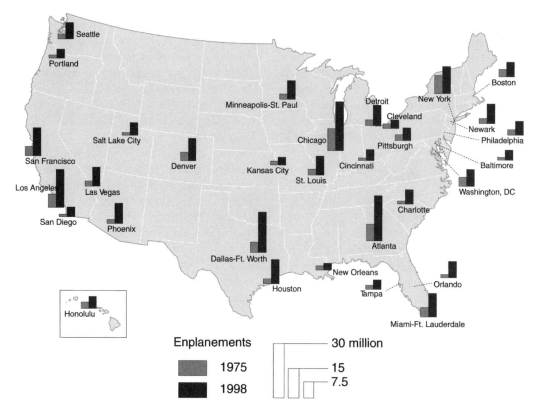

FIGURE 2.10. Enplanements in major markets: 1975 and 1999. Source: U.S. Department of Transportation, Bureau of Transportation Statistics (2000a, p. 2-19).

ber of ways. Perhaps most conspicuous in addition to fare competition was the move away from a linear point-to-point network to a hub–spoke system. This spatial reconfiguration of travel allowed airlines to connect several origins with multiple destinations without having all points directly connected (U.S. Department of Transportation, Bureau of Transportation Statistics, 2000a, p. 2-45).

Air Hubs

The concept of "hubbing" has encouraged carriers to serve more markets without having to increase fleet size and seat capacity, has allowed airlines to cut operating costs, and has enabled airlines to serve many different city pairs more efficiently (Leinbach & Bowen, 2004). In the process of trans-

forming the network structure of operations, the hub has taken on new importance and has allowed airlines to become more entrenched and dominant at their designated hubs. An example is the way in which United Airlines dominates operations at its hubs in Chicago, Denver and—to a lesser extent—San Francisco. Similarly Delta Airlines is the dominant provider in Atlanta and Cincinnati, and Northwest Airlines dominates in Detroit and Minneapolis.

The growth of especially the larger hubs has produced an important instrument of economic development. A prime example is that of Hartsfield International, Atlanta's airport. The airport is not only the major economic generator of Georgia but also arguably a major economic engine in the Southeast. With an estimated 43,000 people tracing employment through the airport (including airlines, ground transportation, concessions, security, federal government, City of Atlanta, and tenant employees), the airport's payroll is in excess of $1.9 billion annually. In 2000, the direct and indirect economic impact of the airport was estimated to be $16.8 billion, or $46 million per day! The success of the regional economy is directly linked to Hartsfield, and over the past 5 years the region served by the airport has been the fastest growing in the United States. The airport is also one of the facilities in the country most severely affected by delays. Under a major development program, a fifth runway expansion costing $1 billion is the first element in a $5.4 billion enhancement planned for the next 10 years.

Air passenger travel and transport of goods and services are now critically important. In fact, airports may shape business locations and urban development in the current century the way in which highways did in the 20th, railroads in the 19th, and seaports in the 18th centuries. Some researchers argue that major international gateway airports are giving rise to an emerging urban form called the "aerotropolis" (Armbruster, 2001; Kasarda, 2001). The aerotropolis can stretch up to 15 miles outward from airports along transportation corridors that radiate from them. Airports can attract various types of business development including time-sensitive manufacturing, e-commerce fulfillment, telecommunications, and third-party logistics firms; entertainment, hotel, retail, and exhibition complexes; and office buildings that house regional corporate headquarters and air-travel-intensive professionals such as consultants, auditors, and high-tech industry executives. Washington Dulles in the northern Virginia region is an excellent example of the way in which an air facility can draw high-tech industry.

Some airports have even assumed the same roles as metropolitan central business districts by becoming major employment, shopping, meeting, and entertainment destinations. An example is Amsterdam's Schiphol, which employs 52,000 people daily. Two main motorways link the airport to downtown Amsterdam, a modern train station operates directly under the terminal, and the terminal itself contains shopping and entertainment opportunities that appeal to both travelers and the general public. Strings of businesses are clustered along Schiphol's motorways; all of the businesses are airport-intensive users. In addition Singapore's Changi International Airport is a prime example of how an airport can serve as a major element of development (Raguraman, 1997).

This new urban form may appear simply as additional sprawl along main airport transportation corridors. Yet the aerotropolis is actually a highly networked system. Accessibility will replace location as the most important business-location and commercial-real-estate organizing principle. Corporate and real estate professionals who understand this fact can select strategic sites near major gateway airports and can position investment to be leveraged by air commerce. As more and more companies depend on time-intensive transportation and quick response, those who choose to locate near airports will be one step ahead of their competitors (Kasarda, 2001).

Yet such developments add to the already existing externalities of noise, pollution, and increased traffic congestion of the airport. The residents that live within the influence of these developments must shoulder these costs. A large airport such as Los Angeles International generates substantial externalities in terms of noise and increased traffic congestion, which depress the value of the immediately surrounding real estate. Some income groups may, however, be more willing to accept the high level of noise pollution in return for the much lower rents they have to pay for housing. Putting a dollar value on the degradation in the quality of life for those residents affected is very difficult and continues to be debated.

The Health of the Airline Industry

Since deregulation, air carriers have earned more than $38 billion in operating revenues and $6.5 billion in net profits. But the health of the airline industry has followed a roller-coaster path. Prosperity from 1984 to 1988, resulting from a strong economy and low fuel costs, was followed by a strong downturn from 1989 through 1993, stemming from the depressed domestic and global economy as well as rising jet fuel prices. Uncertainties during the Gulf War and the threat of terrorism were also factors. Under the Clinton administration, special efforts were made to study and stimulate the industry. As a result of measures then put into place a resurgence occurred.

Yet following the boom-and-bust cycle, the current period is perhaps the darkest thus far in the history of the airlines. High costs of the transition to modern aircraft, excess capacity, strong competition, and union wage demands have produced an unprecedented depressive state. Bringing all of this to a culmination were the events of September 11, 2001, which brought a fear of flying and a steep falloff in enplanements and of course in leisure and business travel. This loss of traffic, even with significant short-term federal financial support, has resulted in record operating losses. As a consequence, U.S. Airways declared hardship and began operations under Chapter 11 bankruptcy protection. And in December 2002, under the pain of daily $1 million loses, United Airlines, the nation's largest carrier, also sought Chapter 11 protection. The financial picture among most air carriers is similar. Despite the apparent success of deregulation, competition has not produced a viable system of air carriers. The future of the airlines is uncertain, particularly in a struggling domestic economy. The answer certainly lies in downsizing, moving more strongly to a no-frills model such as that employed by Southwest, and perhaps further consolidation. The future role of the federal government in subsidizing the airlines is unclear (Leinbach & Bowen, 2004).

The health of the air transport industry has important impacts on urban areas. Perhaps most obvious is the resulting decline in services as carriers cut back, liquidate, or consolidate. Air services are absolutely critical to the vitality of large- as well as small- and medium-sized urban areas. Increasingly, firms examining a potential expansion or relocation site examine air services as one of the highest priorities in a location decision. The proximity of Bluegrass Airport in Lexington, Kentucky, for example, was a major factor in Toyota's decision to locate in Georgetown, Kentucky. The employment effect of even small urban airports is highly significant. Beyond this, the issue of aircraft manufacture and its suppliers is also critical. The impact of Boeing and its local and regional suppliers on the Seattle urban area and beyond is noteworthy. The slowdown in aircraft demand can produce significant repercussions.

Regional and Commuter Carriers

One additional impact of deregulation was the growth of the regional and commuter carriers. Regional/commuter airline revenue enplanements increased sevenfold from 1975 to 1999 (see Figure 2.11). Large air carriers began purchasing their regional

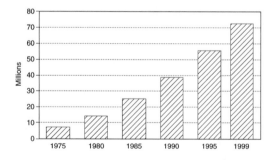

FIGURE 2.11. Regional/commuter airline revenue enplanements: 1975–1999. Source: U.S. Department of Transportation, Bureau of Transportation Statistics (2000a, p. 2-23).

partners as a way of ensuring a source of traffic to their major flights. One example is United Express as a feeder to United Airlines, with subsidiary operations performed by, for instance, Air Wisconsin in the upper Midwest and Atlantic Coastal Airlines in the southeast. Another model is ComAir, an independently owned feeder based in Cincinnati and linked to Delta, operating in the southeast. Serving a range of markets and utilizing a diverse fleet of aircraft has also stimulated the evolution of the commuter/regional airline industry.

A related effect has been the growth of regional jets (RJs) (Taylor, 2000; Yung, 2000). Essentially smaller jet aircraft (with 50 or 70 seats) have gradually begun to replace traditional turboprop aircraft and in some instances larger jets in nodes where demand is fairly low (see Figure 2.12). The pace of this replacement will accelerate rapidly over the next 25 years (see Figure 2.13). Reflecting the expansion of the regional airline market over time, the length of the average trip has gradually increased (see Figure 2.14). The result of this innovation can be immediate access for smaller communities and improved scheduling to larger communities. The 50-seat aircraft (the Canadair Regional Jet produced by Bombardier in Toronto is one example) has many advantages over more traditional air-

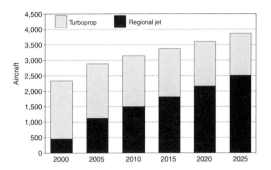

FIGURE 2.13. Regional/commuter commercial passenger aircraft fleet: projected 2000–2025. Source: U.S. Department of Transportation, Bureau of Transportation Statistics (2000a, p. 2-29).

craft. They are cheaper to fly because they cost less to purchase, burn less fuel, and employ lower cost crews. In some situations the use of these aircraft will provide business travelers with an alternative to congested hubs and crowded airports. More important, it allows airlines to cope more easily with fluctuating demand. For example, after September 11 United Airlines was able to develop a more frequent schedule of service and accommodate demand between Lexington, Kentucky, and Chicago by using the RJs. Larger jet aircraft, the British Aerospace 146, formerly used on the same segment, were displaced to western markets where the higher capacity could be better utilized (Leinbach & Bowen, 2004).

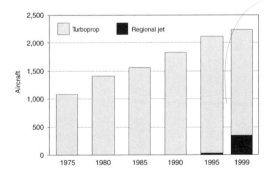

FIGURE 2.12. Regional/commuter passenger airline fleet: 1975–1999. Source: U.S. Department of Transportation, Bureau of Transportation Statistics (2000a, p. 2-24).

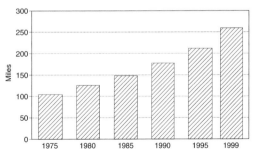

FIGURE 2.14. Average trip length on regional/commuter carriers: 1975–1999. Source: U.S. Department of Transportation, Bureau of Transportation Statistics (2000a, p. 2-24).

Airfreight

Air cargo, while still accounting for a small proportion of total freight movement in the United States, has been growing. In fact, as noted above, from 1990 to 1996 it was the fastest growing mode of freight transport. Air cargo also grew much faster than the passenger sector and increased fivefold from 5 billion revenue ton-miles (RTMs) in 1975 to 25 billion RTMs in 1999 (see Figure 2.15). The expansion is attributed to the new demand in a restructured economy; the emergence of all-cargo carriers such as UPS, FedEx, and Airborne; and the availability of capacity in passenger aircraft. Freight is moved in dedicated all-cargo aircraft as well as in the belly hold of passenger aircraft. For example, a dedicated Boeing 747-400 freighter can hold 124 tons of cargo and fly a maximum range of 8,240 kilometers, while an McDonald Douglas MD11, used by FedEx from Subic Bay, in the Philippines, can hold 85 tons. The belly capacity of a 747 in combi mode is a not-insignificant 16 tons. The new Airbus 380 will accommodate 150 tons in cargo configuration. About 225 747 Freighters currently carry half of all of the world's air cargo freight.

While in 1999 domestic air cargo exceeded international freight for the first time since 1975, the international potential is huge (see Figure 2.15). The trans-Pacific traffic, in particular, is some of the most heavily contested in the global market (Bowen, 2004). In response to this demand, all-cargo carriers have established hubs in the Pacific Rim in order to gain access to lucrative traffic. FedEx, with its domestic base in Memphis, established a now-successful base in Subic Bay, the Philippines in 1995 (Bowen, Leinbach, & Mabazza, 2002). UPS similarly has recently established a Pacific hub at the former Clark Air Force Base in the Philippines.

Air cargo has had a major impact on particular cities. For example, as a result of FedEx's location, Memphis is the world's largest cargo hub. The firm is the city's largest employer, with over 30, 000 employees located in the Memphis area. The total financial impact, both direct and indirect, of the Memphis International Airport on the regional economy amounts to almost $10 billion in annual sales. Research has shown

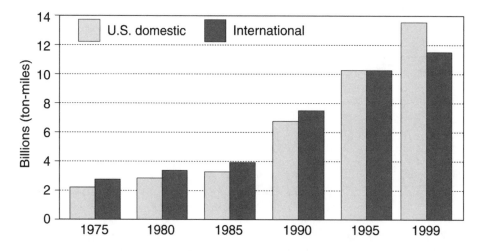

FIGURE 2.15. U.S. domestic and international freight revenue ton-miles: 1975–1999. Source: U.S. Department of Transportation, Bureau of Transportation Statistics (2000a, p. 2-20).

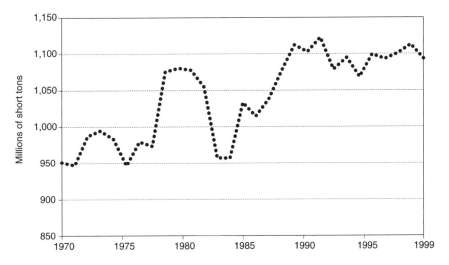

FIGURE 2.16. Waterborne domestic trade: 1970–1998 (annual totals). Source: U.S. Department of Transportation, Bureau of Transportation Statistics (2000a, p. 2-31).

that one in five residents in the Memphis area owe their jobs to the airport and its activities. A recent contract with the U.S. Postal Service will generate an additional 1,500 jobs by 2008. The Memphis Area Chamber of Commerce lists 40 FedEx customers in the last 5 years that have located to or expanded in Memphis because of FedEx, representing $710 million in new investment and 9,813 new jobs. The decision by FedEx to locate a new hub in the Piedmont Triangle near Greensboro is expected to create over 16,000 jobs in a 16-year period and add $7.5 billion to the local economy as a result of the hub itself, existing businesses growing, and new businesses locating in the area. Yet the plan has generated significant debate over the externality effects of congestion that such a hub would create.

In an analogous situation, UPS has made a tremendous impact on the economy of Louisville, where it is responsible for 23,000 jobs and half a billion dollars annually in economic impact. UPS launched its $1.1 billion, 4-million-square-foot global air hub in Louisville in late September 2002, which will serve as a growth engine for the city and region.

MARITIME INTERACTIONS

In addition to movements between cities via air, rail and road, the United States does have a significant maritime system of transportation that consists of waterways, ports, and, most important, their intermodal connections. Of course, these movements are nearly exclusively freight-oriented. In 1975, the shipping industry enjoyed rapid growth as a result of the crude oil trade. But 23 years later growth had stagnated mostly as a result of the increased dependence on pipelines to ship crude oil (see Figure 2.16). U.S. domestic waterborne trade consists of bulk commodities moving on the inland Great Lakes and coastal waterways. This trade suffered during the oil shortages of the mid-1970s, the financial crises of the 1980s, and again during the flooding of the Mississippi River in 1993. Over the past 10 years the major commodities shipped in this fashion have been petroleum products, crude materials, and coal.

In contrast waterborne foreign trade has grown 65% by weight (see Figure 2.17; U.S. Department of Transportation, Bureau of Transportation Statistics, 2000a). In general the growth in waterborne foreign trade re-

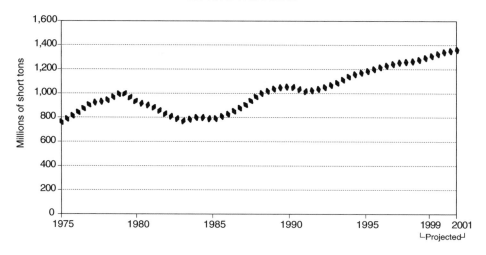

FIGURE 2.17. Waterborne foreign trade: 1975–2001. Source: U.S. Department of Transportation, Bureau of Transportation Statistics (2000a, p. 2-32).

flects the expansion of the global production system as extranational ties in the production and distribution chains are established.

Over the past 25 years the shipping industry, like the other modes, has been under pressure to become more cost-efficient. In fact this has taken place via the application of a variety of innovations and technological measures. Perhaps the most revolutionary change in U.S. shipbuilding has been the production of advanced container ships and roll-on/roll-off vessels. At the end of 1975 the United States held 25% of the world's cargo fleet, which carried 30% of the tonnage. The growth in container ships has generated a huge development in container facilities by ports and shipping lines to compete with noncontainer cargo vessels (U.S. Department of Transportation, Bureau of Transportation Statistics, 2000). As a response to and reflection of this development, world containerized trade almost doubled between 1991 and 1999 (see Figure 2.18). The top container ports for 1987 and 1999 are depicted in Figure 2.19. Most striking are the dramatic increases in container trade out of San Diego and Los Angeles, both of which have surpassed New York, the former leading port. The change

in traffic clearly reflects the growth of the Asian trade partnerships that have evolved over the past decade.

While the surge in container traffic has been a positive development, it has also pointed up infrastructure deficiencies facing U.S. ports and waterways. Deepening of waterways and ports has not been carried out but must be. In addition, containerization and cost-efficiency in cargo movement have emphasized the critical importance of seamless intermodal transportation (Slack, 1990).

It should be mentioned that while New York, Long Beach, California, and Los Angeles are the top container ports in the United States in terms of total volume of cargo, where noncontainer cargo such as oil movements are recorded, the ports of Houston, New Orleans (and southern Louisiana), Corpus Christi, Texas, and Beaumont, Texas, dominate the total volume rankings. While these and other ports have a significant economic impact on their associated urban areas and regions, there are, as with large airports, significant negative externalities. Ports conflict with other urban demands for scarce coastal resources and are major sources of air and water pollution. In addition, nearly all ports generate huge

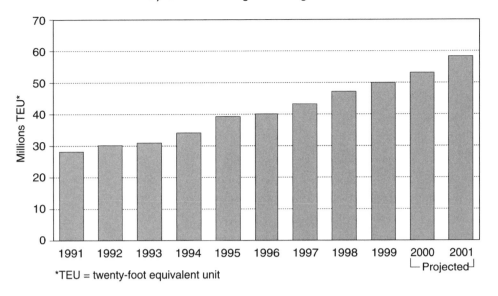

FIGURE 2.18. World containerized trade moves: 1991–2001. Source: U.S. Department of Transportation, Bureau of Transportation Statistics (2000a, p. 2-35).

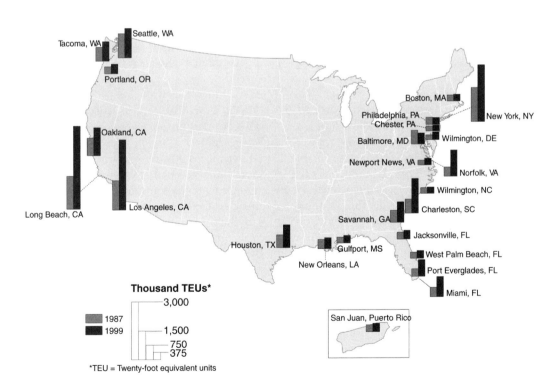

FIGURE 2.19. Top 25 U.S. container ports: 1987 and 1999. Source: U.S. Department of Transportation, Bureau of Transportation Statistics (2000a, p. 2-36).

amounts of truck traffic and are therefore major congestion points and sources of vehicle-generated air pollution. The expansion of ports, and other transport terminals, often entails dilemmas, especially those involving relations between terminals and the communities in which they operate (Hoyle & Pinder, 1992). Increased traffic, congestion, air and noise pollution, and environmental degradation of coastal areas will force communities and the operators of these facilities to explore creative alternatives. These include satellite hubs and inland load centers that can help divert some of the freight-handling activity away from congested urban centers (Goetz & Rodrigue, 1999; Slack, 1999).

DYNAMICS OF THE INTERURBAN TRANSPORTATION SYSTEM

In sum, a variety of forces have influenced city interactions within the United States and beyond. Since its beginning in the 1970s, the impact of liberalization in the forms of deregulation and privatization has been influential. Motor carriers were given authority to set rates, and as a result existing carriers expanded into new services and routes and new carriers emerged. Deregulation impacts have been especially conspicuous in the railroads, where strong consolidation has taken place and nonprofitable routes have been jettisoned. Similarly, as the Civil Aeronautics Board was dissolved, consolidation has taken place in the air transport industry. A major spatial impact of this change has been the move from a linear point-to-point routing pattern to one dominated by a hub system in a newly configured feeder network. Heightened price and route competition has occurred as a result of the entry of new startups, particularly the no-frills carriers. The new flexibility has allowed the airlines to service new and existing markets with a variety of aircraft. In particular, the regional jet has strengthened the air services of smaller cities and enhanced their development potential.

Globalization has also shaped the transportation system. U.S. multinational firms import materials from extranational locations, and indeed many firms have international bases either because of access to materials, markets, or more frequently a lower cost source of labor. This international production milieu affects the transport of goods to and from Canada and Mexico as well as to and from Japan and China, to name only four significant trading partners. While all modes have been affected, it is air that is most conspicuous in driving our interactions abroad.

The types of commodities moved in the freight system are directly related to the structure of our economy. In the mid-1970s the U.S. economy was transitioning from manufacturing to services. Today the service sector is larger than all other sectors in output and growth potential. This new economy is shifting from mass manufacturing and distribution toward custom manufacturing and retailing. These changes have transformed the nature of freight movements. More freight is being moved over longer distances and increasingly to international destinations. This freight is typically lighter and higher in value.

At the same time, critical changes between consumer and producer have occurred. Logistical decisions are required to optimize performance, customer service, and profits. Contemporary needs require higher priced and higher quality transport services to assure timely delivery with minimum damage. With the transformation of the economy, the change might be characterized as a shift away from a "*push*" system of logistics where manufacturing, distribution, and retailing were dedicated to a mass-production purpose with associated warehousing and retailing. In this system centralization of design, production, and marketing was used to achieve economies of scale. Long production runs were warehoused and large inventories developed in anticipation of consumer demand (U.S. Department of Transportation, Bureau of Transportation Statistics, 2000a, p. 2-55).

Today the logistics system is increasingly one defined by "*pull*." In this environment retailers, distributors, manufacturers, and suppliers closely track customer demand on a daily, or sometimes hourly, basis. Orders and purchase patterns pull goods through the supply chain. It is now common for industries not to produce parts and final goods until an order is placed. This gave rise to the "just-in-time" manufacturing concept. Indeed, this "just-in-time" concept is already a dated term as "total logistics solutions" is now used to describe a seamless process of demand and supply (U.S. Department of Transportation, Bureau of Transportation Statistics, 2000a, p. 2-56).

This transition from a "push" to a "pull" logistics system has been made possible by the information revolution brought about by enhanced computerization and the application of new technologies (Leinbach & Bowen, 2004). Given the importance of time and the higher value of shipments, time-definite services have become much more important. Carriers have been pressed to provide integrated and intermodal services that are reliable and cost-effective. This new environment has encouraged the growth of new logistical relationships such as third party logistics (3PLs), where an intermediate firm will act on behalf of a client to store, package, and ship goods to a customer on demand. Indeed 4PLs are an emerging trend where two intermediaries function as logistical agents and relieve producers and consumers of the need to deal with these aspects of operations (U.S. Department of Transportation, Bureau of Transportation Statistics, 2000a, p. 2-56).

At the same time, a variety of other factors affecting our economy and society have spatial interaction implications. Among these are the long-felt democratization of mobility. This equalizing of mobility will continue to be critical to the search for employment and the fuller use of discretionary income and leisure time. The aging of the U.S. population, a changing pattern of immigration, and growing affluence all have

implications for transportation. Off-peak travel with an older population and the location of immigrants settling in metro suburbs rather than cities are important. Larger incomes mean generally that access to automobiles will increase and trips per household and trip lengths will increase. It is clear that transportation will continue to serve to disperse population. Airport congestion may drive development to those areas where capacity is available without the negative externalities of congestion and pollution. A major trend is the application of technology and displacement of transportation. The growth of the Internet, e-commerce, and telework all will expand and have new impacts on society and the economy.

FUTURE PROSPECTS

Cities will be affected in a variety of ways by changing transportation forms and connections. Corridors of many urban-anchored regions of the United States will be provided with HSR service (Figure 2.7 above). The diffusion of such services potentially will produce important benefits to other modes, namely, reductions in highway and airport congestion. Forecasts suggest that in the Northeast Corridor HSR could generate 3.5 billion passengers annually by 2025. Such services in already-designated corridors would mean that nearly 75% of the nation's metropolitan population, over 150 million people, would have access (U.S. Department of Transportation, Bureau of Transportation Statistics, 2000a, p. 2-17).

Major changes are also likely among the airlines. Greater efficiency will have to be initiated through a variety of cost-cutting measures. While deregulation has clearly had some important positive impacts, some areas have not felt lower fares and better schedules. Low-cost convenient airline service is a major factor in the viability of all communities regardless of size. Competition among communities for such services will become more intense. In addition, budget carriers will increasingly utilize alternate, less-congested hubs with important consequences for the associated cities. It is certain that air travel, while affected by economic downturns, will continue to expand for both leisure and business purposes.

Perhaps most critical are the changes and the ongoing dynamics in the realm of freight transportation. As a result of firms' search for profitability, the press of competition, and advances in technology, the industry is dramatically different from that of even a decade ago. The umbrella concept over this period has been *intermodalism*, the term used to describe the movement of freight using two or more modes that interconnect and interchange, allowing cost-efficient delivery. One of the most significant continuing developments has been the recognition of the efficiencies to be achieved through containerization (Slack, 1990). Over the past 25 years, urban-based motor carriers, railroads, and ports have invested in hub container facilities. The motor carrier industry is a critical aspect of the trend toward intermodal supply chains. Logistics providers who manage and create these chains may have their own fleets of trucks and will use them to interface with their air- and sea-based fleets to provide time-definite services in a truly global environment. These investments have altered the urban landscape and culture. Increasingly important are the air hub complexes associated with both private- and public-sector operations. In truth, freight movement is increasingly becoming "mode invisible" with performance (time, cost, and reliability) determining the choice of modes (U.S. Department of Transportation, Bureau of Transportation Statistics, 2000a, pp. 2-50–2-51). The significance of such intermodality and the options for integrating modes is that it provides flexible responses to changing supply chain requirements in global markets. In this perspective, supply chain management through transport is becoming more important in the success and survival of firms and their urban bases in a global context.

REFERENCES

Amtrak derailed. (2002, June 26). *Financial Times*, p. 14.

Armbruster, W. (2001, July 9–15). Aerotropolis. *Journal of Commerce Weekly*, p. 10.

Bowen, J. T. (2004, March). The geography of freighter aircraft operations in the Pacific Basin. *Journal of Transportation Geography, 12*, 1–11.

Bowen, J. T., & Leinbach, T. R. (1995). The state and liberalization: The airline industry in the East Asian NICs. *Annals, Association of American Geographers, 85*, 468–493.

Bowen, J. T, Leinbach, T. R., & Mabazza, D. (2002). Air cargo services, the state and industrialization strategies in the Philippines: The redevelopment of Subic Bay. *Regional Studies, 36*, 451–467.

Burns, J. B. (1998). *Railroad mergers and the language of unification*. Westport, CT: Quorum.

Cairncross, F. (1997). *The death of distance: How the communications revolution will change our lives*. Boston: Harvard Business School Press.

Debbage, K. (1994). The international airline industry: Globalization, regulation, and strategic alliances. *Journal of Transport Geography, 2*, 190–203.

Field, D. (1996). On 40th birthday, interstates face expensive midlife crisis. *Insight on the News, 12*, 40–42.

Fishman, R. (2000, Winter). The American metropolis at century's end: Past and future influences. *Housing Policy Debate, 11*(1), 199–213.

Fossey, J. (1996, October). Full steam ahead—Privatization has put new life into Canadian National Railway. *Containerisation International*, pp. 72–75.

Gifford, J. (1994). *Planning the Interstate Highway System*. Boulder, CO: Westview Press.

Gillespie, A., & Richardson, R. (2000). Teleworking and the city: Myths of workplace transcedence and travel reduction. In J. O. Wheeler, Y. Aoyama, & B. Warf (Eds.), *Cities in the telecommunications age* (pp. 228–245). New York: Routledge.

Goetz, A. R., & Rodrigue, J. P. (1999). Transport terminals: New perspectives. *Journal of Transport Geography, 7*, 255–261.

Hoyle, B., & Pinder, D. A. (Eds.). (1992). *European port cities in transition*. London: Belhaven Press.

Hu, P. S., & Young, J. R. (1999). *Summary of travel trends: 1995 Nationwide Personal Transportation*

Survey. Washington, DC: U.S. Department of Transportation, Federal Highway Administration.

International Civil Aviation Organization. (2001, June 13). *Growth in air traffic projected to continue: Long term forecasts.* Montreal: Author.

Janelle, D. G., & Hodge, D.C. (Eds.). (2000). *Information, place and cyberspace: Issues in accessibility.* Berlin: Springer.

Kasarda, J. D. (2001, August–September). From airport city to aerotropolis. *Airport World, 6,* 42–47.

Kay, J. H. (1997). *Asphalt nation: How the automobile took over America and how we can take it back.* New York: Crown.

Leinbach, T. R., & Bowen, J. (2004). Airspaces: Air travel, technology and society. In S. Brunn, S. Cutter, & J. W. Harrington (Eds.), *Geography and technology* (pp. 285–313). Amsterdam: Kluwer.

Leinbach, T. R., & Bowen, J. (in press). Transport services and the global economy: Toward a seamless market. In J. Bryson & P. Daniels (Eds.), *The handbook of service industries.* Cheltenham, UK: Edward Elgar.

Leinbach, T. R. (2001). Emergence of the digital economy and E-commerce. In T. R. Leinbach & S. D. Brunn (Eds.), *Worlds of E-commerce: Economic, geographical and social dimensions* (pp. 3–26). Chichester, UK: Wiley International.

Lewis, T. (1997). *Divided highways: Building the interstate highways, transforming American life.* New York: Viking Press.

Machalaba, D. (2001, December 26). High speed grain terminals bode change for rural U.S. *Wall Street Journal,* p. 5.

Mehndiratta, S. R., & Hansen, M. (1998). *Because people sleep at night: Time-of-day effects in inter-city business travel.* Washington, DC: Transportation Research Board.

Mitchelson, R. L., & Fisher, J. S. (1987). Long distance commuting and population change in Georgia, 1960 to 1980. *Growth and Change, 18,* 44–65.

Moon, H. (1994). *The Interstate Highway System.* Washington, DC: Association of American Geographers.

National Household Travel Survey. (2001). Washington, DC: U.S. Department of Transportation and Federal Highway Administration. Available online at http://www.bts.gov/nhts/

Norris, D. A. (1987). Interstate highway exit morphology: Nonmetropolitan exit commerce on I-75. *Professional Geographer, 39,* 23–32.

Park, J. J., Babcock, M. W., & Lemke, K. (1999). The impact of railway mergers on grain transportation markets: A Kansas case study. *Transportation Research, Part E, Logistics and Transportation Review, 35,* 269–290.

Perl, A. (2002). *New departures: Rethinking rail passenger travel in the 21st century.* Lexington: University of Kentucky Press.

Raguraman, K. (1997) International air cargo hubbing: The case of Singapore. *Asia Pacific Viewpoint, 38,* 55–74.

Rose, M. H. (1990). *Interstate express highway politics, 1939–1989* (Rev. ed). Knoxville: University of Tennessee Press.

Rubenstein, J. M. (1992). America's "just-in-time" highways: I-65 and I-75. In D. G. Janelle (Ed.), *Geographical snapshots of North America* (pp. 432–435). New York: Guilford Press.

Sharkey, J. (2002, June 14). Rail projects are a sign of a quiet revolution. *International Herald Tribune,* p. 3.

Slack, B. (1990). Intermodal transportation in North America and the development of inland load centers. *Professional Geographer, 42,* 72–83.

Slack, B. (1999). Satellite terminals: A local solution to hub congestion? *Journal of Transport Geography, 7,* 241–246.

Spear, B. D., & Weil, R. W. (1999). Accessibility to intercity transportation services from small communities: A geospatial analysis. In *Transportation Research Record 1666* (pp. 65–73). Washington, DC: Transportation Research Board.

Taylor, A. III. (2000, September 4). Little jets are huge! *Fortune,* pp. 274–278.

U.S. Department of Transportation. (2001). *National household travel survey.* Washington, DC: Author. Available online at http://nhts.ornl.gov/2001/index.shtml

U.S. Department of Transportation, Bureau of Transportation Statistics. (1997). *Transportation statistics annual report* (BTS97-S-01). Washington, DC: Author.

U.S. Department of Transportation, Bureau of Transportation Statistics. (2000a). *The changing face of transportation* (BTS00-007). Washington, DC: Author.

U.S. Department of Transportation, Bureau of Transportation Statistics. (2000b). *North American transportation in figures* (BTS00-05). Washington, DC: Author.

U.S. Department of Transportation, Bureau of Transportation Statistics. (2000c). *Transportation statistics annual report* (BTS01-02). Washington, DC: Author.

U.S. Department of Transportation, Bureau of Transportation Statistics. (2001). *Airport activity statistics of certificated air carriers: Summary table twelve months ending December 31, 2000.* Washington, DC: Author.

U.S. Department of Transportation, Bureau of Transportation Statistics. (2002). *U.S. international travel and transportation trends* (BTS02-03). Washington, DC: Author.

U.S. Department of Transportation, Federal Aviation Administration. (2000). *Strategic plan.* Washington, DC: Author.

U.S. Department of Transportation, Federal Highway Administration. (2002, October). *Freight analysis framework* (FHWA-OP-03-006, DL 13694). Available online at http://ops.fhwa.dot. gov/freight/publications/faf.html

Yung, K. (2000, June 11). Regional jets are airlines' new competitive arena. *Dallas Morning News,* p. 1H.

Zuckermann, W. (1991). *End of the road: The world car crisis and how we can solve it.* Post Mills, VT: Chelsea Green.

CHAPTER 3

Transportation and Urban Form

Stages in the Spatial Evolution
of the American Metropolis

PETER O. MULLER

As the opening chapter demonstrated, the movement of people, goods, and information within the local metropolitan area is critically important to the functioning of cities. In this chapter, I review the U.S. urban experience of the past two centuries and trace a persistently strong relationship between the intraurban transportation system and the spatial form and organization of the metropolis. Following an overview of the cultural foundations of urbanism in the United States, I introduce a four-stage model of intrametropolitan transport eras and associated growth patterns. Within that framework it will become clear that a distinctive spatial structure dominated each stage of urban transportation development and that geographical reorganization swiftly followed the breakthrough in movement technology that launched the next era of metropolitan expansion.[1] Finally, I briefly consider the contemporary scene, both as an evolutionary composite of the past and as a dynamic arena in which new forces may already be forging a decidedly different future.

CULTURAL FOUNDATIONS OF THE U.S. URBAN EXPERIENCE

Americans, by and large, were not urban dwellers by design. The emergence of large cities between the Civil War and World War I was an unintended by-product of the nation's rapid industrialization. Berry (1975), recalling the observations of the 18th-century French traveler Hector St. Jean de Crèvecoeur, succinctly summarized the cultural values that have shaped attitudes toward urban living in the United States for the past two centuries:

> Foremost . . . was a *love of newness*. Second was the overwhelming desire to be *near to nature*. *Freedom to move* was essential if goals were to be realized, and *individualism* was basic to the self-made man's pursuit of his goals, yet *violence* was the accompaniment if not the condition of success—the competitive urge, the struggle to succeed, the fight to win. Finally, [there is] a great *melting pot* of peoples, and a manifest *sense of destiny*. (p. 175)

As the indigenous culture of the emergent nation took root, its popular Jefferso-

59

nian view of democracy nurtured a powerful rural ideal that regarded cities as centers of corruption, social inequalities, and disorder. When mass urbanization became unavoidable as the Industrial Revolution blossomed after 1850, Americans brought their agrarian ideal with them and sought to make their new manufacturing centers noncities. For the affluent, this process began almost as soon as the railroad reached the city in the 1830s; by midcentury, numerous railside residential clusters had materialized just outside the built-up urban area. But middle-income city dwellers could not afford this living pattern because of the extra time and travel costs it demanded. With cities becoming increasingly unlivable as industrialization intensified, pressures mounted after 1850 to improve the urban transportation system to permit the burgeoning middle class to have access to the high-amenity environment of the periurban zone.

The necessary technological breakthrough—in the form of the electric (streetcar) traction motor—was finally achieved in the late 1880s. By the opening of the final decade of the 19th century, the city began to spill over into the much-desired surrounding countryside. By 1900, the decentralization of the middle-income masses was no longer a trickle but a widening migration stream (which has yet to cease its flow) that rapidly spawned the emergence of the full-fledged metropolis, wherein a steadily increasing multitude of urban dwellers shunned the residential life of the industrial city altogether. Hardly had this initial transformation of the U.S. city been completed when the automobile introduced mass private transportation in the 1920s for all but the poorest urban dwellers. As the intrametropolitan highway network expanded in the interwar period, successive rounds of new peripheral residential development were launched, and the urban perimeter was pushed ever farther from the downtown core. But these centrifugal forces still operated at a rather leisurely

pace, undoubtedly slowed after 1930 by a decade and a half of economic depression and global war.

Following the conclusion of World War II, however, all constraints were removed, and a massive new wave of deconcentration was triggered. Spurred by a reviving economy, widespread housing demands, federal home loan policies that favored new urban development, copious highway construction, and more efficient cars, the exodus from the nation's cities reached unprecedented proportions between 1945 and 1970. The proliferation of urban freeways (introduced in Southern California in the late 1930s) heightened the centrifugal drift. With the completion of these high-speed, limited-access, superhighway networks in the 1960s and 1970s came the elimination of the core-city central business district's (CBD's) regionwide centrality advantage, as superior intrametropolitan accessibility became a ubiquitous spatial good available near any expressway interchange location. As entrepreneurs swiftly realized the consequences of this structural reorganization of the metropolis, nonresidential activities of every variety began their own massive wave of intraurban deconcentration. Manufacturing and retailing led the way.

By 1980, the erstwhile ring of bedroom communities that girdled the aging central city had become transformed into a diversified, expanding *outer city* that was increasingly home to a critical mass (i.e., more than half) of the metropolitan area's industrial, service, and office-based business employers. Moreover, major new multipurpose activity centers have been emerging in the outer city during the past three decades, attracting so many high-order urban functions that residents of surrounding areas have completely reorganized their lives around them. Thus the compact industrial city of the recent past has today turned inside-out. The rise of downtown-type centers in the increasingly independent outer city has also forged a decidedly polycentric metropolis, the product of both the cumulative spatial processes outlined above and the emerging forces of a postindustrial society that are shaping new urbanization patterns that represent a clean break with the past.

The legacy of more than two centuries of intraurban transportation innovations and the development patterns they etched on the landscape of metropolitan America is *suburbanization*—the growth of the edges of the urbanized area at a rate faster than that in the already developed interior. Since the spatial extent of the continuously built-up urban area has, throughout history, exhibited a fairly constant time–distance radius of about 45 minutes' travel from the center, each breakthrough in higher-speed transport technology extended that radius into a new outer zone of suburban residential opportunity. In the 19th century, commuter railroads, horse-drawn trolleys, and electric streetcars each created their own suburbs— *and thereby also created the large industrial city,* which could not have been formed without absorbing these new suburbs into the pre-existing compact urban center. But the suburbs that developed in the early 20th century began to assert their independence from the enlarged, ever more undesirably perceived central cities. Few significant municipal consolidations occurred after the 1920s, except in postwar Texas and certain other Sunbelt locales (a trend that ended during the 1980s). As the automobile greatly reinforced the intraurban dispersal of population, the distinction between central city and suburban ring grew as well. And as freeways greatly reduced the friction effects of intrametropolitan distance for most urban functions, nonresidential activities deconcentrated to such an extent that by the mid-1970s the emerging outer city became at least the coequal of the neighboring central city that spawned it— making the word *sub*urb an oxymoron.

This urban experience of the United States over the past two centuries is the product of uniquely U.S. cultural values and contrasts sharply with modern urbanization trends in Europe. The European metropolis, though also now experiencing decentralization, retains a more tightly agglomerated spatial structure, and the historic central city continues to dominate its immediate urban region. Sommers (1983) has summarized the concentrative forces that have shaped the cities of postwar Europe:

> Age is a principal factor, but ethnic and environmental differences also play major roles in the appearance of the European city. Politics, war, fire, religion, culture, and economics also have played a role. Land is expensive due to its scarcity, and capital for private enterprise development has been insufficient, so government-built housing is quite common. Land ownership has been fragmented over the years due to inheritance systems that often split land among sons. Prices for real estate and rent have been government controlled in many countries. Planning and zoning codes as well as the development of utilities are determined by government policies. These are characteristics of a region with a long history, dense population, scarce land, and strong government control of urban land development. (p. 97)

Further evidence of the persistent dominance of the European central city is shown in Figure 3.1. The density gradient pattern of the North American metropolis (Figure 3.1A) is marked by progressive deconcentration, whereas the counterpart European pattern (Figure 3.1B) exhibits sustained intraurban centralization. The newest metropolitan trends of the past decade in Europe do show accelerating suburbanization, but the central city remains the intraurban social and economic core.

THE FOUR ERAS OF INTRAMETROPOLITAN GROWTH AND TRANSPORT DEVELOPMENT

The evolving form and structure of the U.S. metropolis, briefly outlined in the previous section, may be traced within the frame-

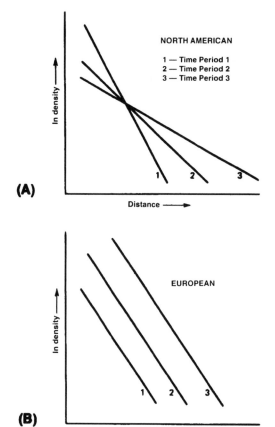

(A)

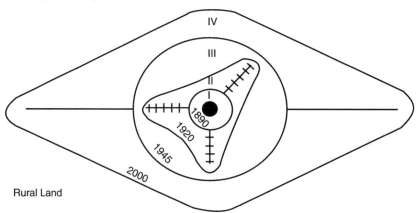

work of four transportation-related eras identified by Adams (1970). Each growth stage is dominated by a particular movement technology and network expansion process that shaped a distinctive pattern of intraurban spatial organization:

1. Walking–Horsecar Era (1800–1890)
2. Electric Streetcar Era (1890–1920)
3. Recreational Automobile Era (1920–1945)
4. Freeway Era (1945–present)

This model, diagrammed in Figure 3.2, reveals two sharply different morphological properties over time. During Eras 1 and 3 uniform transport surface conditions prevailed (as much of the urban region was similarly accessible), permitting directional freedom of movement and a decidedly compact overall development pattern. During Eras 2 and 4 pronounced network biases were dominant, producing an irregularly shaped metropolis in which axial development along radial transport routes overshadowed growth in the less inaccessible interstices.

A generalized model of this kind, while organizationally convenient, risks oversimplification because the building processes of

FIGURE 3.1. Density gradients over time in the North American and European metropolis. Source: Hartshorn (1992, p. 230). Copyright 1992 by John Wiley & Sons. Reprinted by permission.

FIGURE 3.2. Intraurban transport eras and metropolitan growth patterns: (I) Walking-Horsecar Era, (II) Electric Streetcar Era, (III) Recreational Auto Era, and (IV) Freeway Era. Source: Adams (1970, p. 56). Copyright 1970 by The Association of American Geographers. Adapted by permission.

several simultaneously developing cities do not fall into neat time–space compartments (Tarr, 1984, pp. 5–6). An examination of Figure 3.3, which maps Chicago's growth for the past 150 years, reveals numerous empirical irregularities, suggesting that the overall urban growth pattern is somewhat more complex than a simple, continuous, centrifugal thrust. Yet, when developmental ebb-and-flow pulsations, leapfrogging, back-filling, and other departures from the normative scheme are considered, there still remains a reasonably good correspondence between the model and historical-geographical reality. With that in mind, each of the four eras is now examined in detail.

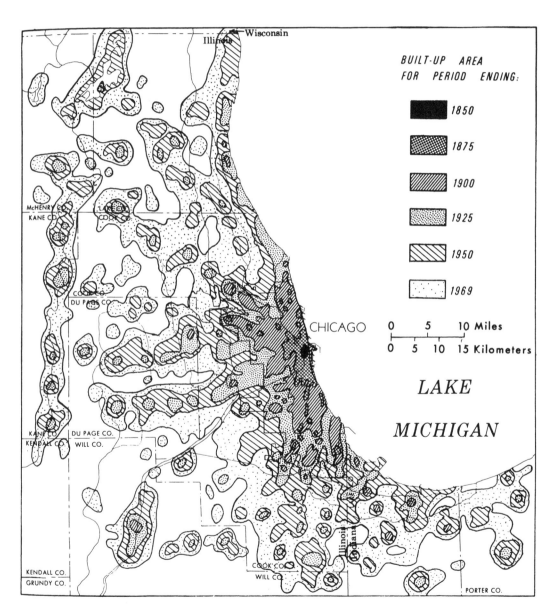

FIGURE 3.3. The suburban expansion of metropolitan Chicago from the mid-19th century through 1970. Source: Berry et al. (1976, p. 9). Copyright 1976 by B. J. L. Berry. Reprinted by permission.

Walking-Horsecar Era (1800–1890)

Prior to the middle of the 19th century, the U.S. city was a highly agglomerated urban settlement in which the dominant means of getting about was on foot (Figure 3.2, Era I). Thus people and activities were required to cluster within close proximity of one another. Initially, this meant less than a 30-minute walk from the center, later extended to about 45 minutes when the pressures of industrial growth intensified after 1830. Any attempt to deviate from these mobility constraints courted urban failure: Washington, D.C., struggled enormously for much of its first century on L'Enfant's 1791 plan that dispersed blocks and facilities too widely for a pedestrian city, prompting Charles Dickens to observe during his 1842 visit that buildings were located "anywhere, but the more entirely out of everyone's way the better" (Schaeffer & Sclar, 1975, p. 12).

Within the walking city, there were recognizable concentrations of activities as well as the beginnings of income-based residential congregations. The latter behavior was clearly evinced by the wealthy, who walled themselves off in their larger homes near the city center; they also favored the privacy of horse-drawn carriages to move about town—undoubtedly the earliest U.S. form of wheeled intraurban transportation. The rest of the population resided in tiny overcrowded quarters, of which Philadelphia's now-restored Elfreth's Alley was typical (see Figure 3.4).

The rather crude environment of the compact preindustrial city impelled those of means to seek an escape from its noise as well as the frequent epidemics that resulted from the unsanitary conditions. Horse-and-carriage transportation enabled the wealthy to reside in the nearby countryside for the disease-prone summer months. The arrival of the railroad in the early 1830s soon provided the opportunity for year-round daily travel to and from elegant new trackside suburbs. By 1840 hundreds of affluent businessmen in Boston, New York, and Philadelphia were making these round-trips every weekday. The "commutation" of their fares to lower prices, when purchasing tickets in monthly quantities, introduced a new word to describe the journey to work: *commuting*. A few years later, these privileges extended to the nouveau riche professional class (well over 100 trains a day ran between Boston and its suburbs in 1850), and a spate of planned rail suburbs, such as Riverside near Chicago, soon materialized.

As industrialization and its teeming concentrations of modest, working-class housing increasingly engulfed the mid-19th-century city, the worsening physical and social environment heightened the desire of middle-income residents to suburbanize as well. For those unable to afford the cost and time of commuting—and with the pedes-

FIGURE 3.4. Elfreth's Alley in the heart of downtown Philadelphia. Its restoration provides a good feel for the lack of spaciousness in the Revolution-era city. Photo courtesy of R. A. Cybriwsky.

trian city stretched to its morphological limit—these middle-class yearnings intensified the pressure to improve intraurban transport technology.

As early as the 1820s, New York, Philadelphia, and Baltimore had established *omnibus* lines. These intracity adaptations of the stagecoach eventually developed dense networks in and around downtown (see Figure 3.5); other cities experimented with cable-car systems and even the steam railroad, but most efforts proved impractical. With omnibuses unable to carry more than a dozen or so passengers or to attain the speeds of people on foot, the first meaningful breakthrough toward establishing in-

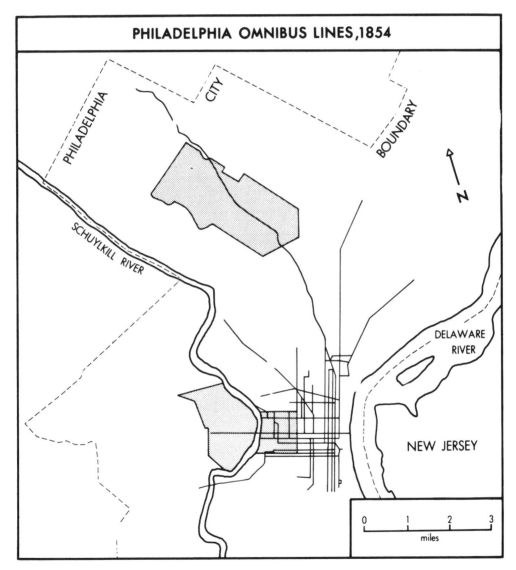

FIGURE 3.5. Philadelphia's omnibus routes in 1854, the year the city annexed all of surrounding Philadelphia County. Source: Miller (1982, p. 364). Copyright 1982 by The Association of American Geographers. Reprinted by permission.

FIGURE 3.6. The horsecar introduced mass transportation to the teeming U.S. city. This dictograph was taken in downtown Minneapolis in the late 1880s. Source: Minnesota Historical Society.

tracity "mass" transit was finally introduced in New York City in 1852 in the form of the horse-drawn streetcar (see Figure 3.6). Lighter street rails were easy to install, overcame the problems of muddy unpaved roadways, and allowed horsecars to be hauled along them at speeds slightly faster (ca. 5 miles per hour) than those of pedestrians. This modest improvement in mobility allowed a narrow band of land at the city's edge to be opened for new home construction. Middle-income urbanites flocked to these *horsecar suburbs,* which proliferated rapidly after 1860. Radial routes were usually the first to spawn such peripheral development, but the steady demand for housing required the construction of crosstown horsecar lines, thereby filling in the interstices and preserving the generally circular shape of the city.

The nonaffluent remainder of the urban population was confined to the old pedestrian city and its bleak, high-density, industrial appendages. With the massive influx of unskilled laborers, increasingly of European origin after the Civil War, huge blue-collar neighborhoods surrounded the factories, often built by the mill owners themselves. Since factory shifts ran 10 or more hours 6 days a week, their modestly paid workers could not afford to commute and were forced to reside within walking distance of the plant. Newcomers to the city, however, were accommodated in this nearby housing quite literally in the order in which they arrived, thereby denying immigrant factory workers even the small luxury of living in the immediate company of their fellow ethnics. Not surprisingly, such heterogeneous residential patterning almost immediately engendered social stresses and episodic conflicts that persisted until the end of the century, when the electric trolley would at last enable the formation of modern ethnic neighborhood communities.

Toward the end of the Walking-Horsecar Era, the scale of the city was slowly but inexorably expanding. One by-product was the emergence of the downtown central business district (CBD). As needs intensified for specialized commercial, retailing, and other services, it was quickly realized that they could best be provided from a single center at the most accessible urban location. With immigrants continuing to pour into the all-but-bursting industrial city in the late 19th century, pressures redoubled to improve intraurban transit and open up more of the adjacent countryside.

In retrospect, horsecars had only been a stopgap measure, relieving overcrowding temporarily but incapable of bringing enough new residential space within the effective commuting range of the burgeoning middle class. The hazards of relying on horses for motive power were also becoming unacceptable. Besides high costs and the sanitation problem, disease was an ever-present threat—for example, thousands of horses succumbed in New York and Philadelphia in 1872 when respiratory illnesses swept through the municipal stables (Schaeffer & Sclar, 1975, p. 22). By the late 1880s that desperately needed transit revolution was at last in the making. When it came, it swiftly transformed both city and suburban periphery into the modern metropolis.

The Electric Steetcar Era (1890–1920)

The key to the first urban transport revolution was the invention of the electric traction motor by one of Thomas Edison's technicians, Frank Sprague. This innovation surely must rank among the most important in U.S. history. The first electric trolley line opened in Richmond, Virginia, in 1888, was adopted by two dozen other major cities within a year, and by the early 1890s was the dominant mode of intraurban transit. The rapidity of the diffusion of this innovation was enhanced by the immediate recognition of its ability to mitigate the urban transportation problems of the day: motors could be attached to existing horsecars to convert them into self-propelled vehicles,

powered via easily constructed overhead wires. Accordingly, the tripling of average speeds (to over 15 miles per hour) now brought a large band of open land beyond the city's perimeter into trolley-commuting range.

The most dramatic impact of the Electric Streetcar Era was the swift residential development of those urban fringes, which expanded the emerging metropolis into a decidedly star-shaped spatial entity (Figure 3.2, Era II). This morphological pattern was produced by radial trolley corridors extending several miles beyond the compact city's limits; with so much new space available for home building within easy walking distance of these trolley lines, there was no need to extend trackage laterally. Consequently, the interstices remained undeveloped.

The typical "streetcar suburb" around the turn of the 20th century was a continuous corridor whose backbone was the road carrying the trolley tracks (usually lined with stores and other local commercial facilities), from which gridded residential streets fanned out for several blocks on both sides of the tracks. This spatial framework is illustrated in Figure 3.7, whose map reconstructs street and property subdivisioning in a portion of streetcar-era Cambridge, Massachusetts, just outside Boston. By 1900, most of the open spaces between these streets were themselves subdivided into small rectangular lots that contained modest single-family houses.

In general, the quality of housing and prosperity of streetcar suburbs increased with distance from the central-city line. As Warner (1962) pointed out in his classic study, however, these continuous developments were home to a highly mobile middle-class population, finely stratified according to a plethora of minor income and status differences. With frequent upward (and local spatial) mobility the norm, community formation became an elusive goal, a process further inhibited by the relentless grid-settlement morphology and the heavy dependence on distant downtown for em-

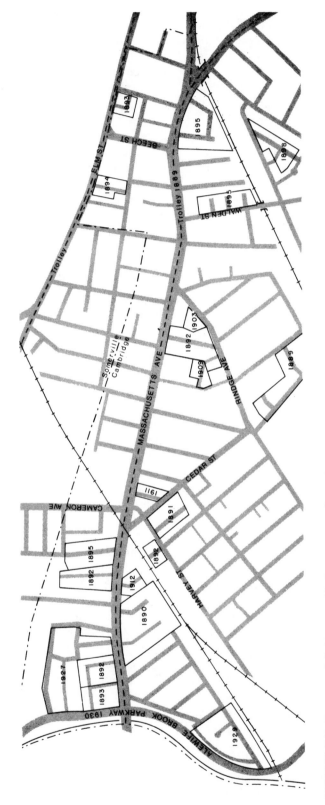

FIGURE 3.7. Streetcar subdivisions outside Boston in North Cambridge, Massachusetts, 1890–1930. Source: Krim et al. (1977, p. 44). Copyright 1977 by the Cambridge Historical Commission. Reprinted by permission.

ployment and most shopping. As Warner put it so aptly, this kind of a society generated "not integrated communities arranged about common centers, but a historical and accidental traffic pattern" (1962, p. 158).

The desire to exclude the working class also shaped the social transportation geography of suburban streetcar corridors. "Definitional" conflicts usually revolved around the entry of saloons, with middle-income areas voting to remain "dry" while the blue-collar mill towns of lower status trolley and intercity rail corridors chose to go "wet" (Schwartz, 1976, pp. 13–18).Within the city, too, the streetcar sparked a spatial transformation. The ubiquity and low fare of the electric trolley now provided every resident access to the intracity circulatory system, thereby introducing truly *mass* transit to urban America in the closing years of the 19th century. For nonresidential activities, this new ease of movement among the city's various parts quickly triggered the emergence of specialized land use districts for commerce, industry, and transportation as well as the continued growth of the multipurpose CBD—now abetted by the elevator, which permitted the construction of much taller buildings. But the widest impact of the streetcar was on the central city's social geography, because it made possible the congregation of ethnic groups in their own neighborhoods. No longer were these moderate-income masses forced to reside in the heterogeneous jumble of rowhouses and tenements that ringed the factories. The trolley brought them the opportunity to "live with their own kind," enabling the sorting of discrete groups into their own inner-city social territories within convenient and inexpensive travel distance of the workplace.

The latter years of the Electric Streetcar Era also witnessed additional breakthroughs in public urban rail transportation. The faster electric commuter train superseded steam locomotives in the wealthiest suburban corridors, which had resisted the middle-class incursions of the streetcar sec-

tors because the rich always seek to preserve their social distance from those of lesser status. In some of the newer metropolises that lacked the street-rail legacy, heavier electric railways became the cornerstone of the movement system; Los Angeles is the outstanding example, with the interurban routes of the Pacific Electric network (Figure 3.8) spawning a dispersed settlement fabric in preautomobile days, and many lines forging rights-of-way that were later upgraded into major boulevards and even freeways (see Banham, 1971, pp. 32–36). Finally, within the city proper elevated ("els") and underground rapid transit lines (subways) made their appearances, the "El" in New York as early as 1868 (using steam engines—the electric elevated was born in Chicago in 1892) and the subway in Boston in 1898. Such rapid transit was always enormously expensive to build and could be justified only in the largest cities that generated the highest traffic volumes. Therefore, els and subways were restricted to New York, Boston, Philadelphia, and Chicago, and most construction concluded by the 1920s. Rapid-transit-system building did not resume until the 1960s with metropolitan San Francisco's Bay Area Rapid Transit (BART) network, followed in the 1970s and 1980s by projects in Cleveland, Washington, D.C., Atlanta, Baltimore, Miami, and Los Angeles.

The Recreational Automobile Era (1920–1945)

By 1920, the electric trolleys, trains, interurbans, els, and subways had transformed the tracked city into a full-fledged metropolis whose streetcar suburbs and mill-town intercity rail corridors, in the largest cases, spread out to encompass an urban complex more than 20 miles in diameter. It was at this point in time, many geographers and planners would agree, that intrametropolitan transportation achieved its greatest level of efficiency—that the burgeoning city truly "worked." How much closer the U.S. metropolis might have approached optimal workability for all its millions of residents, however, shall never be known because the second urban transportation revolution was already beginning to assert itself through the increasingly popular automobile.

Whereas many scholars have vilified the automobile as the destroyer of the city, Americans took to cars as completely and wholeheartedly as they did to anything in the nation's long cultural history. More balanced assessments of the role of the automobile (see, e.g., Bruce-Briggs, 1977) recognize its overwhelming acceptance for what it was: the long-hoped-for attainment of private transportation that offered users almost total freedom to travel whenever and wherever they chose. Cars came to the metropolis in ever greater numbers throughout the interwar period, a union culminating in accelerated deconcentration—through the development of the bypassed streetcar-era interstices and the pushing of the suburban frontier farther into the countryside—to produce once again a compact, regular-shaped urban entity (Figure 3.2, Era III).

Although it came to have a dramatic impact on the urban fabric by the eve of World War II, the automobile was introduced into the U.S. city in the 1920s and 1930s at a leisurely pace. The first cars had appeared in both Western Europe and the United States in the 1890s, and the wealthy on both sides of the Atlantic quickly took to this innovation because it offered a better means of personal transport. It was Henry Ford, however, with his revolutionary assembly-line manufacturing techniques, who first mass-produced cars; the lower selling prices soon converted them from the playthings of the rich into a transport mode available to a majority of Americans. By 1916, over 2 million autos were on the road, a total that quadrupled by 1920 despite wartime constraints. During the 1920s, the total tripled to 23 million and increased another 4½ million by the end of the depression-plagued 1930s; passenger car registrations paralleled these increases (Fig-

ure 3.9). The earliest flurry of auto adoptions had been in rural areas, where farmers badly needed better access to local service centers; accordingly, much of the early paved-road construction effort was concentrated in rural America. In the cities, cars were initially used for weekend outings—hence the *Recreational* Auto Era—and some of the first paved roadways built were landscaped parkways that followed scenic waterways (such as New York's Bronx River Parkway, Chicago's Lake Shore Drive, and the East and West River Drives along the Schuylkill in Philadelphia's Fairmount Park).

In the suburbs, however, where the overall growth rate now for the first time exceeded that of the central cities, cars were making a decisive penetration throughout the economically prosperous 1920s. Flink (1975, p. 14) reported that, as early as 1922, 135,000 suburban dwellings in 60 metropolises were completely dependent on motor vehicles. In fact, the subsequent rapid expansion of automobile suburbia by 1930 so adversely affected the metropolitan public transportation system that, through significant diversions of streetcar and commuter-rail passengers, the large cities started to feel the negative effects of the car years before accommodating to its actual arrival. By encouraging the opening of unbuilt areas lying between suburban rail axes, the automobile effectively lured residential developers away from densely populated traction-line corridors into the now-accessible interstices. Thus the suburban home-building industry no longer found it necessary to subsidize privately owned streetcar companies to provide cheap access to their trolley-line housing tracts.

Without this financial underpinning, the modern urban transit crisis soon began to surface. Traction companies, obliged under their charters to provide good-quality service, could not raise fares to the level necessary to earn profits high enough to attract new capital in the highly competitive money markets. As this economic squeeze

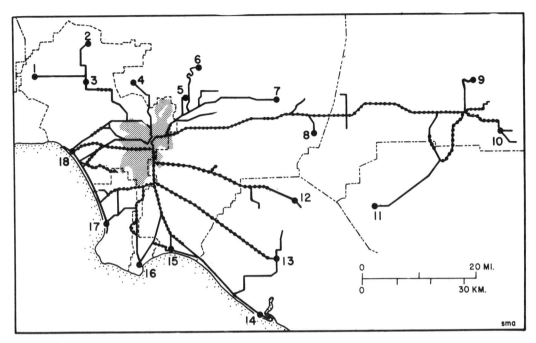

FIGURE 3.8. Interurban railway routes of the Pacific Electric system at their greatest cumulative extent in the mid-1920s. Shading denotes portions of central Los Angeles that were situated within a half-mile of narrow-gauge streetcar lines in the early 1920s. Selected interurban destinations are numbered in clockwise sequence as follows: (1) Canoga Park; (2) San Fernando; (3) Van Nuys; (4) Burbank; (5) Pasadena; (6) Mount Lowe; (7) Glendora; (8) Pomona; (9) Arrowhead Springs; (10) Redlands; (11) Corona; (12) Yorba Linda; (13) Santa Ana; (14) Newport Beach; (15) Long Beach; (16) San Pedro; (17) Redondo Beach; and (18) Santa Monica. Source: Steiner (1981, p. 95). Copyright 1981 by Kendall/Hunt. Reprinted by permission.

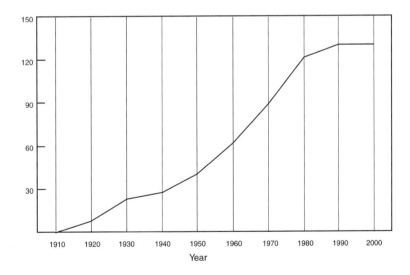

FIGURE 3.9. Passenger car registrations in the United States, 1910–2000. Source: *Statistical Abstract of the United States* (various years).

intensified, particularly during the Great Depression of the 1930s, local governments were forced to intervene with subsidies from public funds; eventually, as transit ridership continued to decline in the postwar period, local governments assumed ownership of transit companies when lines could not be closed down without harming communities. Several additional factors also combined to accelerate the interwar deterioration of the superlative trolley-era metropolitan transit network: the growing intrasuburban dispersal of population that no longer generated passenger volumes great enough to support new fixed-route public transportation facilities; dispersion of employment sites within the central city, thereby spreading out commuter destinations as well as origins; shortening of the work week from 6 to 5 days; worsening street congestion where trolleys and auto traffic increasingly mixed; and the pronounced distaste for commuting to the city by bus, a more flexibly routed new transit mode that never caught on in the suburbs.

Ironically, recreational motorways helped to intensify the decentralization of the urban population. Most were radial highways that penetrated deeply into the suburban ring; those connecting to major new bridges and tunnels—such as the Golden Gate and Bay Bridges in San Francisco, the George Washington Bridge and the Holland and Lincoln Tunnels in New York—usually served to open empty outer metropolitan sectors. Sunday motorists, therefore, had easy access to this urban countryside and were captivated by what they saw. They responded in steadily increasing numbers to the home-sales pitches of developers who had shrewdly located their new tract housing subdivisions beside the suburban highways. As more and more city dwellers relocated to these automobile suburbs, by the end of the interwar era many recreational parkways were turning into heavily traveled commuter thoroughfares—especially near New York City, where the suburban parkway network devised by planner Robert Moses reached far

into Westchester County and Long Island, and in the Los Angeles Basin, where the first "freeway" (the Arroyo Seco, now called the Pasadena) was opened in 1940.

The residential development of automobile suburbia followed a simple formula that was devised in the prewar years and perfected and greatly magnified in scale in the decade after 1945. The leading motivation was developer profit from the quick turnover of land, which was acquired in large parcels, subdivided, and auctioned off. Accordingly, developers much preferred open areas at the metropolitan fringe where large packages of cheap land could readily be assembled. As the process became more sophisticated in the 1940s, developer-builders came to the forefront and produced huge complexes of inexpensive housing—with William J. Levitt and his Levittowns in the vanguard. Silently approving and underwriting this uncontrolled spread of residential suburbia were public policies at all levels of government that included the financing of highway construction, obligating lending institutions to invest in new home building, insuring individual mortgages, and providing low-interest loans to Federal Housing Administration (FHA) and Veterans Administration (VA) clients.

Although the conventional wisdom view of U.S. suburbanization holds that most of it occurred after World War II, longitudinal demographic data indicate that intrametropolitan population decentralization had achieved sizeable proportions during the interwar era. Table 3.1 reveals that suburban growth rates began to surpass those of the central cities as early as the 1920s, and that after 1930 the outer ring took a commanding lead (which has not ceased widening to this day). With an ever larger segment of the urban population residing in automobile suburbs, their spatial organization was already forming the framework of contemporary metropolitan society. Because automobility removed most of the preexisting movement constraints, suburban social geography now became dominated by lo-

TABLE 3.1. Intrametropolitan Population Growth Trends, 1910–1960

Decade	Central-city growth rate	Suburban growth rate	Percent total SMSA[a] growth in suburbs	Suburban growth per 100 increase in central-city population
1910–1920	27.7	20.0	28.4	39.6
1920–1930	24.3	32.3	40.7	68.5
1930–1940	5.6	14.6	59.0	144.0
1940–1950	14.7	35.9	59.3	145.9
1950–1960	10.7	48.5	76.2	320.3

[a]SMSA, Standard Metropolitan Statistical Area, constituted by the central city and county-level political units of the surrounding suburban ring.
Source: U.S. Census of Population.

cally homogeneous income-group clusters that isolated themselves from dissimilar neighbors. Gone was the highly localized stratification of streetcar suburbia; in its place arose a far more dispersed, increasingly fragmented residential mosaic that builders were only too happy to cater to, helping shape this kaleidoscopic settlement pattern by constructing the most expensive houses that could be sold in each locality.

The long-standing partitioning of suburban social space was further legitimized by the widespread adoption of zoning (legalized in 1916). This legal device gave municipalities control of lot and building standards, which, in turn, assured dwelling prices that would only attract newcomers whose incomes at least equaled those of the existing population. For the middle class, especially, such exclusionary economic practices were enthusiastically supported because it now extended to them the capability that upper-income groups had enjoyed to maintain their social and geographic distance from people of lower socioeconomic status.

Nonresidential activities were also suburbanizing at a steadily increasing rate during the Recreational Auto Era. Indeed, many large-scale manufacturers had decentralized during the previous streetcar era, choosing suburban freight-rail locations that rapidly spawned surrounding working-

class towns. These industrial suburbs became important satellites of the central city in the emerging metropolitan constellation (see Taylor, 1915/1970). The economic geography of the interwar era reflected an intensification of this trend, as shown in the curves of activity gradients in Figure 3.10. Industrial employers accelerated their intraurban deconcentration in this period as more efficient horizontal fabrication methods were replacing older techniques requiring multistoried plants—thereby generating greater space needs that were too expensive to satisfy in the high-density inner central city. Newly suburbanizing manufacturers, however, continued their spatial affiliation with intercity rail corridors, because motor trucks were not yet able to operate with their present-day efficiencies and the highway network of the outer ring remained inadequate until the 1950s.

The other major nonresidential activity of interwar suburbia was retailing. Clusters of automobile-oriented stores had first appeared in the urban fringes before World War I. By the early 1920s, the roadside commercial strip had become a common sight in many Southern California suburbs. Retail activities were also featured in dozens of planned automobile suburbs that sprang up in the 1920s, most notably in outer Kansas City's Country Club District where builder Jesse Clyde Nichols opened the nation's first

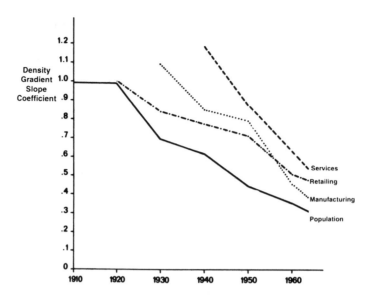

FIGURE 3.10. Intrametropolitan population and activity density gradients, 1910–1963. The lower the slope coefficient, the more dispersed the spatial distribution of an activity. Source: Mills (1970, p. 14). Copyright 1970 by *Urban Studies*. Reprinted by permission.

complete shopping center in 1922. But these diversified retail centers spread rather slowly before the 1950s; nonetheless, such chains as Sears & Roebuck and Montgomery Ward quickly discovered that stores situated alongside main suburban highways could be very successful, a harbinger of things to come in post–World War II metropolitan America.

The central city's growth reached its zenith in the interwar era and began to level off (Table 3.1) as metropolitan development after 1925 increasingly concentrated in the urban fringe zone that now widely resisted political unification with the city. Whereas the transit infrastructure of the streetcar era remained dominant in the industrial city (see Figure 3.11), the late-arriving automobile was adapted to this high-density urban environment as much as possible, but not without greatly aggravating existing traffic congestion.

The structure of the U.S. city during the second quarter of the 20th century was best summarized in the well-known *concentric-ring*, *sector*, and *multiple nuclei* models (re-

viewed in Harris & Ullman, 1945), which together described the generalized spatial organization of urban land usage. The social geography of the core city was also beginning to undergo significant change at this time as the suburban exodus of the middle class was accompanied by the arrival of southern blacks. These parallel migration streams would achieve massive proportions after World War II. The southern newcomers were attracted to the northern city by declining agricultural opportunities in the rural South and by offers of employment in the factories as industrial entrepreneurs sought a new source of cheap labor to replace European immigrants, whose numbers were sharply curtailed by restrictive legislation after the mid-1920s. But urban whites refused to share residential space with in-migrating blacks, and racial segregation of the metropolitan population swiftly intensified as citywide dual housing markets dictated the formation of ghettoes of nonwhites inside their own distinctive social territories.

FIGURE 3.11. The automobile did not become a major force in the central city until the post-World War II era, but its presence can already be detected by the 1930s. This photograph was taken in St. Paul, Minnesota, in 1932. Source: Minnesota Historical Society.

Freeway Era (1945–Present)

Unlike the two preceding eras, the post-World War II Freeway Era was not sparked by a revolution in urban transportation. Rather, it represented the coming of age of the automobile culture, which coincided with a historic watershed as a reborn nation emerged from 15 years of economic depression and war. Suddenly the automobile was no longer a luxury or a recreational diversion: it quickly became a necessity for commuting, shopping, and socializing—essential to the successful exploitation of personal opportunities for a rapidly expanding majority of the metropolitan population. People snapped up cars as fast as the reviving peacetime automobile industry could roll them off the assembly lines, and a prodigious highway-building effort

was launched, spearheaded by high-speed, limited-access expressways.

Given impetus by the 1956 Interstate Highway Act, these new freeways would soon reshape every corner of urban America as the new suburbs they engendered represented nothing less than the turning inside-out of the historic metropolitan city. In retrospect, this massive acceleration of the deconcentration process "cannot be considered a break in longstanding trends, but rather the later, perhaps more dynamic, evolutionary stages of a transformation which was based on a pyramiding of small scale innovations and underlying social desires" (Sternlieb & Hughes, 1975, p. 12).

The snowballing effect of these changes is expressed spatially in the much-expanded metropolis of the postwar era (Figure 3.2, Era IV), whose expressway-dominated in-

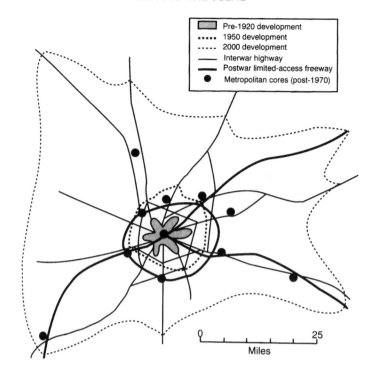

Pre-1920 development
..... 1950 development
- - - - 2000 development
—— Interwar highway
━━ Postwar limited-access freeway
● Metropolitan cores (post-1970)

0 ⊢——┴——┴——┴——┴ 25
Miles

FIGURE 3.12. The spatial pattern of growth in automobile suburbia since 1920. Source: Muller (1981, p. 257). Copyright 1982 by Charles E. Merrill. Adapted by permission.

frastructure again produced a network-biased development pattern reminiscent of the Electric Streetcar Era (Figure 3.2, Era II). A more detailed representation of contemporary intraurban morphology is seen in Figure 3.12, showing the culmination of eight decades of automobile suburbanization. Most striking is the enormous band of growth that was added between 1950 and 2000, with freeway sectors pushing the metropolitan frontier deeply into the surrounding zone of exurbia. The huge curvilinear *outer city* that arose within this new suburban ring was most heavily shaped by the circumferential freeway segments that girdled the central city—a universal feature of the metropolitan expressway system, originally designed to allow long-distance interstate highways to bypass the congested urban core. Today, more than 100 of these expressways form complete *beltways* that are the most heavily traveled roadways in their regions. The prototype high-speed circum-

ferential was suburban Boston's Route 128, completed in the early 1950s; by the 1980s, such freeways as Houston's Loop, Atlanta's Perimeter, Chicago's Tri-State Tollway, New York–New Jersey's Garden State Parkway, Miami's Palmetto Expressway, and the Beltways ringing Washington and Baltimore had become some of the best-known urban arteries in the nation.

The maturing freeway system was the primary force that turned the metropolis inside-out after 1970, because it eliminated the regionwide centrality advantage of the central city's CBD. Now *any* location on that expressway network could easily be reached by motor vehicle, and intraurban accessibility swiftly became an all-but-ubiquitous spatial good. Ironically, large central cities had encouraged the construction of radial expressways in the 1950s and 1960s because they appeared to enable downtown to remain accessible to the swiftly dispersing suburban population. As one economic activity after another discovered its new locational footlooseness in the freeway metropolis, however, nonresidential deconcentration greatly accelerated. Much of this suburban growth has gravitated toward beltway corridors; Figure 3.13 displays the typical sequence of land use development along a segment of circumferential I-494 just south of Minneapolis.

As high-speed expressways expanded the radius of commuting to encompass the entire dispersed metropolis, residential loca-

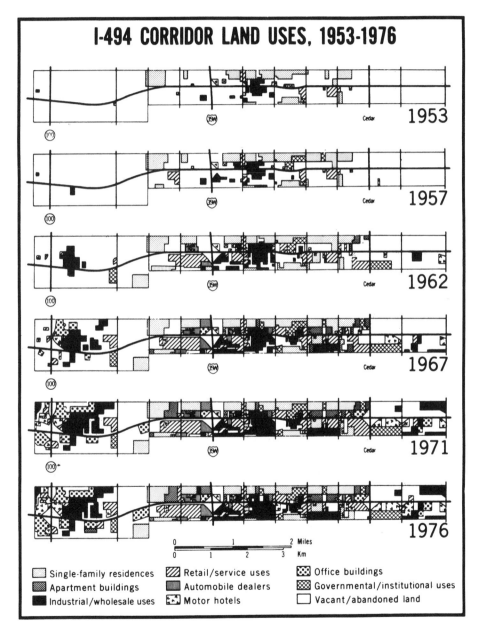

FIGURE 3.13. Land use change in the Interstate-494 corridor south of Minneapolis, 1953–1976. Source: Baerwald (1978, p. 312). Copyright 1978 by The American Geographical Society. Reprinted by permission.

tional constraints were relaxed as well. No longer were most urbanites required to live within a short distance of their job. Instead, the workplace had now become a locus of opportunity offering access to the best possible residence that a household could afford anywhere within the urbanized area. Thus the heterogeneous patterning of sociospatial clusters that had arisen in prewar automobile suburbia was writ ever larger in the Freeway Era—giving rise to a *mosaic culture* whose component tiles were stratified not only along class lines but also according to age, occupational status, and a host of minor lifestyle differences (Berry, 1981, pp. 64–66).

These developments fostered a great deal of local separatism, thereby intensifying the balkanization of metropolitan society as a whole:

With massive auto transportation, people have found a way to isolate themselves; . . . a way to privacy among their peer group. . . . They have stratified the urban landscape like a checker board, here a piece for the young married, there one for health care, here one for shopping, there one for the swinging jet set, here one for industry, there one for the aged. . . . When people move from square to square, they move purposefully, determinedly. . . . They see nothing except what they are determined to see. Everything else is shut out from their experience. (Schaeffer & Sclar, 1975, p. 119)

After more than a half-century of Freeway Era change, certain structural transformations have emerged from what was, in retrospect, one of the most tumultuous upheavals in U.S. urban history. Figure 3.12 reveals the existence of several new outly-

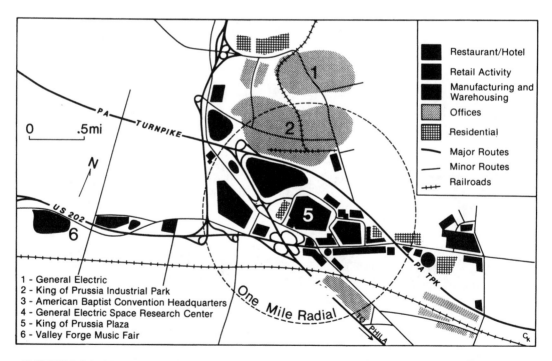

1 - General Electric
2 - King of Prussia Industrial Park
3 - American Baptist Convention Headquarters
4 - General Electric Space Research Center
5 - King of Prussia Plaza
6 - Valley Forge Music Fair

FIGURE 3.14. The internal structure of nonresidential activities in a typical suburban downtown: King of Prussia, Pennsylvania. Source: Muller (1976, p. 41). Copyright 1976 by The Association of American Geographers. Adapted by permission.

FIGURE 3.15. An aerial view of Tyson's Corner in Fairfax County, Virginia, outside Washington, D.C., one of the nation's largest suburban downtowns. Source: County of Fairfax (Virginia), Office of Comprehensive Planning.

ing metropolitan-level cores. Today, such downtown-like concentrations of retailing, business, and light industry have become common landscape features near the major highway intersections of the outer city that now encircles every large central city. A representative *suburban downtown* of this genre is mapped in Figure 3.14, revealing the array of high-order activities that have agglomerated around the King of Prussia Plaza shopping center at the most important expressway junction in Philadelphia's northern and western suburbs.[2] Dozens of such diversified activity cores have matured since the 1970s, and suburban downtowns

such as Washington's Tyson's Corner (Figure 3.15), Houston's Post Oak Galleria, Los Angeles' South Coast Metro, and Chicago's Schaumburg have now achieved national reputations.

In his book-length survey of suburban downtowns—which he calls *edge cities*—Garreau (1991) set forth some minimum requirements that an activity center must meet in order to be classified as an edge city:

1. At least 24,000 jobs
2. 5,000,000-plus square feet of leasable office space

3. 600,000-plus square feet of leasable retail space
4. More jobs than bedrooms
5. An identity as a single place
6. No significant structure more than 30 years old.

By the mid-1990s, nearly 200 of these new urban agglomerations had been identified nationwide, "most at least the size of downtown Orlando (in contrast, fewer than 40 [central-city] downtowns are Orlando's size)" (Garreau, 1994, p. 26).

As the suburban downtowns of the outer city achieve economic–geographical parity with each other (as well as with the CBD of the nearby central city), they provide the totality of urban goods and services to their surrounding populations, and thereby make each sector of the metropolis an increasingly self-sufficient functional entity. This transition to a polycentric metropolis of *realms*—the term coined by Vance (1964) to describe the ever-more-independent areas served by new downtown-like activity cores—requires the use of more up-to-date generalizations than are provided by the "classical" concentric-zone, sector, and multiple nuclei models of urban form. Such an alternative to these obsolete core-periphery models of the interwar metropolis is seen in Figure 3.16.

Another useful contemporary model (which builds on the work of Baerwald [1978], Erickson [1983], and others) was developed by Hartshorn and Muller (1989) to interpret the evolution of the suburban spatial economy. Accordingly, the developments of the Freeway Era are generalized as five growth stages. First was the *bedroom community* stage (1945–1955), dominated by a massive postwar residential building boom but accompanied by only a modest expansion of suburban commercial activity. This was followed by the *independence* stage (1955–1965), during which suburban economic growth accelerated, led by the first wave of industrial and office parks and, after 1960, by the rapid diffusion of regional

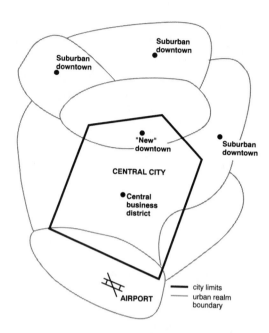

FIGURE 3.16. The generalized layout of urban realms in relation to the central city and suburban downtowns of the polycentric metropolis. Source: Hartshorn and Muller (1989, p. 378). Copyright 1989 by V. H. Winston & Son. Reprinted by permission.

shopping centers. These malls came into their own during the third stage, *catalytic growth* (1965–1980), attracting a myriad of office, hotel, and restaurant facilities to cluster around them, and sparking the swift maturation of the suburban economic landscape in these expanding cores and many freeway corridors that connected them. The fourth stage, *high-rise/high-technology*, spanned the 1980s and saw the flowering of scores of suburban downtowns increasingly dominated by high-rise office buildings; simultaneously, these burgeoning activity cores attracted high-technology research-and-development facilities, and by 1990 the outer suburban city had become the leading geographic setting for the nation's new postindustrial service economy. The post-1990 period constitutes the fifth stage in which the trends of the 1980s have continued unabated, and suburban downtowns are now evolving into *mature urban centers* as their land use complexes steadily diversify and perform ever more important economic, social, and cultural functions that are increasingly international in scope.

URBAN TRANSPORTATION IN THE POSTINDUSTRIAL METROPOLIS

As the nation completes its transition to a postindustrial economy and society in the opening years of the 21st century, intra-urban movement patterns will continue to adjust to new geographical circumstances. Quaternary (information-related) and quinary (managerial- and decision-making-based) economic activities, which increasingly dominate the U.S. labor force, are demonstrating locational preferences that assume the outer metropolitan ring will continue to be the essence of the contemporary U.S. city for a long time to come. Above all, these activities seek the most prestigious metropolitan sites. The leading concentrations of the pacesetting electronics/computer industry—which overwhelmingly prefer high-amenity suburbs from California's Silicon Valley to North Carolina's Research Triangle Park to outer Boston's Route 128 corridor—provide a classic example.

Today there is much evidence that these research-and-development/manufacturing complexes have become cornerstones in the reorganizing urban landscape shaped by the digital revolution. Silicon Valley has proven to be a prototype, spawning dozens of imitators throughout the world that function as *technopoles*—planned techno–industrial agglomerations that create the hardware and software products of the new informational economy (Castells & Hall, 1994). Technopoles are increasingly regarded as the most successful places in shaping "the geographic importance of future cities and communities," and the biggest winners have been "self-contained high-end suburbs that . . . service the needs of both the burgeoning high-technology industries and their [supremely skilled] workers" (Kotkin, 2000, pp. 6–7, 9).

Spearheaded by technopoles, the expanding development of high-order suburban activity centers is also driven by changes in the U.S. economy. One set of forces involves the dynamic economic and social networks of *globalization*, whose expanding international linkages have triggered new investment flows and entrepreneurial opportunities that have revitalized aging CBDs in a number of central cities (Wilson, 1997). But less often realized is that globalization forces simultaneously work to intensify and accelerate the suburban transformation of the U.S. metropolis in a number of ways: (1) suburban centers participate strongly when their urban region becomes a "world city," and some even establish their own direct international ties; (2) improvements in a region's telecommunications network creates a grid of local nodes that links suburban downtowns as well as the CBD to the rest of the world; (3) the transnational corporations that control much of the global economic system are increasingly headquartered in suburban locations, particularly in the Greater New York region; and (4) the foreign presence in suburban America is constantly growing, ranging from the ownership of businesses large and small to the rise of thriving new ethnic communities dominated by affluent professionals (Muller, 1997).

The other major set of forces reshaping the U.S. economic landscape is associated with the relentless *expansion of the services sector*, a term understood to also include the quaternary and quinary activities mentioned above. The growth of this sector has been at the forefront in the suburbanization of employment since the mid-20th century (see Figure 3.10). That trend picked up speed during the 1960s, and by 1973 the nation's suburban rings surpassed the central cities in total number of jobs. This gap has widened steadily, and the author's calculations show that a quarter-century later the

1998 intrametropolitan split for the 35 largest metropolitan areas averaged 28% of the jobs in the central city and 72% in the suburbs.

A more detailed picture of this services-led shift is presented in Table 3.2, which traces the suburban percentage of metropolitan-area employment by major sector from 1970 to 1998. Philadelphia was selected from the eight metropolises for which data are available because it is a microcosm of the national metropolitan economy and precisely reflects the 28/72% percent split exhibited by the 35 largest urban areas. During the final three decades of the 20th century, overall employment in Philadelphia's suburbs advanced from just under 50% of the metropolitan total to a position of critical mass (greater than half) during the 1970s, and then to the level of dominance (greater than two-thirds) by the mid-1980s. Sectorally, this pattern was mirrored by manufacturing, wholesaling, and retailing, and all three today are more than 80% suburbanized. The two services sectors shown in Table 3.2 were slower to decentralize, but both approached suburban dominance as the century ended. The finance/insurance/real estate sector is a particularly good barometer of spatial change in the services sector because of its traditionally strong ties to the CBD and relative independence from intraurban population shifts. Nonetheless, its detachment from Center City Philadelphia between 1970 and 1990 is starkly apparent, and if present rates persist nearly 75% of the region's jobs in this sector will be located in the suburbs by the end of this decade. Moreover, this trend is paralleled by another pacesetting services subsector: professional/scientific/technical services. As of 1998, the first year such data were reported, the employment split for this triad of specialized-services professions stood at central city—34.5%/suburban ring—65.5%.

The changes just described strongly point toward continuing development of the outer city and even greater suburban dominance of intrametropolitan employment geography. Thus the leading urban transportation challenges of the early 21st century focus on the efficiencies of moving people about the dispersed, polycentric city of realms. Although urban freeways spawned the new multinodal metropolis, it is unlikely that many more will be built in the foreseeable future. Local resistance is intensifying, and governments at all levels are increasingly unable to afford the enormous construction costs that are often heightened by the need to conform to stricter environmental regulations (see Bae, Chapter 13, this volume). Moreover, there is consider-

TABLE 3.2. Suburban Percentage of Major Employment Sectors, Metropolitan Philadelphia, 1970–1998

Employment sector	1970	1978	1988	1998
Manufacturing	**54.5**	**66.0**	**75.0**	**81.5**
Wholesale trade	39.7	**60.5**	**73.6**	**82.0**
Retail trade	**55.9**	**68.0**	**73.0**	**80.3**
Finance/insurance/real estate[a]	31.0	45.9	**59.2**	**66.5**
Health services[b]	46.1	**51.8**	**57.4**	**62.5**
Total employment	48.8	**60.2**	**68.3**	**72.2**

[a]Pre-1998 data also include Real Estate employment in Finance/Insurance sector.
[b]Pre-1998 data do not include Social Assistance employment in the Health Services sector.
Note. Percentages shown in boldface exceed the *critical mass* level (50%); percentages shown in underlined boldface exceed the *suburban dominance* (66.7%) level. Source: U.S. Bureau of the Census, *County Business Patterns* (annual).

able evidence that building new express-ways does not improve the flow of traffic: metropolitan Los Angeles and Houston, for instance, both possess extensive superhigh-way networks, yet each new road link that is added creates more congestion as traffic from other routes is quickly attracted to fill the new highway's vehicle-handling capac-ity (some of the reasons for this are outlined in Giuliano, Chapter 9, this volume).

As an alternative to additional highways, the construction of new public mass transit systems is still being pursued as a possible solution to urban transportation problems (a subject treated in Pucher, Chapter 8, this volume). Since the 1960s, heavy-rail, elec-tric-train systems have been started in met-ropolitan San Francisco, Washington, D.C., Cleveland, Atlanta, Miami, Baltimore, and Los Angeles. Elsewhere, less expensive light-rail trolley lines have been constructed in San Diego, Buffalo, Portland (Oregon), Dallas, and more than a dozen other cities; major bus system improvements have been undertaken in Detroit, Indianapolis, and Dallas. Nonetheless, although total ridership grew during the 1990s, transit's market share has been declining in these cities over the past two decades. One major reason is that transit lines are incapable of serving even a significant minority of the increas-ingly dispersed travel demands in the low-density, automobile-oriented outer subur-ban city.

The rapid proliferation of suburban downtowns and specialized activity centers magnifies two additional mobility problems. At the local level, infrastructure develop-ment usually lags behind the pace of growth in these mushrooming cores, thereby spawn-ing traffic congestion nightmares at peak travel hours that contribute to the rising clamor for density controls in these areas. (To some extent these new suburban cen-ters resemble the old CBDs, but their spatial organization is different and requires new strategies for travel demand management.) Planners have responded with a number of imaginative short-term policies and long-

term master plans to counter the trend to-ward "suburban gridlock" (see Cervero, 1986, 1989), but implementation of strict measures is often resisted by local political leaders.

The other mobility problem engendered by the reorganization of the metropolitan space-economy is the growing geographical mismatch between job opportunities and housing. Glittering suburban downtowns are invariably surrounded by upper-income residential areas, thereby requiring most of the people who work there to commute considerable distances to the nearest com-munities with affordable housing. This not only heightens the level of lateral suburb-to-suburb commuting on already over-burdened highways for middle-income employees; it also increases the flow of ex-tended-distance sectoral commuting—from both the inner central city in one direction and the outlying exurban fringe in the other—for less-skilled workers.

With the U.S. metropolis now all but turned inside-out, and with most of its continuing growth occurring in peripheral zones, ever greater reliance on the automo-bile becomes unavoidable despite the rising costs of such dependency. In much of the United States today, people already pay more for transportation than for clothing, entertainment, and health care combined; and, in a steadily rising number of urban re-gions, residents spend more on transporta-tion than on *housing* (Mencimer, 2002). With so many urbanites willing to pay this steepening price of access to the bur-geoning opportunities of the outer city, policymakers, planners, and public officials will be hard-pressed to keep up as the pro-liferation of cars threatens to overwhelm their abilities to manage the metropolitan circulatory system.

ACKNOWLEDGMENTS

In the preparation of the three editions of this chapter, the comments, suggestions, and assis-tance of the editors and the following are gratefully acknowledged: Raymond A. Mohl of Florida Atlantic University (now of the University of Alabama at Birmingham), Thomas J. Baerwald of the Science Museum of Minnesota (now of the National Science Foundation), Ira M. Sheskin of the University of Miami, Roman A. Cybriwsky of Temple University, John B. Fieser of the Economic Development Administration of the U.S. Department of Commerce, Sterling R. Wheeler of the Fairfax County (Virginia) Office of Comprehensive Planning, and Sharyn Caprice Goodpaster of the City of Houston's Department of Planning and Development. I am particularly indebted to Truman A. Hartshorn of Georgia State University, my research colleague of the past two decades, for allowing me to use materials that were jointly developed in our studies of suburban business centers.

NOTES

1. This approach, while emphasizing the key role of transportation, does not mean to suggest that movement processes are the only forces shaping intraurban growth and spatial organization. As will be demonstrated throughout this chapter, urban geographical patterns are also very much the products of social values, land resources, investment capital availability, the actions of private markets, and other infrastructural technologies.
2. Since this map was compiled, this huge suburban complex has continued its growth, spearheaded by a doubling in size of the superregional mall at its heart (labeled *5* in Figure 3.14).

REFERENCES AND FURTHER READINGS

Adams, J. S. (1970). Residential structure of midwestern cities. *Annals of the Association of American Geographers, 60,* 37–62.

Baerwald, T. J. (1978). The emergence of a new "downtown." *Geographical Review, 68,* 308–318.

Banham, R. (1971). *Los Angeles: The architecture of four ecologies.* New York: Penguin Books.

Berry, B. J. L. (1975). The decline of the aging metropolis: Cultural bases and social process. In G. Sternlieb & J. W. Hughes (Eds.), *Postindustrial America: Metropolitan decline and interregional job shifts* (pp. 175–185). New Brunswick, NJ: Rutgers University, Center for Urban Policy Research.

Berry, B. J. L. (1981). *Comparative urbanization: Divergent paths in the 20th century* (2nd ed.). New York: St. Martin's Press.

Berry, B. J. L., Cutler, I., Draine, E. H., Kiang, Y.-C., Tocalis, T. R., & de Vise, P. (1976). *Chicago: Transformations of an urban system.* Cambridge, MA: Ballinger.

Bruce-Briggs, B. (1977). *The war against the automobile.* New York: Dutton.

Castells, M., & Hall, P. (1994). *Technopoles of the world: The making of twenty-first century industrial complexes.* London and New York: Routledge.

Cervero, R. (1986). *Suburban gridlock.* New Brunswick, NJ: Rutgers University, Center for Urban Policy Research.

Cervero, R. (1989). *America's suburban centers: The land use–transportation link.* Winchester, MA: Unwin Hyman.

Erickson, R. A. (1983). The evolution of the suburban space-economy. *Urban Geography, 4,* 95–121.

Flink, J. J. (1975). *The car culture.* Cambridge, MA: MIT Press.

Garreau, J. (1991). *Edge city: Life on the new frontier.* Garden City, NY: Doubleday.

Garreau, J. (1994, February). Edge cities in profile. *American Demographics,* pp. 24–33.

Harris, C. D., & Ullman, E. L. (1945). The nature of cities. *Annals of the American Academy of Political and Social Science, 242,* 7–17.

Hartshorn, T. A. (1992). *Interpreting the city: An urban geography* (2nd ed.). New York: Wiley.

Hartshorn, T. A., & Muller, P. O. (1989). Suburban downtowns and the transformation of metropolitan Atlanta's business landscape. *Urban Geography, 10,* 375–395.

Hartshorn, T. A., & Muller, P. O. (1992). The suburban downtown and urban economic development today. In E. S. Mills & J. F. McDonald (Eds.), *Sources of metropolitan growth* (pp. 147–158). New Brunswick, NJ: Rutgers University, Center for Urban Policy Research.

Jackson, K. T. (1985). *Crabgrass frontier: The suburbanization of the United States.* New York: Oxford University Press.

Kotkin, J. (2000). *The new geography: How the digital revolution is reshaping the American landscape.* New York: Random House.

Krim, A. J. (1977). *Northwest Cambridge* (Survey of architectural history in Cambridge, Report 5). Cambridge, MA: Cambridge Historical Commission and MIT Press.

Lang, R. E. (2002). *Edgeless cities: Exploring the elusive metropolis.* Washington, DC: Brookings Institution Press.

Leinberger, C. B., & Lockwood, C. (1986, October). How business is reshaping America. *Atlantic Monthly, 258,* 43–52.

Mencimer, S. (2002, April 28). Economics of transportation: The price of going the distance. *New York Times Magazine,* p. 34.

Miller, R. (1982). Household activity patterns in nineteenth-century suburbs: A time–geographic exploration. *Annals of the Association of American Geographers, 72,* 355–371.

Mills, E. S. (1970). Urban density functions. *Urban Studies, 7,* 5–20.

Muller, P. O. (1976). *The outer city: Geographical consequences of the urbanization of the suburbs* (Resource Paper No. 75-2). Washington, DC: Association of American Geographers, Commission on College Geography.

Muller, P. O. (1981). *Contemporary suburban America.* Englewood Cliffs, NJ: Prentice-Hall.

Muller, P. O. (1997). The suburban transformation of the globalizing American city. *Annals of the American Academy of Political and Social Science, 551,* 44–58.

Pacione, M. (2001). *Urban geography: A global perspective.* London and New York: Routledge.

Schaeffer, K. H., & Sclar, E. (1975). *Access for all: Transportation and urban growth.* Baltimore: Penguin Books.

Schwartz, J. (1976). The evolution of the suburbs. In P. Dolce (Ed.), *Suburbia: The American dream and dilemma* (pp. 1–36). Garden City, NY: Anchor Press/Doubleday.

Sommers, L. M. (1983). Cities of Western Europe. In S. D. Brunn & J. F. Williams (Eds.), *Cities of the world: World regional urban development* (pp. 84–121). New York: Harper & Row.

Stanback, T. M. Jr. (1991). *The new suburbanization: Challenge to the central city.* Boulder, CO: Westview Press.

Statistical Abstract of the United States. (various years). Washington, DC: U.S. Government Printing Office.

Steiner, R. (1981). *Los Angeles: The centrifugal city.* Dubuque, IA: Kendall/Hunt.

Sternlieb, G., & Hughes, J. W. (Eds.). (1975). *Postindustrial America: Metropolitan decline and inter-regional job shifts.* New Brunswick, NJ: Rutgers University, Center for Urban Policy Research.

Tarr, J. A. (1984). The evolution of the urban infrastructure in the nineteenth and twentieth centuries. In R. Hanson (Ed.), *Perspectives on urban infrastructure* (pp. 4–66). Washington, DC: National Academy Press.

Taylor, G. R. (1970). *Satellite cities: A study of industrial suburbs.* New York: Arno Press. (Original work published 1915)

U. S. Congress, Office of Technology Assessment. (1995). *The technological reshaping of metropolitan America* (OTA-ETI-643). Washington, DC: U.S. Government Printing Office.

Vance, J. E., Jr. (1964). *Geography and urban evolution in the San Francisco Bay Area.* Berkeley: University of California, Berkeley, Institute of Governmental Studies.

Vance, J. E., Jr. (1990). *The continuing city: Urban morphology in Western civilization.* Baltimore: Johns Hopkins University Press.

Ward, D. (1971). *Cities and immigrants: A geography of change in nineteenth-century America.* New York: Oxford University Press.

Warner, S. B., Jr. (1962). *Streetcar suburbs: The process of growth in Boston, 1870–1900.* Cambridge, MA: Harvard University Press and MIT Press.

Wilson, D. (Special Ed.). (1997). Globalization and the changing U.S. city. *Annals of the American Academy of Political and Social Science, 551,* 8–247.

CHAPTER 4

Impact of Information Technologies

DONALD G. JANELLE

Information technologies tie together computers, software, and information management systems. In transportation, these technologies are integrated into a large and growing number of significant applications. Some of these operate invisibly—for example, microprocessors and various sensors embedded in engines to enhance vehicle operation and to inform operators of vehicle performance levels and maintenance concerns. Others, based on linkages among computers that host large databases, telecommunication systems, and Internet capabilities, promise to revolutionize the way transportation agents and travelers gather, store, and represent information to facilitate the movement of goods and people. Applications include automated billing for system use and system designs to increase safety, speed, and synchronization of movement—all areas where information technologies offer some scope for improved transportation services in metropolitan regions. Nonetheless, even the most technically advanced regions in the world are a long way from creating fully integrated and automated intelligent transportation systems (ITS). As promising as these systems appear, there is considerable ap-prehension and a lack of knowledge about their ultimate impact on the spatial and temporal patterns of behavior that shape the structures and functions of cities and urban regions.

Another important link between information and transportation technologies concerns the practices and prospects for virtual mobility. *Virtual mobility* refers to the substitution of electronic transfers and exchanges for physical transport activities. It includes electronic commerce, or e-commerce (e.g., downloading the latest music instead of purchasing it at the store); telecommuting (e.g., working from home via computer and telecommunication facilities); conducting business from the cell phone in one's car; and other forms of virtual exchange: for example, telemedicine, tele-education, electronic business (e-business), e-leisure, and so forth. Although one can reject the prospects of Cairncross's (1997) death of distance, the implications of nearly ubiquitous connectivity among individuals and societal networks might significantly alter the foundations for transportation in the future, a prospect that demands careful consideration.

Many questions arise in addressing the roles of these new information and communication technologies in transportation and their likely impacts on the development of urban areas. How will new and more widespread technologies influence land use patterns in metropolitan regions and in their exurban hinterlands? Will they encourage new functional and spatial relationships between the home and the workplace that will alter the journey to work, the frequency of trips, or the choice of travel mode? In general, too little is known about these changes and the responses to them to permit even simple approximations based on the standard modeling techniques used in urban and regional planning. Answers may lie in understanding the complex and revolutionary changes that affect the very nature of the interdependence between individuals and the socioeconomic and technological structures that both hinder and facilitate their activities.

This chapter focuses on the implications of information and communications technologies (ICTs), using primarily North American examples. It describes (1) how structural and functional changes in the metropolitan economy have placed greater reliance on the quality and accessibility of information needed by corporate and public decision makers, (2) how technological changes have altered the ways that information needs are met, (3) how firms have changed their organizational structures and locational patterns, (4) how new locational and functional relationships between the home and the workplace have developed, (5) how the nature of the work task controls the extent to which communications may be substituted for transportation, and (6) how the changing transportation–communication relationship poses general planning and policy issues for cities and their metropolitan regions. Before considering these matters, it is useful to develop a conceptual framework that relates the structure and function of urban regions to changes in the significance of distance brought about by the technological and social innovations of the information society.

URBAN STRUCTURE AND DISTANCE

Urbanization is the very embodiment of communication. By concentrating a wide variety of creative specialists in a region of limited extent and of high connectivity, cities minimize the need for costly movement of goods and people. This is most evident in the central business areas of large cities where the chief executives of major firms and public agencies have unsurpassed opportunities for face-to-face exchanges. These urbanization benefits are achieved, however, at the price of crowding, congestion, excess demands on the natural environment, and other social and economic costs that may propel a dispersion of activities and people to outlying areas. The root of this dilemma (i.e., the need for specialized interaction and information vs. the desire for "elbow room" and amenity) is, in part, the product of temporal and spatial constraints on human behavior.

In their elemental form, these constraints are internal, defined by one's biological clock (e.g., the need for sleep, food, etc.) and lifespan, and by the inability to be in more than one place at a time. The importance of these constraints, however, is as much due to social and technological factors as it is to biological traits. Whereas biological traits are fixed by a genetic code, social and technological factors are subject to modification through human innovation.

Social and Technological Constraints on Distance

Social practices prescribe the precise timing and locations of events (such as work and school) that confine human movement to set rhythms and that result in general maximum distance limits between activity sites (e.g., home and work). Technology can extend these distance limits by altering the

TABLE 4.1. Constraints on Individual Behavior

	Temporal	Spatial
Biological	Prescribes essential life-support events (eating, sleeping, hygiene) according to distinct patterns of time	Sets limits on the geographical range of movement
Social	Prescribes events in time (church attendance, work hours, school, holiday periods)	Allocates activities in space and the distances between them
Technological (transportation, communications, production)	Frees time for other activities, allows faster response times for meeting needs, permits storage of information for use at any time	Alters the distance that can be covered per unit of time, allows linkages between distant places, permits different levels of separation between production units

amount of time required for given tasks and by lessening the time required to move between places—time that one might spend in other activities or in traveling even greater distances between activity sites. These interrelationships are listed in Table 4.1. Any critical discussion of the transportation–communication trade-off and telecommuting options must be related to elemental spatial and temporal constraints and their associations with biological, social, and technological factors.

Geographers Spencer and Thomas (1969) referred to transportation and communication innovations as "space-adjusting technologies" that change the significance of distance and allow for higher degrees of accessibility to specific locations. Such technologies can also be "time-adjusting," however, in that time is freed for alternative use in a variety of activities, not just in increasing the number of kilometers logged per day.

Increasing Space Standards and Distance

While much of the time freed from work and travel has gone to other creative efforts and leisure, some of it has been used to achieve more space per person, or higher space standards (such as larger house lots,

extra space for potential factory and store expansion, space for parking more and more cars, etc.). In addition, new technologies (like single-floor industrial plants) have increased the average distances between activities in urban regions. These increases have not been systematically documented, but a well-known Greek planner, Constantinos Doxiadis (1968), has speculated that, in the 40 years before 1968, average densities in several major cities dropped by two-thirds, implying that each individual was using three times the space used by people in the 1920s. Ivan Illich (1974), the late social critic, argued that increased speeds create distance by allowing for increased separations between related activities. An implication of such dispersal has been the growing reliance on the automobile and the growing difficulty in sustaining public transportation systems that depend on higher densities for generating traffic.

The Distance between Home and Work

One of the most basic distance constraints on human activity is the separation between the home and the workplace. This distance is crucial to interpreting urban transportation patterns and to gauging the possible transfer of many work-related activities to the home, via communications technologies. The home is the base for satisfying most of one's daily and lifetime biological needs and for meeting most of one's critical social obligations. Increasing the home's distance from the workplace incurs a time expense that could go to activities other than commuting. National Personal Transportation Surveys (NPTS; Hu & Young, 1999; U.S. Department of Transportation, 1992) document long-term trends of increasing average home–work travel distances in the United States. While average home–work trip lengths in kilometers decreased for the nation as a whole from 15.2 to 13.7 between 1969 and 1983, they increased to 17.7 in 1990 and to 18.7 in 1995. In addition, the 1995 NPTS data reveal average work–trip speeds of 55 kilometers per hour in 1995 in contrast to average speeds of only 45 kilometers per hour in 1983. Notably, this increase was accompanied both by an increase in average distances and by increases in the average duration of commutes (from 18.2 to 20.7 minutes). Furthermore, a preliminary analysis of the 2001 National Household Travel Survey (NHTS) and the 2000 Census (McGuckin & Srinivasan, 2003) indicates a shift to longer commutes. For example, one-way commutes of over 30 minutes accounted for 33.9% of workers in 1980, increasing to 36.4% in 1990, and to 40.5% in 2000.

Such evidence suggests interesting questions about human behavior and motivations. What are the social and technological factors that permit greater or lesser separation between residences and workplaces? Why do people spend their travel-speed gains on home locations that require longer journeys to work? What are the principal constraints that set the outer limits of this distance relationship? And how might uses of communications technologies alter these limits? Two factors that offer partial interpretation of these trends are changes in the terms of employment and time–space convergence.

Terms of Employment

I have already mentioned that social factors and technologies for overcoming distance play important roles in determining the temporal and spatial constraints on human activities. Gottmann (1977a), the geographer who coined the term "megalopolis" to describe the massive coalescence of urban development along the East Coast of the United States, saw the terms of employment as probably the most significant social determinant of urban spatial patterns. These terms represent contractual agreements between the employee and the employer concerning such matters as compensation, hours of work, time for holidays, work location, and so forth. Historical trends to shorter work weeks, longer paid vacations, flexible work hours, and even job sharing and early retirement have given employees greater control over how they spend their discretionary time. In the early period of the Industrial Revolution, 14-hour workdays and 6-day workweeks constrained workers and their families to live within short distances of the factories (described in Muller, Chapter 3, this volume). By the early 20th century, concessions granted to labor provided some release from these temporal constraints and permitted employees to use some of their freed time to commute greater distances. Gottmann saw this as contributing to a broader range of living environments and to the potential for massive suburbanization. Some of the free time could be channeled into creative and pleasurable activities, such as sports, hiking, and the arts. But these then generate needs for space—for parks, stadia, museums, and other cultural and recreational facilities. Two-day weekends permitted day trips into the countryside and to weekend cottages (i.e., secondary residences), and paid vacations helped to promote seasonal tourism, the growth of specialized resort centers, and urbanization in the coastal and mountain zones of high amenity.

Concurrent with these social changes, the advent of the automobile and the paving of roads ushered in new possibilities for expanding the commuter fields of cities, the residential options for households, and the geographical range for leisure activities. Such space-adjusting consequences of transportation innovation may be assessed directly by the concept of time–space convergence (Janelle, 1968).

Time–Space Convergence

If it required 74 hours to travel by stagecoach from Boston to New York in 1800 and only 5 hours to travel the same distance by automobile in 2000, then these two places have been approaching each other in time–space at the average rate of 20.7 minutes per year. Such measures provide a convenient way of characterizing the degree of transportation innovation between any pair of places and for determining relative changes in their overall accessibility with other places in the urban system. Needless to say, time–space divergence is also a possibility, reflecting the deterioration of transportation service between locations. These measures of convergence and divergence

objectively express what has long been characterized as a "shrinking world" at regional and global scales and as "congestion" at local scales.

Convergence rates vary depending on the degree and timing of innovations and on the degrees of route deterioration and congestion. Even under conditions of uniform changes in transportation quality, however, the convergence rates between more distant places are always greater than those for closer places. An example illustrates this point.

In Figure 4.1, a simple hypothetical urban highway network is focused on a single city. The travel-time distances for 1960 and 2005 and the resulting time–space convergence rates for each of nine locations within the metropolitan area are shown in Table 4.2. Whereas the 1960 pattern in Figure 4.1 shows a simple crossroads situation, the 2005 pattern represents a typical limited-access beltway with assumed speeds of 80 kilometers per hour. The outer fingers of the network are improved to 90 kilometers per hour by 2005, and, because of the diversion of traffic around the city, speeds within the city increase from 35 to 40 kilometers per hour. For each place, the average dis-

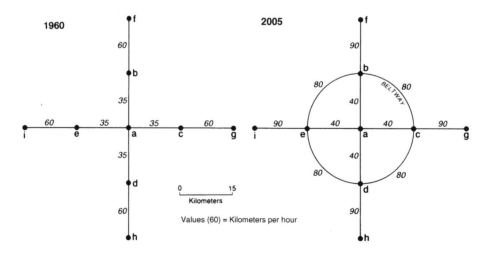

FIGURE 4.1. Hypothetical urban highway network and travel speeds used in calculating rates of time–space convergence (see Table 4.2).

TABLE 4.2. Convergence Rates and Distance in a Hypothetical Metropolitan Setting

Places[a]	Average kilometers (each place to all eight other places)			Average travel times (min) (each place to all eight other places)			Convergence rates, 1960–2005 (average min/yr) (each place to all eight other places)	
	1960 Network	2005 Beltway	Straight line	1960	2005 (using shortest distance paths)	2005 (using shortest time paths)	Using shortest distance paths	Using shortest time paths
a (city center)	22.50	22.50	22.50	33.24	27.54	27.54	0.127	0.127
b, c, d, e	31.80	32.98	28.68	49.26	27.90	25.50	0.475	0.528
f, g, h, i	45.00	46.10	41.80	62.40	36.66	34.26	0.572	0.625

[a]Letters refer to the places shown in Figure 4.1.

tance and the average convergence to all other places are shown in Table 4.2. The greatest accessibility gains have been recorded for the more distant journeys, demonstrating the increasing accessibility of peripheral locations.

Two critical conclusions may be drawn from this simple example. First, because of its central location, the city center has the shortest average distance to all other points in the city. Hence, it stands to make the least absolute gains from overall urban transportation improvements. Second, any overall increase in speed and flexibility of transportation, such as that facilitated by the automobile, means that the urban edge will, in contrast to the central city, on average, converge in time–space more rapidly with other places in the city than either the center or any points intermediate between the center and the edge. Since space limitations are not as critical in the outskirts as in the central city, the urban periphery has taken full advantage of automobile benefits.

Questioning the Costs and Benefits of Exurban Expansion

Most urban transportation improvements over the last half-century have been automobile-oriented, emphasizing a combination of limited-access radial and circumferential highways. These innovations have been of primary advantage to the ur-

ban edge, allowing easier cross-town movements by private vehicles and easier service of both central and regional hinterland positions. Garreau (1991) uses the term "edge city" to describe the expanding role of the urban periphery. In response to its improved accessibility, the emergent dominance of the urban periphery has been reinforced by the inflow of capital investment (see Chapter 3) and by the in-migration of talented residents. From analysis of the U.S. 2000 census, Pisarski (2002) concludes that enhanced mobility (multiple-car households) has permitted an unprecedented expansion of new residences in exurban environments. The locations of airports and other transshipment depots augment still further the accessibility advantages of the urban edge.

Changing patterns of accessibility within urban regions and changes in the terms of employment are seen as significant facilitators for greater separation between home and work and for establishing functional linkages over ever greater distances. Other factors also contribute to such trends. The extended open hours of stores and shopping centers permit people to travel longer distances, since with less scheduling constraint they can more easily budget their trips according to their personal needs and availability of time. In addition, rising incomes and trends toward more than one income per household have lessened the fiscal restraints on the use of household income.

Greater discretionary incomes and higher proportions of the population in the labor force are reflected in an increase in the number of commuting trips, in the number of vehicles per capita, and in vehicle miles traveled. While these are seen as contributing to higher levels of congestion, a study of automobile trip times in the 20 largest metropolitan areas in the United States from 1980 to 1985 (Gordon, Richardson, & Jun, 1991) revealed that average trip durations either declined or stayed the same. The authors referred to this as "the commuting paradox." Their explanation for the paradox was that commuters make adjustments in their places of work and residence to reduce commuting time or to use less-congested roads. In addition, they cited structural adjustments in the spatial forms of cities that have favored shorter commutes. These include residential suburbanization, the decentralization of firms, and the emergence of polycentric metropolitan areas. Furthermore, they see these changes as outcomes of an efficient market mechanism that precludes the need for public intervention. In focusing on the duration of commuting trips, Gordon and colleagues (1991) have sidetracked issues regarding the increasing distances traveled, made possible in part by increases in average travel speeds in the United States.

Bourne (1992) challenged the perspective of Gordon and colleagues (1991), citing the important role of institutional regulatory practices, and questioned their assumptions and values regarding what are preferred spatial forms for metropolitan regions. Is it one with extended low-density development (as is the norm for many U.S. cities) or one of planned high-density development of inner cities and suburbs (as is the case in Vancouver, Toronto, and other Canadian cities)? In turn, Gordon and Richardson (1997) have questioned the value of controlled compact growth, downtown revitalization, protection of agricultural land, and transit promotion as desirable planning goals, thereby questioning most aspects of state policy intervention and challenging some of the ideas of "smart growth."

Transportation Expansion and the Law of Constant Travel Time

In the Netherlands, an analysis of personal transportation for all purposes (not just commuting) and by all modes of movement revealed, in general, that people neither spend more time traveling nor make more trips from decade to decade. The author, Hupkes (1982), sees this as a general tendency and likens it to a "law of constant travel time." This observation has bearing on the issue of urban area expansion and on the synergistic links among transportation, telecommunications, and the spatial form of metropolitan regions.

Time, a finite resource in people's lives, is by definition a scarce resource that must be allocated according to subjective notions of utility, whether measured in dollars per hour or pleasure per hour. Hupkes's investigations indicate that the number of daily trips and the amount of time devoted to travel remains relatively constant over time. He cites such tendencies in many different countries and provides an aggregation of results from time-budget surveys in the Netherlands. Although shifts occur over time in the modes used (biking, walking, busing, driving), a remarkable stability in the overall number of trips and in the total hours devoted to travel was documented. In general, over time, people have used the advantage of extra speed to broaden the geographical range of their activities rather than to increase the number of their trips, while incurring no appreciable increase in the hours they devote to travel per day.

Of course, within any society there are variations across individuals in what might be considered acceptable as a maximum amount of time to spend in average daily travel. Hupkes suggests that, beyond a certain point, extra travel might generate biopsychological stress, which could dictate a cap to the time judged as generally acceptable.

Using the concept of utility, Hupkes offers an interesting explanation for an upper limit of acceptable travel commitments. Figure 4.2A illustrates the *intrinsic utility* that travelers derive from the journey itself (e.g., the sense of speed, the opportunity for fresh air or exercise, or other benefits). But, as trips continue and boredom or fatigue begin to dominate, this utility may decline, and additional increments of distance may even yield negative utility. *Derived utility* is related to the value obtained from activities that traveling makes possible, but this too

will peak at some point beyond which extra travel cuts into the time supply for these activities or is judged as too high a price to pay. These intrinsic and derived utility curves will differ for different individuals and may shift as one moves through the life cycle. Furthermore, technological changes may result in significant variations.

In Figure 4.2B, Hupkes's model is modified slightly to help shed insight on how advanced telecommunications (e.g., picture phones, interactive cable television, computers with modem and wireless linkages to the Internet, etc.) might influence utility values within the context of telecommuting

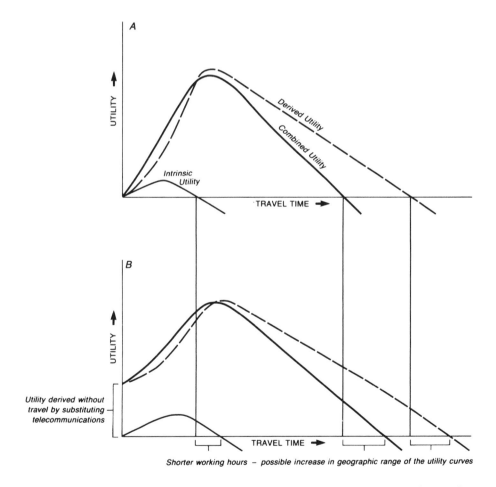

FIGURE 4.2. (A) The utility of travel time. (B) Changes in utility of travel time following the introduction of telecommunication systems and reduced work hours. Source: Hupkes (1982, pp. 42, 45). Copyright 1982 by Elsevier. Adapted by permission.

options. It is assumed that these technologies are cheap, easy to use, and widely distributed among businesses and households, allowing for some substitution for working, shopping, and personal business trips. In this case, the upward shift in the utility curve at the origin reflects the utility derived without trips. In addition, Figure 4.2B illustrates the effects of a reduction in work hours. Following Hupkes's assumption, if travel outside of working hours is evaluated for 20% of leisure time, then a reduction from 8 to 7 working hours per day would permit an extra 12 minutes for travel, some of which might be allocated to broadening one's geographical base of operation; some might also go for more pleasurable but slower trips (such as walking).

The merits of Hupkes's approach require empirical evaluation. There is no definitive proof that an optimal amount of time to devote to travel exists. Although the average value is relatively stable, it is subject to change as a consequence of social and technological forces, and it would be sensitive to policies on such matters as speed limits, transport investments in different modes, the availability of fuel, and support for technological substitutes for transportation. The possible uses of time, freed as a consequence of technical change or of changes in the terms of employment, depend in part on the choices that people make about where to live in relation to jobs and other activity sites. In general, an outcome of these decisions has been the outward migration of the city (a prospect allowed for in Figure 4.2B). With an increase in average commuting speeds from 60 to 70 kilometers per hour, the distance that can be reached within a 30-minute trip from a central point (e.g., the workplace) will increase by 5 kilometers—a 17% gain in distance. But this gain in distance increases the accessible land resource from 2,827 to 3,848 square kilometers—a 36% gain! Hence, a small increment in travel speed produces a considerable expansion of accessible land resources and potentially alters the potential variety of living

environments for commuters. It is important, therefore, to view the substitution of communication for transportation in terms of what motivates people to spread out at lower overall population densities.

Motivations and Opportunities for Exurban Lifestyles

Many factors motivate the migration of populations to regions beyond the edges of outwardly expanding North American cities. The symbolic "newness" of suburbs and the opportunity to start with a clean slate, where land is unencumbered by existing structures, may add a certain existential character to the process that is hard to label. Cheaper land (the possibility of having more of it per capita) and lower taxes have been important inducements for the centrifugal spread of city populations. Ginzberg and Vojta (1981) note that, with rising incomes, an increasing proportion of household income goes for products and activities related to self-identity and lifestyle, and less to basic necessities, comfort, and convenience. One manifestation of this change is the dispersal of exurbanites to the peripheral regions of the urban field—in part, a trend related to the quest for amenities, and one that contradicts standard expectations from location theory. Nonetheless, there appear to be basic regularities in this process that should not be ignored.

The "law of constant trip rates and constant travel time" is one such regularity that suggests outer limits to the time and resources that people will devote to travel. But increasingly the traditional economic argument of a trade-off between more housing and land, on the one hand, and less transportation cost, on the other hand, may be overshadowed by preferences for "lifestyle locations." Zelinsky (1975) calls these "voluntary regions," where people with like concerns stake out and create territories or places that cater to their interests. Residential communities built around golf courses or horseback riding, within the commuting range of large cities, represent obvious examples of such places. The post-1950 development of "dispersed cities" (Burton, 1963), "counterurbanization" (Berry, 1980), and "edge cities" (Garreau, 1991) has resulted in a great variety of specialized communities, a trend that also reflects the relaxation of social and technological constraints on where one chooses to live.

Population densities and travel speeds are also interrelated factors in the determination of the land supply available for urban growth. According to economist Wilbur Thompson (1968, p. 356), if the population of a city doubles and its population density remains constant, the distance from the edge of the city to its center would increase by only 40%. But, if there were a 50% drop in average population density, the size of the urban area would quadruple, and one would have to double the transport speed to maintain the same journey times from the edge to the center. He suggested the possibility that transportation developments in North America may have maintained this constant relationship over the past century. But even in this situation, the areal extent of cities would be expanding faster than the average distances of commuting trips. Thus, as the supply of land within a tolerable commuter range increases, the choice of residential site can be matched more closely with individual lifestyle goals. Such choices become less constrained by the locational selections made by others, and more attention can be given to weighing the amenities to be gained from specific physical and cultural environments.

Amenities and Urban Expansion

The combination of increasing discretionary money and increasing discretionary time gives a strong leisure orientation to many locational decisions. Ullman (1954) equated these trends with the growing significance that amenity plays in shaping the development of urban systems. He noted the rapid growth of Sunbelt cities and of mountain and coastal resort centers at regional and national scales. He also saw amenity as a leading contributor to suburbanization, to the increase of scattered nonfarm residents in countrysides, and to the growth of towns and small cities located within the fields of influence of major metropolitan centers.

Coppack (1985) adds significant evidence of the growing importance of amenity in one of the most urbanized parts of North America: the zone within a 2-hour driving range of Toronto. Coppack defines amenity in terms of the ability of places to satisfy human psychological needs. More pragmatic (but still difficult to define) concerns for "a good place to raise children" and the opportunity to escape from the less desirable features of urban life (congestion, pollution, crime, etc.) are embraced in this concept of amenity. However, intangible as such notions as scenic beauty and country atmosphere might be, Coppack describes amenity as a high-order commodity that people are willing to pay for and to travel long distances to obtain.

Coppack views the outward extension and transformation of urban fields as direct products of amenity-seeking behavior. The resulting mix of permanent nonfarm residents, farm population, and temporary users of the urban periphery (weekend cottagers, day-trippers, and weekend shoppers) contribute to satisfying the market needs of many lower order commercial and service functions. In such a way, amenity (the motivation) is seen as the conceptual link between the expanding urban field and the changing functional offerings of central places, leading to the resurgence of rural places that once bore the brunt of population loss and decay during the earlier era of cityward migration.

In principle, the supply of amenity associated with the land resource does not vary with increasing distance from a city. Travel costs and bid rents for the use of land, however, mean that the amenity resource develops a distinct concentric pattern around the city. Thus Coppack describes an urban core of high cultural amenity; an inner field of low-amenity supply, resulting from the competitive displacement by other land uses; a middle field that possesses a high-amenity value owing to the presence of acceptable bid rents and travel costs; and an outer field, where the evaluation of the amenity is low because of the high travel costs, even though the bid rents are low.

Figure 4.3 expands on Coppack's model to suggest that, through time, the accessibility of amenity resources will shift outward. In part, this may be due to new opportunities provided by transportation and communication developments. In addition, however, the intensification of urban facilities and population growth in centers within the initial (Time 1) area of high-amenity use may convert it into the jumping-off point for a succeeding round (Time 2) of outward expansion. This may be an attempt

to regain the amenity that one sees slipping from one's grasp. Thus, through a cycle of amenity growth and decline, places at increasing distances from the central metropolis become the major foci of amenity attraction and the centers for the progressive outward expansion of urbanization influences. This model suggests two possible scenarios: one driven by a perception of amenity that is rooted in rural and natural environments, and the other rooted in the significance of cities as the base for cultural amenity.

In the first scenario, if amenity is the driving force of expanding exurban settlement patterns, then the perceived outward shift of the zone of high amenity would likely result in one or a combination of the following responses: (1) demands for improved accessibility via transportation investments that permit longer distances without any appreciable increase in travel

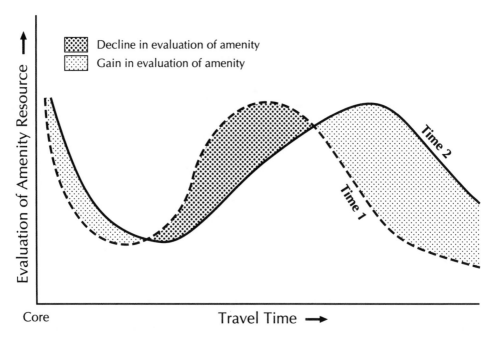

FIGURE 4.3. Perceived value of the amenity resource with increasing travel time from the core of a metropolitan region. See Coppack (1985, pp. 72–77) for a discussion of the principles that underlay the original model upon which this expanded model is based.

times, thus increasing the amenity supply; (2) the substitution of communication technologies for trips, including telecommuting and teleshopping; or (3) the transfer of jobs to the exurban frontier.

The alternative scenario places greater weight on the cultural amenities associated with residence within high-quality and densely populated cities or in their immediate suburbs. In this case, responses are most likely to focus on (1) demands for efficient, cost-effective, and accessible forms of public transportation to combat increasing levels of congestion; (2) opportunities to walk and to cycle in safe and attractive environments to obtain low-order goods and services and for recreation; and (3) the transfer of any personal savings in commuting time attained through telecommuting into higher levels of personal productivity, greater engagement with the social and cultural opportunities associated with cities, or greater internalization of amenity within the household.

The evaluation of these scenarios requires not only an understanding of responses by individuals and households, but also a focus on changes in the employment structure and interaction requirements of corporate and public agents. Telecommuting must be considered both in terms of how it effects the ability of private and public establishments to carry out their objectives and in terms of the locational relationships between these establishments and the households of their employees.

STRUCTURAL AND FUNCTIONAL CHANGES IN THE METROPOLITAN ECONOMY

High levels of employment in primary (fishing, agriculture, and mining) and secondary (manufacturing and construction) economic activities characterized the formative period of urban growth in the United States. In contrast, employment is now dominated by the service, or tertiary, sector, which in 2001 ac-counted for approximately 76% of the national labor force. The significance of this trend is manifested in the growing importance attached to the flow of information within society. Between 1990 and 2000, employment in information technology (IT) industries (e.g., computer production and services, communications) expanded by more than 50% (U.S. Economics and Statistics Administration, 2002), while employees within other sectors of the economy, whose use of IT is central to their job roles, expanded by nearly 30%.

The flow of information is particularly critical to the most rapidly growing component of the service sector, that of producer services—primarily high-level managerial, research, and professional jobs that focus on processing, analyzing, and distributing information that is vital to the increasing globally dispersed secondary (manufacturing) sector. Producer services, which account for more than a quarter of the U.S. gross national product (GNP), include marketing, advertising, legal counsel, accounting, and specialist consultants, whose timely access to highly reliable information is vital to the success of many private and public establishments (Beyers, 2000). Employment in these areas expanded by 50% between 1990 and 2000 (U.S. Economics and Statistics Administration, 1999, 2002). Many of the larger corporate firms compete for information and for the best minds to process and interpret it; they invest in the most sophisticated equipment to store, retrieve, manipulate, and communicate it; and they pay vast sums of money to locate their corporate offices in the most information-rich areas of major metropolitan centers. For example, in Canada these high-growth sectors of the economy are concentrated in the three largest Census Metropolitan Areas (Montreal, Toronto, and Vancouver) that account for half of the country's jobs in producer services (Coffey, 1994). Not surprisingly, the significance of communication linkages can be expected to vary according to the labor composition of cities and to the presence of information-sensitive firms within them. In addition, this rapid growth in the use of computers, the Internet, and other information and communication tools is altering the fundamental character of work and commerce. Kellerman (2002) sees the emergence in the 1990s and 2000s of an information-dominated society leading to significant structural changes in the economy and to related changes in the competitive positions of urban centers.

The spatial impact of information availability and use on the development of urban systems is related, in large measure, to how public and private agents have organized their operations to make effective use of information, labor, and other resources. As businesses increase in scale and in complexity, the controlling decision-making functions of upper-level management have tended to centralize for maximum access to consultation and information, but productive and administrative activities, based more on routine operations, have become more widely dispersed—to peripheral regions on the national and international levels (e.g., globalization).

Although the central business areas of large U.S. cities maintained their share of new office construction through the 1980s, there is now considerable evidence of dispersal at the metropolitan level. While "edge cities" (Garreau, 1991) and the "new suburbanization" (Stanback, 1991) have attracted the majority of U.S. businesses with fewer than 50 employees to the vicinities of beltway interchanges, there is evidence that large employers of information-intensive workers are also moving to these locations. Unlike urban sprawl, edge cities and suburban centers have captured locational and agglomeration advantages that now enable many of them to enjoy the information and contact advantages of central business districts. Recent moves of large employers suggest that the urban edge may no longer be incompatible with the information requirements of large firms.

Information Needs, Communication Linkages, and Contact Requirements

Communication linkages are needed to pass on or to receive information, to seek advice, to make assessments, to negotiate agreements, and so forth. Such task distinctions are critical to the evaluation of information needs and to determining the appropriateness of different means for transferring information. A classification of job-related tasks, used by Thorngren (1970) to analyze the communication needs of Swedish firms, distinguished among "orientation," "planning," and "programming" tasks, each directed to different kinds of objectives and requiring different kinds of contacts in order to be carried out successfully.

Orientation tasks may involve initiating new projects that require complicated discussions, difficult negotiations, and the assessment of prospects. In general, these activities entail considerable nonroutine decision making and coping with novel situations. The inherent uncertainty in new ventures makes face-to-face meetings particularly important, allowing for immediate feedback, flexibility in seeking on-the-spot clarification, and the use of various communication aids (demonstrations, exhibits, discussions, etc.). The complexity of many orientation tasks requires consultation with a variety of specialists. Thus prearranged meetings, often of considerable duration and in many locations, make travel an essential component in the daily work life of many managerial, professional, and technocratic occupations.

Planning tasks represent efforts to implement decisions that have already been made—for example, organizing procedures for production, marketing, finance, and so on. Although each problem to be solved may be unique, established channels of communication are used in resolving them. The contacts are generally less dispersed spatially and of less duration than for most orientation activities. Often, indirect contacts, via telephone or e-mail, are sufficient to carry out planning activities.

Programmed tasks refer to routine (sometimes repetitive) exchanges of information and to decision making that is based on structured and often standardized procedures ("according to the book"). Many production, sales, and clerical tasks fall into this category. At this level, indirect contacts, particularly by telephone and increasingly by e-mail are usually sufficient for most intra- and interfirm communications.

In general, the transition from orientation to programmed activities is toward decisions that require less time and tasks that require less travel. Olander (1979) uses Hägerstrand's (1970) time–geography approach to depict the interrelations among these kinds of tasks, seeing them as integrated activity systems that result in distinct time–space patterns of movements (paths), linking individuals and places, both within and among corporate establishments (see Figure 4.4). This focus on the daily activity patterns of a firm's employees calls attention to the spatial and temporal constraints that firms must consider in choosing the locations of their subsidiaries and corporate headquarters, and in matching their communications requirements with appropriate means of contact. Thus issues concerning

the relocation of a firm or parts of its operations (possibly to the amenity region outside the city), the separation between the home and work locations for different kinds of employees, possibilities for telecommuting, and the possible substitution of new ways to exchange information are seen in relation to the sensitivity of communication needs and to the daily activity systems of individuals.

Some work tasks are more contact-intensive than others. For example, Törnqvist (1970) suggests that the following activities are particularly reliant on face-to-face exchanges: decision making, planning and negotiation, distribution of information, publicity and selling, control and intelligence activities, research, analysis and educational work, construction, and product development and designing. Interestingly, as a proportion of all occupations, these nonroutine activities have increased proportionately over routine-oriented jobs. The importance of face-to-face meetings would suggest that this occupational shift could augment still further the centralization of office functions within the central business areas or suburban business centers of large cities. If advances in telecommunications allow for the substitution of many direct contacts, then two related propositions may be advanced.

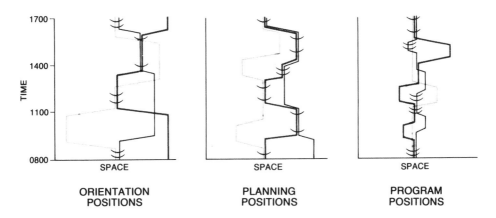

FIGURE 4.4. Activity paths of a firm's employees. Source: Olander (1979, p. 16). Copyright 1979 by John Wiley & Sons. Adapted by permission.

First, telecommunications can substitute for transportation; and, second, these technologies will permit information-based office work to decentralize beyond the central business area and even beyond the city's built-up zone to outlying regions of high amenity. An evaluation of these propositions must focus on two questions. First, to what extent can electrically transmitted communications replace face-to-face meetings? And, second, how might telecommunication linkages alter the relationships between the spatial organization of firms and the locational patterns of employees' households?

TELECOMMUNICATIONS AND TRANSPORTATION

According to Gottmann (1977b), the universal availability of the telephone had a profound impact on the spatial form of the U.S. metropolitan system, on the organization of corporate and public agencies, and on the lifestyles of individuals. If this was true in the mid-1970s, imagine the impact of innovations in telephone and computer uses over the past 30 years: cheaper flat-rate and wide-area services; the ability of fiber optic transmission to accommodate integrated voice, video, and data exchange; the advances in inexpensive high-quality facsimile (i.e., fax) transfer of documents; the locational flexibility associated with rapidly expanding networks for cellular telephones, and their compatibility with fax and computer linkages. Add to this the diffusion of home computers and television cables, the emergence of a rapidly expanding global Internet, and the growing possibilities of wireless access to a broad range of information resources. All of these innovations have promoted a powerful incentive to substitute telecommunications contacts for costly and more time-consuming face-to-face linkages.

To illustrate some of the potential benefits to be gained by substituting communications for transportation, consider the ex-

ample illustrated in Figure 4.5. Figure 4.5a shows individual B traveling directly from home to A's work location for a 30-minute meeting. B has sacrificed alternative activities that could have been carried out in the time required for this trip. Of course, this opportunity cost would have to be evaluated against the benefits of a face-to-face meeting over that of a telephone linkage. In Figure 4.5b a 30-minute telephone link between the work locations of A and B is represented by the horizontal broken lines. In this example, although it is not possible to assess the comparative quality of the communication for the two situations, it is possible to illustrate the general extent of the opportunity cost. Figure 4.5c represents the opportunity cost to B by means of space–time prisms (Hägerstrand, 1970). Space–time prisms (see Hanson, Chapter 1, this volume) define the set of places that are potentially accessible to an individual from a specific location within a given span of time. In this case, the inner prism (the black diamond) represents the space–time opportunity cost (i.e., the set of places to which B could have gone had B not made the telephone call) associated with the telephone call and the outer prism (the grey diamond) encloses the lost opportunities associated with the face-to-face meeting and the required journeys. It is clear that B's trip to meet A in person has preempted a greater range of alternative activities than would a similar exchange by telephone. The significance of this cost, however, will depend on the value that B attaches to time, on the differences in financial costs, and on the value that might be attached to the foregone options. Had B stayed at home as a telecommuter instead of traveling to the work site, then the savings in space–time opportunity cost would have been still greater, freeing an additional 56 minutes (28-kilometer round-trip at 30 kilometers per hour) in commuting time for alternative allocation.

One criterion that is most relevant to the substitution of communication for trans-

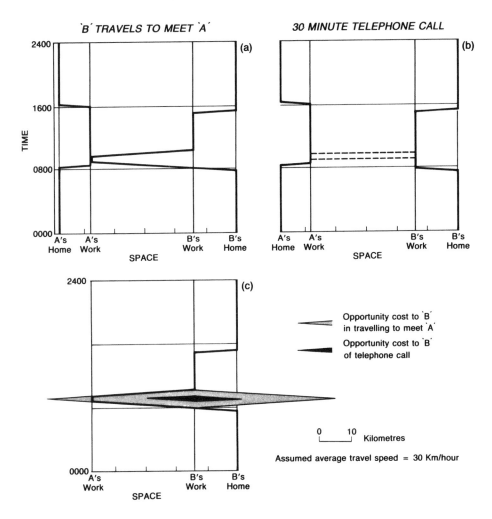

FIGURE 4.5. Comparative space–time opportunity cost of a telephone call and a business meeting.

portation relates to the degree of spatial and temporal constraint on the use of the system. This is summarized in Table 4.3 by whether or not locational and/or temporal coincidence is required between communicants. Face-to-face contacts require that all participants be in the same place at the same time, and, as noted earlier, such contacts entail transportation expenses and the displacement or exclusion of activities that could have been completed during the travel period. In contrast, systems based on the permanent storage of messages (such as tape recordings, books, and e-mail) pose the

lowest levels of constraint, since they can be accessed and responded to at any time that is convenient to the individual; the cost of such systems depends on the seriousness of the delay.

Assessing the Effectiveness of Teleconferencing

The quality of information may be insensitive to the distance over which communication takes place, but, in contrast, it is highly sensitive to the mode of transfer. Decisions on the use of specific means will depend on the suitability of a given mode for the task and on some assessment of the cost-effectiveness. Will the system allow the user to achieve the desired objectives (such as persuasion of the other individual or group)? Is it physically available to others? Can documents be transferred? Cost-effectiveness depends on the price of terminals and related equipment, the tariff for transmission of information, and the time required to prepare for and to carry out the communication. Since some services, such as audiovisual teleconferencing, represent large investments, it is important that private firms and public agencies give careful thought to these concerns.

Research on the effectiveness of audio and audiovisual communication dates to the mid-1970s, much of it carried out in laboratory situations in the United Kingdom. Though the technologies have changed significantly since then, some of the major findings are still relevant (Reid, 1977). Given the choice between teleconferencing and travel, the decision will depend on (1) the task of the meeting, (2) the familiarity among participants, and (3) the extent of travel that would be involved. Certain tasks lend themselves more than others to telecommunications. Thus face-to-face meetings are preferred for situations that require bargaining, negotiation, or the exchange of complex ideas; audiovisual modes are preferred to audio-only modes in situations where it is important to get to know people (e.g., a job interview); and audio media in combination with document transmission facilities are as effective as audiovisual or face-to-face modes for collaborative problem solving between people who know each other. Pye (1979) notes that participants in actual teleconferencing meetings regarded them as suitable for simple tasks (such as delegating work) but as less satisfactory for interpersonal interaction (e.g., disciplinary interviews).

An evaluation of the comparative costs for telephone and videoconferences and for face-to-face meetings was conducted by Salomon, Schneider, and Schofer (1991). They simulated analytically the costs for interaction with Chicago from seven selected cities across the United States. Using cost data for business travel and communications, they considered air miles, driving time, flight and telecommunication costs,

TABLE 4.3. Spatial and Temporal Constraints on Communications Systems

	Geographical coincidence of communicators required	
	Yes	No
Temporal coincidence of communicators required	Face-to-face meeting	Picture phones Telephones Cellular phones Teleconference (audio or audiovisual) Web conferencing and collaboration systems Instant messaging CB radio
Temporal coincidence of communicators not required	Refrigerator notes[a] Hospital charts[a]	Answering and recording machines Computer conferencing and bulletin boards Electronic mail Voice mail Mail Telegrams, telex, fax Printed publications

[a]Suggested by Harvey and Macnab (2000, p. 150).

the value of time, number of participants and number of origin centers, duration of meeting, and set-up time for arranging the meeting. They concluded that travel costs are less than telecommunication costs for short distances, but will vary depending on the number of people involved, the number of origin places, and the duration of the meeting. In contrast, teleconferencing is generally more economical for short meetings where long distances separate the origin centers.

Similar assessments are now underway with regard to web conferencing (or online meeting) technologies (Guly, 2003). Current web conferencing and web collaboration software/hardware requirements are expensive and permit only a few people to interact simultaneously to share and exchange data and documentation, but the expectation is that this technology will improve significantly in the near future. Guly reports that a recent survey by the Travel Industry Association of America found that security and cost concerns have prompted business travelers increasingly to hold meetings over the phone or via video or web conferencing. However, although two-thirds of the business interviewees reported cost savings over air travel, only 20% regarded the telemediated meetings as more effective than face-to-face sessions.

Expanding the Technological Base for Telecommuting

The impact of communication technologies in the long term will depend largely on their accessibility to potential users. Radios, televisions, and wired telephones may be the most widely distributed communication technologies within households, but personal computers, cellular telephones, fiber connectivity for fast Internet access, and wireless Internet are the emergent tools that define the information society. Based on reports available at *http://cyberatlas.internet.com* (a web marketer's guide to online news and information), approximately 66% of U.S.

adults (137 million) were using the Internet (as of March 2002); of these, 55% were accessing it from home and 28% from work. And about 34% of Internet-user households had high-speed broadband, which allows for more sophisticated communication potentials. As Kellerman (2002) observes, the diffusion of these technologies is advancing differentially through the settlement system, favoring connections within and among the large metropolitan markets long before diffusing to rural environments. To assess whether these technologies will bring disproportionate benefits to existing urban regions or allow for increasing dispersal of information-based jobs to remote amenity-rich regions, more data on geographical variations in the market penetration of home computers and other technologies are required. Nonetheless, the rapidity of the change generally is laying the base for intensifying a work-from-home scenario, seen by some as a means of reducing transportation requirements in metropolitan regions. Success in telecommuting is at least partially dependent on the technical capabilities of different modes of communication to allow for effective transactions, but it also requires attention to social issues involved in embedding paid work within the household and in separating workers from the traditional workplace.

Working from Home: Assessing the Transportation Substitution Potentials

Options for telecommuting range from Toffler's (1980) "electronic cottage," serving the elite in their remote hideaways in regions of high amenity, to the "electronic sweatshops" for work-at-home low-paid keypunchers. Between these extremes are dispersed telecommuting work centers that allow people to commute short distances for linkage with the main office via a variety of communications and computer resources. Toffler observes that the electronic cottage may be more suited to some jobs than others. Suitable candidates are those who work in sales and who prearrange visits by phone; consultants and counselors, who have widely dispersed customers; and those who require long periods for reflective work—for example, academic researchers, computer programmers, writers, artists, architects, and designers. In addition, some managerial and professional jobs require periods of intense contact with others and periods to be alone, either to create a large block of uninterrupted time for work on a single-objective task (such as a report), or to withdraw temporarily from the pressures of the job and from "information overload." The selective access to public and private databanks and to consultants via home-based telecommunication systems may permit these and many other activities to be carried out from the distant amenity regions of exurbia. Nonetheless, for most electronic commuters, the reality is working from one's existing principal residence, whether in cities or in surrounding suburbs.

"Telework" and "telecommuting" are terms that are often used interchangeably in the literature, though distinctions are sometimes drawn between any use of IT to substitute for work trips (telework) and their specific use to avoid commuting (telecommuting). Although telecomuting portends independence and economic advantage to many workers who are confined to their homes by virtue of household responsibilities or by disabilities that preclude commuting to distant jobs, there are growing concerns about social problems that accompany telecommuting. Holcomb (1991) cites evidence that the traditional workplace contributes to individual self-esteem, provides companionship, and offers the advantages of direct feedback and motivation. In addition, they raise the possibility that working at home could alter the social meaning of the home and place new burdens on the family unit. Even the journey to work is seen by some as an opportunity to make the psychological transition from family to work roles. Holcomb also notes possibilities for increased social stress resulting from the mix

of family and work lives. She notes how, for women with career aspirations, working from home might deny one the same opportunities enjoyed by those who are in closer personal contact with the job environment. More importantly, she observes that most of the women who work at home are involved in low-pay activities, such as data entry, catalog sales, ticket sales, processing insurance claims, and other routine tasks. Concerns over low job security, few worker benefits, and poor prospects for job advancement are at the root of labor union opposition to widespread adoption of telecommuting. These concerns illustrate how little is understood about the likely social side effects of the significant trend to work from home.

However important these concerns, it is evident that telecommuting offers interesting possibilities for altering transportation demands within urban regions. Various experiments illustrate the likely benefits and pitfalls of telecommuting and offer some insight into the prospects that widespread adoption might either mitigate the traffic problems of metropolitan regions or fuel the flight to exurbia.

Telecommuting and Traffic Mitigation

The impact of telecommuting on traffic levels and flow patterns is clearly a function of the number of telecommuters, their anticipated growth in numbers, and their choices about where to live and what activities to engage in. Estimates on the numbers of telecommuters in the United States can vary quite significantly, based largely on how one defines the term, but also on how the survey was administered. The Telework America 2001 survey by the International Telework Association and Council (ITAC; Davis & Polonko, 2001) estimates 28 million teleworkers, whereas estimates from some other surveys (see, e.g., Langhoff, 2004) record more than 32 million, up from 19 million in 2000 (these estimates include those who work at home occasionally as well as independent self-employed individuals). ITAC estimated 16.5 million regularly employed telecommuters in 2000. The growth in teleworking is fueled, in part, by the diffusion of household computers and fast Internet access, but it is important to recognize that substantial numbers of telecommuters rely on nothing more sophisticated than a telephone to carry out their work.

Jack Nilles (1991, 1998), one of the founders of the telework and telecommuting movement in the United States, estimates that telecommuters could number 50 million or more by 2010 and could save significantly on passenger miles and resultant pollutants. Nilles cofounded one of many private firms that provide consulting services and resources to government agencies and private-sector employers for setting up telework programs. These organizations advise large employers in the government and private sectors on how to implement telework options for employees to work at home and at dispersed telework centers, and on how to integrate these practices within existing agency or corporate structures and cultures.

The first neighborhood-level telecenter in the United States opened in Coronado, California, in October 1993, funded by the state transportation agency, Caltrans, and administered by the University of California, Davis. This neighborhood telecenter (closed in June 1996 following conflicts with the state regarding the ownership of equipment) was accessible from within the community by foot or bicycle. Telecenters are equipped with all of the prerequisite technologies for modern office functions, are usually accessible by automobile and/or public transportation, and provide shorter journeys to work for potential clients. For example, many telework centers (such as those in southern Maryland) specifically target individuals whose commute to work is 30 minutes or more and who would like greater flexibility in their work schedules. Among the advantages of telework centers

over home-based telecommuting (noted by Mokhtarian, 1991) is that they provide workers exposure to social ties that are normally associated with employment and they offer employers greater security over information and resources along with shared access to specialized equipment and more confidence in employee productivity. One of the obstacles to their establishment is the need for businesses and agencies to restructure centralized single-site establishments and fragmented branch-bank-type structures into organizations of dispersed workforces. Workers would travel to the closest one and then telecommute to the main work center through telephones, audio-video conference facilities, and computer terminals. As there is little financial incentive for private firms to develop such facilities, some jurisdictions may offer tax credits for employers who initiate telecommuting and may, at least partially, subsidize their development.

Private organizations with telecommuting programs include firms such as Aetna Life and Casualty Company, IBM, Boeing, AT&T, American Express, Blue Cross/Blue Shield, Apple Computer Company, Levi Strauss, and Control Data Corporation. The example of AT&T may not be typical, but its experience illustrates the lure and potentials of telecommuting. Based on a representative stratified random sample of 1,200 managerial employees, an AT&T telework white paper (Roitz, Allenby, Atkyns, & Nanavati, 2003) notes that one-third of the company's managers now telework at least 1 day per week, four times the level reported in 1992. Moreover, 17% work full time at home or at client locations. Teleworkers cite the increase in their productivity and improved balance between family and work as primary benefits, but 63% of them also mentioned savings in money and personal time, and 59% regard teleworking as beneficial to the environment. Relying on networks of knowledge workers instead of buildings, the white paper reports annual benefits to the firm in excess of $150 million in 2003 arising from increased productivity, reduced overhead costs (e.g., real estate), and enhanced retention and recruitment (2003, p. 1). The average number of days that teleworkers worked at home per month rose to 9.0 in 2003 from 6.7 in 2002. Furthermore, the white paper suggests that AT&T teleworkers saved over 150 million miles and related fuel consumption in not commuting to work in 2002. The concluding paragraph from the report is especially interesting:

> Broadly, as our societal reliance on location-based work shrinks, so do the traditional perceptual and physical barriers—and the inefficiencies and disparities—arising from age, disability, physical appearance or worker locale. Telework offers a powerful way for those who have typically been disenfranchised from the economic and social mainstream, such as seniors, or members of the disabled community, or those in rural areas, to become active participants in the knowledge economy. (p. 3)

Several U.S. states (e.g., Arizona, Florida, Oregon, and Washington) now promote programs in telecommuting as significant components of their transportation policies. The dominant current focus is to integrate telecommunications with strategies to manage the demand for transportation by reducing the need for new infrastructure and cutting congestion and transport-generated air and noise pollution. California has been a leader in this area, with early pilot studies on telecommuting and telework centers targeted at saving office space for state government agencies while contributing to solving problems of traffic congestion and air quality. By 2002, 60 California government agencies were actively engaged in managing or creating formal telework policies and programs (California Department of Personnel Administration, 2003). From an intitial pilot stage of about 200 teleworkers in 1987, these programs accommodated more than 7,000 state employees in 2002.

The Institute of Transportation Studies at the University of California, Davis, was contracted in 1992 to evaluate California's experience with telecenters to assess their potential as a model for demand management. A report based on detailed monitoring of these operations outlines the fate of California's telecommuting centers (Buckinger, Mar, Mokhtarian, & Wright, 1997). Of the 45 that opened between 1991 and mid-1997, only 23 were still operating in June 1997. The average lifespan was slightly less than 2 years, with most of those that closed doing so because of financial problems and insufficient steady clients. Reliance on public funding, their designation as demonstration projects, and conflicting goals all mitigated against long-term self-sustained operations for these centers. Those centers that succeeded tended to establish broader economic linkages and partnerships with local businesses and provided a greater range of business services to clients.

Most of the research on the utility of telecommuting as an instrument of transportation policy has looked at small non-random samples of workers or has focused on the responses for single firms or jurisdictions in comparatively few regions (with a bias toward metropolitan settings in California). While providing useful assessments, many are based on the unique circumstances of employees of a single firm or agency and have monitored the travel behavior of telecommuting participants for relatively short periods (usually for no more than 2 years) following the adoption of telecommuting. In addition, there is need to explore a broader range of interdependencies. For instance, Mokhtarian (1991) cites the need to understand how telecommuting might compete with other traffic-reduction methods (e.g., van pooling and transit). She also has suggested the need for cost–benefit analyses and the refinement of procedures for incorporating the telecommuting mode into standard forecasting and planning models. Other forms of telecommuting have also been neglected—for example, the use of telemediated remote diagnostics used to monitor the operations of elevators or heat and air quality systems in buildings. Yet another form of work arrangement facilitated by communication technologies needs systematic study: mobile work that is neither home- nor workplace-based (see Vilhelmson & Thulin, 2001). Working on the road, in airports, or while eating in restaurants draws on the portability of the latest handheld communication tools.

While U.S. teleworkers may save up to 5,000 or more miles in commuting per year, there are concerns that the freed time might be translated into other activities that require the same or more amount of travel. In general, an overemphasis on automobile travel has especially resulted in a neglect of research on pedestrian trip making and how this might vary for telecommuters and commuters. Because telecommuting has emerged as an increasingly important instrument of demand management, the need to clarify the ambiguity over behavioral responses to its adoption should represent a high priority for transportation research.

While research in this area needs substantial expansion, it is useful to identify some of the general findings. Mokhtarian (1991, pp. 329–330), summarizing some of the early findings on the travel behavior of telecommuters, notes that, in general, commute trips are reduced, noncommute trips do not increase, telecommuters tend to shift their activities to destinations closer to home, and they make fewer linked multipurpose trips than nontelecommuters. The same general findings were identified in Holland, where 7-day diaries were kept by 30 teleworkers for five sessions, 3 months apart, revealing a 15% overall reduction in commuting trips, but a more substantial 26% drop in peak-hour use of automobiles (Hamer, Kroes, & Ooststroom, 1992). More recent studies augment these general find-

ings, while expanding the range of questions explored (Handy & Mokhtarian, 1996; Mokhtarian, 1997, 1998; Salomon, 1998).

Research findings generally support the role of telecommuting as a suitable instrument to reduce the number and length of trips, but fears persist that the telecommuting option may exacerbate urban sprawl and related transport-generated pollution. For example, Hamer and colleagues warn that "introducing more flexible work hours and work locations, for instance, through teleworking, may result in workers accepting even longer commute distances for the remaining commutes" (1992, p. 98). They raise the specter that, in the long term, telecommuting may have the same consequences as mass motorization. A more recent empirical investigation by Ellen and Hempstead (2002) suggests, however, that this fear of decentralization may be unfounded. They analyzed Work Schedule supplements from the 1997 Census Bureau's *Current Population Study* to determine the extent to which telecommuters in the United States are opting for household locations in remote areas and in small, dispersed settlements. The survey covered 50,000 households, 18% of which engaged in some form of telecommuting from home. Contrary to the decentralization hypothesis, their findings show that telecommuting has had very little impact on choice of residential location; indeed, the telecommuters were more heavily concentrated in the larger metropolitan centers than the workforce as a whole.

CONCLUSIONS

In an apparent contradiction to the centralizing forces of information and contact requirements, recent North American trends toward exurban dispersal of residences and jobs indicate the increasing importance of perceived amenities. While the contact requirements of the knowledge industries pose constraints on this process, time–space convergence represents a potential relaxation of such constraints. The penultimate geographic expressions of technological substitution for direct contacts are Webber's (1963, 1964) "community without propinquity" and "nonplace urban realms"; but, in spite of the availability of technological advances that were not envisioned in the early 1960s and the emergence of the "information highway" in the mid-1990s, the prerequisite levels of mobility and access to advanced communications have only recently approached levels of adoption that could lead to fundamental changes in how work and commerce are transacted. This chapter has focused primarily on individual travel behavior, the relationship between home and work locations, and the changing structures of human settlement patterns. Equally important in regard to emergent forms of settlement and to household lifestyles is the movement of freight. In its manifestation as e-commerce, freight movement is also greatly affected by information technologies, a theme that is addressed in an excellent overview article by Golob and Regan (2001).

This chapter has stressed the importance of behavioral constraints that set outer limits on the degree of acceptable physical separation from the centers of metropolitan life and from the cultural amenities that require concentrated populations. Among the most significant are the need for face-to-face contacts and the law of constant travel time and trip frequency. In addition, the related biological and social constraints (such as the need for rest and terms of employment that favor prescribed work hours and workdays) impose limits to the geographical range of daily movements. Concerns for fossil fuel consumption and the environmental consequences of urban sprawl also may impose limits on the exodus to exurbia. On the strength of these constraints, I conclude that transportation technologies, enhanced by information technologies, will remain fundamental to the spatial structuring of urban regions. It seems axiomatic to the material basis of human existence that transportation provides the vital links in the day-to-day system of life support and that the outer limits of transport feasibility set the range of a person's social, economic, and other opportunities. Such a view portends gradual rather than revolutionary change.

Here is a simple example to illustrate how the impact of widespread telecommunications technologies on the spatial organization of metropolitan regions could depend on individual behavioral responses. The day has traditionally been regarded as the fundamental temporal unit for programming the choice, sequence, and duration of events in a household's life. In the future, the week may become a more significant unit of household planning. Of many possible alternatives (if the terms of employment permit), people might decide to allocate only 3 workdays at the office for essential face-to-face meetings and to spend alternate weekdays at home engaged in work that might benefit from greater isolation.

In this example, if the daily/weekly scheduling complications of two-worker households don't intrude on the household's freedom of residential mobility, one could consider relocating the home to an area of higher amenity at greater distance from the city office without increasing the weekly aggregate of time spent in traveling to work. Such a shift in behavior would facilitate an expanded range of opportunities for exurban living, but it might not be supportive of an environmentally sustainable form of settlement structure. For example, a modest average travel speed of 80 kilometers per hour and a time allocation of 30 minutes per day (150 minutes per 5-day workweek) for the one-way journey to work permits an outer distance between home and office of 40 kilometers. If one substitutes telecommunications from home for 2 days per week and allows for the full

150 minutes per week for traveling to work, the range for three 50-minute work trips would be 67 kilometers. Thus, without increasing the weekly allotment of time in work-related travel and without altering the level of linkage to the city, the zone of potential exurbanite expansion increases 180%, from 5,027 to 14,103 square kilometers.

Though the technology that enables decentralized patterns of job locations will expand in the future, many office activities will continue to require central locations, and the most information-sensitive activities will likely remain in metropolitan settings. Decentralized development must recognize the specific information and contact requirements of individuals and establishments.

Nonetheless, even under assumptions of modest travel speeds and current standards of communications capability, the land supply for exurban residential expansion could increase dramatically. Widespread application of information technologies to the provision of transport services (such as Intelligent Vehicle Highway Systems [IVHS], could substantially accelerate the trend toward exurban expansion in the advanced economies of North America, Western Europe, and Japan. One version of this innovation is referred to as "road transport informatics" (RTI) (Hepworth & Ducatel, 1992). Based on these technologies, Svidén (1993, pp. 182–183) projects scenarios for Europe to the year 2100. For example, in 2050 he envisions highly reliable and widely available RTI technologies that allow for

integrated European route guidance, automatic debiting, speed guidance, dynamic speed keeping, lane keeping, car following, and basic collision warning. The tutoring function can compensate for most types of driver handicap. The road-pricing scheme is efficiently distributing the costs to those who benefit, and raises the capital needed for further infrastructure improvement. The intelligent roads define the structure of the dispersed transurbia with its high info-mobility.

If the scenario holds, these and related RTI developments for public passenger transportation, and for freight transport, might greatly reduce the level of stress on drivers and could result in an even wider geographic choice of job-residence relationships.

A high "info-mobility society" is an alluring prospect, but it holds uncertainties. Will environmental consequences of transportation system expansion and fossil fuel depletion put a halt to mobility as a desired social outcome? Will expanded fields of household interaction go beyond the effective range of governance? Or will the new technologies allow for broader participation in the political process at local and regional levels? Will only certain groups of people have access to the new technologies and their associated lifestyles? Will Abler's (1975) fears be realized, that by allowing for more selective patterns of communication, new technologies could promote higher degrees of segregated social networks than currently exist? Appropriate forms of social innovation, structural changes in institutional organizations, even public subsidization for widely distributing the prerequisite equipment and education, may be needed to adapt to the impending changes. No matter how society responds to these and other questions, the future will be as much a product of political choice as one of technological possibility. My best guess for the first quarter of the 21st century is that both mobility and telecommunications systems will expand in their synergistic relationships; that both decentralized exurban development and centralized urban patterns will continue to expand, offering the range of contact requirements essential for a complex postindustrial society; and that transportation planning must accommodate simultaneously the needs to support both patterns of development.

REFERENCES

Abler, R. (1975). Effects of space-adjusting technologies on the human geography of the future. In R. Abler, D. Janelle, A. Philbrick, & J. Sommer (Eds.), *Human geography in a shrinking world* (pp. 35–56). North Scituate, MA: Duxbury Press.

Berry, B. J. L. (1980). Urbanization and counterurbanization in the United States. *Annals, American Academy of Political and Social Science, 451,* 13–20.

Beyers, W. (2000). Cyberspace or human space: Wither cities in the age of telecommunications? In Y. Aoyama, J. O. Wheeler, & W. Barney (Eds.), *Cities in the telecommunications age: The fracturing of geographies* (pp. 161–180). New York: Routledge.

Bourne, L. S. (1992). Self-fulfilling prophecies?: Decentralization, inner city decline, and the quality of urban life. *Journal of the American Planning Association, 58,* 509–513.

Buckinger, C., Mar, F., Mokhtarian, P., & Wright, J. (2001). *Telecommuting centers in California: 1991–1997.* Davis: University of California, Davis, Institute of Transportation Studies. Available online at www.its.ucdavis.edu/tcenters/repts/status/disc.html

Burton, I. (1963). A restatement of the dispersed city hypothesis. *Annals of the Association of American Geographers, 53,* 285–289.

Cairncross, F. (1997). *The death of distance: How the communications revolution will change our lives.* Boston: Harvard Business School Press.

California Department of Personnel Administration. (2003). *The State of California Telework–Telecommuting Program, 1983–21st Century.* Sacramento: California Department of Personnel Administration. Available online at http://www.dpa.ca.gov/telework/teleworkmain.shtm

Coffey, W. J. (1994). *The evolution of Canada's metropolitan economies.* Montreal: Institute for Research on Public Policy.

Coppack, P. M. (1985). *An exploration of amenity and its role in the development of the urban field.* Unpublished doctoral thesis, University of Waterloo, Waterloo, Ontario, Canada.

Davis, D. D., & Polonko, K. A. (2001). *Telework America 2001 Summary.* Rockville, MD: International Telework Association and Council. Available online at http://www. workingfromanywhere.org/telework/twa2001.htm

Doxiadis, C. A. (1968). Man's movement and his city. *Science, 62,* 326–334.

Economics and Statistics Administration (1999). *The emerging digital economy II, appendices.* Washington, DC: U.S. Department of Commerce.

Economics and Statistics Administration. (2002). *Digital economy 2002.* Washington, DC: U.S. Department of Commerce. Available online at http://www.esa.doc.gov/pdf/DE2002r1.pdf

Ellen, I. G., & Hempstead, H. (2002). Telecomuting and the demand for urban living: A preliminary look at white-collar workers. *Urban Studies, 39,* 749–766.

Garreau, J. (1991). *Edge city.* New York: Doubleday.

Ginzberg, E., & Vojta, G. J. (1981). The service sector of the U.S. economy. *Scientific American, 244,* 48–55.

Golob, T. F., & Regan, A. C. (2001). Impacts of information technology on personal travel and commercial vehicle operations: Research challenges and opportunities. *Transportation Research, Part C, 9,* 87–121.

Gordon, P., & Richardson, H. W. (1997). Are compact cities a desirable planning goal? *Journal of the American Planning Association, 63*(1), 95–106.

Gordon, P., Richardson, H. W., & Jun, M.-J. (1991). The commuting paradox: Evidence from the top twenty. *Journal of the American Planning Association, 57,* 416–420.

Gottmann, J. (1977a). Megalopolis and antipolis: The telephone and the structure of the city. In I. de Sola Pool (Ed.), *The social impact of the telephone* (pp. 303–317). Cambridge, MA: MIT Press.

Gottmann, J. (1977b, December 8). *Urbanisation and employment: Towards a general theory.* Paper presented as the Walter Edge Public Lecture at Princeton University, Princeton, NJ.

Guly, C. (2003, March 20). Meetings via mouse clicks. *The Globe and Mail* (Toronto, Canada), p. B11.

Hägerstrand, T. (1970). What about people in regional science? *Papers of the Regional Science Association, 24,* 1–21.

Hamer, R., Kroes, E., & van Ooststroom, H. (1992). Teleworking in the Netherlands: An evaluation of changes in travel behaviour—Further results. *Transportation Research Record, 1357,* 82–89.

Handy, S. L., & Mokhtarian, P. L. (1996). The

future of telecommuting. *Futures, 28,* 227–240.

Harvey, A. S., & Macnab, P. A. (2000). Who's up?: Global interpersonal temporal accessibility. In D. G. Janelle & D. C. Hodge (Eds.), *Information, place, and cyberspace: Issues in accessibility* (pp. 147–170). Berlin: Springer.

Hepworth, M., & Ducatel, K. (1992). *Transport in the information age: Wheels and wires.* London: Belhaven Press.

Holcomb, B. (1991). Socio-spatial implications of electronic cottages. In S. D. Brunn & T. R. Leinbach (Eds.), *Collapsing space and time: Geographic aspects of communications and information* (pp. 342–353). London: HarperCollins Academic.

Hu, P. S., & Young, J. R. (1999). *Summary of travel trends. 1995 nationwide personal transportation.* Washington, DC: U.S. Department of Transportation, Federal Highway Administration.

Hupkes, G. (1982). The law of constant travel time and trip-rates. *Futures, 14,* 38–46.

Illich, I. D. (1974). *Energy and equity.* London: Calder & Boyars.

Janelle, D. G. (1968). Central place development in a time–space framework. *Professional Geographer, 20,* 5–10.

Kellerman, A. (2002). *The Internet on earth: A geography of information.* Chichester, UK: Wiley.

Langhoff, J. (2004). *FAQs about telecommuting.* June Langhoff's Telecommuting Resource Center. Available online at http://www.langhoff.com

McGuckin, N., & Scrinivasan, N. 2003. A walk through time: Changes in the American commute. Available online at http://nhts.ornl. gov/2001/index.shtml

Mokhtarian, P. L. (1991). Telecommuting and travel: State of the practice, state of the art. *Transportation, 18,* 319–342.

Mokhtarian, P. L. (1997). The transportation impacts of telecommuting: Recent empirical findings. In S. P. Stopher & M. Lee-Gosselin (Eds.), *Understanding travel behavior in an era of change* (pp. 91–106). Oxford, UK: Elsevier.

Mokhtarian, P. L. (1998). A synthetic approach to estimating the impacts of telecommuting on travel. *Urban Studies, 35,* 215–241.

Nilles, J. M. (1991). Telecommuting and urban sprawl: Mitigator or inciter? *Transportation, 18,* 411–432.

Nilles, J. M. (1998). *Managing telework: Strategies for managing the virtual workplace.* New York: Wiley.

Olander, L. (1979). Office activities as activity systems. In P. W. Daniels (Ed.), *Spatial patterns of office growth and location* (pp. 159–174). Chichester, UK: Wiley.

Pisarski, A. (2002, March). *Mobility congestion and intermodalism.* Testimony before the U.S. Senate, Committee on Environment and Public Works, Washington, DC. Available online at http://www.senate.gov/~epw/Pisarski_031902.htm

Pye, R. (1979). Office location: The role of communications technology. In P. W. Daniels (Ed.), *Spatial patterns of office growth and location* (pp. 239–275). Chichester, UK: Wiley.

Reid, A. A. L. (1977). Comparing telephone with face-to-face contact. In I. de Sola Pool (Ed.), *The social impact of the telephone* (pp. 386–414). Cambridge, MA: MIT Press.

Roitz, J., Allenby, B., Atkyns, R., & Nanavati, B. (2003, April 11). *AT&T telework white paper. Organizing around networks, not buildings: 2002/2003 AT&T employee telework research results.* Available online at http://www.att.com/telework/article_library/survey_results_2003. html.

Salomon, I. (1998). Technological change and social forecasting: The case of telecommuting as a travel substitute. *Transportation Research, Part C, 5,* 17–45.

Salomon, I., Schneider, H. N., & Schofer, J. (1991). Is telecommuting cheaper than travel?: An examination of interaction costs in a business setting. *Transportation, 18,* 291–317.

Spencer, J. E., & Thomas, W. L. Jr. (1969). *Cultural geography: An evolutionary introduction to our humanized earth.* New York: Wiley.

Stanback, T. M. (1991). *The new suburbanization.* Boulder, CO: Westview Press.

Svidén, O. (1993) A scenario on the evolving possibilities for intelligent, safe and clean road transport. In G. Giannopoulos & A. Gillespie (Eds.), *Transport and communications innovation in Europe* (pp. 175–185). London: Belhaven Press.

Thompson, W. R. (1968). *A preface to urban economics.* Baltimore: Johns Hopkins University Press.

Thorngren, B. (1970). How do contact systems affect regional development? *Environment and Planning, 2,* 409–477.

Toffler, A. (1980). *The third wave.* New York: William Morrow.

Törnqvist, G. (1970). *Contact systems and regional development* (Lund Studies in Geography, Series B, Human Geography, No. 35). Lund, Sweden: Royal University of Lund, Department of Geography.

Ullman, E. (1954). Amenities as a factor in regional growth. *Geographical Review, 44,* 119–132.

U.S. Department of Transportation. (1992). *1990 Nationwide Personal Transportation Survey: Summary of travel trends.* Washington, DC: Federal Highway Administration.

Vilhelmson, B., & Thulin, E. (2001). Is regular work at fixed places fading away?: The development of ICT-based and travel-based modes of work in Sweden. *Environment and Planning A, 33,* 1015–1029.

Webber, M. M. (1963). Order in diversity: Community without propinquity. In L. Wingo, Jr. (Ed.), *Cities and space: The future use of urban land* (pp. 23–54). Baltimore: Johns Hopkins University Press.

Webber, M. M. (1964). The urban place and nonplace urban realm. In M. M. Webber, J. W. Dyckman, D. L. Foley, A. Z. Guttenberg, W. L. C. Wheaton, & C. B. Wurster (Eds.), *Exploration into urban structure* (pp. 79–153). Philadelphia: University of Pennsylvania Press.

Zelinsky, W. (1975). Personality and self-discovery: The future social geography of the United States. In R. Abler, D. Janelle, A. Philbrick, & J. Sommer (Eds.), *Human geography in a shrinking world* (pp. 108–121). North Scituate, MA: Duxbury Press.

PART II

PLANNING FOR MOVEMENT WITHIN CITIES

CHAPTER 5

The Urban Transportation Planning Process

ROBERT A. JOHNSTON

The rapid movement of people and goods is essential to economic growth in the United States and in every one of its urban areas. Funds for building and operating highway and transit systems are a major portion of expenditures for local, state, and federal governments. Traffic congestion is one of the most important issues facing local elected officials everywhere. Suburbs fight for more highway funds, while central cities push for transit improvements. Within all large urban regions, city, suburban, and rural jurisdictions must cooperate in regional transportation planning exercises every 3 years.

Urban transportation planning in the United States began with the development of travel demand models in the 1950s and then became formalized in the 1960s with the U.S. Department of Transportation (U.S. DOT) codifying practice and software. Travel models begin by making 20-year projections of future land uses in small zones throughout the region. These land use projections (residential, commercial, industrial) are then used to project, for each small area, the number per zone of households, which generate trips, and the number

of employees, which attract trips. With computer software, the trips are assigned to a network of roads that connects the zones. As more trips are loaded onto the network in the modeling process, travel speeds slow down and some trips are assigned to faster sets of links in the network. After many iterations where all trips try all possible routes, total traveler time is minimized. Then travel speeds on the links are analyzed to see where congestion is the worst. These congested areas are where additional investment will be needed, to increase road or transit capacity.

I begin this chapter with a review of the history of regional transportation planning in the United States and then identify recent federal laws that have placed additional requirements on transportation modeling and planning. Next, I review the steps in travel modeling, outlining improvements that the larger metropolitan agencies are undertaking and the variety of policy analysis capabilities that regional transportation planning agencies are now pursuing to address the needs of citizen groups and local governments. These "bottom-up" needs, which come from the constituent groups, are in addition to the "top-down" requirements from the federal laws. After this review I outline an ambitious model improvement program in one region to bring all of these pieces together. Finally, I describe foreseeable future modeling advances to indicate the nature of planning practice during the next decade.

HISTORY OF REGIONAL TRANSPORTATION PLANNING

Federal law requires regional transportation planning for medium-sized and large urban areas in order to qualify for federal highway and transit funding. Many states also have similar requirements to get state funding. Transportation is the only remaining type of infrastructure funding for which Congress still requires regional planning. Such planning was formerly necessary to apply for federal funding for hospitals, sewage treatment plants, and water supply systems, but these regulations were dropped in the 1970s. Understanding regional transportation planning is vital because it is the only regional planning process undertaken regularly in most regions and because changes in transportation systems affect many aspects of urban life, including traffic congestion, air quality, transit availability for lower income households, and land development patterns.

Before World War II, transportation planning did not include travel modeling. After the war, rapid increases in automobile ownership, widespread suburbanization, and a vast increase in federal spending for transportation investments led to the need for regional modeling of travel behavior. Early studies were done in Detroit and Chicago, followed by studies in other eastern cities. Many of the early methods were developed by the Chicago Area Transportation Study. The Federal Highway Act of 1963 required urban areas to use a "continuing, comprehensive, and cooperative" transportation planning process (the 3C process) to qualify for federal funding. Urban areas were de-

fined as regions, and so regional modeling became the standard procedure. The Bureau of Public Roads (which later became the Federal Highway Administration [FHWA]) sponsored research to develop standardized methods and published manuals describing travel modeling methods so that transportation modeling and planning practice became standardized nationwide. Specifically, these methods became codified in the Urban Transportation Modeling System (UTMS), or Urban Transportation Planning System (UTPS), which was a standardized set of methods for "four-step" travel modeling. The four steps are (1) *trip generation*, which predicts the number of daily household trips; (2) *trip distribution*, which predicts where each trip goes; (3) *mode choice*, which predicts which travel mode is chosen; and (4) *route assignment*, which predicts travelers' routes on roads, rail lines, or bus lines. Many regions used the software disseminated by the U.S. DOT for use on mainframe computers.

Travel models play a central part in transportation planning. How are they designed? First, the Metropolitan Planning Organization (MPO) divides the urban region into 100–2,000 zones, generally composed of census block groups and census tracts. The MPO develops a set of networks in the travel model software that represent all the major roads and all transit lines. Then the MPO performs a household travel survey, where a few thousand randomly selected households are surveyed on a random midweek day, in order to identify all trips made by time, mode, route, destination, and purpose on that day. These data are then used to estimate several submodels, also known as the "travel model steps."

An ideal travel model would represent all behaviors affected by changes in the transportation system. The model would be estimated on large data sets and the equations in the various submodels would have high explanatory value, statistically. Futhermore, the models would include explanatory variables relevant to all major policy issues, such as road tolls, new road capacity, or compact growth.

The MPO begins with the first of the four main submodels, the *trip generation* model, in which household characteristics (e.g., income and size) determine the number of trips the household makes per weekday. Second, a *trip distribution* model is estimated, where trips are generated by households and attracted by employment. The trips are linked randomly from one zone to another and then a proportionate fitting program rematches origin and destination zones, to get a distribution of trip time lengths which matches that in the survey data. This is done separately for each trip purpose (home-based work, home-based other, and non–home-based are the minimum set). Then, the agency estimates a *mode choice* model, where the travel mode chosen is a function of the modes available to the household, household income, and the cost of the mode for each trip taken (time cost plus auto mileage costs or transit fares). Finally, in the *assignment* submodel, trips are assigned to the network, to get from the origin zone to the destination zone. As links become loaded with travelers, the vehicle speeds slow down. This results in some trips being assigned to other routes, until no trip can be made more quickly.

The whole model set is run on the base year data, and traffic volumes are calibrated to actual traffic counts in certain parts of the region. Various adjustments are made in the submodel parameters until the modeled traffic volumes match the observed volumes fairly well. This calibration is also done for transit line ridership for all transit lines and for regional mode shares and number of trips. After the model is calibrated, it is run for those future years of interest, based on the amount and location of projected population and employment. Generally, models are run on a 20-year horizon for facility planning and are also run on the intermediate years for which there is a vehicle emissions reduction deadline. Figure 5.1 shows the steps in this modeling process.

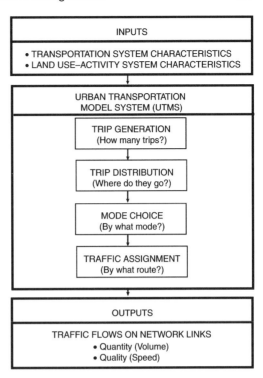

FIGURE 5.1. The Urban Transportation Model System (UTMS). Source: Pas (1995, p. 65). Copyright 1995 by The Guilford Press. Reprinted by permission.

In the 1960s, opposition to freeways in central cities resulted in a statutory requirement for a local freeway agreement to be signed by all cities and counties where the right-of-way would be acquired. Rising environmental concerns led to the adoption of the National Environmental Policy Act (NEPA), in 1969, which requires a detailed environmental assessment of all federally funded projects. In 1975, the U.S. DOT adopted regulations requiring not only a long-range (20-year) plan, but also a short-term element, focusing on low-cost Transportation Systems Management (TSM) policies, such as carpool lanes, ramp metering, and limiting parking. These measures were intended to reduce peak travel demand and to manage peak flows better; they represented the first federal objectives for demand management, instead of simply providing more capacity on roads and transit lines. Also, public participation requirements were strengthened and paying attention to the transportation needs of the elderly and the poor was mandated.

In the early 1980s, the federal agencies reduced their oversight of regional planning somewhat, but the basic planning requirements stayed in place. Federal funding for new freeway and highway capacity began to decline steadily as the Interstate Highway System was almost completed. Greater suburban travel and higher auto ownership in general led to continuing rapid increases in vehicle miles of travel (VMT), a broad measure of travel demand within urban regions. Although tailpipe controls on vehicles continued to improve, the vast majority of larger urban regions did not meet all federal air quality standards. Many urban regions adopted plans with light rail systems, and, in general, there was an increased interest in transit improvements, some of which was directed to urban renewal objectives for central cities.

In the 1950s and 1960s, travel modeling addressed mainly the simple issue of where to provide new freeways and highways. Solving this problem merely requires the agency to determine the most congested corridors in the road system. Only the trip distribution model step (where trips originate and where they go) and the route assignment model step have to be roughly right to determine relative congestion levels on roads. Trip generation and mode choice do not have to be very accurate. Models were generally daily models, in that they represented all daily travel. Vehicle speeds were poorly simulated. Travel models were generally accurate enough for ranking network links by congestion level, however.

In the 1970s and 1980s, travel modeling practice advanced statistically, with the use of disaggregate discrete choice models, mainly for mode choice, advocated at the Williamsburg Urban Transportation Modeling Conference, held by the U.S. DOT in the early 1970s. There have been international travel modeling conferences every few years since then, focusing on making travel models more behavioral and accurate. Many consulting firms are competent in discrete choice model development. Smaller urban areas with, say, less than 200,000 people, still generally do not use mode choice models, but regions above this population level do, as they have significant bus and rail systems, or wish to plan for them.

Let us now review the recent federal laws that have greatly increased the requirements placed on travel modeling and on transportation planning. Agencies are still trying to adjust to these new mandates. The key point is that whereas in the 1960s, 1970s, and 1980s the federal government was primarily concerned about congestion reduction from transportation investments, in the 1990s air quality concerns became equally important.

RECENT FEDERAL PLANNING REQUIREMENTS

The Clean Air Act of 1970 required all states to adopt a State Implementation Plan (SIP) that includes an emissions inventory for each region in the state and a plan for attainment of all ambient federal air quality standards. The Clean Air Act Amendments of 1977 and the subsequent Highway Act both required all regional transportation plans to show attainment of the vehicle emissions reductions specified in the SIP, through the modeling of travel and emissions from on-road vehicles. The emissions modeling is governed by the U.S. Environmental Protection Agency (U.S. EPA) and most states use the EPA emissions modeling software. This program takes daily VMT (in each vehicle speed class) data for the region (taken from the travel model) and other data, such as number of trips, and produces emissions projections. It also relies on vehicle fleets specified by the EPA for all future years (number of vehicles by model and year). Under the 1977 amendments, a region that did not do a transportation plan could have its federal funds for transportation improvements denied (the "funding sanction").

The Clean Air Act of 1990 is the most important policy advance in transportation planning in recent decades. It greatly strengthens the previous requirement for air quality attainment by providing that federal transportation funds can be withheld from regions that adopt transportation plans but cannot show attainment (through modeled emissions reductions) by the deadlines for each pollutant. This was the first time the funding sanction had been tied to actual attainment, and therefore represents a much stronger incentive for regions to reduce emissions. The Clean Air Act of 1990 also strengthens the role of demand-side policies, by requiring that Transportation Control Measures (TCMs) be studied in nonattainment regions, and lists the TCMs that must be studied for each level of nonattainment (moderate, serious, severe, and extreme). The TCM lists include much stronger demand reduction policies than had appeared on the earlier lists of TSM policies. TCMs, for example, include peak-period road tolls.

The Surface Transportation Act of 1991 reinforced the policies in the Clean Air Act of 1990 and emphasized multimodal planning (for all modes, in a connected system). It also strengthened the requirements for public participation and interagency consultation and required that MPOs with more than 200,000 people have their planning procedures recertified by the U.S. DOT every 3 years. In sum, the objectives of transportation planning have evolved from adding road and transit capacity to also managing travel demand, connecting modes, and reducing emissions.

The Clean Air Act air quality conformity rule (40 CFR 93.122[b][1]), adopted pursuant to the act, requires that all regions with "serious" or worse ozone or carbon monoxide nonattainment status run models that are equilibrated across all model steps to fully represent the effects of changes in accessibility on travel demand. This rule was adopted because academics and environmental groups have long contended that adding road capacity speeds up travel and leads to longer trips. There is considerable evidence that increased accessibility on roads leads to increased travel (Cervero & Hansen, 2000, 2002; Transportation Research Board, 1995). Most larger MPOs have now begun to run their travel models in this fashion. Equilibrated, or "full feedback," modeling means that the speeds in the assignment submodel step are fed back to the other submodel steps.

The conformity rule (40 CFR 93.122[b][1]) also requires that the land development effects of regional transportation plans be accounted for, in the "serious" and worse ("severe" and "extreme") air quality regions. It says that the land use patterns and facility plans must be "consistent" in each alternative. This is a much more contentious rule, as it requires MPOs to adopt some sort of land use forecasting committee or model. Because of a lawsuit in the Chicago region in the mid-1990s and this rule, most large MPOs are now developing land use models to run in combination with their travel models. Medium-sized MPOs with "serious" or worse air quality are beginning to do the same. MPOs do not regulate land use, and so they must gain the cooperation of their member cities and counties to allow alternative land use projections to be run with each regional transportation (facility) plan.

More fundamentally, the Clean Air Act of 1990 requires the MPO modeling in these regions to be much more accurate than it has been, so that emissions will not only be ranked correctly across alternative plans, but also be projected accurately absolutely, in order to judge the performance of any plan against emission reduction budgets developed for each nonattainment region. This means that travel must be projected by time of day (peak and off-peak periods), to forecast congestion, vehicle speeds, and emissions in more detail. Travel speeds must also be much more accurate than in the past. Most large, and some medium-sized, MPOs now use speed postprocessors (extra models) to project vehicle speeds more accurately. These models are calibrated against observed vehicle speeds in the base year model. Since MPOs must demonstrate air quality conformity every 3 years, this new level of modeling accuracy is very difficult for MPOs to achieve.

In addition, the Clean Air Act of 1990 requires that all MPOs be able to model the effects of TCMs such as peak-period tolls, parking charges, fuel taxes, flextime, paratransit, transit, land use intensification near to rail stations, bike and pedestrian facilities, park-and-ride lots for transit and for carpooling, ramp metering, and carpool lanes. This requirement means that the travel model must be sensitive to a much greater variety of policies and travel behaviors than in the past. Models need to represent all transit, walk, and bike modes, for example. They also must include prices for travel (time cost, distance cost, transit fares, tolls, and parking charges) for all modes, in all model steps. The more advanced large MPOs' models do include all of these variables to make them sensitive to such policies, but the medium-sized MPOs need to catch up by adopting substantial model improvement programs.

The Civil Rights Act of 1964 has resulted in a series of transportation discrimination cases that are making MPOs consider the equity effects of their plans more seriously. A very important settlement in Los Angeles County in the 1990s, for example, requires much greater investment in bus system improvements, because poor people ride buses, not rail (see Deka, Chapter 12, this volume). The MPO for that region now includes simple equity measures in its regional plans. The 1998 EPA guidelines on environmental justice are being implemented by all federal transportation agencies (U.S. EPA, 1998). Some states also have adopted their own statutes requiring

the analysis of such equity issues. It is worth noting that, to get theoretically complete economic equity measures, you have to run economic models within your travel model or land use model. This is the least understood and least developed area of MPO practice, even though such models are in widespread use in other developed countries.

In some regions, NEPA is increasingly being enforced (through lawsuits) regarding the growth-inducing impacts of highways. The MPOs will need to model induced travel (VMT) and induced land development (type and location) for major road and passenger rail improvements. Since cities and counties control land use planning, MPOs often treat these projections of land development as "impact evaluations," not land use plans.

One useful way to think of all of these requirements for modeling is to list all of the behaviors we wish to simulate. In conventional travel models these behaviors are only crudely represented:

1. Trip generation is represented poorly. Nonmotorized modes are omitted, only a few trip purposes are identified, and there are no land use and accessibility explanatory variables. This means, for example, that the effects of land use density on trip generation are not modeled and the effects of land use policies on walking and biking are also omitted.

2. Trip distribution (specifying the number of trips from each residence zone to each workplace zone) is rudimentary. Trips are not linked in sequences, but are all modeled separately. Workers from households are not matched to job types by income. The monetary costs of travel are not represented. The whole model set is not run to equilibrium, and so the effects of vehicle speeds on trip lengths is neglected. Equilibrium modeling takes the link speeds from the assignment

step and feeds them back to the trip distribution and mode choice steps and the whole model set is run (iterated) until speeds no longer change. All model steps now have the same link speeds.

3. Mode choice, at least in the larger MPOs, is done fairly well with disaggregate models. Nonmotorized modes are missing in most models, however. Small MPOs omit this step.

4. Travel assignment is quite inaccurate, as road capacities are often inaccurate and the relation between vehicle flows (road volumes) and speeds are poorly represented. Generally, quite inaccurate speeds result, especially in daily models. As a consequence, projected congestion levels are unreliable. Emissions estimates are therefore inaccurate.

Behaviors that are generally missing completely (for which new submodels must be created) include:

1. Auto ownership (autos per household), which strongly affects trip generation and mode choice.

2. Trip chaining, which increases as congestion gets worse (drivers link more trips together). Chaining also reduces the number of cold engine starts, which are a critical input to an emissions model.

3. Time of travel, which is affected by congestion. Peak spreading occurs as congestion worsens, and this affects speeds and emissions (more people travel at shoulder times and off-peak times).

4. Location of land development, which is affected by congestion levels in subareas in the region.

5. Location of firms. As congestion levels change, firms move around in the existing building stock, to be close to their workers, suppliers, and customers.

6. Location of households. As congestion changes, households move around in the available dwellings, in order to have access to employment, shopping, and schools.

To summarize, MPOs are now required to greatly improve their capabilities for modeling travel and land development as well as the effects of the resultant travel and land use patterns on the economy, environment, and social equity. Box 5.1 summarizes the many federal requirements that now affect modeling.

We can see that MPOs are under great pressure to improve their travel models. We will identify improvements that can be made on all of these submodels, below. To do this, we review all of the steps in the overall transportation planning process. We will propose improvements in this process, as well as for the travel models, which are a major part of the planning process.

OVERVIEW OF TRAVEL MODELS AND LAND USE MODELS

Following Pas (1995), we will divide the planning process into three phases and describe past practice. Drawing from Beimborn, Kennedy, and Schaefer (1996), Deaken, Harvey, and Skabardonis (1993), and the recent model development programs for Oregon and the Sacramento region, we outline at the end of each section the improvements generally agreed on for medium-sized and large MPOs. We will not discuss goods movement, owing to limited space. Other good sources include U.S. Department of Transportation (1994c) and Johnston and Rodier (1994). The best recent set of recommendations for travel models and land use models is Miller, Kriger, and Hunt (1999).

Preanalysis Phase

The preanalysis phase consists of (1) problem identification, (2) formulation of objec-

BOX 5.1. Summary of Federal Legal
Requirements Affecting
Travel Modeling

Clean Air Act of 1990

Serious and Worse Air Quality Regions:

- MPO must show attainment by deadlines
- Run network-based regional travel demand model
- Run model to equilibrium (to show induced travel)
- Land development patterns must be consistent with facility plans
- Peak and off-peak time periods
- Travel costs must be included in all model steps

All Other Areas:

Must use any of the above methods, if the MPO has them available

Surface Transportation Act of 1991

- Agencies must plan for all travel modes
- MPOs must consult with other agencies and interest groups on scenarios, modeling methods, and indicators
- Planning process must be recertified every 3 years

Civil Rights Act of 1964

- Agencies cannot discriminate against minorities in transit services
- Agencies must consider the effects of all projects and plans on minorities and on lower income households (Executive Order on Environmental Justice)

National Environmental Policy Act

- MPOs must consider the growth-inducing impacts of projects (including the land development effects of major projects)

tives, (3) data collection, (4) generation of alternatives, and (5) definition of evaluation measures. The most important aspect of this phase is to define problems broadly and then identify broad objectives. In this way, many solutions can be sought, not just increases in road capacity. So, rather than simply aiming to "widen congested roadways," an MPO should define the objective to reduce congestion as "maximize accessibility in the region, especially for lower income households." This definition focuses on those travelers with few options and so encompasses equity. This approach is preferable to defining a second objective for equity, which can be downplayed in the analysis. My definition also states the problem as accessibility, not improved level-of-service (speeds) on roads, because higher speeds results in higher VMT (which then increases emissions and can increase congestion). As described in Chapter 1, accessibility is simply access to activities, which is what the traveler is seeking. Travel is primarily a means to get to activities, such as work, shopping, or visiting friends, not a good in itself. So we can increase accessibility with a policy for higher density mixed-use land use centers, by themselves or with better transit. More freeway capacity will also increase accessibility, and so this definition is mode-neutral. Given the broad policy charge in the most recent transportation acts and the need to examine TCMs in the air quality law, we need to define transportation planning objectives in a holistic fashion.

The generation of alternatives is another especially weak step in transportation planning. Most MPOs initially analyze several roadway and transit investment schemes, but do this in-house, not in published documents. In the official plan, the MPOs usually analyze only the "Preferred Plan" and the "No Action Alternative," which is required by NEPA. The preferred plan, in many instances, seems to be the maximum investment in roads where the plan can just meet the future emissions requirements. Transit is added to reduce emissions, as necessary, and to satisfy interest groups advocating for transit and for lower income households in general. MPOs almost never identify and evaluate all-transit alternatives (the author has reviewed most of the plans from the largest 15 MPOs). The main reason for this appears to be that transit investments will generally focus on the central cities in a region and so will not meet the chief political test of MPO decision making, which is geographic equity (i.e., spreading the money around all areas within the urban region). Each MPO board member is an elected local government official and so he or she wants to bring the dollars into his or her jurisdiction. In some states, such as California, transportation funds are legally allocated by population, and so outlying counties get their "share" of funds. These funding rules make transportation planning more a process of spreading projects around than of seeking to meet regional objectives for economic efficiency, equity, or even congestion reduction. This is a good example of how "planning is politics." Nevertheless, it is important for MPOs to examine several alternative plans, including ones that focus on transit and land use in central cities, as well as on typical freeway widenings, to see the results. Several MPOs have engaged in visioning processes over the last 20 years, in which they examined broadly different alternatives, projected for 40 or 50 years. This is an essential part of transportation planning and should be done every 10 years or so by MPOs, outside of the regular planning cycle. These broad analyses may affect the funding process in ways that benefit the whole region.

The third weakness in this phase is inadequate plan evaluation criteria. The outputs from travel models are measures of congestion, such as level-of-service (A–F) on each network link and person-hours of travel delay. They also send their VMT-by-speed-class projections to emissions models that estimate the number of tons of each pollutant produced per weekday. MPOs, however, seldom evaluate equity outcomes well and they seldom evaluate aggregate economic

welfare at all. These measures are urged, but not required by, the Surface Transportation Act of 1991. They are, however, essential to an informed and accountable planning process. Easily applied measures of comparative aggregate economic welfare (net benefits to travelers) do exist. These measures can come directly from the mode choice model or, for smaller MPOs without such models, from travel costs calculated off-model using travel times and distances from the travel model. These measures can also be calculated for travelers by income class and so can give a vertical equity measure (i.e., economic welfare by household income class). Since few MPOs in the United States use these measures, economic welfare and social equity are not given much weight in plan evaluation. Such measures are used in most developed countries and in many developing ones; multilateral banks require them for major transportation investments.

Last, ongoing data collection is fundamental to accurate travel modeling. All MPOs should perform a household travel survey every decade, coordinated with the national census. In addition to travel questions asked of a random sample of the whole region's population, surveys can be done on the same households over time (panel surveys), and surveys can be used to support land use modeling by asking questions about household and firm location behavior. One should also survey firms to determine goods movements by commodity type. MPOs need to perform a survey of employment to supplement national sources such as InfoUSA. If an MPO is going to develop a land use model, then it must also gather land use data. Most MPOs now have geographic information system (GIS) capabilities and so can use parcel data sets, if all of their constituent counties have these data (see Nyerges, Chapter 7, this volume). Parcel data, supplemented with air photos and satellite data, are the only way to get spatially accurate land use maps. All of these data-gathering efforts are expensive and time-consuming, and so each MPO needs to have a model development program where the data gathering and data cleaning is part of an ongoing process.

In general, these issues of narrow objectives, narrow range of alternatives, inadequate evaluation criteria, and insufficient data gathering are a consequence of inadequate MPO budgets for transportation planning. MPOs that spend $1 billion per year on transportation projects will spend only a few million dollars on transportation planning, which is a few tenths of a percent of the budget. Private firms spend a much greater proportion of their overall costs on planning. Federal funding rules permit the use of several categories of funds for MPO planning, but most MPOs prefer to spend those funds primarily on facilities. This generalization, that MPOs don't spend enough on planning in the preanalysis phase, applies to the next two phases of planning, as well. Figure 5.2 outlines the three phases of transportation planning.

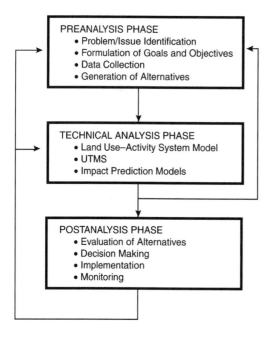

FIGURE 5.2. A general representation of the urban transportation planning process. Source: Pas (1995, p. 60). Copyright 1995 by The Guilford Press. Reprinted by permission.

Technical Analysis Phase

The next phase of transportation planning is the technical analysis phase, which is primarily travel modeling but may also include a land use model. I noted earlier that a complete set of models would represent all behaviors in the urban system that are affected by changes in the transportation system. Now we will define an ideal model in more detail from the perspective of urban modeling, where all urban systems are represented. According to Wegener (1994), urban models should represent all subsystems. He defines these systems by the speed with which they change. *Slow change systems* include transportation networks and land uses (permitted economic activities). *Medium-speed change systems* include workplaces and housing, by which he means the construction of buildings. *Fast change systems* include employment and population, meaning the movement of workers and households among existing buildings. *Immediate change systems* include goods transport and personal travel. It is widely accepted that changes in the transport networks will affect travel and goods movement and, subsequently, employment and population locations, and, finally, the construction of workplaces and housing. This broader urban modeling perspective is now becoming accepted in transportation planning and modeling. When the U.S. DOT had four teams of consultants develop outlines of advanced travel forecasting models in the early 1990s, three of the proposals included land use models as part of the package to ensure accuracy of travel forecasts (U.S. Department of Transportation, 1994a).

With this broader perspective, let us now review again the typical MPO travel modeling process, derived from the 1960s and still in use today, called the "four-step" modeling process. This typical process does not include a land use model.

In recent years, several conferences and publications have called for improved travel models and for the addition of land use models to MPO practice. A compilation of recommendations from peer reviews of travel models, done during MPO recertifications, was published in 1994 (Peer Review Panel Functions and Organization, 1994). This report lists desirable modeling characteristics for observed data, demographic and economic forecasts, model system design, general forecasting issues, trip generation, trip distribution, mode share, assignment, and other details. The U.S. DOT published *Short-Term Travel Model Improvements* in 1994 (U.S. Department of Transportation, 1994b). Good papers from this period were by Stopher (1993a, 1993b) and Deakin, Harvey, and Skabardonis, Inc. (1993). Hundreds of papers have been published and presented at the Transportation Research Board annual meetings over the last 20 years on how to improve travel models.

Activity Forecasts

These forecasts are done before the four-step model is run. The MPO needs data on the location of different types of households and employment for the base year and for the forecast years. Base-year data are gotten from the decennial census and from the MPO's household travel survey and other local surveys and GIS data sets (from the census: households, number of workers per household, persons per household, household income, number of autos per household, all by zone; from other data sources: number of employees by type of employment by zone). Then, population and employment forecasts for the counties in the region are taken from a state agency. These are usually derived from national and state input–output and econometric models of the national and state economies. We will not concern ourselves with the accuracy of these projections, since they are exogenous, except to note that the error is, on average, at least 1% for each year of the forecast, meaning that the typical 20-year projections can be high or low by 20% or more.

In addition, the MPO must get or make forecasts for household income and size, the other most important variables in the travel model. Many states provide these forecasts for counties. Errors in demographic inputs to travel models are unavoidable. This is one reason that MPOs are required to redo their transportation plans every 3 years.

The MPO then must allocate the firms and households to zones. This is usually done for a few hundred to 2,000 travel analysis zones (TAZs), which are generally the base-year block groups or census tracts. Only a few large MPOs use some sort of land use model to allocate households and firms to zones in future years. The vast majority of MPOs use a judgmental process in which the staff examines vacant lands in each county and what the land use designations are for them in each county's land use plan. By hand, or using a simple GIS system, the MPO staff assigns future households by density class and income (income correlates with density and with existing incomes). Future household auto ownership level is also generally correlated with existing auto ownership, which comes from census data. Types of firms (generally manufacturing, retail, other, for small MPOs, with more categories for larger MPOs) are allocated by land use categories in local plans.

Most local land use plans, however, are for a 20-year period and are redone about every 12 years. This means that the average plan is about 6 years old and so may have as little as 14 years of land remaining in it designated for urban land uses. So the MPO staff will run out of land for some land use types in many cities and counties in the region when doing 20-year projections. As a result, the MPO staff must project where lands will be designated for future growth by interviewing local planners. The local planners often will not commit to any specific areas, so the MPO staff must do this on their own.

Jobs/housing balance also presents a problem to the MPO staff. Most localities want more employment than housing, because housing units pay only property taxes whereas retail uses pay property and sales taxes. So if the MPO staff uses the local projections in the land use plans, or uses the local land use maps literally, they will get far more employment than there will be workers in households for all future years. That is, local governments project and plan for too much employment, to keep such lands in oversupply, which makes these lands less costly, to attract firms. The MPO, however, cannot run a travel model with excessive employment, as the model must balance jobs with workers. So the MPO staff must reduce employment projections for many of its jurisdictions, and they do so by negotiating with the local planners. This procedure is inaccurate since employment can still be projected in zones where firms would probably not locate, in terms of access to markets, workers, and related firms.

MPOs should match forecasted land use patterns to each facility scenario to be accurate because land development is affected by changes in accessibility. This can be done by using a land use model or an expert panel. These projections must be approved by all the member cities and counties because they control land uses in the region. This process really just formalizes the currently poorly documented procedures, where projections do not conform to local land use plans. The difference is that the MPO would use different land use patterns for each of its transportation alternatives.

Demographic trends must also be accurately forecasted. Average household incomes, for each income class, should not be simply forecasted to stay the same in real dollars. This masks the changes that have occurred in the past two decades, where in most parts of the United States the lower income groups have experienced a large drop in real income. In some parts of the country, the middle-income groups have also lost income. The MPO must project these trends, in order to capture their strong effects on auto ownership, trip generation, and mode choice.

In the four model steps below, it is important to observe that the first three submodels are estimated on data for individuals and households. This gives the maximum statistical power to the models, as all variation in the data are available. After the equations are estimated for each household and employment type, however, the model is applied in an aggregate fashion, on categories of households and employment in each zone. The four-step model is sometimes called an "aggregate travel model" because of the zonal averages used in application, but the submodels are estimated in a disaggregate fashion.

Trip Generation

This is the first step of the four-step travel modeling process. This simple statistical model (a cross-classification table or regression model) projects the number of weekday trips a household will produce, based on household income, number of autos owned, number of workers, and household size. It is insensitive to the level of road congestion in the region and is also insensitive to transit accessibility. This is theoretically incorrect, as regions with good transit and congested roads would tend to get a higher transit mode share and lower auto ownership for each household type. It is also insensitive to land use density in the residence zone, which is known to be inversely related to trip generation. The model also projects trip attractions, which are the destinations for the trips. Attractions are determined by the number and types of employees in each zone, again using simple models. There are standard books with trip attraction data for each type of employee. Trip productions (from households) and attractions (to employment) must be equal, and to do this the travel model iterates to convergence on tables of all trips, by origin and destination zones. Usually, the model is set to make attractions match productions, as agencies feel that their local data for trip generation are reasonable.

An auto ownership step should be added to the model set, so that household number of autos is dependent on accessibility for auto travel and to land use variables in the home zone, such as density and mix. Trip generation should be made sensitive to auto ownership and to origin (home) zone land use density and mix and to parking costs in the home zone and employment zone. The pedestrian and bike modes need to be added, as these are the subject of much policymaking. Trip purposes need to be elaborated to include at least home-based shop trips, home-based school trips, and work-related trips. More trip purposes results in more accurate estimates for trip generation, trip distribution, and mode choice, as these submarkets behave differently.

Trip Distribution

The second step matches trip origins at households (by zone) to trip destinations at firms (by zone). All trips are considered to be separate, even though in reality trips are linked in tours (e.g., home, to work, to shop, then to home). Trip distribution is done separately by trip purpose, with the simplest models having home-based work, home-based other, and non-home-based trip purposes. The trips are distributed across the production/attraction pairs (now called origins and destinations) according to a gravity-type model, in which the number of trips falls off as a function of trip time length, raised to a power greater than 1. The gravity model captures the fact that trips are more likely to be short than they are to be long. The other determinant of the number of trips placed in each cell in the zonal trip distribution table is the number of trips produced in each zone and the number of trips attracted in each zone, which is taken from the previous model step. To summarize, all trips are considered separately and are assigned to origin and destination zone pairs in a large table, according to the time length of the trips, using an iterative fitting algorithm to get all trips distributed.

The main issue with this model step is whether trip distances are fixed in the model or not. If an MPO runs through the four model steps just once, one can have different vehicle speeds and road volumes in the trip distribution step, the mode choice step, and the network assignment step. This is, of course, illogical and inaccurate. The very earliest UTPS manuals from the U.S. DOT warned against this practice, but it became routine in MPOs, ostensibly because of (mainframe) computer runtimes and costs. If all of the model steps are run through iteratively until all steps have the same travel speeds or volumes on a sample of links, or the trip distribution table does not change significantly, then one has an equilibrated model. This is the only logical and legally defensible method, and it is the method recommended in all academic textbooks and articles. It is called "full-feedback" modeling. Several academic articles discuss methods for this procedure, as does a U.S. DOT-commissioned report on methods for model equilibration (COMSIS, Inc., 1996).

Not iterating the whole model set is not only inaccurate, it is biased toward adding road capacity. With a properly equilibrated model, when you widen roads, travel speeds go up and people make trips of longer distances in about the same amount of time as before. This is based on the observation that the household travel time budget (daily total travel times) is very stable over time and so can be treated as almost fixed in any given model year. With travel time budgets that are almost constant, if road travel becomes faster, people will travel longer distances on average, and if road travel slows, they will make shorter distance trips. So, in a properly functioning model with full feedback, the No Action Alternative will show high levels of congestion in future years, but this congestion will result in shorter distance trips and so the congestion will be reduced somewhat by this behavioral change. In a model set with a "fixed trip table" (the model steps are only run through once), there is no such feedback to reduce trip lengths, VMT, and congestion, and so congestion levels are overprojected. Exaggerated congestion levels make road expansion projects appear more urgent than they really are. Several lawsuits have attacked this practice.

This same bias applies to the Proposed Plan alternative, in which a properly equilibrated model will show that adding to road capacity speeds up travel, leading to longer distance trips, which then adds to VMT, which in turn leads to greater congestion. So the benefits of adding road capacity in reducing travel times is diminished somewhat in the model, as the extra travel adds to congestion. A model run with a fixed trip table, however, does not represent this important behavior and shows greater congestion reductions than will really occur. This makes the plan alternative look better than it will be.

This issue of "induced travel" has been the subject of dozens of papers in the last few years, and there is now a strong consensus that adding capacity speeds up travel, inducing longer distance trips (Cervero & Hansen, 2000, 2002; Noland & Lem, 2000). One of the critical improvements MPOs can make to their travel models is to run them to full equilibration. The air quality conformity rule requires this for "serious" and worse air quality regions. This method is necessary, however, to pass even a threshold test of basic scientific soundness in any lawsuit that questions an MPO's modeling. Virtually all MPOs run their models on desktops or workstations and so computer runtime costs are irrelevant. Most travel modeling systems software (the management shell that runs all the submodels [model steps] in sequence) have the ability to equilibrate the model steps, and so this is a small effort, except in the largest MPOs that have very complex models. These MPOs, however, have the resources to write the code to iterate the model steps.

Large MPOs now equilibrate their model sets. More recently, they have also gone to a more accurate type of trip distribution model, at least for the home-based work trip, which is associated with the peak travel period, where congestion is the worst. This is the destination choice model, which is a discrete choice model where the destination zones are the choice set. Some large MPOs have joint mode-destination choice models for work trips, where both are chosen simultaneously. This is the theoretically best model type, but it is more difficult to estimate. In principle, all trips should be distributed using destination choice models, as they are based in microeconomics (traveler utility), whereas the current method of trip distribution by iterative fitting of the trip table is ad hoc.

Mode Choice

In this model step, the trips in the trip distribution table are allocated to modes with a statistical model that includes trip cost (distance cost, time cost, fares, tolls, and parking charges), household income, household auto ownership (number of cars), and accessibility. The whole model set must be run to get the travel times, which then are fed back to the mode choice step. Small MPOs generally do not have a mode choice step, and so run what are called "traffic models" (only auto travel is represented). Some medium-sized MPOs use diversion curves (or mode split curves) or run simple mode split models that are sensitive only to auto travel times, where slower times create more demand for transit. Often, they are manipulated by hand, that is, mode percentages are changed judgmentally. Most medium-sized and all large MPOs use discrete choice models, which are much more accurate than mode split models and, more importantly, are behaviorally based. They also are more policy-sensitive. That is, they can project changes in mode choice that result from the construction of a new passenger rail line or the widening of a freeway.

The mode choice model is the strongest submodel in terms of its behavior, which is firmly based in microeconomic theory. A major weakness of many mode choice models, however, is that they do not include nonmotorized modes, such as walk and bike. The walk mode share is much larger than the transit share in most medium-sized urban regions, for example, and so must be modeled, in order to forecast shares for other modes accurately. Also, improving transit generally shifts some travelers from the walk and bike modes to transit, so they must be represented. Most large MPOs have recently included the walk and bike modes in their models. Another weakness is that mode choice models are not sensitive to land use policies, such as increased densities and land use mix, which are expected to reduce auto mode shares. So, many larger MPOs have recently added land use variables in their mode choice models, usually for the transit, walk, and bike modes. Medium-sized MPOs need to also make these two types of model improvements in order to be able to evaluate transit options.

Assignment or Route Choice

The fourth step involves assigning the vehicles to the network of roads and the transit passengers to the rail and bus lines. Most MPOs use capacity-restrained assignment, where a computer program assigns vehicles to the shortest distance routes for each origin–destination pair of zones. The trips are loaded onto the network and the travel speeds are then calculated from the traffic volumes, related to the capacity of each link. For each volume/capacity ratio, there is a related speed in a table. This permits the creation of a trip time table (time lengths of trips). As vehicles on links slow down, the model assigns some vehicles to other faster pathways to link the origin zone to the destination zone, until such new assignments cannot decrease travel times in the system.

In most MPOs, the road and transit network files need to be "cleaned" (i.e., checked for errors) thoroughly to reduce errors in link capacities and to fix nonconnected links and nodes in the spatial graph. Link capacities should be carefully recalculated with additional data on road geometry (e.g., width, slope, sight lines). Carpool lanes must be a separate link type, as speeds will be higher in them. Transit access links (from parking lots and streets) should also be coded, especially for rail stations.

Intersection delays need to be represented in the networks, and delays for links and intersections must be calibrated to local data. Postprocessing of speeds is also becoming a common practice. In this procedure, modeled vehicle speeds are made to match observed speeds. Time-of-day must be represented, at least as peak and off-peak periods. Even more time periods (peak, shoulder, off-peak, nighttime) will make the model more accurate in projecting speeds and emissions.

Advanced practice for the large MPOs is to develop tour-based models that incorporate time-of-day explicitly in trip generation. Portland, Oregon, and San Francisco adopted such models in 2001. Trip chaining in tours also permits more accurate estimates of the number of cold, warm, and hot vehicle starts, which is important to emissions analysis. Very advanced models will also represent changes in household "activity allocation," such as working at home or deferring an out-of-home activity. Household travel surveys must be improved to match these new modeling requirements, and so multiday activity and travel diaries are coming into use.

Emissions Modeling

After the four-step model is run, the MPO then calculates the on-road vehicle emissions. As noted above, a standard emissions model is used in all states, except California, which has its own model. We will not go into the details of these models, except to say that the past emissions factors for each vehicle type and age were much lower than were the empirically observed emissions, and so the most recently revised emissions models have higher emissions for any given fleet with any given regionwide VMT. This change will make it more difficult for many MPOs to show air quality attainment in regional transportation plans completed after 2002. The new emissions models also show higher emissions at high speeds (over 55 miles per hour), which will make building more freeway lanes produce more emissions in the models. Last, we note that new ambient air quality standards for small particulates and for ozone have been adopted by the U.S. EPA recently, and the new standards will cause many regions that are now in attainment to go out of attainment for these new standards.

A very important change required by the air quality conformity rules is that travel speed in the base-year models must be calibrated (matched) to actual measured speeds in the base year. This means MPOs cannot continue their past practice of capping highway speeds in the model at 55 miles per hour, which resulted in large underestimates of emissions on highways. Models now should include speed categories up to 75 miles per hour, since substantial proportions of vehicles travel at over 65 miles per hour in most regions.

Postanalysis Phase

This phase includes plan evaluation, plan implementation, and monitoring of the results. We will only discuss evaluation here because of limited space.

Plan evaluation should be comprehensive, that is, it should include all major impact categories. These many measures can be summarized as economic, equity, and environmental issues. Regarding economic issues, MPOs generally evaluate congestion levels with level-of-service on links and also

with person-hours of delay for all travelers. These are narrow measures, as they depict only road congestion and do not include the overall economic utility of the travelers. A utility measure can be derived from most mode choice models, so that a complete measure is readily available for large MPOs and for the medium-sized ones that have mode choice models. Smaller MPOs can get a similar measure, based on travel costs derived from their model runs. Social equity can be measured in many ways, but the most complete measure is traveler utility by household income class, which is measured by using the logsums from the logit mode choice model. All mode choice models require that households be categorized by at least three income classes so that one can perform this equity analysis (Rodier & Johnston, 1998).

For environmental measures, MPOs give the on-road emissions of several pollutants and can project energy use by vehicles. It is fairly simple also to project the emissions of greenhouse gases, using published research that relates fuel use to greenhouse gases. NEPA requires that the MPO also evaluate runoff from all roads in the plan, but most MPOs do not do this adequately. Recent legal cases and changes in EPA rules require local governments to evaluate and treat nonpoint runoff (water pollution), that is, runoff from impervious surfaces in the region. A recent Clean Water Act case in Southern California, for instance, requires California DOT and U.S. DOT to treat all runoff from federally funded and state-funded roads. MPOs need to improve practice in this area of analysis. Many runoff models are available.

Habitat damage is a big issue in the southeastern, southwestern, and Pacific Coast states, where large numbers of listed species occur, and in all wetlands and riparian lands in the United States. With recent advances in mapping habitat types and in GIS methods in general, the analysis of habitat impacts of transportation plans is much easier than it was a decade ago. Many MPOs are beginning to adopt one or more of the many methods in the conservation biology literature. (Bae, in Chapter 13, this volume, treats environmental impacts in detail.) A key aspect of evaluating habitat damage, however, is that the MPO must project the land use effects of the transportation system improvements. A new freeway may induce more rapid urban or suburban development in some part of the region. NEPA clearly requires the consideration of the "growth-inducing impacts" of all major transportation projects. Federal transportation rules also require a NEPA evaluation of regional transportation plans in order to speed up subsequent project analyses. Virtually no MPOs have modeled the land use effects of their transportation plans in the past, but several large and medium-sized MPOs are now developing land use models. We will return to this subject, below. Some state DOTs now model the land development effects of new highways, using statewide input–output models. These models project economic growth in counties (or townships), partially based on transport costs. They are available in all states. Box 5.2 summarizes our discussion of the technical analysis phase so far.

As land use modeling is a new and controversial requirement, we review it briefly now.

LAND USE MODELING

Land use modeling has a long history in the United States. Early academic models were developed in the 1950s and 1960s by researchers such as Lowry and Alonso. The federal government was supportive of these models as they could be used in both transportation planning and in urban redevelopment. A U.S. DOT request for proposals in 1971 states:

The following issues have been of concern to transportation planning agencies:

BOX 5.2. Summary of Good Modeling Practice for Medium-Sized and Large MPOs

Time Representation
- Peak and off-peak periods

Data Gathering
- Household travel survey every decade with tours
- Vehicle speed surveys
- Data for urban model

Activity Forecasts
- GIS land use model or economic urban model

Auto Ownership
- Discrete choice model, dependent on land use, parking costs, and accessibility by mode

Trip Generation
- Walk and bike modes
- More trip purposes
- Dependent on auto ownership
- Three or more time periods

Trip Distribution
- Full model equilibration
- Composite costs used (all modes, all costs)
- All-day trip tours represented

Mode Choice
- Discrete choice models used
- Land use variables in transit, walk, and bike models

Goods Movement
- Fixed trip tables

Assignment
- Capacity-restrained
- Cleaned-up link capacities
- Speeds calibrated
- Three or more time periods

1. In urban areas, it is necessary to achieve a balance between the need for mobility and the preservation of environmental quality and social stability. How can this process be incorporated into the transportation/land use planning program beginning with plan development?

2. To what extent can transportation system planning and facility capacity planning be used to influence land development? What are the available controls for accomplishing this, and how can they be applied effectively?

3. Providing too high a level of service of the auto-highway system in urban areas might result in travel demand exceeding the supply that can feasibly be provided. What should determine the minimum or maximum level of service (speed), which should be used as a criterion for planning transportation facilities in specific parts of the urban area?

4. What would be the consequence of controlling transportation facility capacities as a means of directing land-use development even though such controls could mean congestion with more pollution and higher travel costs?

This study should determine the feasibility of the balanced development of land use and transportation facilities. It should also determine whether or not land development and travel can be brought into a satisfactory balance such that any transportation improvements made can provide a continuing high level of service over time. (Putman, 1983, pp. 39–40)

This U.S. DOT contract led to the development of the earliest integrated urban model in the United States, ITLUP (Integrated Transportation and Land Use Planning), by Putman in the early 1970s, which has since been used by about two dozen MPOs in the United States. This is a Lowry-type model with land use types dependent on accessibility and past demand for land, but without the use of land (or floor space) prices as a feedback. This type of model requires the MPO to gather data

only on land consumption (acres of each land use type) by firms and households in each traffic analysis zone or district. All the other data are already in use for travel modeling in each region.

Echenique developed a more complete urban modeling system in the late 1960s and 1970s, one that included the supply and demand for floor space, mediated by price (Hunt & Simmonds, 1993). Again, the U.S. DOT supported the development of the generalized software package for this model (Echenique, M. H. & Partners, and Voorhees, A. M., & Associates, 1980, in Echenique et al., 1990). This package later became MEPLAN (Marcial Echenique and Partners Planning Model), which has been applied in more than 50 urban regions in the world. The U.S. DOT software was apparently never used by any region in the United States. This class of models requires floor space lease value data and floor space consumption data, and it is more difficult to implement than the simpler type of land use model. Calibration of urban models is quite difficult, but new software that substantially automates this process is now becoming available (Abraham & Hunt, 2000).

There is a long-established trend of improved travel modeling and land use modeling, both in the United States and abroad. The Federal Highway Administration (FHWA) has a website that reviews the various types of urban models (*fhwa.dot.gov/ planning/toolbox/land_develop_forecasting.htm*). Dissemination of these methods into agency practice has been quite slow, however, but new legal mandates are putting pressure on MPOs to advance their models. In addition, these agencies are now expected to evaluate a great range of land use and transportation policies that require land use models and much better travel models.

Modeling the induced land development effects of transport improvements can change VMT projections substantially (Rodier, Abraham, Johnston, & Hunt, in press). We compared a future No Build scenario with a future Maximum Freeway Expansion scenario, using both a travel model and an urban model. We found that the land use effects added about 15% to the VMT difference between the scenarios. In another paper, we found that using a land use model along with a travel model changed the direction of change in emissions (future Build case minus future No Build case) and in travel speeds for highway scenarios, compared to running a travel model only (Rodier, Johnston, & Abraham, 2002).

Planners in the Portland, Oregon MPO compared an analysis of their 20-year transportation plan with their new integrated urban model (travel and land use models, run in sequence) with their previous typical analysis, using only their travel model and found: (1) using the land use model makes the travel model results more reasonable, in terms of economic theory and common sense; and (2) when both models were used, congestion levels on some roads were significantly lower because firms moved outward to take advantage of less-congested conditions in the countercommute (Conder & Lawton, 2002). In other words, not using a land use model can result in exaggerated projections of congestion, because the effects of congestion on land development are not represented.

A National Academy of Sciences review panel concluded that both induced travel and induced land development are real behaviors (Transportation Research Board, 1995). This work was partly based on much more detailed research previously done by a national panel in the United Kingdom. Many published papers show the induced travel effect, using historical data; for a review, see Noland and Lem (2000), and for a detailed empirical study see Cervero and Hansen (2000, 2002). A review of textbook sections on model equilibration is in Purvis (1991). This area is fairly settled now, given the recent papers with strong statistics. A few recent papers show the land development effect (Boarnet & Haughwout, 2000; Cervero, 2003; Nelson & Moody, 2000).

The FHWA disseminates several travel models and impact models that include provisions for entering induced travel factors.

State governments also recognize that added road capacity affects travel and land use. The National Governors Association (NGA) policy position NR-13 states that "the impact that highway decisions have on growth patterns and, in turn, on the environmental health of communities must also be appropriately analyzed in advance" (2001, p. 1). A report by the NGA also says that "rapid suburbanization and urban decay are mirror images of the same phenomenon" (National Governors Association, 2000). The latter statement recognizes the "hollowing out effect" that sprawl induces, where central city business and residential properties are abandoned when growth moves outward.

Federal agencies are encouraging the use of land use models by supporting research on them. The U.S. Environmental Protection Agency has published a reference work on land use models (U.S. EPA, 2000). The National Center for Geographic Information and Analysis has a review of GIS-based land development models on their website at *ncgia.ucsb.edu/conf/landuse97/summary.html*. A review of a great variety of land use forecasting methods was done for the FHWA and covers both simple and complex models (Parsons Brinckerhoff Quade & Douglas, Inc., 1999). A detailed review of integrated urban models was done for the Transit Cooperative Research Program (Miller, Kriger, & Hunt, 1998). The U.S. DOT has published recommendations for improving urban models (U.S. Department of Transportation, 1995).

Many MPOs have been using land use models of various sorts for many years, but not in iteration with their travel model. That is, they use the models for their base case demographic forecasts, which are then used for all transportation scenarios (Porter, Melendy, & Deakin, 1995; SAI International, 1997). A more recent survey shows a few MPOs using land use models for projecting different scenarios (MAG, 2000). Full microeconomic urban models were recently developed for Honolulu, Salt Lake City, Eugene (Oregon), Sacramento, New York City, and the state of Oregon. Similar models are being developed for Calgary, Edmonton, the state of Ohio, and the Seattle and Chicago regions. The San Diego, Atlanta, and San Francisco regions have had zonally aggregate urban models for many years.

Having now reviewed the recent "top-down" federal laws requiring better modeling methods and having also gone through the steps in a modeling process and identified best practices for each step, we now review the "bottom-up" policy analysis capabilities being espoused by citizens groups and local governments. MPOs should respond to these needs too.

NEW POLICY ANALYSIS NEEDS

It is important to emphasize that, regardless of federal laws, lawsuits, and political pressures for more elaborate modeling, the increased policy analysis demands being placed on MPOs by their constituent local governments and by citizens groups also require better modeling. Recent recertification reviews, ongoing MPO experience with the strong consultation requirements, and agency surveys during model design programs reveal a growing list of policy concerns expressed by local member jurisdictions and citizens groups.

For example, the Sacramento region MPO surveyed the staffs of local agencies in 2001 and held several public workshops to get feedback on policy concerns. The survey revealed that the following issues were hot (in order of popularity): (1) land use and smart growth; (2) pricing of parking and roads; (3) automated traveler information systems; (4) paratransit, bus rapid transit; (5) environmental justice, social equity; (6) induced land development; (7) induced and suppressed travel; (8) peak spreading, departure time choice; (9) effects of land use and design on travel; (10) sidewalks, bike lanes; (11) air quality conformity, NEPA documents, traffic impact studies, land use planning; (12) models useful for subregion and subarea studies (fine spatial detail); (13) models useable by other agencies, standard modules, GIS; (14) making all assumptions explicit; (15) interregional travel; (16) open space planning and habitat protection; (17) useful for sensitivity analyses of policies; (18) including lots of understandable performance measures; (19) representing all travel behaviors; (20) representing nonmotorized modes and telecommuting; (21) representing land markets, not just local land use plans; (22) representing multimodal trips in tours, by time of day; and (23) useful for broad scenario testing. This is a long and demanding list. Most of these needs came from local transportation and land use planning staffs. So we can see that the higher standards for MPO modeling are coming from both external legal mandates and also from internal needs among member cities and counties.

There is great variation in MPO modeling practice, even among large MPOs. While many MPOs use models from the 1960s that are no longer scientifically defensible, many other MPOs regularly improve their capabilities and engage in good practice, even as expectations continue to increase. Some large and medium-sized MPOs recently have gone to full feedback in their travel models, an inexpensive improvement. Many have added mode choice models, some with walk and bike represented. Many MPOs have developed peak and off-peak time periods in their travel models. Most recently, some MPOs are developing land development models of various kinds. The great variation in quality of modeling, even among large MPOs, seems to reflect a lack of detailed guidance from the federal agencies and from professional organizations.

Another interesting aspect of this issue is that a few MPOs, even small ones, have developed good methods that are quite inexpensive. The Anchorage MPO, for example, has walk and bike modes, equilibrates their model set, and has a spreadsheet land use allocation model that includes accessibility from the base-year network as an attraction. Whereas this MPO currently uses the land use model to project only one land use pattern and then uses this pattern as input for all transportation plans, it could easily adapt this model to be used for projecting land uses appropriate for each future scenario, simply by using the future networks to get accessibility.

Another example is the Merced County Association of Governments in California (population 211,000 in 2000), which is running a simple GIS-based land use allocation model to design land use alternatives and then taking those land uses into the travel model. Many MPOs have developed similar GIS-based land use allocation models.

To show how agencies are trying to respond to both the "top-down" legal mandates and the "bottom-up" needs being expressed within their regions, we will review one agency's model improvement program.

AN EXAMPLE MODEL IMPROVEMENT PROGRAM

The Sacramento region (population 1.9 million in year 2000) has a high growth rate and many important resources in need of protection, including prime agricultural lands, valley and foothill riparian corridors, native grasslands, and oak woodlands. It is an interesting region for model development as it includes heavy passenger rail, light rail, and several bus systems. Lower income households in this region have seen drops in their real incomes over the last 30 years, and so the need for transit services is increasing.

The region is in nonattainment for ozone, and the regional transportation plan may have difficulty showing conformity with the State (air quality) Implementation Plan (SIP) for future years. California has recently adopted a new emissions model that will likely make it more difficult for this region to demonstrate conformity after 2002. Because of these concerns about air quality and concerns about sprawl and loss of agricultural lands and of habitats, rising traffic congestion, lack of transit service for many lower income people in the central cities, and a public dislike for homogeneous (monotonous) segregated land use patterns, the MPO recently decided to improve its modeling capabilities and to design and evaluate broad regional scenarios.

The current model improvement program in this region is starting from a strong base, owing to past model upgrades. In the early 1990s, the Portland, Oregon MPO upgraded its travel model and the Sacramento agency adopted the same improvements in the mid-1990s. The 1994 model was developed with a 1991 household travel survey conducted in the Sacramento region. The resulting SACMET (Sacramento Metropolitan Travel Model) is based on the four-step modeling framework but has the following improvements:

1. Full model feedback (trip lengthening is represented);
2. Auto ownership and trip generation steps with accessibility variables (more road capacity results in more autos and more auto trips);
3. A joint destination-mode choice model for work trips (more accurate work trip mode choice and distribution);
4. A mode choice model with separate walk and bike modes, walk and drive transit access modes, and two carpool modes (more accurate mode share projections and better sensitivity to land use policies);
5. Land use, travel time and monetary costs, and household attribute variables included in the mode choice models (can evaluate land use policies and road tolls, fuel taxes, and parking charges);

6. All mode choice equations in disaggregate form (more accurate mode shares); and

7. A trip assignment step that assigns separate A.M. peak, P.M. peak, and off-peak periods (more accurate volumes, speeds, and emissions).

The SACMET model has considerable detail in the description of conditions, with 11,159 roadway links and 8,403 roadway lane miles. The zone system is also finely detailed, with 1,077 transportation analysis zones. There is an explicit representation of truck movements in the model with a fixed trip table for heavy trucks. Heavy-duty trucks constitute a small proportion of urban trips, but their size and slow acceleration gives them a significant effect on congestion. Land use is a fixed exogenous input in SACMET, however.

This travel model meets all the conditions in the air quality conformity rule, except the requirement for "consistent" land use and facility scenarios. This model set also does not permit the evaluation of the growth-inducing effects of expanded highways on land development, as required by NEPA. This travel model, however, presumably produces more accurate vehicle speeds and emissions for each time period, as speeds are now calibrated in the base-year model. The model is also sensitive to land use intensification policies, which are a major issue in this region.

More recently, the MPO has come under increasing pressure to model the land use effects of its transportation plans. The agency has been sued once and, as we saw above, land use impacts are a concern to its member governments and to citizens groups. During the mid-1990s, the author led a project that implemented and compared three land use models in the region (Hunt et al., 2001). We went on to implement an improved version of the MEPLAN land use model in this region, funded by the U.S. DOT, the California DOT, the U.S. EPA, and the California Energy Commission.

This model was then, and still is, the easiest-to-apply full urban model with land markets represented. With a team from the University of Calgary, we implemented a model with floor space explicitly represented, in quantity and price. Several policy analyses were completed with these models, and the results were broadly reasonable (Rodier et al., in press; Rodier, Johnston, & Abraham, 2002). The cost to the MPO of developing and applying this 60-zone model would have been between $100,000 and $200,000 for consultants and about $100,000 for internal staff time. These estimates would apply to other large MPOs that have fairly complete land use data.

Under pressure for more accurate travel modeling and better policy analysis capabilities, the MPO designed a staged model improvement program in 2001. In 2002, it adopted our 60-zone version of MEPLAN and updated the base-year data sets from 1990 to 2000. It is run in 5-year steps. In 2004, the MPO will implement the PECAS (Production, Exchange, and Consumption Allocation Model), recently developed by Oregon DOT. This is a model similar to MEPLAN, but it includes a more detailed representation of goods exchange and so gives a valid measure of economic welfare, called "producer surplus" for firms and households. This is the most comprehensive measure of urban economic welfare, as rents and travel and goods movement costs comprise about 40% of an urban economy. This state-of-the-practice model will be run in 1-year steps on 600 zones.

During 2003, the MPO will engage in a scenario testing process with citizens and interest groups. Using PLACES, a public-license (free) GIS software package, the MPO will generate 50-year land use maps in real time in workshops, according to interest group and citizen wishes. This will be done for each city and county in the region and for the whole region. The staff will summarize these scenarios into a few for further analysis. They will also add in relevant transportation facilities. These generalized land use scenarios will then be taken into PECAS as changes in the local land use designations, which control potential land use types and densities. PECAS will then be run to give market-based projections for the 600 zones. Then PLACES will be used to disaggregate the zonal land use projections to parcels. The SACMET travel model will be run on these data, to get travel, emissions, congestion, and other transportation measures. In addition, several GIS-based impact models will be run on the land use data, to assess effects on habitats, flood damage, local service costs, and soil erosion.

Concurrently, in this near-term period, the SACMET travel model will be improved by adding trip tours, using 15-minute time periods, and simulating all travelers individually. Modeling tours allows a more accurate representation of trip chaining. Travelers will possess many demographic characteristics, and so equity analyses can be done by income, race, and the like, and by zone. Implementing the initial PECAS model in 2003 will cost about $600,000. Other regions implementing PECAS include the State of Ohio and Calgary.

In the 2004–2005 period, the MPO intends to improve the land use model further by using parts of the Oregon Generation 2 model, which uses the PECAS economic structure but requires the microsimulation of all firms and households, located by street address. The model will be run in 1-year steps. Matching of workers to jobs will be improved with more data on individuals' educational status and job experience. Goods movement will be represented by individual shipments of commodities. The travel model is integrated into this structure by using microsimulation of all households and individuals in them, by address. Household activity allocation will be represented, followed by trip tours. Time of travel will be represented in smaller increments and departure time choice represented with more variables. The cost of going to the improved, more-disaggregate models will be

between $1 and $2 million, depending on the extent of the disaggregation of firms. All of this disaggregation in time and space should result in more accurate projections of travel and emissions. These disaggregate models also permit detailed equity analysis, because each traveler and locator has a rich set of characteristics attached to it.

The Sacramento MPO's plan for model development is ambitious and it will keep the agency at the forefront, with state-of-the-practice models. This will help insulate the MPO from lawsuits, as well as permit it to address the policy issues listed above. Because of the more detailed representation of travelers, trip tours, types of firms, and households, the models will be more accurate, as well as more policy-sensitive. Last, because the models will include theoretically sound indicators for economic welfare, social equity, and environmental impacts, all important evaluation issues can be represented. This comprehensive modeling will then allow the MPO and its constituencies to evaluate the trade-offs among the scenarios being examined. The 50-year time horizon in the scenario process is necessary in order to see significant effects from land use policies and from transit-building policies too. The fact that MPOs do 20-year plans biases them against such policies. The Sacramento region's model improvement program is similar to those in many large MPOs now, but it is more comprehensive. Its scenario testing is also exemplary; only a few other regions have done 40-year or 50-year scenario exercises.

Now that we have reviewed travel models and land use models in some detail and seen the improvements that are being made, we will look a bit further ahead to see what future developments will include.

THE FUTURE OF MODELING TRAVEL AND LAND USE

In general, modeling is moving toward the microsimulation of travel, goods shipments, households, and firms with land, developers, and floorspace demand all represented. Firms and households will be simulated by point locations. Large MPOs are progressing to disaggregate travel modeling, including tour-based joint mode-destination choice models initially and eventually microeconomic models of household activity allocation. Household vehicle holdings are represented. These models are discrete in time and space with households represented by address and firms by block or tract. The travel model will be based in GIS for data management and display (see Nyerges, Chapter 7, this volume). The goods movement model will represent shipments of commodities by truck type and time of day, with terminal costs included. Many of these methods recently have been applied in Portland, Oregon, the State of Oregon, and San Francisco. A household travel survey that asks about trip tours is needed, and several MPOs are currently doing these.

For land use modeling in the future, large MPOs will develop similarly disaggregate land use models, where households are represented by address and firms by block or tract. Floorspace quantity and price will be represented. Developer decisions will be represented with a discrete choice model, based on historical parcel data. Reuse of buildings and redevelopment of parcels are included, as are the production and consumption of all goods and services. The land use model is based in GIS for data management and display. All local services and regulations are represented.

Data are handled in a data library, which all submodels address directly. The travel and land use models are closely linked, so that person and goods movements are derived directly from the spatial flows between sectors for persons and goods in the land use model. The models will be run with time periods appropriate to each behavior (minutes, hours, days, months). UrbanSim has been developed around these concepts and recently applied to the Eugene–Springfield, Oregon, region, the Salt Lake region, and the Honolulu area.

UrbanSim is an open-code model, available on the World Wide Web for download, and was supported by the National Science Foundation, the Oregon DOT, and the U.S. DOT (*www.urbansim.org*). Digital census data are available for 1990 and 2000, making the estimation and calibration of urban models more tractable than in the past. Also, parcel data are increasingly being used by MPOs for various other purposes. MPOs do need to gather data on the locations of firms, number of employees, and floorspace lease values. Some of these data are available from national vendors.

In general, MPOs need to budget for continuous model improvements over each decade. Some data sets take a year or two to gather, clean, and assemble, before a new model can be estimated and calibrated. As member jurisdictions make more demands on the MPO models for various types of policy analysis, the models will have to keep pace. Many MPOs are making the models available to all of their member agencies and better user interfaces are being developed for these local staff people. A good framework for seeing how models can evolve over time is presented in Miller, Kriger, and Hunt (1999). It is also important to understand that the interagency consultation process required by the surface transportation act intends for the partner agencies to influence the evaluation and choice of models to be used by MPOs. Local governments and partner state and local agencies can use this joint decision-making process to require better modeling practices by MPOs.

REFERENCES

Abraham, J. E., & Hunt, J. D. (2000). Parameter estimation strategies for large-scale models. *Transportation Research Record, 1722*, 9–16.

Beimborn, E., Kennedy, R., & Schaefer, W. (1996). *Inside the black box: Making transportation models work for livable communities.* Milwaukee, WI: Citizens for a Better Environment.

Boarnet, M. G., & Haughwout, A. F. (2000, August). *Do highways matter?: Evidence and policy* implications of highway's influence on metropolitan development (Brookings Institution Discussion Paper). Washington, DC: Brookings Institution.

Cervero, R. (2003). Road expansion, urban growth, and induced travel: A path analysis. *Journal of the American Planning Association, 69*, 145–163.

Cervero, R., & Hansen, M. (2000). *Road supply–demand relationships: Sorting out causal linkages* (University of California Transportation Center, Working Paper No. 444). Berkeley, CA: University of California Transportation Center.

Cervero, R. & Hansen, M. (2002). Induced travel demand and induced road investment: A simultaneous equation analysis. *Journal of Transport Economics and Policy, 36*, 469–490.

COMSIS, Inc. (1996). *Incorporating feedback in travel forecasting: Methods, pitfalls, and common concerns* (DOT-T-96-14). Washington, DC: Federal Highway Administration, Office of Environmental Planning.

Conder, S.,& Lawton, K. (2002). Alternative futures for integrated transportation and land use models contrasted with trend-delphi models: Portland, Oregon, metro results. *Transportation Research Record, 1805*, 99–107.

Deakin, Harvey, & Skabardonis, Inc. (1993). *Manual of regional transportation modeling practice for air quality analysis* (Prepared for the National Association of Regional Councils). Washington, DC: U.S. Department of Transportation.

Echenique, M. H., Flowerdew, A. D. J., Hunt, J. D., Mayo, T. R., Skidmore, I. J., & Simmonds, D. C. (1990). The MEPLAN models of Bilbao, Leeds, and Dortmund. *Transportation Reviews, 10*, 309–322.

Echenique, M. H., & Partners & Voorhees, A. M., & Associates. (1980). *Functional specification for ITPM: Final report*. Washington, DC: U.S. Department of Transportation.

Hunt, J. D., Johnston, R. A., Abraham, J. E., Rodier, C. J., Garry, G., Putman, S. H., & de la Barra, T. (2001). Comparison from the Sacramento model testbed. *Transportation Research Record, 1780*, 53–63.

Johnston, R. A., & Rodier, C. J. (1994). Critique of Metropolitan Planning Organizations' capabilities for modeling transportation control measures. *California. Transportation Research Record, 1452*, 18–26.

MAG (Maricopa Association of Governments). (2000). *Survey of modeling practices*. Phoenix, AZ: Rita Walton and Anubhar Bagley.

Miller, E. J., Kriger, D., & Hunt, J. D. (1998). *Integrated urban models for simulation of transit and land-use policies* (TCRP Project H-12, Final Report). Washington, DC: U.S. Department of Transportation.

Miller, E. J., Kriger, D. S., & Hunt, J. D. (1999). Research and development program for integrated urban models. *Transportation Research Record, 1685*, 161–170.

Nelson, A. C., & Moody, M. (2000). Effect of beltways on metropolitan economic activity. *Journal of Urban Planning and Development, 126*, 189–196.

National Governors Association. (2000). *Growing pains: Quality of life in the new economy* (Issue brief). Available online at http://www.nga.org/center/divisions/1,1188,C_ISSUE_BRIEF_D609,00.html

National Governors Association. (2001). *NG-13. Principles for Better Land Use Policy* (Policy Position Detail). Available at http:www.nga.org/nga/legislativeUpdate/1,1169,C_POLICY_POSITION_D662,00.html

Noland, R. B., & Lem, L.. (2000, January). *Induced travel: A review of recent literature and the implications for transportation and environmental policy*. Paper presented at the annual meeting of the Transportation Research Board, Washington, DC.

Parsons Brinckerhoff Quade & Douglas, Inc. (1999). *Land use impacts of transportation: A guidebook* (NCHRP Report No. 423A). Washington, DC: National Academy Press.

Pas, E. I. (1995). The urban transportation planning process. In S. Hanson (Ed.), *The geography of urban transportation* (2nd ed., pp. 53–77). New York: Guilford Press.

Peer Review Panel Functions and Organization. (1994). *Travel Model Improvement Program*. Washington, DC: U.S. Department of Transportation, U.S. Department of Energy, U.S. Environmental Protection Agency.

Porter, C., Melendy, L., & Deakin, E. (1995). *Land Use and Travel Survey data: A Survey of the Metropolitan Planning Organizations of the 35 largest U.S. metropolitan areas*. Berkeley: University of California Transportation Center.

Purvis, C. (1991, November 19). *Review of transportation planning textbooks and other material on feedback and equilibration* (Memorandum).

Oakland, CA: Metropolitan Transportation Commission.

Putman, S. H. (1983). *Integrated urban models: Policy analysis of transportation and land use.* London: Pion.

Rodier, C. J., Abraham, J. E., Johnston, R. A., & Hunt, J. D. (in press). Anatomy of induced travel: Using an integrated land use and transportation model in the Sacramento region. *Transportation Research, A.*

Rodier, C. J., Johnston, R. A., & Abraham, J. E. (2002). Heuristic policy analysis of regional land use, transit, and travel pricing scenarios using two urban models. *Transportation Research, D, 7,* 243–254.

SAI International. (1997). *MPO use of modeling tools* (Draft report). San Rafael, CA: Arlene Rosenbaum.

Stopher, P. R. (1993a). Deficiencies of travel forecasting methods relative to mobile emissions. *Journal of Transportation Engineering, 119,* 723–741.

Stopher, P. R. (1993b). Predicting TCM responses with urban travel demand models. In T. F. Wholley (Ed.), *Transportaton and Air Quality II, Conference proceedings.* New York: American Society of Civil Engineers.

Transportation Research Board. (1995). *Expanding metropolitan highways: Implications for air quality and energy use* (Transportation Research Board, Special Report No. 245). Washington, DC: National Academy Press.

U.S. Department of Transportation. (1971). *Request for proposal on inter-relationships of trans-portation system development and land development* (GS-10, RFP-26, FHWA). Washington, DC: Office of Highway Planning, Urban Planning Division.

U.S. Department of Transportation. (1994a). *New approaches to travel forecasting: A synthesis of four research proposals* (DOT-T-94-15). Washington, DC: Travel Model Improvement Program.

U.S. Department of Transportation. (1994b). *Short-term travel model improvements.* Washington, DC: Travel Model Improvement Program.

U.S. Department of Transportation. (1994c). *Summary of comments prepared by travel forecasting peer review panels.* Washington, DC: Travel Model Improvement Program.

U.S. Department of Transportation. (1995). *Land Use Modeling Conference Proceedings.* Washington, DC: Travel Model Improvement Program.

U.S. Environmental Protection Agency. (1998, April). *Final guidance for incorporating environmental justice concerns in EPA's NEPA Compliance Analysis.* Washington, DC: Arthur Totten, et al.

U.S. Environmental Protection Agency. (2000). *Projecting land use change: A summary of models for assessing the effects of community growth and change on land-use patterns* (USEPA.EPA/600/R-00/098). Washington, DC: Systems Applications, Inc.

Wegener, M. (1994). Operational urban models: State of the art. *Journal of the American Planning Association, 60,* 17–29.

CHAPTER 6

Reflections on the Planning Process

MARTIN WACHS

Transportation systems provide one of the most essential human services and comprise a large proportion of our everyday urban environment. Although transportation is rarely desired for its own sake, we travel in order to get to work, obtain medical care, become better educated, and participate in recreational, social, cultural, and religious pursuits. The ability to access places and activities provides choices that are the essence of human freedom. Low-cost, reliable transportation is also essential to the missions of manufacturing and service firms and public organizations. In addition to the social and economic benefits of travel, we also face many costs associated with mobility. There are the obvious monetary costs of automobiles, trucks, and public transit, and costs to society that are important even though more indirect and difficult to measure, including those associated with traffic congestion, environmental pollution, resource depletion, and injuries and deaths due to crashes.

Transportation is of central interest to geographers because it is a fundamental determinant of human interaction in space. During the 20th century societies changed continually as the ease of movement from place to place increased. In 100 years we evolved from walking and animal power, to railroads, and eventually to increasingly universal automobile access and to a global community reliant on intercontinental jets. Such rapid changes in access impart economic and social value. Transportation investments order the space in which human interactions take place, creating prosperous communities by reducing the friction of overcoming space and creating economic opportunity by lowering the cost of people's and firms' interactions with one another. Uneven change across space and time also leaves some places and groups behind, ignored and disadvantaged. Increasingly reliant on transportation systems, people live in suburbs while working in cities; resources are moved at low cost and high speed from mines and fields to factories and markets, companies open branches 10,000 miles from their headquarters, and people decide at a moment's notice to visit the natural wonders of the world. Others, however, still depend on walking and animal-powered movement, and live their entire lives within one community. No wonder, then, that transportation is essential to our lives and at the same time it is at the heart of many political arguments.

This chapter examines these complex issues by looking first at the criteria by which scholars evaluate and compare transportation networks and why technical criteria, such as those described in Chapter 5, give way in policymaking situations to political struggles. We turn next to an analysis of the role of governments in transportation and how in the U.S. context the roles of different governmental bodies are simultaneously complementary and competing. Some of the formal products of transportation decision making are examined, in the form of physical plans and prioritized public expenditure programs. This is followed by a discussion of how citizens and communities can influence policymaking. Because the focus of many policy debates over transportation systems in the United States has been their impacts on environmental quality, we also discuss environmental impacts of transportation and the concepts of environmental justice and sustainability. Lastly, we consider how transportation policy debates and policymaking will look in the 21st century.

CRITERIA BY WHICH TRANSPORTATION POLICIES ARE JUDGED

Because economic and social well-being are enhanced by improved access to opportunities, but the direct and indirect costs of transportation can be very high, the purpose of transportation policy is to provide society with the benefits of transportation services while minimizing their costs as much as possible. Achieving a balance between benefits and costs requires consideration of effectiveness, efficiency, and equity and the trade-offs among them. *Effectiveness* refers to how well the transportation system meets its intended objectives. Holding other conditions constant, the effectiveness of an investment in transportation is proportional to its capacity to deliver service. *Efficiency*,

dealt with further in Box 6.1, refers to how much a transportation investment costs in relation to the benefits it provides. Resources are always limited, and where more service is provided per dollar of expenditure society is better off. *Equity* refers to the fairness with which benefits are delivered and to the costs imposed on the many communities that comprise society. If some groups obtain a disproportionate share of benefits and others incur a disproportionate share of monetary or environmental costs, such disparities heighten the inequalities that exist in our society.

THE POLITICAL NATURE OF TRANSPORTATION POLICY

It is not easy to achieve effectiveness, efficiency, or equity. The modeling approaches discussed in Chapter 5 are intended to address these criteria using formal methods of analysis that are taught at universities and encapsulated in widely distributed software. Yet different people and interest groups define the terms "effectiveness," "efficiency," and "equity" differently. Even if we could define each term unambiguously, it is still difficult to balance all three concepts whenever an important decision must be made. This problem, inherent in making most decisions about transportation systems, is one of the principal reasons that transportation is always highly political.

People often disagree for many reasons regarding what should be done in any particular situation. Should a new road be built or not? Should an existing road be widened or not? Should cyclists be encouraged on a given route or not? Should investments be made in public transit facilities or in highways or in both? Should social service agencies trying to reduce unemployment help poor people to buy cars or help them to use public transit more effectively? Should we pay for transportation services through property and sales taxes or through tolls, fuel taxes, and other fees imposed on those who use the system? Answers to these

BOX 6.1. Cost–Benefit Analysis and Transportation Policy

Detailed technical studies by professional analysts and debates in public sessions and committee meetings by members of the public, staff, and consultants can identify alternative approaches to solving transportation problems or meeting a stated need. After options are identified there is a systematic method for comparing them called the "cost–benefit analysis." A cost–benefit analysis identifies all of the costs and benefits due to a project and assigns monetary values to them. This allows analysts and decision makers to gauge whether the benefits from a project exceed its costs and if the amount of the benefits make the project worth undertaking. Although this may seem straightforward, it often is not. There are many opportunities for value judgments and there is much debate about how to quantify some types of benefits and costs. Common criticisms of cost–benefit analyses are that:

- Costs (construction, operation, maintenance, etc.), although they are generally easier to quantify than benefits (shorter travel times, improved safety, better air quality, etc.), are typically underestimated, especially in the long term.
- Many costs and benefits are hard to quantify—for example, the costs of noise that make sitting in the backyard unpleasant, or the costs of speeding traffic that make it dangerous for children and pets to be outside unsupervised, or the benefits of an improvement in neighborhood appearance. These costs and benefits are often left out of typical cost–benefit analyses.
- In order to complete the analysis decision makers have to agree on how to quantify benefits (such as travel time savings) and costs (such as loss of life in traffic accidents).
- Even if overall benefits exceed overall costs, there are individuals and/or groups impacted by the project for whom the costs will exceed the benefits.
- Cost–benefit analyses sometimes do not look at the distribution of benefits and costs, which is a key issue in environmental justice.

Even though the cost–benefit analysis takes place within a technical framework, it is inherently political and can be influenced by people involved in the evaluation of a project, whether they be politicians, staff, or members of the general public.

Source: Greig, Cairns, and Wachs (2003).

questions always depend on how different interests perceive the effectiveness, efficiency, and equity associated with the issues these questions raise. That means that such questions are inherently political. Indeed, transportation decision making is always intensely political.

Citizens often cannot agree on such local transportation issues as whether a stop sign is needed at their corner, or whether a speed bump would improve safety on their street. No wonder, then, that there is always controversy over whether or not to build a regional subway or a new airport. Large infrastructure projects have high costs and extensive positive and negative impacts. These are likely to be distributed unequally across space and across population groups, so they naturally have very high political visibility. In one well-known case, authorities have been attempting to build a freeway through the communities of El Sereno and South Pasadena near Los Angeles for well over 50 years. Well-organized and thoughtful community groups who oppose the freeway have been able to repeatedly delay the project that would benefit many other constituencies by using a variety of tactics, from suits to repeated studies to mobilizing different public agencies. As the alignments and design details are reconsidered and redesigned, the cost of building the freeway rises, commitments of funds are made to other projects, and the probability of ever building the proposed freeway is reduced owing to the cumulative effects of the many delays. (For a more comprehensive discussion, see Wachs, 1995.)

One of the most interesting and important characteristics of transportation systems is that they almost always depend on the interaction of many different individuals and organizations. Vehicles such as automobiles and bicycles are owned and operated by individuals, families, or organizations, and trucks are owned and operated by companies. A furniture dealer may make deliveries in her own vehicle, while a small appliance manufacturer might contract for his deliveries with a separate delivery service. Vehicles are manufactured by private companies, domestic and international, which purchase components from other companies and labor from individuals. Many vehicles are insured by still other private companies and financed by loans from banks, while the performance characteristics of vehicles and the ways in which we operate them are regulated in many different ways by governmental bodies. In the United States, speed limits, for example, are typically set by state governments, while allowable air pollutant emissions and fuel economy standards for vehicles are set by the federal government. Governments at different levels are also the primary providers of streets and highways. Local streets and rural roads usually are built, maintained, and regulated by towns, cities, and counties, while arterial highways and freeways are often provided by states, many with financial support and within regulatory constraints set by the federal government. And, while many public transit systems were traditionally privately owned, in recent decades transit services have also come to be provided by government agencies, usually different ones from those providing streets and highways. Regional transit authorities are typically overseen by boards composed of elected officials representing the jurisdictions within which service is offered.

In addition to the myriad companies and agencies that formally own and operate transportation facilities and services, many other groups work hard to influence transportation policies and choices. Automobile clubs are among the largest membership organizations in the United States, and they try to influence policies to the benefit of their members. Environmental organizations are committed to the regulation of transportation systems to achieve better air quality and reduce the production of greenhouse gases. Important membership associations are committed to improving highway safety. Labor unions devote enormous energy to improving the welfare of those working in industries that provide transportation. Industry associations represent the providers of transit services, taxicab operators, truckers, and automobile manufacturers. Membership associations of various sorts, such as the American Association of Retired People, speak up regarding the transportation needs of their members, while other groups call for measures that would better meet the mobility needs of disabled people, children, and war veterans.

Because these diverse groups have different members and pursue competing purposes, they more likely disagree than agree over the effectiveness, efficiency, and equity of different measures, and which should be the highest priority for transportation policy; in short, they disagree over what the solution might be to a particular problem. Beyond that, transportation is often seen as a tool by which our society can accomplish other important social or economic objectives like the creation of jobs or the reduction of economic inequalities. We sometimes say that successful transportation systems are the result of coalitions or partnerships among many public agencies, private companies, and voluntary associations.

The other side of the same coin leads us sometimes to say that transportation failures are the result of institutional complexities, turf battles, inadequate leadership, or organizational conservatism. But it is clear that the transportation enterprise is always inherently political in nature. In order to understand transportation, we must come to grips with its institutional complexity. Intellectual analyses and research can help us set priorities, but policy is actually made by legislatures, regulatory agencies, and the sometimes vigorous interaction among organizations advocating on behalf of competing interests.

Change does occur, but usually change is gradual because the complexity of the system is itself a source of inertia. Initiatives stall when even a small number of influential interest groups say no, whereas to be successful such initiatives may require that many groups say yes. In a complex political system that addresses the competing positions of many interests, Altshuler (1965) noted that to be acceptable a change must benefit many interests, which consequently would decide to support the change. It must also concentrate the costs or disbenefits narrowly, to minimize the number of interests who would oppose the change. In other words, changes in transportation systems or policy are most likely to occur when groups perceiving themselves to be "winners" far outnumber and exceed in influence those who think the policies would make them "losers." Even this is often not sufficient, as those who consider themselves losers often seek redress through the courts. This conflict is the essence of U.S. politics, and transportation is one of the most common settings for political interaction.

GOVERNMENT INVOLVEMENT IN URBAN TRANSPORTATION

Governments at many levels interact with one another to build and manage the many public elements of the transportation system. While some would represent the different governments involved as a "layer cake" with local governments at the bottom and the federal government at the top, it is probably more realistic to think that transportation politics more resembles a "marble cake," with local, regional, state, and federal interests all mixed together through multiple programs in which different governments cooperate, compete, regulate, and represent their unique interests and concerns. As indicated in Chapter 11, the availability of funding from national or state government can strongly influence the types of projects favored by local political leaders. For example, if bus transit appears to be cost-effective locally but federal dollars are more readily available for rail investments, local preferences can often be changed.

Local Governments

Urban streets and local county roads make up the vast majority of the transportation system when measured in terms of lane miles or surface area, but outside of busy activity centers they are often characterized by low traffic volumes and so they account for a small minority of all travel. They are critically important because they provide access to homes, businesses, and institutions. Local roads and streets enable us to make the first and last part of each trip; they also support postal and parcel deliveries; emergency visits by police, fire, and ambulance services; trash collection; and a wide range of similar services. Local streets are the rights-of-way for telephone and electric power lines and pipes that provide gas, water, and sewerage services to our homes and businesses. Because both accessibility to property and access to services impart value to land, local governments most typically require developers of land to build streets and cede them to local governments. Citizens also support the maintenance and operation of local streets by paying real estate taxes to local governments on residential, commercial, and industrial property. In recent years, many local governments have used similar financial support to provide local public transit services, which can be viewed as providing basic accessibility similar and complementary to that provided by local streets.

State Governments

Under the U.S. Constitution, most government functions not specifically assigned to the federal government are the responsibility of the states. In the early part of the 20th century, states' roles in transportation grew dramatically as automobile and truck travel grew much faster than population and the provision of intercity highway connections became necessary. Getting rural farmers "out of the mud" was an enormous and expensive undertaking. Most heavily traveled roads in the United States, and many transportation facilities serving other modes of travel, are under the jurisdiction of the states. They are typically owned and operated by state departments of transportation and overseen by commissions of citizens appointed by the governor and/or the legislature. State fuel taxes and vehicle registration fees finance many transportation projects, and major policy debates about transportation often take place in state legislatures and involve the application of gubernatorial authority.

National Government

While most transportation facilities were formally built, owned, and operated by the states, the national government's role in transportation has also grown significantly throughout the last century, and in many ways actions undertaken by the states reflect national transportation and environmental policies. At first, the federal government provided roads only on federal lands, but the U.S. Constitution specifically notes a federal responsibility for "interstate commerce." Gradually, the national government provided support for roads and oversight of railroads that moved mail; and eventually

the economies of the states became more integrated and the federal government provided financial support to connect major population centers and military facilities. A federal gasoline tax was enacted in the 1930s that enabled national sponsorship of intercity facilities. After World War II, the federal government became a major player in all realms of transportation. The National System of Interstate and Defense Highways, consisting of more than 40,000 miles of high-quality highways, was made possible by a substantial increase in federal fuel taxes and excise taxes on vehicles and their components like tires. Congress provided that the Interstate program make available nine federal dollars for each state dollar when a state built a component of that vital road system. Gradually, while the states were still ostensibly making the most important decisions, the federal government required them to fulfill certain requirements in order to receive these valuable federal funds. Thus the states must plan their highways and transit facilities in accordance with federal planning guidelines and must meet federal environmental protection requirements. Because the federal government is responsible for the health and well-being of all citizens of the nation, it also regulates the safety features, energy consumption, and pollution production of vehicles that are produced by private companies for private owners.

Regional Agencies

Most Americans are far more aware of their local, state, and federal governments than they are of regional agencies, yet, as was described in Chapter 5, planning and systematic resource allocation to accommodate travel must be addressed at the regional level. Metropolitan areas contain many government jurisdictions, and trips commonly cross local boundaries, as do the troublesome environmental and social impacts of transportation facilities. Box 6.2 shows how many different levels of government interact over transportation decision making.

BOX 6.2. The Government and Transportation Decision Making

Transportation decision making is carried out by several governmental levels:
- State departments of transportation (DOTs) are the largest units of government that develop transportation plans and projects. They are responsible for setting the transportation goals for the state. To do so, they work with all of the state's transportation organizations and local governments. They are responsible for planning safe and efficient transportation between cities and towns in the state.
- Metropolitan Planning Organizations (MPOs) represent areas with a population of 50,000 people or more. An MPO may have "council of governments" or "regional planning commission" in its official name. Each MPO is different because individual metropolitan areas are so different. A policy board, which is comprised of local elected officials, sets an MPO's policy; but other groups, such as nonprofit organizations, community organizations, or environmental organizations, can influence the direction an MPO follows. The MPOs' mission is to provide short- and long-term solutions to transportation and transportation-related concerns.

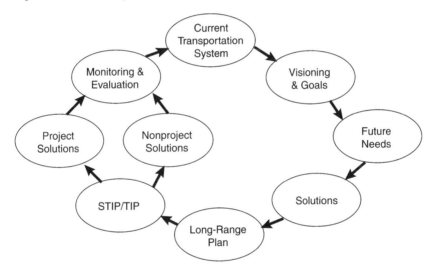

- Local governments carry out many transportation planning functions, such as scheduling improvements and maintenance for local streets and roads.
- Transit agencies are public and private organizations that provide transportation for the public. Public transportation includes buses, subways, light rail, commuter rail, monorail, passenger ferryboats, trolleys, inclined railways, and people movers.
- The federal government (U.S. DOT) oversees the transportation planning and project activities of the MPOs and state DOTs. The federal government also provides advice and training on transportation topics, ranging from pavement technology, to design, to efficient operations of highway and transit systems. The federal government also supplies critical funding needed for transportation planning and projects. At least every 2 years, the federal government approves a program of projects submitted by state DOTs that includes projects proposed for federal funds.

Source: Federal Highway Administration (http://www.fhwa.dot.gov/planning/citizen/citizen4.htm)

When the commissioners of Cuyahoga County, Ohio, in 1927 solicited the help of the Federal Bureau of Public Roads (since renamed the Federal Highway Administration) to deal with growing problems of traffic congestion in Cleveland, they sought to establish a "scientific plan of highway improvement." The bureau responded that it would cooperate and provide funds for the study only under the condition that it extend to the area principally controlled by the City of Cleveland and surrounding suburban areas *without regard to political boundaries*, and that all governmental agencies in the area having jurisdiction over highways and traffic cooperate in the establishment of a general highway development plan and agree to carry out this plan cooperatively when it is completed. Cleveland was not unique. At exactly the same time, a group in Boston under the direction of Robert Whitten was developing a relatively advanced traffic analysis and forecast based on an "origin and destination survey" that extended to 39 cities and towns of the metropolitan area (Heightchew, 1979). These early efforts, based on voluntary cooperation among cities, towns, and counties, were important precursors to the metropolitan organizations that today perform similar though more extensive functions.

Starting in the 1960s, after the creation by act of Congress of a U.S. Department of Transportation, a series of federal funding laws gradually required more and more coordination between highway and transit planning and programming and between the many jurisdictions that make up each metropolitan area. Eventually, transit and highway planning were unified under a single Metropolitan Planning Organization (MPO) in each large urbanized area, designated by the governor and often overseen by a board or commission of elected officials representing the many local governments of the region. About half of all MPOs are "regional councils of governments" made up of members who are the municipalities comprising the region. Other

MPOs are regional economic development organizations, transportation planning agencies created explicitly for that purpose by state statutes, or specific divisions within state highway or transportation departments (Yaro, 2000, p. 74). These MPOs can sometimes encompass a single county, and sometimes they extend over many counties with combined populations as large as several states. These agencies usually have professional staffs who have become increasingly responsible for planning transportation capital investment programs and for allocating federal and state transportation funds within the metropolitan area.

While regional agencies rarely have the authority to levy taxes or raise their own funds, they have in some cases gained formal authority over state funds spent within their jurisdictions, and this gives them considerable power (Wachs & Dill, 1999). For decades, regional MPOs decided how federal transportation funds would be spent within fairly tight limits as determined by federal and state laws and within narrowly specified funding categories. These agencies gained considerable influence with the passage of the Intermodal Surface Transportation Efficiency Act (ISTEA) of 1991. This law reduced federal limitations and increased the flexibility with which regional agencies could allocate funds among different types of projects. For example, while previously federal funds came as *either* highway dollars *or* transit dollars which had to be spent on projects devoted exclusively to one mode or the other, the Surface Transportation Program (STP) and Congestion Mitigation and Air Quality (CMAQ) Program under current legislation allow regional agencies to determine their particular mix of expenditures among transit and highway programs. Flexibility was retained when federal transportation funding was reauthorized in 1998 by a more conservative Congress that passed the Transportation Equity Act for the Twenty-First Century (TEA-21) and allocated $217 billion dollars to transportation spending over the next 6 years. States have also acted in recent years to increase the role of regional planning agencies. In California, for example, three-fourths of federal and state highway and transit funds are designated by statute to be spent at the discretion of the MPOs in that state.

THE REGIONAL TRANSPORTATION PLAN AND TRANSPORTATION IMPROVEMENT PROGRAM

One of the most important responsibilities of an MPO is the preparation, approval, and updating of the Regional Transportation Plan (RTP), required by federal law and by many states as well. The RTP is a blueprint to guide the region's transportation development for the coming 25 years; it is updated every 3 years based on projections of growth in the region's population, economic activity, patterns of housing development, and the resulting projections of travel volumes. Under federal law, the RTP must also be "fiscally constrained," meaning that all new transportation programs and projects included in the plan must be capable of being financed under reasonable projections of revenue from various governmental sources.

The RTP is intended to be a guide to ongoing regional expenditures on transportation projects, so, in general, local and state governments are required by law and federal regulations to invest only in projects that are consistent with the stated objectives of the RTP. This is usually accomplished through the development of a Transportation Improvement Program (TIP), which is a list of projects to be implemented during a specified time period, including delineation of the agency responsible for implementing that project. Planning regulations adopted by the U.S. Department of Transportation in 1975 require each MPO to officially adopt a TIP which consists of a staged, 3- to 5-year program of transportation improvements based on realistic

estimates of the projects' total costs and revenues from various sources during the investment period. The TIP must be updated every 2 years, but projects included in the TIP must always be consistent with the RTP (Meyer & Miller, 2001, p. 567).

True "regionalism" in transportation planning has proven to be elusive, and regional agencies can often more accurately be described as a forum in which individual counties, cities, transit operators, and interest groups battle over turf and struggle for larger shares of regional resources. In a number of regions, for example, law suits have recently been filed asserting that the MPOs have been inequitable in the distribution of resources among the many jurisdictions that make up the regions. One common source of disagreement arises from the fact that suburban communities often have higher incomes than the central city, yet public transit is often most effective and efficient where central-city residents live at higher densities, own fewer cars, and travel shorter distances. The central cities urge the MPOs to redistribute tax revenues collected from suburban communities for the purpose of expanding central-city transit. At the same time, representatives of the suburbs usually demand "equity," which they define as spending resources in the jurisdictions in which they were collected. These disagreements often lead to overcrowded transit vehicles in the inner city which receive fewer subsidy dollars per rider, while buses and rail lines operate in the suburbs at high cost and large subsidies per user, but with lower ridership.

CITIZEN INVOLVEMENT IN TRANSPORTATION: THE RISE OF CITIZEN ACTIVISM

The complexity and extent of the transportation system are enormous; highway and transit systems shape and form the city and determine the quality of daily life. It is not surprising, then, that in a democratic society diverse interest groups take direct and competing interest in regional transportation policymaking. Thus, in addition to federal, state, local, and regional governments that all participate in regional transportation policymaking, businesses, trade associations, environmental organizations, representatives of inner-city and suburban communities, spokespersons for transit users and automobile clubs, and numerous other groups want a voice in determining which projects will be undertaken. These groups must all be accommodated in the regional transportation planning process.

During the 1960s, when most urban regions planned dramatic expansions of their transportation systems, diverse interests began to participate in transportation policy debates. In many communities, despite federal requirements to consider the social, economic, and environmental effects of transportation plans and systems prior to submission of requests for federal aid, minority ethnic groups and citizens representing neighborhoods became exasperated that their views were less frequently influential in the planning process than were those of business interests and well-established political groups (Weiner, 1997, p. 69). Citizens could voice their opinions at public hearings prior to the adoption of planned projects, but when hearings were held late in the planning process they provided insufficient opportunity for the diverse voices of the populace to be heard. As early as 1969, the Federal Highway Administration addressed this concern by establishing a "two hearing process," so that citizens' views could first be heard at a "corridor" public hearing, while planning concepts were being considered in general terms, and then again at a "highway design public hearing," when specific project designs were being considered. While this requirement was a small step in the right direction, it hardly stemmed community objections to certain transportation investments.

In many U.S. cities, including New Orleans, Chicago, St. Louis, and New York, groups formed to oppose regional transportation plans that they believed would adversely affect their communities. The results of many transportation studies of the 1960s and 1970s called for major expansions of regional highway networks, expansions that often threatened the integrity of particular communities. Residents and local business owners objected formally at the public hearings, but increasingly took to the streets to hold disruptive demonstrations and to deride public officials. When the long-range transportation plan for the Boston region was published in 1969, opposition was so vocal and sustained that Governor Francis Sargent ordered a moratorium on new highway construction in 1970 and called for a review of regional plans that was much more open to the positions of a wide variety of interest groups. The Boston Transportation Planning Review led to a new regional transportation plan, which greatly reduced highway construction through inner-city neighborhoods and gave much more emphasis to public transit. Just as important as its effects on the region's transportation plan were its influence on the planning process. The review occurred in a highly charged atmosphere, welcomed organized input from a wide range of diverse groups, addressed social and economic impacts as well as travel, and focused on community-level as well as regional impacts of the transportation plans under consideration (Gakenheimer, 1976; Humphrey, 1974).

Since the 1970s, regional transportation planning authorities have struggled continuously with their responsibility to balance technical responsibility and responsiveness to federal and state statutes against their responsibility to represent and incorporate the views and needs of citizens in the planning process. Among the many approaches adopted by transportation planning agencies in efforts to provide clearer voices to citizens and diverse interest groups have been the publication of guidebooks for citizens that outline their opportunities to influence regional decision making, the creation of multiple special-purpose advisory boards and task forces, the use of specialized consultants for the purpose of organizing citizen response processes, the use of surveys of citizen opinion, and the creation of two-way electronic communications facilities that link citizens to planning agencies (Meyer & Miller, 2001, p. 73). Despite all of these efforts, it has proven much easier to incorporate into decision-making processes the views of well-organized, well-educated, and articulate middle-class constituencies than the views of low-income, localized, or minority groups. Many citizens remain distrustful of regional transportation planning agencies and view their efforts to achieve consensus as falling far short of true collaborative decision making.

ENVIRONMENTAL JUSTICE

Today, environmental justice is one of the most important elements of policymaking in transportation. It is not specific to any mode of transportation, particular community, or single policy issue. It is fundamentally about fairness toward the disadvantaged and often turns on issues of racial and ethnic minorities who feel excluded from influence in policymaking. The federal government has identified environmental justice as an important goal for public spending in transportation; local and regional governments must comply with requirements to provide environmental justice in transportation programs.

Environmental justice issues arise most frequently when (1) some communities get the benefits of improved accessibility and congestion relief while others experience fewer benefits from transportation spending programs; (2) some communities suffer disproportionately from such negative impacts of transportation programs as air pollution and community disruption; (3) some communities have to pay higher transportation taxes or higher fares than others in relation to the services they receive; (4) some communities' members receive a disproportion-

ately small share of the jobs that are created by the construction and operation of transportation facilities; and (5) some communities are less represented than others when policymaking bodies debate and decide what should be done with transportation resources (Forkenbrock & Schweitzer, 1999; Greig, Cairns, & Wachs, 2003).

One of the best- known disputes illustrating the complexity of environmental justice in transportation began in the early 1990s in Los Angeles. The Metropolitan Transportation Authority (MTA), a regional transportation construction and operating agency, was building a network of rail transit lines, many of which were focused on lower income communities in the central city. As planning and construction proceeded, the MTA experienced rising costs and shortages of funds and consequently decided to reduce some bus service in order to conserve resources. Eventually, when the agency proposed a substantial increase in bus fares in order to raise needed funds, a group known as the Bus Riders Union brought suit against them, alleging environmental injustices and violations of Title VI of the Civil Rights Act. The plaintiffs claimed that the agency was reducing bus service, raising bus fares, and subjecting minority bus riders to intense crowding while vigorously pursuing its rail construction program. They argued further that rail users would include a higher proportion of white and higher income travelers, who would receive higher subsidies than bus riders. Though the agency believed that its rail investment program would benefit many minority citizens and refused to acknowledge any discrimination, an agreement was eventually reached between the parties in which bus fares were stabilized for a certain period of time and bus crowding would be relieved by the addition of buses to some routes. The MTA continued to disagree with the plaintiffs, and 10 years after the suit was initiated it challenged the judge's order to buy or lease 248 new buses to reduce overcrowding on certain bus routes. The U.S.

Court of Appeals upheld the judge's order, and the U.S. Supreme Court refused to hear the case.

REGIONAL TRANSPORTATION PLANNING: BALANCING ANALYSIS AND POLITICS

The regional transportation planning process is complex; it is simultaneously highly technical and highly political. Important roles are played by professional social scientists, engineers, and planners working with large data sets, and important roles are played by the representatives of stakeholder groups who express their concerns and preferences at public meetings. Box 6.3 summarizes the "planning factors" that federal law requires be included in the planning process. As explained in the previous chapter, analysts use data that describe regional population, economic growth and change, land use and urban form, trip making, and environmental and social impacts of transportation. They often employ complex mathematical models to forecast changes in travel over fairly long time horizons and to evaluate the potential impacts of different transportation investment strategies. For example, regional transportation plans often result from the systematic evaluation of alternative ways of accommodating forecast growth that range from increasing movement capacity through the expansion of public transit, to the expansion of highways, to improvements in the operation and management of existing facilities, to redirection of growth into denser urban corridors within the region. Quantitative measures, such as benefit–cost ratios, are often used to compare the social efficiency of different plan elements, and these ratios are reviewed by committees of citizens and stakeholders including environmental organizations, chambers of commerce, and representatives of affected industries. Plans are revised and reevaluated, alternative investment strategies are suggested and tested, and

---

BOX 6.3. Transportation Planning Factors

The Transportation Equity Act for the Twenty-First Century (TEA-21) requires that seven "planning factors" be included in regional transportation plans. The plans must:

1. Support the economic vitality of the metropolitan planning area, especially by enabling global competitiveness, productivity, and efficiency;
2. Increase the safety and security of the transportation system for motorized and non-motorized users;
3. Increase the accessibility and mobility options available to people and for freight;
4. Protect and enhance the environment, promote energy conservation, and improve the quality of life;
5. Enhance the integration and connectivity of the transportation system, across and between modes, for people and freight;
6. Promote efficient system management and operation;
7. Emphasize the efficient preservation of existing transportation system.

Source: Transportation Equity Act for the Twenty-First Century.

---

eventually a plan is arrived at through a process of modeling, review, comment, debate, and voting that might extend over several years.

Fifty or more years ago, when regional transportation planning was a new undertaking, long-range transportation plans were most often characterized by the inclusion of major new highway capacity investments such as radial and circumferential freeways and bridges or tunnels across major topographical barriers. Improved access to ports, airports, and railroad yards have also been important, especially with respect to the accommodation of anticipated growth in freight movements along with increased personal travel. In many metropolitan areas, as highway networks have been completed and some potential expansion projects have become controversial, more recent plans have emphasized transit improvements such as extensions of bus routes and construction of new rail transit lines. While RTPs can also frequently provide for improvements to the operational efficiency of existing systems—for example, through better coordination between highways and transit systems or through widespread implementation of computerized traffic signal timing—it is fair to say that RTPs have most often emphasized expansions of the physical capacity of the transportation system.

Planners have long understood that the arrangement and intensity of activities in space affects the levels and patterns of urban travel, and their models have to the extent possible incorporated relationships between urban form and transportation. In earlier years, urban development was forecast, and transportation plans were developed to accommodate, the travel that would arise as the forecasts came to pass. In recent years, planners have tried, where possible, to incorporate into regional transportation plans greater recognition that land use and transportation can be mutually and jointly planned. Transit-oriented development and an orientation toward "smart growth" are gradually becoming central features of larger numbers of regional transportation plans, and federal laws such ISTEA, passed in 1991, and TEA-21, passed in 1998, have encouraged that type of orientation within regional transportation plans.

Most observers agree that recent federal legislation and changing planning doctrine are creating more regional autonomy and a gradual shift in regional transportation plans from a focus on accommodating forecast growth toward an emphasis on shaping and limiting regional growth and travel. Yet great differences of opinion exist as to how desirable this change is and how quickly a change in orientation is occurring . Some have called recent trends a "radical and visionary transformation of the nation's transportation policy" (Rusk, 2000, p. 78). Some metropolitan areas, among which Portland, Oregon, has gained a national reputation for innovation, are self-consciously attempting to slow suburbanization by emphasizing investments in transit rather than in highway capacity and encouraging clusters of dense and mixed-use development. Critics believe that these approaches will fail because suburban development and preferences for automobile-dependent lifestyles will continue despite these efforts (O'Toole, 2001). Others see many regional transportation organizations as bastions of protection of the "status quo," responding predictably to the wishes of the states and of the local elected officials serving on their boards who pursue their constituents' particular local priorities in preference to the development of creative visions by which entire regions might be transformed (Yaro, 2000).

In truth, regions differ dramatically from one another and their regional transportation plans should undoubtedly reflect their differences. Some U.S. metropolitan areas are growing very slowly while others are booming; some are focused on providing services and catering to tourists while others are manufacturing centers; some are plagued by congestion while others have capacity to absorb growth. Some, especially in the Northeast and the Midwest, have high-density transit-oriented cores that predate universal ownership and use of automobiles, while many Sunbelt cities grew to prominence after the widespread adoption of automobiles. Some areas have serious air quality problems that must be addressed through regional transportation planning, while other areas do not have such problems. For all of these reasons, it seems appropriate that regions should differ in their approaches to transportation planning, and these differences should be encapsulated in their RTPs.

ENVIRONMENTAL DIMENSIONS OF TRANSPORTATION PLANNING

Transportation systems shape and form the urban environment in which we live in many different ways. While the regional transportation planning process is intended to provide an efficient and aesthetically pleasing transportation system, that system significantly affects the quality of the natural environment in many unintended ways. Transportation facilities cover a reasonably large proportion of the surface area of urban communities; they affect watersheds by changing watercourses and producing many contaminants that enter streams and rivers. Roads and their by-products, such as salt used to melt winter ice and snow, intrude upon animal and plant habitats. Transportation vehicles are among the most common sources of urban noise, and vehicle exhaust gases contribute significantly to urban air pollution. The manufacture, operation, and scrappage of transportation vehicles create additional environmental challenges. The planning process has repeatedly been modified to make it more responsive to the goal of preserving the natural environment. Today a large proportion of the resources devoted to transportation planning and management is focused on minimizing or mitigating the environmental impacts associated with transportation.

Section 4(f) Provisions

As the process of building urban portions of the Interstate Highway System accelerated in the 1960s, environmentalists became alarmed that in their efforts to avoid locating highways in populated communities, transportation planners were routing many new roads through public parks, wetlands, and other sensitive open spaces. In response, federal planning requirements were changed in an effort to preserve sensitive habitats and aesthetically pleasing environments and prevent their destruction by the construction of transportation facilities.

Section 4(f) of the federal law governing regional transportation planning was enacted in 1966, requiring "special effort . . . to preserve the natural beauty of the countryside and public park and recreation lands, wildlife and waterfowl refuges, and historic sites." The U.S. secretary of transportation is permitted to approve construction of a transportation facility in such a sensitive environment only if "there is no feasible and prudent alternative," and only if planning has been done to "minimize harm" to such sites. Numerous legal decisions have resulted in the current policy that planners must fully document all of the environmental costs of such projects and that mitigation measures must be genuine and effective. In some cases, for example, highway agencies have restored two or three times as much acreage of wetland as that affected by a new road project in order to mitigate the impacts of that project in conformance with Section 4(f) (Federal Highway Administration, 1989).

National Environmental Policy Act

Section 4(f) of the federal highway program was in many ways the model for the most widely applied federal environmental legislation, the National Environmental Policy Act (NEPA) in 1969, which applies to all federally funded transportation projects as well as many other public and private projects. This demanding law requires that any project to be supported by federal funds undergo an "environmental impact assessment process," which is summarized in Boxes 6.4 and 6.5. The agency responsible for the proposed project must publish and widely circulate a draft environmental impact statement (EIS) and invite public comments in writing and through public hearings. In the case of large projects, these statements can be massive, consisting of thousands of pages of detailed analysis. The responsible agency must respond in writing to comments and suggestions made by the public and eventually must issue a revised

---

BOX 6.4. Environmental Impact Statements

The following are the basic steps in preparing an EIS based on "Final Guidance for Incorporating Environmental Justice Concerns in EPA's NEPA Compliance Analyses, April 1998" which is a comprehensive discussion of NEPA and environmental justice requirements that was produced by the U.S. Environmental Protection Agency:

- *Scoping* is the first step when an agency must file an EIS and is the first opportunity for public input into the EIS. At this stage, the lead agency invites representatives from all government agencies that might be involved, the project's supporters, and interested members of the public to a meeting to identify all of the issues involved with the project that could have a significant impact. Alternatives for a project can be developed at this stage. These meetings are advertised in local newspapers and on the lead agency's website, and announcements are sent to people who have been involved with the agency's activities in the past or are on their mailing list. Getting on this mailing list is a good way to hear about scoping sessions and other public meetings.
- *Draft EIS* is the first document produced; it discusses the impact of each alternative on the human and natural environment and how serious the impacts are. In cases where the effects of a project are significant but they can be reduced, a mitigation strategy is presented. The draft is circulated to all involved parties, interested individuals and organizations, and is available to the public at libraries and other public offices.
- *Public Comment* is the second major opportunity for public involvement in the EIS process. At this stage stakeholders or members of the general public can voice concerns with the technical analyses, elimination or inclusion of specific alternatives, mitigation strategies, or anything else addressed in the draft EIS. Comments can be made in writing to the lead agency or orally at a pubic hearing. This comment period lasts for 180 days from the time the draft is issued.
- *Response to comments.* All comments on a draft EIS must be addressed either by modifying an alternative, developing and evaluating additional alternatives, improving the analysis, making corrections, or documenting why no action was taken.
- *Final EIS* is the resulting document after all comments on the draft EIS have been addressed.

---

and final EIS. This must be approved by several agencies within the federal government, including those responsible for the funding of the project and others responsible for environmental policies and programs. Approval of the statement does not ensure that the project or program will go ahead without negative environmental impacts, but it does pressure the planning and construction agencies to be careful and responsible when taking actions that affect the environment. Tens of thousands of environmental reviews have taken place since the enactment of NEPA, including many dealing with highways and transit projects. The law has withstood many attempts to defeat it in court actions and efforts to weaken it through amendment, and a number of states have adopted similar legislation requiring environmental reviews of projects that do not depend on federal funding (Bear, 1989).

Clean Air Conformity Requirements

In no area have environmental concerns had more influence over transportation planning than the demanding requirements placed on transportation planning agencies with respect to air quality. Becoming gradually aware of the growing health problems associated with urban air pollution, Con-

BOX 6.5. NEPA Graphic

Project Development Process

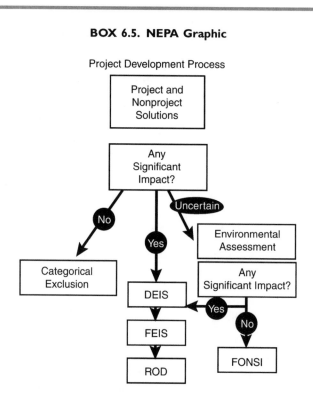

In preparing an Environmental Impact Statement, Categorical Exclusion or Environmental Assessments for projects, the responsible agency must consider all of the relevant socioeconomic and environmental impacts and pursue public involvement. Although the size and complexity of the three levels of NEPA documentation are different, they all serve the same purpose: to achieve better decisions by making the impact of choices known and by involving the public in making transportation decisions.

- Environmental Impact Statements (EIS) are prepared for federal actions that have a significant effect on the human and natural environment.
- Draft EIS (DEIS) and Final EIS (FEIS) are disclosure documents that provide a full description of the proposed project, the existing environment, and analysis of the anticipated beneficial and adverse environmental effects of all reasonable alternatives.
- Categorical Exclusions (CE) are prepared for federal actions that do not have a significant human and natural environmental effect.
- Environmental Assessments (EA) are prepared for federal actions where it is not clearly known how significant the environmental impact might be. If, after preparing an Environmental Assessment, it is determined that the project's impact is significant, an Environmental Impact Statement is then prepared. If not, a finding of "no significant impact" is documented.
- Record of Decision (ROD) is a concise decision document for an Environmental Impact Statement that states the decision (selected alternative or choice), other alternatives considered, and mitigation adopted for the selected alternative or choice.
- Finding of No Significant Impact (FONSI) is a statement indicating that a project was found to have no significant impacts on the quality of the human environment and for which an environmental statement will therefore not be prepared.

gress passed the Clean Air Act of 1955, a first step designed to foster cooperation between federal, state, and local governments. National concern over deteriorating air quality in many cities has led to substantial amendments of this act on five separate occasions since 1967 (Garrett & Wachs, 1996, chap. 1). The latest federal Clean Air Act Amendments (CAAAs) and the most recent federal transportation spending bills, ISTEA and TEA-21, require that air quality policies and transportation planning be more systematically integrated than ever before. Air quality laws have set standards limiting the production of several major pollutants by motor vehicles, and over time Congress has strengthened these standards and extended them to classes of vehicles, such as heavy trucks, that had previously not been regulated. Federal regulations also address air pollution by strictly regulating the content of fuels used by those vehicles. In combination, these regulations have induced technological changes in transportation that have substantially reduced air pollution even as the amount of vehicular travel has grown steadily (see Chapters 1 and 14).

Despite this improvement, many metropolitan areas fail to meet the National Ambient Air Quality Standards (NAAQS) specified in current legislation. Because "mobile sources" continue to account for a substantial proportion of the air pollution in most of those regions, it is necessary that air quality impacts be explicitly considered in regional transportation planning. This, in turn, requires agencies responsible for regional transportation planning to cooperate with those responsible for regional air quality attainment. Under federal air quality laws, regions that fail to meet air quality standards must produce plans (often called "State Implementation Plans," or SIPs) by which they propose to meet those standards within time periods specified in the CAAAs. And, because transportation systems are the source of a major proportion of the air pollution in those regions, these air quality plans often specify transportation

measures by which air quality should be improved. Transportation Control Measures (TCMs), such as those listed in Box 6.6, are explicitly enumerated in air quality plans for implementation in order to demonstrate that a region will bring its air quality into compliance with federal requirements.

In order to ensure that progress is made with respect to air quality, federal air quality and transportation laws require consistency between regional transportation plans and air quality plans in areas that have failed to attain the NAAQS. A complex "conformity process" is the mechanism by which consistency is achieved. In a "nonattainment area," the regional air quality planning agency prepares a plan by which it intends to meet the NAAQS by the date at which they are required to do so. The agency uses mathematical models that quantify relationships between travel volumes and pollutant emissions, and conducts analysis using other models of the dispersion of pollutants given local topographic and atmospheric conditions. If the plan is approved by state and federal air quality agencies based on their expectation that its implementation would bring about attainment of the standards, the plan is said to be a "conforming" plan. To achieve consistency between air quality and transportation planning, federal law mandates that the MPO and the regional air quality agency employ the same data to estimate air pollutant emissions. Very importantly, the transportation and air quality plans must be consistent with each other. The RTP must be consistent with the regional air quality plan, the Transportation Improvement Program (TIP) must include projects that are part of both plans, and pollution estimates associated with projects must be consistent in the two plans. To achieve conformity, the RTP and the TIP must conform to the air quality plan's purposes of eliminating or reducing the number and severity of violations of the NAAQS and achieving expeditious attainment of the standards. To achieve conformity, no transportation project or plan may

**BOX 6.6. Transportation Control Measures
Enumerated in the Clean Air Act Amendments of 1990**

- Programs to improve public transit;
- Restriction of certain roads or lanes to use by buses, vanpools, and carpools;
- Employer-based transportation management programs, including incentives for using alternatives to driving alone;
- Trip reduction ordinances;
- Traffic flow improvement programs that reduce emissions;
- Fringe and transportation corridor parking facilities serving high-occupancy vehicles and transit;
- Programs to limit or restrict the use of vehicles downtown and in other activity centers in which emissions are concentrated;
- Programs that promote high-occupancy vehicle and shared-ride services;
- Programs that limit portions of the road network to the use of nonmotorized vehicles and pedestrians at certain times;
- Programs that provide bicycle lanes and secure bicycle storage facilities in public and private areas;
- Programs to control the extended idling of vehicle engines;
- Reducing emissions from extreme cold-start conditions;
- Employer-sponsored programs to encourage flexible work hours;
- Programs and ordinances that facilitate travel by mass transit and alternatives to the automobile and to reduce travel by single-occupant vehicles at shopping centers, special events, and other centers of vehicular activity;
- Programs for new construction and major reconstruction of paths, tracks, or areas for the sole use of pedestrians and nonmotorized means of transportation.

be approved if it conflicts with the air quality goals of the SIP by causing new violations of the NAAQS, worsening existing violations, or by delaying timely attainment of the statutory deadlines (Garrett & Wachs, 1996, pp. 23–24).

The requirements linking transportation planning to air quality goals are complex, demanding, and intensely political. They require technical cooperation between air quality and transportation agencies, the sharing of databases, and agreement on mathematical models. Often, imprecise models and databases complicate this process even when all participants have the best of intentions. Transportation planners must evaluate long-range plans based on tons of air pollutant emissions, yet the models that are used to forecast travel and emissions often yield errors that are larger than the differences in emissions that would result from

the selection of different alternatives (Transportation Research Board, 1995). Also required are negotiation and agreement between planning and air quality agencies on planning strategies that will enhance mobility while addressing regional air quality needs. Furthermore, these negotiations must take place openly at public meetings in which competing constituencies argue aggressively for different regional solutions.

INCREASED FUTURE FOCUS ON ENVIRONMENTAL IMPACTS OF TRANSPORTATION

The requirement that transportation agencies anticipate and enumerate the many environmental impacts resulting from new projects and programs has, over the more than 30 years since the law was enacted, gradually but substantially changed the nature of transportation planning. At first, environmental impacts were recognized only indirectly and defensively by transportation authorities. The addition of 4(f) and NEPA requirements, and numerous law suits in which courts have upheld their intent and requirements, have empowered and emboldened environmental interest groups which now routinely play active roles in transportation planning. Regional, state, and federal transportation agencies today employ many environmental analysts and devote significant resources to studying and documenting the environmental impacts of potential transportation investments. The provision of mobility and the protection of the environment are sometimes complementary and sometimes conflicting, yet it is fair to say that transportation agencies no longer take environmental issues lightly.

For the last 30 years the single most pressing environmental concern affecting regional transportation institutions has been air quality at the local and regional levels. Emphasis has been placed on the reduction of lead, oxides of nitrogen, reactive organic gases, carbon monoxide, and particulates. While a great deal of progress has been made, the problem of improving urban air quality remains a serious concern for transportation officials. As the health impacts of pollutants have gradually become understood, emphasis has shifted. Fine particulates, for example, now receive increasing attention as their damaging effects on human health have been documented.

In part bolstered by a growing understanding of the role of transportation in achieving better urban air quality, other environmental concerns are just emerging that will certainly complicate the transportation planning process in coming decades. Society has begun to comprehend the risks of producing increasing quantities of carbon dioxide (CO_2) and other "greenhouse gases" and the fact that global climate change is a significant problem (see Chapter 9). Estimates show that transportation is responsible for as much as 20% of worldwide CO_2 and that motor vehicles in the United States account for 20–25% of worldwide transportation emissions of greenhouse gases that are produced by human activity rather than natural sources (Transportation Research Board, 1997, pp. 210–211). Based on global near-surface temperature measurements, temperatures are estimated to have increased by about 0.6 degrees Celsius during the 20th century. More importantly, perhaps, are observations that daily minimum temperatures are increasing at a rate approximately twice that of daily maximum temperatures. Measures of the extent of glaciers and snow cover reinforce concerns that these trends could require enormous changes in policy to fend off permanent irreversible damage (Easterling, 2002).

Transportation systems can dramatically alter the nature of water systems by altering the courses and volumes of flows in waterways and can change natural drainage patterns. Vehicles, highways, and transit routes are also sources of substantial amounts of liquid, solid, and gaseous pollutants that can settle on water surfaces and be carried as runoff into watercourses. The U.S. Army Corps of Engineers has jurisdiction over navigable waterways in the United States and must review and provide permits for transportation projects that will affect the character and content of flows on those waterways. Also, provisions of the Clean Water Act and regulations of the U.S. Fish and Wildlife Service often do limit the routing and design of transportation facilities (Wachs, 2000, p. 107).

Transportation facilities can substantially affect fragile habitats, for example, by placing barriers between sheltered habitats and sources of food and water, disturbing animal or insect migration routes, and polluting local watercourses. Road salt, used in colder regions to make roads more passable in winter by melting ice and snow, can damage certain plant, animal, and insect species. A recent survey of state highway and transportation departments revealed that biodiversity had been raised by participants in the transportation planning process in 21 of the 30 responding states (Herbstritt & Marble, 1996).

Regional transportation agencies are recognizing that the planning challenge of the 21st century will be the provision of improved accessibility and mobility while making transportation systems more "sustainable." Sustainability will require that we not present future generations with declining quality of life as a result of depleted energy reserves and intractable pollution problems. Gradually, efforts are underway to merge transportation planning with regional land use planning. Starting with inventories of fragile habitats and species, planners are increasingly able to derive planning requirements for land conservation and protection simultaneously with those for mobility. Although our ability to foresee the future, respond to the past, and manage human activity in the present is always limited by imperfect collective human understanding, transportation will be at the center of global politics because it plays a critical role in every society and is essential to our individual and collective well-being.

TRANSPORTATION POLICY IN THE 21ST CENTURY

Transportation planning is a technical and professional activity embedded in a constantly changing context of local, regional, state, national, and global politics. In the 21st century, the principal mission of transportation policy will continue to be the enhancement of mobility of people and goods in pursuit of a wide variety of economic and social purposes.

As has been the case in the past, the nature of mobility will constantly evolve in response to changes in technology. It is reasonable to expect fundamental shifts in the technology of fuels and propulsion systems, materials that comprise vehicles, and advances in telecommunications that change the patterns by which people interact with

one another over space, and consequently the way we organize and carry out travel by all modes. As has always been the case, technological change will continue to have intended and unintended consequences, some of which will be beneficial and some of which will be harmful.

Transportation planning at the regional level traditionally has emphasized the construction of facilities and the analysis of the environmental impacts of their construction and of the forecasted flows on those facilities. It is likely that in the coming decades transportation plans will consist to a decreasing extent of facilities plans and that planning goals will increasingly be pursued through a mix of policies that are much broader and more focused on management than on facility construction. For example, strategies that include the promotion of higher density residential communities and mixed land uses in residential and commercial areas are being promoted as environmentally responsible and efficient. Some believe that these approaches can reduce the geographic expansion of metropolitan areas, reduce the pace of urbanization of agricultural land, promote more walking and transit use, and require less automobile travel.

Increasing reliance on telecommunications and the complementarities between telecommunications and travel will permit a wider variety of living environments (see Chapter 4). Telecommunications innovations are already influencing transportation plans through the addition of "intelligent transportation systems." Electronic toll collection, electronic information about current traffic conditions that guides travelers toward less-congested routes, information on the arrival time of the next bus or train, and advanced information about the current availability of parking spaces at the destination are examples of improvements in transportation system management that result from more ubiquitous and reliable innovations in telecommunications. Over time, we can expect regional transportation planners to invest increasingly in such management tools because they provide many of the benefits of increased system capacity without some of the negative social and environmental consequences of capacity expansion.

Transportation is critical to citizens who carry out their daily activities and to communities that must balance the needs of different social and economic groups. Increasingly, the scope of economic and environmental concerns is becoming global, and transportation plays a central role in reducing the separation between people and places throughout the world and in addressing the environmental and sustainability concerns that have also become central on the world stage. Transportation planning will continue to be an intensely political meeting ground on which the efforts of many different public and private interests will intersect.

REFERENCES

Altshuler, A. A. (1965). *The city planning process: A political analysis.* Ithaca, NY: Cornell University Press.

Bear, D. (1989). NEPA at 19: A primer on an "old" law with solutions to new problems. *Environmental Law Reporter, 19,* 10060–10069.

Easterling, D. R. (2002, October 1–2). *Observed climate change and transportation.* Paper presented at the Federal Research Partnership Workshop, Brookings Institution, Washington, DC.

Federal Highway Administration, U.S. Department of Transportation. (1989). *Section 4(f) policy paper.* Available online at http://fhwa.dot.gov/environment/guidebook/vol2/4fpolicy.htm

Federal Highway Administration, U. S. Department of Transportation. (1999, April). *The Congestion Mitigation and Air Quality (CMAQ) Program under the Transportation Equity Act for the 21st century (TEA-21): Program guidance.* Washington, DC: Author.

Forkenbrock, D. J., & Schweitzer, L. A. (1999). Environmental justice and transportation planning. *Journal of the American Planning Association, 65,* 96–111.

Gakenheimer, R. A. (1976). *Transportation planning as response to controversy: The Boston case.* Cambridge, MA: MIT Press.

Garrett, M., & Wachs, M. (1996). *Transportation planning on trial: The Clean Air Act and travel forecasting.* Thousand Oaks, CA: Sage.

Greig, J., Cairns, S., & Wachs, M. (2003). *Environmental justice and transportation: A citizen's handbook.* Berkeley: University of California, Berkeley, Institute of Transportation Studies.

Heightchew, R. E. (1979). TSM: Revolution or repetition? *ITE Journal, 48*(9), 22–30.

Herbstritt, R. L., & Marble, A. D. (1996). Current state of biodiversity impact analysis in state transportation agencies. *Transportation Research Record, 1559,* 51–63.

Humphrey, T. F. (1974). Reappraising metropolitan transportation needs. *Transportation Engineering Journal, Proceedings of the American Society of Civil Engineers, 100*(TE2).

Meyer, M. D., & Miller, E. J. (2001). *Urban transportation planning: A decision-oriented approach* (2nd ed.). New York: McGraw-Hill.

O'Toole, R. (2001). *The vanishing automobile and other urban myths: How smart growth will harm American cities.* Banden, OR: Thoreau Institute.

Rusk, D. (2000). Growth management: The core regional issue. In B. Katz (Ed.), *Reflections on regionalism* (pp. 78–106) Washington, DC: Brookings Institution Press.

Transportation Research Board. (1995). *Expanding metropolitan highways: Implications for air quality and energy use* (Special Report No. 245). Washington, DC: National Research Council.

Transportation Research Board. (1997). *Toward a sustainable future: Addressing the long-term effects of motor vehicle transportation on climate and ecology* (Special Report No. 251). Washington, DC: National Research Council.

Wachs, M. (1995). The political context of transportation policy. In S. Hanson (Ed.), *The geography of urban transportation* (2nd ed., pp. 269–286). New York: Guilford Press.

Wachs, M. (2000). Linkages between transportation planning and the environment. In *Refocussing transportation planning for the 21st century, conference proceedings, 20,* 102–112. Washington, DC: Transportation Research Board.

Wachs, M., & Dill, J. (1999). Regionalism in transportation and air quality: History, interpretation, and insights for regional governance." In A. Altshuler, W. Morrill, H. Wolman, & F. Mitchell (Eds.), *Governance and opportunity in metropolitan America* (pp. 296–323). Washington, DC: National Academy Press.

Weiner, E. (1997). *Urban transportation planning in the United States: An historical overview* (U.S. Department of Transportation, Office of the Assistant Secretary for Transportation Policy, Report DOT-T-97-24). Washington, DC: U.S. Department of Transportation.

Yaro, R. D. (2000). Growing and governing smart: The case of the New York region. In B. Katz (Ed.), *Reflections on regionalism* (pp. 43–77). Washington, DC: Brookings Institution Press.

CHAPTER 7

GIS in Urban–Regional Transportation Planning

TIMOTHY L. NYERGES

WHY BOTHER USING A GIS IN TRANSPORTATION PLANNING?

As discussed in previous chapters, laws passed by the U.S. Congress and several states since 1990 have mandated a more systematic and comprehensive approach to transportation planning, programming, and operations than in previous decades. Among such laws at the federal level are the Intermodal Surface Transportation Efficiency Act of 1991 (ISTEA), the Clean Air Act Amendments of 1990 (CAAAs), the Americans with Disabilities Act of 1990 (ADA), and the Transportation Equity Act for the 21st Century of 1998 (TEA-21). In the State of Washington, relevant laws include the Growth Management Act (GMA) of 1990 and revisions to that act passed in 1991, 1995, and 2001. TEA-21 mandates that every metropolitan area in the United States organize, in cooperation with state transportation agencies, a Metropolitan Planning Organization (MPO) to coordinate plans, programs, and projects within a region. Such coordination is required if the MPO is to be eligible to receive federal funds for transportation improvement in the region. In Washington State, the GMA mandates that certain contiguous counties growing in population must organize, in cooperation with the Washington State Department of Transportation, a Regional Transportation Planning Organization (RTPO) to coordinate plans, programs, and projects across the counties to be eligible to receive state funds for transportation improvement. To avoid duplication, the Puget Sound Regional Council (PSRC) is both the MPO and the RTPO for the central Puget Sound region of Washington State with lead responsibility to coordinate transportation improvements across King, Kitsap, Pierce, and Snohomish Counties (see Figure 7.1). This region is coincident with the Seattle metropolitan region of central Puget Sound and is the focus of the regional, county, and city transportation planning case studies presented in this chapter. Under GMA planning, the PSRC coordinates county plans and the counties coordinate city plans within their boundaries; thus the GMA motivates an across-scale look at transportation planning.

Although transportation concerns have always influenced and been influenced by economic, social, and environmental con-

FIGURE 7.1. Four-county jurisdiction of Puget Sound Regional Council in Washington State.

cerns, federal legislation and state legislation within the State of Washington now mandate explicit consideration of those influences as part of transportation planning. A geographic information system (GIS) is an effective way to integrate the information needed to support the many facets of transportation planning. The principal reason for the effectiveness is that a GIS helps capture, store, analyze, and display geographical information based on the locational, descriptive attribute, and temporal character of phenomena; all are significant to realistic description of transportation information. More efficient, effective, and equitable use of transportation information can promote connections among transportation plans, programs, and operations at multiple scales.

This chapter presents GIS used in transportation planning within organizations at three scales of governance: (1) Puget Sound Regional Council; (2) Metro King County Department of Transportation, Road Services Division; and (3) City of Seattle Department of Transportation, Major Projects and Neighborhood Planning Division. We focus on the similarities and differences in long-range transportation planning based on this difference in geographic scale.

We also treat transportation improvement programming (i.e., transportation project selection, which is also called "short-range planning") because readers should be aware of how long-range decisions relate to decision making in the shorter term.

We begin by defining "GIS," or "GIS for transportation (GIS-T)," hereafter simply referred to as GIS. A frequent complaint from GIS novices is that there is no simple definition of a GIS. Simple definitions tend to emphasize one aspect over another, providing a singular perspective, whereas a multifaceted one is needed. A comprehensive definition is the following:

A GIS is a combination of data, software, hardware, personnel, procedures, and institutional arrangements (Dueker & Kjerne, 1989) meant to capture, store, manipulate, analyze, and display spatially oriented information for addressing complex planning and management tasks (Federal Interagency Coordinating Committee on Digital Cartography, 1988), as part of the policymaking, planning, design, construction, maintenance, management, operations, and evaluation of transportation systems. (Nyerges & Dueker, 1988)

The length of the definition speaks to the sophistication, as well as the potential, of the technology. Embedded in that definition are the following three perspectives, each providing an important key to understanding why a GIS can be so useful:

1. The basic components (generally speaking) of a GIS are *data, software, hardware, personnel, procedures,* and *institutional arrangements.* Each of these six components interacts in various ways that can enhance information flows within and among organizations.
2. The above components, and especially the softwares, are *meant to capture, store, manipulate, analyze,* and *display spatially oriented information.* When data are processed in a broad-based, meaningful context, a GIS information user can better understand issues in complex problems.
3. Lastly, and perhaps most importantly, the purpose of a GIS is to *address complex tasks in policymaking, planning, design, construction, maintenance, management, operations,* and *evaluation of transportation systems.* The purpose of a GIS should match the mandates and missions of organizations that implement them; hence GIS should provide information to organizations responsible for addressing concerns about transportation efficiency (as in ISTEA), effectiveness, and equity (as in TEA-21).

The above definition indicates a wide spectrum of possibilities. Several recent publications treat some aspect of GIS in transportation planning. In a special issue of *Transportation Research, C: Emerging Technologies,* Thill (2000) assembled 22 articles to elucidate a wide range of transportation research topics about transportation modeling and GIS, some six of which deal at least indirectly with planning. In a textbook about urban transportation planning, Meyer and Miller (2001) provide insight about the decision phases of transportation planning and describe several places wherein GIS can be

helpful. Miller and Shaw (2002) provide a comprehensive overview of the concepts, methods, and techniques about GIS for transportation and characterize planning as being among the more important application topics. Some of those GIS applications are being used to foster public involvement in the planning process in face-to-face public meetings (Craig, Harris, & Weiner, 2002; National Research Council, 2002; Talen, 2000). In addition, some of those applications are being redeveloped as WebGIS (i.e., accessed through a standard World Wide Web browser) and Internet GIS (i.e., developed with a customized browser) applications to foster citizen (public) participation in planning (Peng, 1999, 2001). There are many "publics" with rather diverse interests (Taylor, 1998) to be considered in transportation planning (Smith, 1999).

Taken together, those developments and uses of GIS are broadening availability and access to transportation information, as the usefulness of GIS technology grows to support transportation planning. Unfortunately, only certain aspects of GIS in transportation planning can be treated within the space of this chapter. We focus on long-range plan making because it is one of the most widespread applications of GIS and transportation. Despite this focus, the case studies presented later in this chapter will show that GIS use, even within this limited focus, can be different because of scale effects.

As described in Chapter 5, transportation planning involves four decision-oriented stages: visioning, plan making, programming, and system monitoring (Meyer & Miller, 2001). *Visioning* deals with establishing goals and objectives for what a community (taken broadly) might want in regards to a future transportation system over the long range. Visioning within transportation organizations has taken the form of "sketch maps," that is, expressions of future preferred scenarios of land use and/or the built environment. *Plan making* (called "debate and choice" by Meyer and Miller [2001]) refers to the process of creating a map (or maps) of future transportation needs. These maps are educated guesses (called "forecasts") of what the transportation needs of the future might be—or what we might like them to be—given certain socioeconomic and land use–driven population trends over a 20- or 30-year time frame. *Programming* identifies a select set of projects rotating in 2-year increments of scope, design, and construction, hence extending over a 6-year time frame, for which transportation investment monies are directed. The resulting set of projects is called a "capital improvement program." Capital improvement programming is viewed as implementing long-range plans. *Monitoring* deals with reflecting on system performance, and provides feedback for plan making and programming. Most GIS applications in transportation planning have involved plan making; the material in this chapter reflects that emphasis. Visioning, programming, and monitoring can each take advantage of a geographic perspective, however, and represent considerable potential for GIS work as well. A major goal in using GIS for transportation data management, analysis, and display is to provide transportation planners, policymakers, decision makers, citizen groups, and the general public with easier access to information about important geographic concerns as part of the transportation planning process.

This chapter is intended to give you an appreciation for multiscale differences in GIS use in transportation plan making. First, a review of some fundamental technical and institutional issues introduces you to GIS for transportation. Then I describe how GIS has been used in a growth management transportation planning context in Washington State at three scales: the Puget Sound Regional Council, Metropolitan King County, and the City of Seattle. This chapter concludes with a discussion of fundamental potentials and challenges for GIS use in transportation planning.

GIS AS A MIX OF TECHNICAL AND INSTITUTIONAL ISSUES

A number of technical advantages motivate use of GIS in transportation planning. GIS provides an opportunity to broaden and deepen investigation into transportation concerns through data-processing methods not previously possible. A GIS can integrate the locational, temporal, and descriptive attribute characteristics of transportation activity in terms of a data-structuring concept called "data layers" (Sinton, 1978) and more recently "object categories" because of newer technologies (Miller & Shaw, 2002). The term "data layer" originates from the process of laying plastic map layers on top of one another, a process that GIS makes easier because the layers are digital rather than plastic. Consequently, we can examine the relationships among data layers (object categories), such as transportation networks, land uses, water resources, and habitat, in a variety of ways. A highway network data layer can be examined in relation to a land use data layer that generates vehicular movement over that network. In addition, that vehicular movement contributes to air emissions and deposition of oil and gas on the highway. Runoff from highways could find its way into streams and lakes, potentially damaging the habitat of wildlife and fish. The data must share a common spatial and temporal framework in order for the information to be meaningful—that is, spatial coordinate registration is required to put one layer on another and to match locations. Each relationship between phenomena mentioned above can be the basis of a GIS application. From where do these advantages derive? Is GIS only a tool, or is it a technology that facilitates a way of thinking that will help us address issues that in the past were too complicated to consider?

Fundamentally, GIS software technology is an integration of three software technologies: (1) data management for large stores of data, (2) spatial analysis to derive relationships among those large data stores, and (3) map visualization software to provide renderings of complex relationships made easier to understand through a map. *Data management* provides for incremental development of data layers. With every application involving several data layers, an organization can accumulate an enterprise-wide perspective on data, with each unit of the organization contributing to and drawing from a common data store (often called a "data warehouse"). The *data analysis* capability allows us to draw upon those data layers in different combinations, enabling exploration of relationships among phenomena not previously investigated. The *map visualization* capability is used for generating map displays in shorter times than was previously possible, so audiences have an opportunity to gain broader and deeper geographic perspectives. These three software technologies taken together are what make GIS so powerful. Without advanced data management, a GIS would only be transportation modeling software. Without analysis capabilities, a GIS would only be computer-mapping software. Without visualization capabilities, a GIS would only be analytical software portraying numbers from tables.

Communication and decision analysis software technologies have recently been added to the capabilities of GIS to support groups of people working with GIS data. Adding *communications technology* provides a basis to support dialogue about maps, as maps are conversation generators. Adding *decision analysis technology* provides enhanced capabilities for decision support, although a GIS has been touted as a decision support system for some time now (Cowen, 1988). The five software technologies are essential in combination to support data integration to set priorities in addressing transportation concerns. In this chapter, however, we treat only the first three of these technologies in the case studies, as these three are in common use. The other two are relatively new

research topics in GIS (see Jankowski & Nyerges, 2001; Nyerges, Montejano, Oshiro, & Dadswell, 1998) not yet fully integrated into commercial GIS.

What is not new with GIS? The technology is by no means a panacea for transportation issue exploration, assessment, and evaluation. Problem formulation is still as complex as ever, and few suggestions have been put forth in GIS for addressing the complexity of problem scoping. GIS must be able to address traditional information needs in new ways to make the implementation of the technology worthwhile. For example, GIS must address plan making in the traditional transportation planning process, as in accessibility for home-based and non-home-based trips; but a GIS should also be capable of providing the detailed character of those trips as in activity-based modeling (Kwan, 2000). In addition, a GIS must be able to incorporate other variables into the analysis process (e.g., air and water quality concerns), as well as to deal with intermodal issues for accessibility (Miller & Shaw, 2002). Furthermore, a GIS should be useful from a policy perspective—that is, it should be able to facilitate the capture, manipulation, and presentation of data for variables that are relevant to decision makers who set policy for directing urban–regional transportation plans and improvements.

Common to the software technologies in GIS is the need for data representation. Each of the technologies provides a way of representing data as part of data-processing activity. Data become information when they are processed in a meaningful context; but not all people or organizations use the same reference frameworks (whether in their heads or in computers) to make data meaningful in the same way. One of the basic challenges in a GIS is to translate the language of transportation issues, problems, and analysis into a GIS language within an organizational (social) context. One does this through the basic *geospatial data con-*

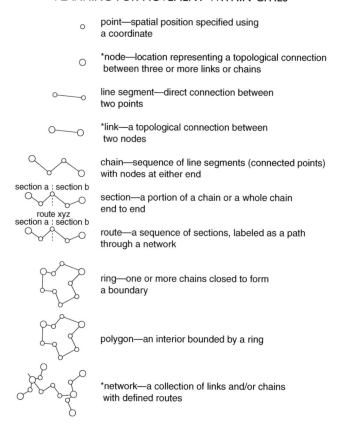

- point—spatial position specified using a coordinate

- *node—location representing a topological connection between three or more links or chains

- line segment—direct connection between two points

- *link—a topological connection between two nodes

- chain—sequence of line segments (connected points) with nodes at either end

- section—a portion of a chain or a whole chain end to end

- route—a sequence of sections, labeled as a path through a network

- ring—one or more chains closed to form a boundary

- polygon—an interior bounded by a ring

- *network—a collection of links and/or chains with defined routes

FIGURE 7.2. Geospatial data constructs commonly found in a GIS for transportation.

structs, such as those listed in Figure 7.2 (in part based on National Institute of Standards and Technology, 1994). The constructs in Figure 7.2 are the *data-structuring* component of software programs. The geospatial constructs are only the spatial dimensions for representing transportation phenomena; attribute and temporal dimensions are also important. Spatial, attribute, and temporal dimensions together form the basic elements of what become transportation networks, residential sites, employment sites, and shopping sites enumerated in terms of "worker activity" that connect residences and employment sites, with shopping stops in-between. People within organizations mediate these contexts when they identify transportation problems and the potential solutions for addressing those problems.

For years, forecasting transportation demand has been thought of as a four-step process: estimate demand through trip productions between origins and destinations, identify modal split for the trips, assign mode split to network links, and balance distribution of trips across the network links (see Chapter 5). Transportation demand modeling software—for example, a package called EMME/2 used by the agencies mentioned in this chapter—has been developed to address those steps, but not much else. One major difference between GIS and transportation modeling software is that in order to process the data efficiently, the modeling software simplifies the spatial representations of the world more than does a GIS. The constructs in Figure 7.2 preceded by an asterisk are those commonly found in travel demand forecast modeling software, whereas all of them are in a GIS. It is worth noting that many travel demand modeling software packages do not contain a construct for polygons—for example, for representing traffic analysis zones (TAZs). Designers of modeling software have assumed that trip productions for origins and destinations in TAZs are coded to the center points (centroids) of a zone; hence there is no need for the polygon construct. To GIS designers and analysts, this is a rather limiting assumption, constraining opportunities for analysis. Nonetheless, there is a need for both GIS and transportation modeling software, as a GIS supports the data management capability of transportation modeling software (McCormack & Nyerges, 1997).

Before 1990, most GIS software could not represent transportation (linear) features effectively because GIS software vendors had recognized only one of two important spatial referencing systems in transportation applications (Nyerges, 1990). In addition to a two-dimensional spatial reference system (such as a latitude–longitude coordinate system or a state plane coordinate system), transportation data require a linear reference system—for example, addresses along a street, mile points along a state route, or time points along a transit route—in order to represent distance along a route, and subsequently represent that same distance as stored along connected chains in a database. Although spatial data are the core of a GIS, the spatial data must be linked to attribute data to realize the GIS advantage. *Attribute data* are those characteristics of phenomena other than space and time. For example, the attribute data of a roadway route include average daily traffic volumes, modes, highway capacities, traffic crashes, and water volume carried through culverts. Linear referencing of data allows for the nonredundant storage of transportation attributes along routes or streets. For example, lane width and speed limit are continuous dimensions

of roadways. Without a linear referencing strategy, a separate segmentation of roadways (i.e., network lines) would be needed to represent each of the attribute data values along the roadway, because speed limit and lane width change at different locations. For example, if you have 100 or more such different attributes, you would need 100 or more different line networks, with each network segmented according to the character of attribute data values associated with sections of roadway. To eliminate the redundant storage of lines, a linear referencing strategy associates reference points to the road (line). Each reference point has attached to it as many attributes as are necessary to describe the changes along the road at that point: number of lanes, surface paving material, speed limit, or the like.

In addition to space (location) and attributes, time is the third aspect in geographic data management. At present, most commercial GIS handle time inadequately for most applications. Interest in representing time in a GIS has become a research topic of significance over the past decade, but little had been done to implement a general GIS approach (Koncz & Adams, 2002; Langran, 1992). To date, time has been handled as an attribute, without a robust means to support efficient data representation, data manipulation, and analysis. Thus entirely separate databases have been stored for different time periods (e.g., 1980, 1990, 2000), making it difficult to determine how the network has changed. A linear referencing system together with dynamic segmentation processing can be applied to such a problem. The temporal issue, making this a multidimensional problem about spatiotemporal data structuring, has been outlined conceptually (Koncz & Adams, 2002) and is at the beginnings of an implementation in object relational database structures such as UNETRANS (Environmental Systems Research Institute, 2002). UNETRANS is a transportation object model developed by a consortium of GIS users for the ArcGIS geodatabase that promises to bring together space, attribute, and time enhancements.

Although these technical issues are often recognized as the main issues of GIS implementation, institutional and cultural issues are important also when working with a GIS (Chrisman, 1987). Of concern are the ramifications of introducing information technology into a work context and the concomitant social impacts on an organization. GIS engenders a cooperative atmosphere because its core lies in data integration around the concept of location. But cooperation has not always been sought out because of turf battles among groups when a common topic arises. With GIS, work relations change with changes in the way information is treated. Many organizations (people) are reluctant to change unless there is a clear benefit in doing so. Organizations are increasingly recognizing that sharing data within and between organizations can yield cost savings and enhance the effectiveness of analysis. GIS tends to get people to work together because of the need to develop comprehensive databases to suit ever more comprehensive transportation analysis built upon the common ground of the spatially referenced database. The institutional and social issues related to GIS are growing in significance as the diffusion of the technology proceeds. Such concerns have broadened to social implications for society (see Pickles, 1995, for an extensive discussion).

Having introduced a few general technical and organizational issues related to GIS use, we now describe three case studies as particular applications. The following three case studies are differentiated primarily in terms of geographic scale (i.e., region, county, and city).

USING GIS IN URBAN–REGIONAL TRANSPORTATION PLANNING AT THREE SCALES

How is GIS being used to support urban–regional transportation planning? To answer this question, we explore three case studies

describing GIS use in the central Puget Sound (Seattle metropolitan) region. The three case studies are at regional, county, and city (together we call them "local") transportation jurisdiction scales, in which transportation agencies work independently and together in growth management planning. Although this context is idiosyncratic to the Seattle, Washington, metropolitan region, the multilevel jurisdictional scales are common in many other areas of the United States, but of course the organizational contexts differ. Two other jurisdictional scales—the state and the federal levels—have a major impact on transportation at the local level, particularly through funding; but they are not considered here at the local level. In addition to the three jurisdictional scales treated here, we identify six functional scales: intersection, link, corridor, subarea, urbanwide (citywide), and regionwide. These functional scales were identified as part of an information needs analysis conducted for metropolitan planning organizations in Washington State (Nyerges & Orrell, 1992).

We treat three jurisdictional scales here to point out some of the transportation planning coordination challenges that can be addressed by implementing a GIS. These challenges have both policy and technical perspectives. From a policy perspective, the three scales challenge organizations to identify and agree on the levels of impact likely to occur with implementation of particular policies, for example, regional projects with metropolitan impacts versus local projects with (mostly) neighborhood impacts. But, of course, even regional projects have neighborhood impacts. Consequently, coordination among the scales is rather important. Under GMA planning, the Puget Sound Regional Council has responsibility for checking the consistency among the four county plans within its state RTPO (federal MPO) jurisdiction. King County is responsible for checking the consistency of the cities' plans within the county jurisdiction. Such a workload would be next to impossible to perform if it were not for GIS. From a technical perspective, organizations across the three scales share data to implement plans as policy directions. Consequently, the technical capability of GIS facilitates policy coordination.

Although GIS has been used for several years to foster coordination among scales, challenges remain. One is the growing recognition that plan making and programming (the two major work activities in planning) treat the concept of "transportation project" differently. Not only is this a problem *within* each jurisdictional scale, but it is even more of a problem *across* jurisdictional scales. Different organizations conceptualize the planning process within different geographic domains, even though the domains are overlapping. Long-range plans contain the concept of a project in a general sense with general impacts, whereas programming requires more specificity to implement an "improvement project." In part, the growing recognition about project inconsistency stems from differences between GIS databases for long-range planning and those for short-term improvement programming. That is, the same projects are not described in the same way across databases because the databases are developed under different mandates and by different units within transportation organizations. This problem will become increasingly apparent as organizations implement GIS applications for transportation improvement programming.

Considerable potential exists for GIS applications in transportation programming, beyond the mere posting of maps of improvement projects on the World Wide Web (WWW), but these applications are not in common use. Prototype software has been implemented to show that it is feasible to use group-based GIS for decision support in transportation programming (Jankowski & Nyerges, 2001; Nyerges et al., 1998). Those demonstration applications were implemented with the data from the transportation agencies mentioned in the case studies below. The challenges in transportation planning that develop due to different organizations implementing GIS and different scales underlay much of the differences in the three cases described in the following subsections.

Transportation Planning at the Puget Sound Regional Council

The Puget Sound Regional Council (PSRC) is a regional planning body with membership including King, Kitsap, Pierce, and Snohomish Counties, plus 68 of the 82 cities and towns in the region.[1] In 2000, the population of the four-county region was 3,275,800, with an area covering 6,295 square miles (Puget Sound Regional Council, 2002). The population in 2030 is expected to be more than 4.5 million, for a 40% 30-year growth of over 1 million residents. Growth in average weekday vehicle miles traveled (VMT) has been outpacing population growth over the past 20 years; during the 1980s growth in VMT was 5.3% greater, but in the 1990s was only 0.8% more. Now the difference in growth rates is a couple of tenths of a percent (1.6% for VMT versus 1.8% for population). Because the base upon which that rate applies has grown, however, substantial traffic congestion will likely continue. As noted in the *Regional Trend* newsletter, "The construction of new roads will barely keep pace with this growth, even if the current VMT rate can be maintained. Congestion at current levels or higher will continue into the foreseeable future" (Puget Sound Regional Council, 2002, p. 1). Such a trend is most likely going to keep the central Puget Sound region on the list of top-10 traffic-congested areas nationwide. Consequently, regional planning in general and regional transportation planning in particular has taken on more significance than ever before, and GIS is playing a major role in this planning.

PSRC, as the MPO and RTPO, coordinates local government transportation plan making by approving (or not) the four county plans. Consequently, the scale of transportation plan making in the organization is regionwide. The principal difference in the transportation plans created by PSRC in relation to those of King County and the City of Seattle is that of scale. The county and the city examine the transportation needs of their respective areas by focusing on specific projects within their jurisdictions or cross-jurisdictions. The PSRC attempts to balance transportation needs across the four-county area, and hence employs a broad, coordinating perspective.

As previously mentioned, TEA-21 and GMA legislation mandate the MPO and RTPO responsibility, respectively, of the Puget Sound Regional Council. That responsibility includes both regional transportation plan making and programming. A regional transportation plan-making process generally follows that outlined by Johnston (Chapter 5, this volume), involving preanalysis scoping, technical analysis, and postanalysis plan selection phases. These phases are at the core of a process used to create and adopt a metropolitan transportation plan (MTP).

Transportation modeling is the technical analysis underpinning of transportation plan making. The three levels of jurisdictions—regional council, county, and city—follow a version of the four-step urban transportation modeling system (UTMS) process described in Chapter 5 (Johnston, this volume). However, the regional planning process differs from the traditional view of transportation modeling in three important ways that relate to policy considerations:

1. Explicit links to land use characterization at a disaggregate level, in terms of both population change and economic activity, are included. The latter issue includes a concern for transportation demands from truck and tourism traffic in addition to typical employment or journey-to-work concerns. In addition, nonmotorized (pedestrian and bicycle) modes are included in the model. In the traditional models, land use categories were fairly broad and special transportation activities were usually not considered.

2. Air and water environmental criteria as well as concurrent developments are now included, in addition to the common financial, safety, and system performance criteria usually used in planning.

3. The analysis phase is explicitly linked with system improvement planning, least-cost financial planning and, at a policy level, with transportation system needs. This linkage implies that the planning process is iterative rather than linear. It also suggests that GIS technology has an integral role in supporting interactive problem solving and decision making.

Every MPO receiving federal funds is responsible for developing a 20-year (or more) MTP. Destination 2030 (Puget Sound Regional Council, 2001), the MTP adopted on May 24, 2001, for the central Puget Sound region, contains a GIS map to overview its intent (see Figure 7.3). The MTP contains information about the regional growth strategy symbolized on the map in terms of urban growth areas and urban centers (21 in mid-2002). The idea of urban growth areas/centers is articulated in VISION 2020. VISION 2020, adopted in 1990 and updated in 1995, is the long-range growth management, economic, and transportation strategy for the central Puget Sound region. The urban growth areas and urban centers are areas were development is being explicitly encouraged through policy, and where significant transportation improvements are likely to be encouraged as well. The MTP is the transportation component of VISION 2020. Because of the causal (and circular) links between land use and transportation, these links must be considered explicitly if a long-term growth strategy is to be successful. More specifically,

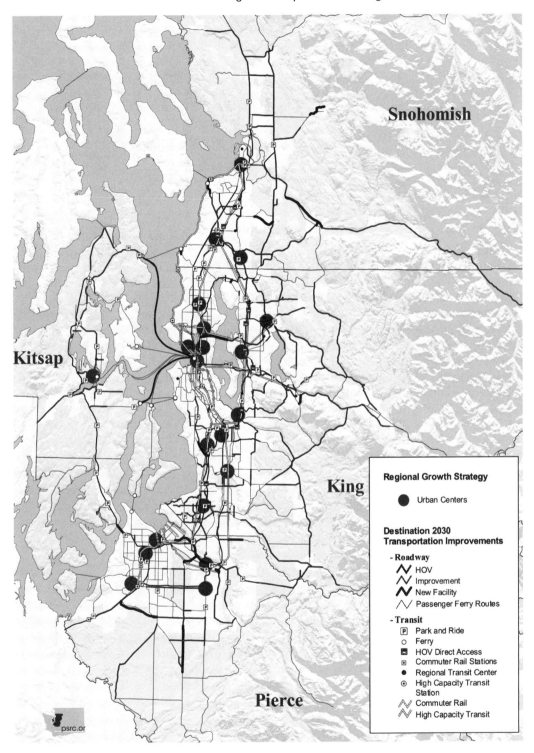

FIGURE 7.3. Overview map for Destination 2030, the Metropolitan Transportation Plan (MTP) for the central Puget Sound region. Source: Puget Sound Regional Council (2001).

the MTP overviews what and where the roadway and transit improvements will enhance the flow of people and goods among urban areas and centers in the four-county region.

The MTP map (the original is in color) shows that the region is focusing considerable effort on high- occupancy vehicle (HOV) lanes. Vehicles with at least three people in the high-density parts of the region, and at least two people in the medium-density parts of the region, are considered "high occupancy." Both roadway improvements and new roadway facilities are also designated on the map.[2] Because of the amount of navigable water in the region, ferries are an important mode, and thus they are depicted as part of the intermodal network.

The second major category of improvements on the map is transit. Among these improvements are park-and-ride lots, ferry terminals, HOV direct access points, commuter rail stations, regional transit centers, high-capacity transit stations, the commuter rail lines, and high-capacity transit. At the current time a considerable effort is underway to route a high-capacity transit (light-rail) line through the core area of the region. Debate centered on the length of the rail line (i.e., north and south of downtown vs. just south of downtown) and how useful the lines would be at mitigating congestion in those corridors. Nonetheless, those planned improvements are moving forward.

PSRC has used an aggregate approach in travel demand modeling, which is the analytic core of the plan development process. Although a change to activity-based transportation modeling is underway, and not yet in place, we describe the four-step UTMS process that was used to create the forecasts for the current MTP.

The basic steps in UTMS at the PSRC are supported by GIS, as depicted in Figure 7.4. The information technology used is a combination of transportation modeling software, GIS, and desktop mapping capabilities; as the databases get larger, desktop mapping capabilities are being moved to

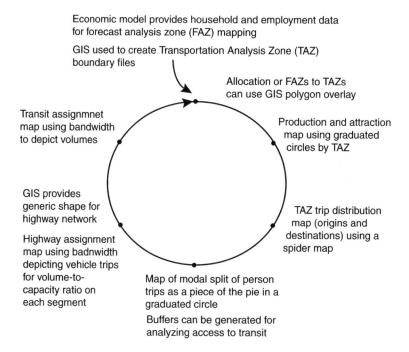

FIGURE 7.4. GIS support for regional travel demand forecast modeling.

GIS. The regionwide perspective makes use of a considerable amount of data. Data management would be difficult under current staffing conditions without the use of a GIS.

The transportation analysis process for long-range plans starts out with an econometric model to provide forecasts of household and employment growth on a regionwide basis for the years 2010, 2020, and 2030. This is the land use-activity system step (described by Johnston, Chapter 5, this volume). After the growth estimates are made, they are allocated to forecast analysis zones (FAZs) based on the previous decade's allocations, the new transportation accessibilities, and the acres of land devoted to current land uses. FAZ boundaries coincide with census tract boundaries available from the Census Bureau (U.S. Bureau of the Census, 2000). The historical FAZ data set has been developed over the years from surveys of local communities. Lately, PSRC has made use of U.S. Geological Survey land use and land cover data to verify the current land use data set. More detailed data set development is underway in the form of county assessor data for land parcels. Thus GIS will continue to play a major role in managing the data and editing the data for inclusion in transportation models.

Allocations to 219 FAZs are performed using EMPAL (EMPloyment Allocation Module) and DRAM (Disaggregate Residential Allocation Module). The employment forecasts are assigned to the FAZs based on employment growth. The household forecasts are assigned to the individual FAZs for residential growth. A GIS is used to create FAZ thematic maps from output of DRAM and EMPAL for residential and employment growth, respectively.

TAZ boundaries delineate relatively homogeneous areas of employment and residence based on data for census block groups, which are smaller than the FAZs. The 956 TAZs (938 internal to the four counties and 18 large external zones) were developed manually, using maps that overlay one another to split census tracts along block boundaries. A GIS is used to convert the FAZ data to TAZ data by distributing the forecasted employment and residential levels at the FAZ level to the more disaggregate level of TAZs. Because land use data will come from land parcel information in the future, the process of estimating the employment and residential levels will involve aggregation, rather than disaggregation, making a GIS an even more significant part of the data preparation.

As outlined in Johnston (Chapter 5, this volume), the first step in the traditional UTMS is estimating trip generation—that is, computing the trip productions and attractions of different areas by trip type and TAZ using EMME/2 travel demand forecast modeling software. As an adjunct, the GIS has been used to create thematic maps of the productions and attractions by TAZ.

The second step is the use of EMME/2 software to compute trip distributions. The desktop GIS can be used again to depict the output from this stage, focusing either on the productions or the attractions (i.e., the 956 × 956 TAZ matrix of origins and destinations).

The third step in the process is estimating mode choice. Variables such as travel time and the cost and accessibility of transit and parking are used to compute the proportion of trips made by each mode (auto, transit, HOV). Mode split is computed on the basis of person trips. The GIS can be used to create production maps and attraction maps as graduated circles, with pie slices symbolizing the proportion of trips made by each mode.

The fourth step is network assignment. The GIS and the EMME/2 software are used in tandem to complete this task. Vehicle trips are assigned to links in the network. The current regional network has 6,000 nodes and 19,000 links. The network is maintained in the GIS and loaded into the EMME/2 software for analysis. After volume assignment is made and volume-to-capacity (V/C) ratios are computed for each link, volume and V/C results are sent back to the GIS for portrayal. The difficulty with the EMME/2 software is that it provides "stick-figure" links, rather than shaped roadway links from intersection to intersection. Shaped roadway links represent a roadway geometry that most map viewers can recognize. Maps of volumes and V/C are generated in the GIS using graduated bandwidth symbolization for the shaped links. It is also possible to determine accessibility to employment based on mode. These results can be depicted as travel-time contours for employment centers.

Another output from the highway assignment step shows volumes and speeds by link, which can be aggregated to TAZ to compute carbon monoxide and ozone levels for the clean air attainment area, which is slightly different from the PSRC jurisdiction. The EPA model MOBILE5A is used to compute air pollution levels based on vehicle miles traveled, speed, mode, vehicle mix within a mode, and other attribute values passed to it from the software.

The last step in assignment analysis is transit assignment. It is performed in the same way as highway assignment, except that transit assignment uses only the transit network. The GIS can create map displays of variables such as number of buses per hour or speed on links using line bandwidth depictions.

The overall four-step analysis is an iterative process. In the travel demand forecasting process, assignment followed by mode choice is performed as many times as necessary to get the model to come to equilibrium—that is, to distribute the modes to links to balance the flows on the links according to a particular set of estimates for mode split. After that, trip distribution analyses are performed to redistribute trips using the new accessibility estimates, creating new inputs to mode choice. This latter cycle is also performed a couple of times. Maps can be made at each stage to see how the equilibrium process is progressing. The analysis is performed for current (2000) and

then for forecasted amounts for 2010. The forecast for 2010 becomes the base level for travel demand forecasts for 2020. The 2020 forecast becomes the base for the 2030 travel demand. These estimates are then used as the basis of generating entities (flow links, stations, etc.) in the transportation network on an iterative basis. The final iteration becomes an MTP alternative plan, defined in principle by the policy conditions (e.g., air pollution levels, congestion levels, and funding) that establish a scenario. Several scenarios are run, each one generating an alternative plan. One of them is eventually designated as the preferred alternative, which is the one that best meets objectives for TEA-21 and GMA policy conditions—for example, mobility needs, environmental protection, and land use change based on population and employment growth.

The MTP was presented to the citizens of the central Puget Sound region at public meetings, first, after the development of the environmental impact statement characterizing plan alternatives, and then again once a preferred alternative was identified. Requests for comments were made before and during the meetings. Comments were integrated into the draft planning material and then considered at decision meetings. Although the maps were created with GIS, a GIS was not interactively available at the meetings. Maps presented to meeting participants were hardcopy. More recently, however, the PSRC has provided World Wide Web maps with an online commenting capability for the transportation improvement programs. This same strategy is likely to be employed for the update of the MTP in 2005.

Transportation Planning at the King County Department of Transportation, Road Services Division

In 2000, King County covered 2,130 square miles with 1,685,000 residents. Seattle is the largest of 39 cities within the incorporated portion, having 563,374 residents, and Sykomish and Beaux Arts were the smallest, each with less than 300 residents. The unincorporated portion of King County contained 341,000 residents. Although King County is the most populous county in the state, a large amount of land is still rural on the eastern side of the county (the western slopes of the Cascade Mountains). The county population is expected to grow to about 1,875,000 by 2012. The Washington State Growth Management Act was passed and updated in part to help rationalize urban growth by protecting rural areas from urban encroachment. The King County Comprehensive Plan of 1994 incorporated those mandates, and in particular established an urban growth boundary to protect rural areas from urban encroachment (King County, 2000). Early in 2002, the county executive announced at a press conference that growth management policies have been successful in limiting urban development in rural areas.

Over the past decade, the transportation planning strategy at the county level has changed to accommodate growth management planning. The strategy addresses development of transportation facilities and land use development, while taking into consideration environmental and cultural concerns. The Washington State Growth Management Act and the county's adopted Comprehensive Plan require the county to specify transportation levels of service (LOS) and enforce them through a concurrency management system. The concurrency management policy ensures that new land use development occurs only with appropriate transportation improvements.

The county is undertaking more coordination with the cities and the Puget Sound Regional Council, all of which have some impact on the quality of life in the region. Because of the urban, suburban, and rural character of King County, transportation planning involves several functional scales: intersection, link, corridor, subarea, urban-

wide, and regionwide (the last of which has an emphasis on the urban–rural transition). Although intersection and link scales are included in the Transportation Improvement Program (TIP) that identifies year-to-year projects, the corridor, subarea, and regionwide scales are emphasized in long-range plans because of the complexity of transportation problems at these scales.

To assist with refining the planning effort, three sets of subareas have been defined over the years, each with a functional focus. The Comprehensive Plan identifies eight urban unincorporated subareas of the county, each of which has a unique service/financing and transportation service strategy. Those subareas were developed based upon community planning considerations, but these areas are no longer the formal basis of transportation planning because of growth changes over the past decade.

As a second, more general, strategy for transportation project enumeration, a Transportation Needs Report (TNR) identifies five subareas as an index for the projects (see Figure 7.5). The areas range from the dense activity centers, as in North King County, to mature residential communities, as in South King County, to the rural areas, as in East King County and Vashon Island. They cover the planning areas from the set of eight mentioned above, essentially covering both urban and rural areas of the county.

In a third approach, five transportation service areas (roughly coinciding with the urban unincorporated areas within each of the index subareas of Figure 7.5) are now used as a basis for compiling projects in a GIS database. This set of subareas is the functional compromise between the other two subarea sets, and the basis for travel demand forecast modeling. Transportation service areas are further disaggregated into TAZs for modeling, roughly compatible with the PSRC TAZs.

The TNR is the transportation element of the King County Comprehensive Plan (King County, 2001). The TNR represents the long-term view of capital projects for

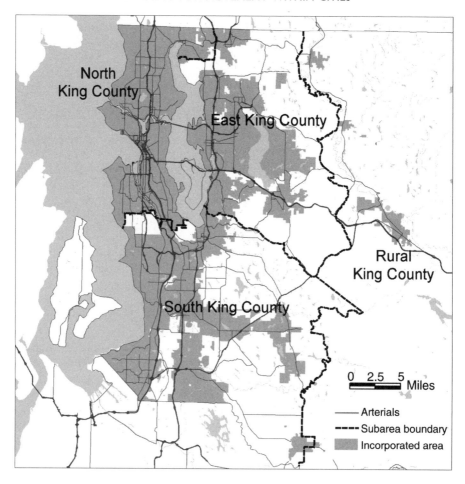

FIGURE 7.5. Transportation planning subareas (gray shading depicts incorporated area and unshaded depicts unincorporated area) of King County, Washington. Source: King County (2001).

transportation improvement, and there is a list of projects for each of the subareas depicted in Figure 7.5. The TNR incorporates the policy direction of the Comprehensive Plan to identify and prioritize transportation projects as the supply side of transportation capacity. The TNR is updated and submitted to the county council for review and adoption on an annual basis. There is a roadway element (Figure 7.6) and a transit element (Figure 7.7), with both data sets now managed in a GIS.

The TNR Priority Process scores and ranks all King County road and transit projects for consideration in the transportation

capital improvement program; hence it establishes a process link between long-range planning and transportation improvement. There are three major steps in the TNR Priority Process. The first step is the identification and screening of potential needs. Proposed projects are compiled from various sources and then screened to eliminate proposals that are noncapital in nature, infeasible, or inappropriate because they conflict with county policies. The second step is a technical evaluation and ranking of all eligible projects. The projects are evaluated on the basis of 20 relevant criteria. A series of rankings and weights is used to develop

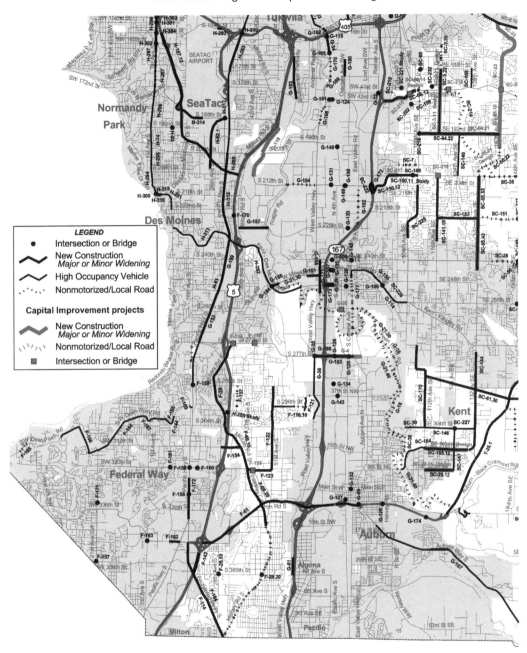

FIGURE 7.6. Transportation element of the Comprehensive Plan—a portion of roadway projects from roadway map in the 2001 *Transportation Needs Report*. Source: King County (2001).

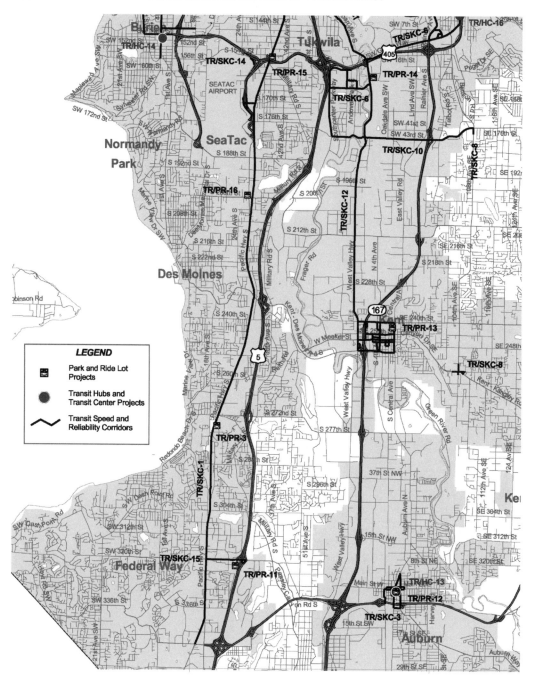

FIGURE 7.7. Transportation element of the Comprehensive Plan—a portion of transit projects from the transit map in the 2001 *Transportation Needs Report.* Source: King County (2001).

individual project scores. The third and final step is an evaluation of nonquantifiable factors to incorporate important considerations that cannot be measured or quantified, such as emergencies, project scheduling, categorical funding, and commitments with other jurisdictions.

Having previously used community planning subareas and the links between them, the King County Department of Transportation now makes use of a corridor perspective based on identification of a regional arterial network. Arterials, which connect the neighborhood streets with freeways, shopping malls, and businesses, are the workhorse of a regional (urban/suburban) transportation system. A main criterion for development of the regional arterial network is the functional classification of the road system, and thus the network starts with the principal arterial system. The principal arterials mainly provide mobility *through* areas rather than *access* to specific sites. The regional arterial network is continually under review to establish a better set of corridors as a foundation for county transportation planning.

The regional corridor perspective provides planners with a foundation for understanding the regional impacts of transportation development; it allows planners to link transportation improvements across city and county jurisdictions as well as with the PSRC regionwide perspective. Each of the corridors can be used as the basis of a transportation partnership for enhancing accessibility and mobility. As of late 2002, there were more than 80 regional corridors identified by name and number for growth management purposes (see Figure 7.8). The map produced on a GIS shows the monitored corridors depicted as critical links to watch. A GIS is used to characterize and maintain this set of corridors (links), a task that would be overwhelming without GIS. The significance of monitored corridors is that they are known to be close to capacity and receive special attention. All corridors have an operational transportation and land

use connection, however, through concurrency management, as described below.

Because of the diversity of development in the county, a major focus in King County's transportation planning process is determining the adequacy of transportation facilities in meeting projected auto and transit travel for activity centers and suburban–rural areas of the county. To establish consistency for transportation planning in cooperation with the PSRC, the county makes use of the TAZ coverage provided by the PSRC as the basis of its transportation modeling. With the use of a GIS, the county splits or aggregates TAZs where necessary into small area zones (SAZs) to reflect the character of subareas requiring more refined attention than at the PSRC four-county scale. This is accomplished on the basis of area, network density, and total population in a manner similar to the creation of TAZs. The SAZs are used by the county's transportation modeling group to forecast travel demand for auto trips. The forecasts are then used to create a transportation adequacy measure (TAM) for each SAZ as well as for monitored corridors. Consequently, the TAM establishes an operational monitoring link between land use development and transportation improvements.

The TAM is used to assess the amount of congestion in small areas (rather than just on road segments), as well as in corridors, because land use generates trips. Making the connection operational through policy comes in the form of a concurrency management program. The King County Council adopted a Transportation Concurrency Management requirement in Ordinance 11617, effective January 9, 1995, and revised under Ordinance 14375, effective June 28, 2002 (King County, 2002b). The ordinance establishes a concurrency management system, which ensures that adequate transportation facilities are available to meet the requirements of new development in King County. A concurrency review must be completed by anyone who intends to apply

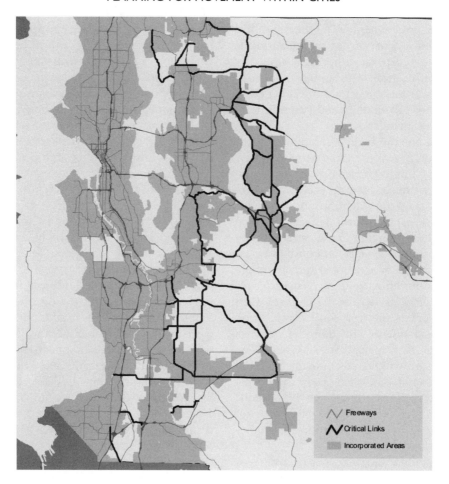

FIGURE 7.8. King County transportation planning corridors—existing monitored critical links. Source: King County (2001).

for a land development permit in unincorporated King County.

Many of the numbers to implement a TAM are computed using the EMME/2 software, with data managed in GIS. The TAM is computed using trips, vehicle miles traveled, and the volume-to-capacity ratio. The number of trips on a segment is computed for each zone relative to the total number of trips across the entire county. Computed trips depend on how close or far vehicles are from a zone. The proportion of trips grows smaller as distance from a zone increases.

We use the trips to then compute VMT, as follows (King County Department of Transportation staff, personal communication, August 15, 2002).

VMT = Segment length × number of trips on a segment

The capacity of a highway is based on the number of lanes, speed, and composition (although speed and lanes are usually sufficient). The volume is the actual number of vehicles at the afternoon/evening peak load time on the segment. Thus, we compute the volume-to-capacity ratio (V/C) as:

V/C = Volume on a segment ÷ capacity on that segment

Thus the TAM for a SAZ is computed as a relationship between VMT and V/C:

$$\frac{\sum VMT \times \frac{V}{C}}{\sum VMT}$$

A threshold TAM is used as a policy constraint, such that SAZs and monitored corridors should not exceed a certain amount of traffic congestion. Exceeding the threshold amount means that there is congestion (lower mobility) in the area because of the level of land use development or the absence of transportation improvements. The threshold TAM is based on a level of service (LOS) recommended by the Federal Highway Administration for urban highways. LOS is commonly graded by a letter (A–F), whereby A is best for freely flowing traffic, and F is worst as in almost standing still on a highway link that is fully capacitated (Transportation Research Board, 1985). LOS is essentially a recommended flow volume-to-capacity ratio. Consequently, in the case of urban traffic, King County specifies E as 0.99, D as 0.89, and C as 0.79 volume-to-capacity, whereas for rural volume-to-capacity, B is always specified as 0.69. Thus the TAM is to be interpreted as a performance measure; it is an indication of people's mobility on the transportation system. Areas expected to be mobility-deficient can be computed based upon changes in land use that generate trips—the concurrency link described below. It is important to note that the TAM is an area-based transportation summary that is computed for SAZs as well as for corridors. It is translated to the same level-of-service coding scheme that is commonly used in assessing links, but the county feels that link-based congestion interpretations miss a broader planning concern, namely, that of adjacent land uses in small areas feeding link congestion (Transportation Research Board, 1985).

The TAM (and associated LOS) as compared to a threshold TAM is the operational concurrency link between land use development and transportation development; it guides the approval process for land de-velopment permits as part of growth management planning. Applicants for development permits must obtain a certificate of transportation concurrency (or availability) prior to obtaining a development permit. The certificate confirms and establishes the availability of transportation facilities (or supply) to serve the development and commits the capacity to the development. A certificate is not issued if the development causes a violation of transportation level of service and if no financial commitment is in place to complete the improvements within 6 years.

The King County Concurrency Management Program updates a GIS map of TAM levels for SAZs and corridors twice a year (see Figure 7.9 for LOS-based TAM for SAZs). The light gray shade (green color in original) indicates that the computed TAM is less than the threshold TAM, and thus developing land is not likely to affect transportation flow very much. A medium gray shade (yellow color in original) indicates that a precautionary level of development has been reached. A dark gray shade (red color in original) indicates that land development permits will not likely be issued because the roads are already at, or over, capacity.

There is actually a two-stage test for concurrency: the SAZ check and the corridor check. If a development permit is denied after checking the TAM in the first phase—that is, against the TAM for an SAZ—then the permit is denied overall. If, however, the permit passes the first-phase TAM check, then the second-phase test is administered: to check the TAM in the corridor. Assuming the test passes this stage, then the permit moves through the regular permitting process for approval/denial consideration.

A GIS is used to characterize corridors and transportation service areas (based on SAZs) in terms of road and transit service, whether they are inside or outside the urban growth boundary, and whether they are in an incorporated or unincorporated area.

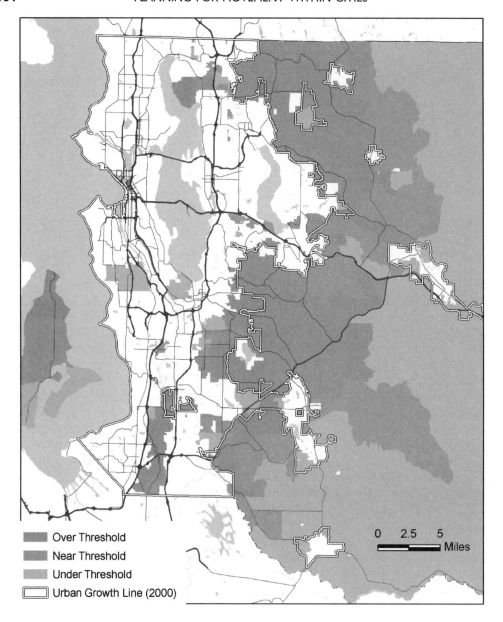

Over Threshold
Near Threshold
Under Threshold
Urban Growth Line (2000)

0   2.5   5
Miles

FIGURE 7.9. Concurrency transportation management in King County, Washington, using a transportation adequacy measure based on level of service. Source: King County (2002c).

Together these characteristics help direct strategy for transportation improvement program projects as part of the capital improvement program.

The King County Department of Transportation (KCDOT) makes use of Community Advisory Boards when creating the TNR. Furthermore, the advisory boards have provided input to the transportation improvement program projects for their particular areas as well. Interactive GIS has not been used at these meetings, but maps from GIS have been. The KCDOT is considering using GIS to enhance public involvement, in regards to obtaining public input on the link between the TNR and the TIP. A strategy to implement that feedback has not yet been established, however.

Transportation Planning at the City of Seattle, Department of Transportation, Division of Major Projects and Neighborhood Planning

The City of Seattle covers approximately 83 square miles in the northwest corner of King County. In 1994, the Washington State GMA office and the City of Seattle estimated the city's 2000 population to be between 540,000 and 548,000. The 2000 census found that the City of Seattle had 563,374 residents, and had grown at a rate of 9% since 1990. The city is expected to grow in population by more than 50,000 by the year 2010.

Appropriate functional scales for performing transportation planning in urban areas are intersection, link, corridor, subarea, and urbanwide. Emphasis here is on the urbanwide and the subarea geographic scales. The urbanwide scale in this case is the City of Seattle (although it is only a portion of the total urbanized area in the county). The subarea scale refers to communities within the city; subareas are larger than neighborhoods but smaller than large sections of the city.

In July 1994, the Seattle City Council adopted the mayor's recommended Comprehensive Plan (City of Seattle, 1994), and in 1998 it adopted amendments to that plan. The Comprehensive Plan makes basic policy choices and provides a flexible framework for adapting to real conditions over time. It is a collection of the goals and policies the city uses to guide decisions about how much growth Seattle should take and where it should be located. The plan also describes in a general way how the city will address the effects of growth on transportation and other city facilities.

The initial building blocks of the Comprehensive Plan are the elements required by the state's Growth Management Act: land use, transportation, housing, capital facilities, and utilities. King County's Countywide Planning Policies require the addition of an economic development element, and the Seattle Framework Policies (Resolution 28535) inspired the inclusion of a neighborhood planning element and a human development element. The ideas in the plan were developed over 5 years through discussion and debate and the creative thinking of thousands of Seattle citizens working with city staff and elected officials (City of Seattle, 2002).

In line with growth management and community development, a major goal of the plan deals with how to reduce single-occupancy auto use. One strategy relies on encouraging the development of urban centers/villages as the basis of the overall land use strategy in the city (see Figure 7.10). Centers/villages were identified using GIS maps in an iterative process undertaken by elected officials, staff, and the affected communities. This is one of the many data layers that would appear in the future land use map generated by a GIS and is integral to the land use–transportation link. Periodically, the urban centers/villages boundaries are reviewed as a matter of implementing land use densification policies.

Urban centers/villages are designated at three scales: urban centers, hub urban villages, and residential urban villages. Zoning ordinances encourage higher residential and commercial densities in urban villages than exist currently. *Urban centers* are intended to accommodate a broad mix of activities and will receive most of the future residential and commercial growth. *Hub urban villages* are somewhat less densely developed, with concentrated mixed-use cores, diverse residential areas, and excellent transit access. *Residential urban villages* are primarily compact residential neighborhoods, each with a small, locally oriented business district. Inside the village the transportation circulation technology will focus on multipassenger vehicles to fit the nature of the develop-

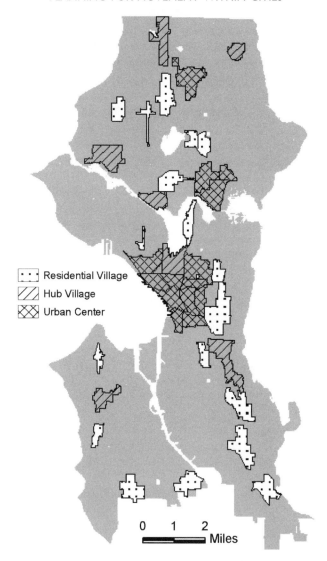

FIGURE 7.10. Urban centers/villages in the City of Seattle, Washington, using the 2000 Comprehensive Plan. Source: City of Seattle (2000a).

ment. Linking urban villages using transit vehicles is a high priority in the transportation plan; the relationship between land use and transportation is made explicit in the urban village development policy.

The locations of these urban villages were worked out over a couple of years. The chosen criteria allowed certain areas to be considered and others not. The primary goals for definition were establishing functional boundaries for centers/villages, local growth targets, and open space availability. Using a GIS enabled rapid turnaround of the conceptual ideas. Each future land use map was of high quality, and the use of sharp boundary lines required the planning department to clarify its ideas. This future land use pattern is the driving factor in the development of the transportation plan.

Whereas the PSRC focuses on the transportation links primarily among urban centers, the City of Seattle focuses on transportation links among *all* centers and villages. Transportation modeling software and GIS software have been used to develop the transportation element of the Comprehensive Plan, much like the process for PSRC and King County. The mobility focus here, however, is on connecting centers and villages, emphasizing regional corridor and neighborhood improvements. The city used a GIS to develop the principal arterial network (see Figure 7.11). This arterial network is similar to the regional arterial network developed by King County. At this time it is unknown if the city arterial network is a subset of the county arterial network, per se. Like the county, the city must monitor the network capacity for congestion, particularly in regard to the concurrency link between transportation and land use. The city uses a screenline level of service measure to implement concurrency. A *screenline* is an average daily traffic measurement taken at a line drawn across a segment on a GIS (database) map such that traffic volumes crossing the lines are monitored and compared against level of service standards to identify congestion on links. Screenlines (using traffic counters) are placed at strategic locations in the city to assess congestion on the network.

Because of the amount of congestion on Seattle roadways, the current emphasis in the urbanwide transportation planning effort at the City of Seattle is on providing intermodal transit linkages among the urban centers and villages. Convenience, safety, and efficient transit linkages are needed to reduce the use of single-occupant vehicles across the city. To do this the city has identified a transit priority network, whereby provision of transit mobility takes priority over other vehicular mobility (see Figure 7.12). This policy perspective in the form of a map plus transportation goals and policy guidelines explicitly appears as part of the transportation element of the comprehensive plan.

To refine and implement that long-term, urbanwide perspective, considerable emphasis is being placed on corridor and subarea planning efforts. In regards to corridor planning, a program called Intermediate Capacity Transit (ICT) was initiated in June 2000. A team bringing together the city, transit agencies, and consultants is examining the feasibility and performance of a network of transit lanes separated from automobiles and trucks. The network would support bus rapid transit, streetcars, trams, and elevated transit (like Seattle's current monorail). ICT service would connect neighborhoods directly to each other, to major destinations, and to transit transfer stations (as part of the regional network). ICT would have fewer stops than regular bus service, so travel would be faster, and would reduce the dependency on cars. As part of planning effort for the ICT network, 47 routes were considered using a GIS. Corridor feasibility was examined in terms of (1) ability to carry existing and new transit riders, (2) cost of implementing technology, and (3) potential impacts within the corridors. Corridor routes were identified using GIS-based neighborhood plans (38 across the city), previous transportation studies, public suggestions, and the work of city staff. From those 47 routes, seven corridors were selected using GIS for further examination, and five were found to be reasonably feasible (see Figure 7.13). Based on that feasibility, the five corridors can link to the regional transit corridors, and together they would provide significant transportation mobility across the city. The corridors are under consideration for formal specification as part of the Comprehensive Plan.

As of late 2002, GIS has not yet been used for subarea planning. Instead, computer-aided drafting has been used from time to time. In computer-aided drafting, a map is purely a picture—that is, no attributes from a database are linked to transportation features. As in many GIS application

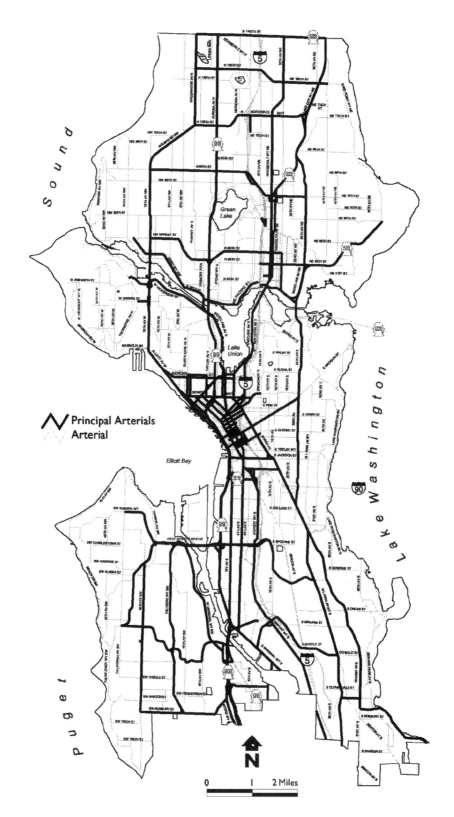

FIGURE 7.11. Transportation element—Principal Arterials Map—of the Comprehensive Plan 2000. Source: City of Seattle (2000b).

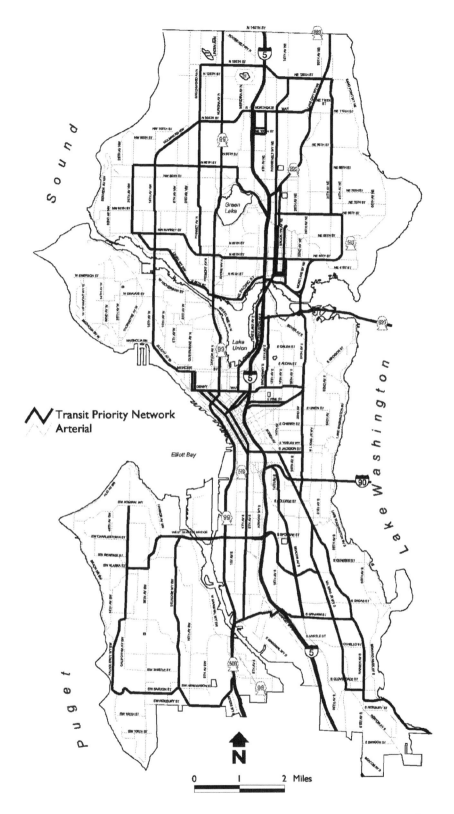

FIGURE 7.12. Transportation element—Transit Priority Network Map—of the Comprehensive Plan. Source: City of Seattle (2000c).

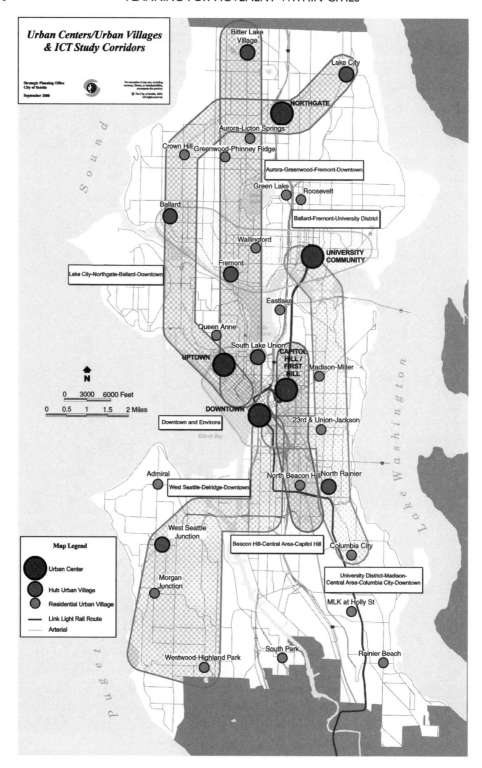

FIGURE 7.13. Recommended Intermediate Capacity Transit System Vision, 2000. Source: City of Seattle (2000d).

topics over the past two decades, GIS is a next step beyond computer-aided drafting to increase flexibility in map creation. A GIS database approach would provide access to attributes, thereby supporting spatial analysis of the network and allowing analysts/planners to examine relationships to other features such as land use and population. Since there are many subareas to plan, and since planning is a continuous endeavor, the potential for GIS use is considerable, particularly as most of these maps are hosted on the World Wide Web for presentation to the public.

Based on subarea studies, projects are collected to provide a basis for the Transportation Strategic Plan (TSP) for the City of Seattle.

> The TSP . . . outlines strategies and actions to help achieve the City's Comprehensive Plan goals: to make Seattle a city where streets and bridges are well-maintained where transit, walking, and bicycling are convenient and attractive, and where we are less dependent on cars for our transportation needs. The TSP also aims to protect the character and livability of our neighborhoods, and to improve our ability to move freight and goods. The intent was that the TSP assists in long-range transportation planning and decision making, and that it be an evolving and "living" document, updated regularly to ensure that it stays relevant. (City of Seattle, 2001, p. 2)

The 1998 TSP, with updates in 1999, 2000, and 2001, establishes a link—at least on paper—between the Comprehensive Plan and the transportation capital improvement program. The TSP provides a strategic direction for project proposals, more operational than the Comprehensive Plan, and more conceptual than the transportation improvement programming that brings the projects to implementation. Linking Comprehensive Plan transportation elements and improvement programming processes receives a growing interest in the city and the region. Such a linkage is critical to monitor growth management progress in transportation.

CONCLUSION

Recognized needs for transportation system improvements in the central Puget Sound region promote an interest in transportation regionwide as one of the major community issues in this current decade. Federal (e.g., TEA-21) and state (e.g., GMA) mandates guide the planning and programming policies and the processes that attempt to implement those policies for transportation system improvements. Such processes, in turn, drive the implementation of GIS technology, as there is a widespread need for access to information about transportation efficiency, effectiveness, and equity across the region. GIS technology is continuing to mature—for example, object-oriented structuring of information allows for more expressive representations (Miller & Shaw, 2002), including community values, assuming they exist (Taylor, 1998). In addition, the World Wide Web was only a glimmer in the mid-1990s. Now the World Wide Web is an integral component of computing to support access to information, but it is still in the beginning stages of supporting interactive dialogue in transportation planning at the local level (Lebeaux, Meheski, Stuart, & Christians, 2002; Peng, 2001).

Although transportation planning at the three jurisdictional scales described here is governed by similar mandates, the outcomes of planning differ because of scale. PSRC is a regional organization, creating regional transportation plans and coordinating TIP across four counties and 68 cities. The projects are regional, working under mandates from federal and state laws, with benefits that accrue across as many of the four counties and all cities as much as possible. The network, and the land use data associated with it, are more extensive but less detailed than those for King County.

King County creates a transportation plan for unincorporated King County, coordinates with counties to the north and the south, and coordinates plans among 39 cities. The County Comprehensive Plan (hence transportation elements) is required to be consistent with the Regional Plan created by PSRC and thus to set the stage for consistency across King County. The transportation element in the King County Comprehensive Plan is a list of projects published as a Transportation Needs Report (TNR). King County creates a capital improvement program by identifying projects within the TNR, while considering the status of concurrency management. The concurrency management system ensures (or at least attempts to ensure) that land use development is consistent with transportation facility improvements. The transportation adequacy measure, as a small-area measure of congestion based on a level of service expected on links in that area, is a strategy King County uses to comply with concurrency requirements of the Washington State GMA.

The City of Seattle creates a transportation element as part of a Comprehensive Plan as required by the GMA and organizes a transportation capital improvement program as well. That plan is mandated to be consistent with King County countywide planning policies. The city planning is performed at urbanwide to neighborhood scales, and corridor planning helps to coordinate the plans across those scales. A screenline level-of-service measurement is used to assess congestion for concurrency purposes. That assessment feeds into the capital improvement program analysis.

At the current time, although GIS is being used at all scales, it is mostly for data management and map display, rather than for analysis and policy investigation. As such, there is still considerable potential for planning use, since analysis is all about exploring relationships about people and goods movement within and among places. For example, accessibility studies are demonstrating considerable potential (Kwan, 2000), but more work definitely needs to be done, particularly in relation to projects that are part of long-range plans and capital improvement programs.

Across the United States, long-range plans are often made without an explicit link to capital improvement programs, at least as implemented in GIS. A major challenge in linking plan making with programming is that the two sets of projects are conceptualized at different levels of detail, reflecting the different spatial and temporal scales involved—that is, long-term general perspectives versus short-term specific perspectives. The different project conceptualizations mean that from an information perspective, plans and programs are loosely coupled at the current time, but from an information technology perspective, they are not coupled at all. A clearer conceptualization of the similarities and differences (as in a conceptual database design) would strengthen the link between the planning activities and the programming activities. Plans would take on grounded direction and programs would obtain a longer term justification with some flexibility for direction.

Recognition that long-range projects eventually become short-range projects is beginning to work its way into GIS implementation. The PSRC staff is now addressing this concern at the level of GIS database design for MTP and transportation improvement program projects. The county and city are using TSPs to bridge the gap between Comprehensive Plan transportation elements (a 20-year perspective) and the short-term capital improvement program (a 2-year perspective revolving over 6 years). This intermediate (10-year or thereabout) perspective provides planners, citizens, and elected officials with a more effective temporal transition for decision making. Unfortunately, no systematic effort across the jurisdictional scales yet addresses the transportation plan-to-program link. That institutional concern is one of the fundamental issues in developing successful GIS, particularly when the topic is growth management. However, data framework efforts at understanding comprehensive approaches to transportation data organization are underway in Washington State (Dueker

& Bender, 2002). Such efforts are being spurred on by continued work in implementing the National Spatial Data Infrastructure (NSDI) under the coordination of the Federal Geographic Data Committee (FGDC), with direction from the FGDC Secretariat at the U.S. Geological Survey (Association of American Geographers, 1994). As GIS technology continues to mature for both workstation and World Wide Web–based environments, providing easier access to transportation information, more complex transportation issues will be tackled with the support of GIS. This means that the geography of urban transportation will be examined in more detail than ever before. Such investigations should lead to more efficient, effective, and equitable alternatives to improve our transportation choices.

ACKNOWLEDGMENTS

This chapter would not have been possible in its current form without the assistance of many people in central Puget Sound transportation agencies providing information and/or organizing permissions for materials use. Regarding the Puget Sound Regional Council, I would like to thank Mary McCumber, Mark Gulbranson, Karen Richter, Jerry Harless, Larry Blain, and Andrew Norton. Regarding the King County Department of Transportation, my thanks go to Jennifer Lindwall, Harry Clark, David Gualtieri, Richard Warren, Rebecca Campeau, and Lisa Schafer. Regarding the City of Seattle Department of Transportation, I extend my thanks to Ann Sutphin and T. J. Moore. Regarding the Seattle Public Utilities, I offer my thanks to Elaine Eberly. I also offer a special thanks to Guirong Zhou in the Department of Geography at the University of Washington for assisting with map graphics conversion and production.

NOTES

1. The other cities are too small to be involved in regional activity—so to speak. It is their choice. Since there is a membership fee involved, they are "free riders" in regards to regional transportation planning benefits.

2. Roadway improvements include lane additions such as adding a center lane for turning (on both two- lane and four-lane roads). New roadway facilities include the building of new links between centers because an upgraded roadway is simply not sufficient. The latter also includes building a new bridge to add four additional lanes over the Tacoma Narrows where there is one bridge of four lanes at this time.

REFERENCES

Association of American Geographers. (1994). President Clinton signs NSDI executive order. *AAG Newsletter, 29,* 2.

Chrisman, N. R. (1987). Design of geographic information systems based on social and cultural goals. *Photogrammetric Engineering and Remote Sensing, 53,* 1367–1370.

City of Seattle. (1994). *Comprehensive plan for the City of Seattle.* Seattle, WA: City of Seattle Planning Department.

City of Seattle. (2000a). *Based on GIS Layer: Urban villages.* Seattle, WA: City of Seattle Planning Department.

City of Seattle. (2000b). Transportation Figure 1: *Principal Arterials.* In *City of Seattle Comprehensive Plan.* Seattle, WA: City of Seattle Planning Department.

City of Seattle. (2000c). Transportation Figure 4: *Transit Priority Network Arterials.* In *City of Seattle Comprehensive Plan.* Seattle, WA: City of Seattle Planning Department.

City of Seattle. (2000d). Urban Centers/Urban Villages & ICT Study Corridors. In *Seattle Transit Study for Intermediate Capacity Transit.* Seattle, WA: City of Seattle Planning Department.

City of Seattle. (2001). *Transportation strategic plan annual report 2001.* Available online at http://www.cityofseattle.net/td/tsp.asp.

City of Seattle. (2002). *City of Seattle comprehensive plan.* Available online at http://www.cityofseattle.net/dclu/planning/comprehensive/homecp.htm

Cowen, D. (1988). GIS versus CAD versus DBMS: What are the differences? *Photogrammetric Engineering and Remote Sensing, 54*(11), 1551–1555.

Craig, C., Harris, T., & Weiner, D. (Eds.). (2002). *Community participation and GIS.* London: Taylor & Francis.

Dueker, K., & Bender, P. (2002). *White paper on issues and strategies for building a state transportation framework.* Portland, OR: Portland State University, College of Urban and Public Affairs, Center for Urban Studies.

Dueker, K., & Kjerne, D. (1989). *Multipurpose cadastre: Terms and definitions.* Bethesda, MD: American Congress on Surveying and Mapping. (Also published in *Technical Papers, ASPRS/ACSM Annual Convention, Baltimore, MD, April 2–7, 1989, 5,* 94–103)

Environmental Systems Research Institute. (2002). ArcGIS data models. Available online at http://www.esri.com/software/arcgisdatamodels/index.html

Federal Interagency Coordinating Committee on Digital Cartography. (1988). *A process for evaluating geographic information systems* (Open File Report 88–105). Reston, VA: U.S. Geological Survey.

Jankowski, P., & Nyerges, T. (2001). *Geographic information systems for group decision making.* London: Taylor & Francis.

King County. (2000). *King County Comprehensive Plan update.* King County, WA: Office of King County Executive. Available online at http://www.metrokc.gov/exec/orpp/compplan/2000/

King County. (2001, January). *2001 Transportation needs report: Transportation element of the King County Comprehensive Plan.* Seattle, WA: King County Department of Transportation, Transportation Planning.

King County. (2002a). King County monitored corridors depicted as critical links. Seattle, WA: King County Department of Transportation. Available online at http://www.metrokc.gov/kcdot/tp/concurr/monitored_corridors.pdf

King County. (2002b). Concurrency management program. Seattle, WA: King County Department of Transportation. Available online at http://www.metrokc.gov/kcdot/tp/concurr/conindex.htm

King County. (2002c).Concurrency management program. King County, WA. Available online at http://www.metrokc.gov/kcdot/tp/concurr/attachmentA32002.pdf

Koncz, N., & Adams, T. (2002). A data model for multi-dimensional transportation location referencing systems. *URISA Journal, 14*(2), 27–41.

Kwan, M. P. (2000). Interactive geovisualization of activity-travel patterns using three-dimensional geographical information systems: A methodological exploration with a large data set. *Transportation Research, C, 8*(1), 185–204.

Langran, G. (1992). *Time in geographic information systems.* London: Taylor & Francis.

Lebeaux, P., Meheski, J., Stuart, J., & Christians, R. (2002). *Internet outreach in statewide long-range planning: The New Jersey experience* (Transportation Research Record 1817, pp. 120–129). Washington, DC: Transportation Research Board.

McCormack, E., & Nyerges, T. (1997). What does transportation modeling need from a geographic information system? *Journal of Transportation Planning, 21,* 5–23.

Meyer, M., & Miller, E. (2001). *Urban transportation planning: A decision oriented approach.* New York: McGraw-Hill.

Miller, H., & Shaw, S. (2002). *Geographic information systems for transportation: Principles and applications.* London: Oxford University Press.

National Institute of Standards and Technology. (1994). *Spatial data transfer specification: Federal Information Processing Standard (FIPS) 173.* Springfield, VA: National Technical Information Service.

National Research Council. (2002). *Community and quality of life.* Washington, DC: National Academy Press.

Nyerges, T. (1990, March). Locational referencing and highway segmentation in a GIS for transportation. *ITE Journal,* pp. 27–31.

Nyerges, T., & Dueker, K. (1988). *Geographic information systems in transportation.* Washington, DC: Federal Highway Administration, Planning Division.

Nyerges, T., Montejano, R., Oshiro, C., & Dadswell, M. (1998). Group-based geographic information systems for transportation improvement site selection. *Transportation Research, C, 5,* 349–369.

Nyerges, T., & Orrell, J. (1992). *Using GIS for regional transportation planning in a growth management context* (Final Report, WA-RD 285.1). Olympia, WA: Washington State Department of Transportation.

Peng, Z-R. (1999). An assessment framework for the development of internet GIS. *Environment and Planning, B: Planning and Design, 26,* 117–132.

Peng, Z-R. (2001). Internet GIS for public

participation. *Environment and Planning, B: Planning and Design, 28,* 889–905.

Pickles, J. (Ed.). (1995). *Ground truth: The social implications of geographic information systems.* New York: Guilford Press.

Puget Sound Regional Council. (2001). *Destination 2030* (CD-ROM). Seattle, WA: Author.

Puget Sound Regional Council. (2002). *Regional view, August 2002.* Available online at http://www.psrc.org/datapubs/pubs/view/0802.htm#Five

Sinton, D. (1978). The inherent structure of information as a constraint to analysis: Mapped thematic data as a case study. In G. Dutton (Ed.), *Harvard papers on GIS* (Vol. 7, pp. 1–19). Reading, MA: Addison-Wesley.

Smith, S. (1999). *Guidebook for transportation corridor studies: A process of effective decision-making* (National Cooperative Highway Research Program, No. 435). Washington, DC: Transportation Research Board.

Talen, E. (2000). Bottom-up GIS: A new tool for individual and group expression in participatory planning. *Journal of the American Planning Association, 66*(3), 279–294.

Taylor, N. (1998). Mistaken interests and the discourse model of planning. *Journal of the American Planning Association, 64(1), 64–75.*

Thill, J. C. (2000). Geographic information systems for transportation in perspective. *Transportation Research, C, 8,* 13–36.

Transportation Research Board. (1985). *Highway capacity manual.* Washington, DC: National Academy Press.

U.S. Census Bureau. (2000). *TIGER/Line Files technical documentation.* Available online at http://www.census.gov/geo/www/tiger/tiger2k/tiger2k.pdf

PART III

POLICY ISSUES

CHAPTER 8

Public Transportation

JOHN PUCHER

Most Americans have only vague ideas about public transportation, or transit, as it is also called. That is not surprising, since only a small percentage of Americans actually use any form of public transportation on a regular basis. Most Americans probably think of public transportation as the buses they occasionally see on the street, or the subways they see in films, or perhaps the new light-rail systems the media have recently focused on.

In fact, public transportation comprises a wide range of travel modes with considerable variation in technology, cost, operating characteristics, geographical extent, market share, and the socioeconomic makeup of their riders. The most common form of public transportation in the United States is the bus, but there are considerable variations even within that modal category: diesel, electric, and natural gas buses; high-floor and low-floor buses; long articulated buses, standard coaches, midsize buses, and minibuses; local buses and express buses; buses on shared rights-of-way versus buses in reserved lanes or on exclusive bus ways. Light-rail transit includes both the older streetcar systems (trolleys or tramways) that often run on city streets as well as modern

light-rail systems, usually with separate rights-of-way, less frequent station stops, and raised boarding platforms. Metro systems (also called "heavy rail," "rapid transit," "subway," or "elevated") always have separate rights-of-way, boarding platforms, higher capacities and faster travel speeds than light rail. Suburban rail systems (also called "commuter rail" or "regional rail") have much in common with metro systems but with much longer distances between station stops, faster speeds, and larger, more comfortable rail cars. (For detailed analysis of the characteristics of transit modes and their classification, see American Public Transportation Association, 2002; Vigrass, 1992; Vuchic, 1992.)

Those are the main modes of public transportation in U.S. cities, but there are many more. Five cities have downtown people movers, similar in technology to the automated shuttle systems in so many airports and amusement parks. Only San Francisco has maintained its historic cable car lines, but five cities have inclined-plane systems (funiculars), cable-drawn rail vehicles on extremely steep slopes. Five cities have trolley buses (trackless trolleys), which operate on city streets but draw power for their electric motors from overhead wires. Seattle and Las Vegas have monorail systems. New York City has an aerial tramway. Finally, 26 cities have ferry boat systems for crossing bays, rivers, and lakes (American Public Transportation Association, 2002).

In addition to those public transportation modes, some experts would include various forms of paratransit: van pools, car pools, taxis, and demand-responsive dial-a-ride systems. Moreover, school bus systems provide important group transportation, with more buses than the total fleet of public transport agencies in the United States. The main difference between the two groups of modes is that public transportation modes are generally defined to include only fixed-route services open to the public at large. Paratransit and school bus services are either special-purpose in nature, limited to special groups, or offer flexible routing and schedules depending on passenger demands. In fact, however, many public transit systems operate some form of paratransit service. Federal law requires them to provide such special services for disabled passengers who cannot use or access fixed route transit services.

Clearly, public transportation comprises a family of travel modes whose precise boundaries are uncertain and changing over time. As documented in Chapter 3, public transportation has changed dramatically since 1800, with old technology being superseded by improved technology. In the mid-19th century, the horse-drawn omnibus was replaced by the horse-drawn streetcar, which in turn, was replaced by the electric streetcar toward the end of the 19th century. Perhaps most strikingly, the electric streetcar, which provided most public transportation in U.S. cities for five decades, was almost entirely replaced by the diesel bus during the 1940s and 1950s. As technologies develop further, new modes are added to public transportation, and older modes are phased out.

The relative importance of the main public transport modes in the United States currently is shown in Table 8.1. Bus services carried 58% of all transit passenger trips in 2000, followed by metro systems (30%),

FIGURE 8.1. San Francisco's Market Street corridor features trackless trolleys, surface streetcars, and diesel buses, in addition to underground metro and light-rail services, plus ferryboat services from the terminal shown in the background. Photo courtesy of San Francisco Municipal Railways and the City of San Francisco; Carmen Magana, photographer.

TABLE 8.1. Public Transportation Usage and Costs in the United States by Mode, 2000

Public transport mode	Unlinked passenger trips (millions)	Passenger miles (millions)	Average trip length (miles)	Average speed (miles per hour)	Unlinked passenger trips per vehicle revenue hour	Operating cost per passenger mile ($)
Bus	5,040	18,807	3.7	12.8	37	0.59
Light-rail transit	316	1,339	4.2	15.3	94	0.45
Metro (subway)	2,632	13,844	5.3	20.5	93	0.28
Suburban rail	413	9,400	22.8	28.5	47	0.29
Demand-response	73	588	8.0	14.8	2	2.09
Other[a]	245	1,222	4.6	NA	37	0.49
Total public transport[b]	8,720	45,100	5.2	NA	40	0.44

[a]The "other" category includes cable cars, inclined planes, ferry boats, aerial tramways, automated guideway transit (people movers), monorails, as well as some van pools and jitney services.
[b]The passenger and passenger mile totals reported in the Federal Transport Administration's (FTA's) *National Transit Database* only include urban transit systems that receive federal subsidies, either directly or indirectly, and thus are required to submit detailed annual reports to the FTA. They understate the total public transport usage in the United States by about 7% because they exclude both rural systems and those systems that do not receive federal aid. The American Public Transportation Association (APTA) includes more systems and reports 9, 363 million unlinked passenger trips for 2000.
Source: Federal Transit Administration (2002). Available online at *www.bts.gov/ntda/ntdb.*

suburban rail (5%), and light rail (4%). In terms of passenger miles traveled, however, the rail modes dominate, accounting for 55% of all transit passenger miles, compared to 42% by bus. That is due to the longer trip lengths on rail transit, especially on suburban rail and metro systems (see Table 8.1).

GEOGRAPHIC VARIATIONS IN PUBLIC TRANSPORT

These nationwide aggregate totals hide the enormous geographic variation in both the extent and type of public transportation among different regions of the country and by city size. For example, most of the nation's public transportation service and ridership is concentrated in the Northeast. Indeed, 43% of all transit trips in the United States take place in the three states of New York, Pennsylvania, and New Jersey, although they account for only 14% of the country's population (Federal Transit Administration, 2002). The most transit-oriented region, the Mid-Atlantic, has a transit modal share of 5.8%, which is twice as high as those of the next most transit-oriented regions (2.2% in the Pacific region and 1.8% in New England) and 15 times higher than in the least transit-oriented region (0.4% in the East South Central region) (Pucher & Renne, 2003). Even more striking, the New York Metropolitan Area alone has 60% of all rail transit riders in the country (compared to 25% of the country's bus riders and only 7% of the country's population) (Pucher, 2002).

Regional variations in walking are also striking, and strongly correlated with transit modal share. Thus, the highest walk modal share is also in the Mid-Atlantic States (15.8%), followed by the Pacific region (10.6%) and New England (10.3%). The correlation between transit use and walking is probably due to the crucial role of compact, mixed-use land use patterns in older, denser cities, which facilitate both transit use and walking. Thus, the same sort of land use pattern encourages both modes. In addition, walking is the most important access mode to transit stops, which also contributes to the correlation between the modal split shares of walking and transit.

Bicycling has a somewhat different regional pattern, with the highest level in the Pacific (1.1%), but roughly the same levels in the rest of the country (0.7% to 0.9%), except for the East South Central, which has a much lower level (0.4%). Thus, the East South Central has the lowest levels of transit use, walking, and cycling, and is the most dependent on the auto for all travel.

In addition to the regional concentration of transit on the East and West Coasts, public transportation is also concentrated in the largest of U.S. metropolitan areas, with transit use strongly correlated with city size. The 10 largest U.S. urbanized areas (with 23% of the country's total population) are home to 76% of the country's transit riders (Federal Transit Administration, 2002). Indeed, per capita transit use in 2000 was 10 times higher, on average, in those 10 largest metropolitan areas than in the rest of the country (112 vs. 11 trips per year).

Corresponding to the spatial concentration of public transportation passengers in large cities, transit's market share of travel is also higher in large cities. As shown in Table 8.2, residents of metropolitan areas with 3 million or more inhabitants made 3.4% of all their trips by transit in 2001 (including all trip purposes, not just the work trip). That market share (or "modal split share," as it is often called) drops to 1.1% for metropolitan areas with 1–3 million population and only 0.6% for metropolitan areas with population between 250,000–1 million.

The spatial variation in public transportation use is equally striking within each individual urbanized area. Residents of central cities are, on average, over seven times more likely to use public transit than are residents of the suburbs. For all U.S. urbanized areas in aggregate, transit accounted for 8.3% of all trips made by central-city residents in 1995, compared to only 1.2% of trips by suburbanites (see Table 8.2). Even within the largest city-size category (3-million-plus), central-city residents made six times as high a percentage of their trips by transit as suburban residents (9.7% vs. 1.6%).

TRENDS IN PUBLIC TRANSPORT USAGE, 1945–1975

In the 30-year period after World War II, public transportation in the United States experienced a sharp decline in both passenger and service levels. Total annual passenger trips fell by 69%, from 22.3 billion in

TABLE 8.2. Spatial Variation in Public Transportation's Market Share by Size and Portion of Metropolitan Area, 1995 and 2001 (Percent of Total Trips by Transit)

Metropolitan area population size	1995			2001
	Central city	Suburb	Total	Total
3 million +	9.74	1.62	3.77	3.41
1.00–2.99 million	3.10	0.61	1.01	1.08
0.50–0.99 million	2.71	0.64	0.88	0.63
0.25–0.49 million	NA	0.39	0.48	0.58
All metro areas	8.26	1.16	1.81	1.69

Note. Unlike the decennial census data, these NPTS and NHTS statistics include all trip purposes, not just work trips. The totals for 2001 are for the metropolitan area totals only, since the central city versus suburban breakdowns are not yet available.
Sources: Pucher and Renne (2003, Table 9), based on data from the 2001 National Household Travel Survey; and Center for Urban Transportation Research (1998, Table A-1), based on data from the 1995 Nationwide Personal Transportation Survey.

1945 to only 8.0 billion in 1975. Using 1945 as the baseline for those calculations, however, exaggerates the decline. Transit use was at its all-time high during World War II, artificially inflated by gas and tire rationing, a halt to new car production, restricted incomes, and a patriotic appeal to use transit to save resources needed in the war effort. Nevertheless, even relative to 1939 (12.9 billion passengers), the fall in transit use by 1975 was substantial: 46% fewer transit trips.

A number of factors contributed to public transportation's decline in the United States. Rising per capita incomes, together with the improved quality and reduced price of cars, led to rapid growth in auto ownership. At the same time, the urban and suburban roadway network expanded rapidly, further encouraging auto use and suburban living. In particular, the metropolitan portions of the Interstate Highway System, built mainly in the 1960s and 1970s, provided an important new impetus to decentralization (Transportation Research Board, 1998a). They opened up vast new suburban areas for low-density homes and workplaces. Most new shopping centers, office complexes, warehouses, and manufacturing plants were located on or near Interstate freeways.

In addition, the federal government provided enormous subsidies to homeownership through income tax deductions and exemptions as well as through loan guarantees (Aaron, 1972). Most of that subsidy was used to finance low-density, single-family homes in the suburbs. Although homeowner subsidies provided an additional impetus to decentralization, many residences and workplaces would have moved to the suburbs anyway. With rising per capita incomes, single-family homes became increasingly affordable. In addition, the rapid increase in population stemming from the postwar baby boom ensured an outward expansion of U.S. cities at any rate, simply to accommodate the additional households. Moreover, technological changes in manu-facturing and materials handling led to a shift from vertical to horizontal plant layouts, thus requiring the more extensive and cheaper land in the suburbs.

All of these inducements to decentralization hurt public transportation, which operates most efficiently in dense, high-volume corridors focused radially on a central downtown area. As discussed in Hanson (Chapter 1, this volume), virtually all of the growth in metropolitan travel has been in the nonradial trips either within or between suburbs. Until 1975 public transportation systems hardly even tried to adapt to changing land use and travel patterns. Thus they almost entirely lost that growing suburban portion of the market.

One reason for the failure to provide service to the rapidly growing suburbs was the severe underfunding of public transportation in the three decades following World War II. Before 1960, most public transit operators were for-profit, privately owned companies, and as such did not receive government subsidies, nor were they heavily regulated by government (Altshuler, Womack, & Pucher, 1979; Pucher, 1995). Their response to increased competition from the automobile was to raise fares, cut services, and delay maintenance and capital investment. As their services deteriorated and became more expensive, public transit systems lost more and more passengers, curtailing revenues further, and forcing yet more service cutbacks and fare hikes. Although increasing auto ownership, rising per capita income, decentralization of homes and jobs, and lack of government financial assistance were largely outside the control of public transit systems themselves, transit companies had no coherent strategy to retain their remaining passengers, let alone to tap the new suburban market. Incompetent, unmotivated management was also a problem in some cities, with managers simply playing the end game of waiting for bankruptcy and public takeover. Innovation of any sort was rare, and marketing was virtually nonexistent (Salzman, 1992).

TRENDS IN PUBLIC TRANSPORT USAGE SINCE 1975

As transit companies went bankrupt, government subsidies and public ownership proliferated, with a virtually complete government takeover of public transit systems by 1980. Thanks to a huge infusion of government funds since the mid-1970s, public transportation systems throughout the United States have been resuscitated. The total annual subsidy from all government levels (federal, state, and local)—and for both operating and capital expenditures—rose from $3.2 billion in 1975 to $22.8 billion in 2000 (American Public Transportation Association, 2002; U.S. Department of Transportation, 2002). Even adjusting for the declining value of the dollar caused by inflation, that subsidy growth amounts to a tripling in the real, constant-dollar value of government assistance over the last quarter of the 20th century.

In most cities, public transportation services were consolidated into metropolitan-wide transit authorities that coordinate fares, routes, and schedules throughout the city and suburbs. Since the governing boards of such transit authorities usually include both state and local government officials, they are highly politicized. They focus more on satisfying political constituencies than on achieving economic efficiency or engineering optimality. That has led to some ma-jor distortions in transit decision making (Altshuler et al., 1979).

Federal subsidies were key to sparking the long-term growth in government assistance to transit. Initially, federal funds were restricted in their use. They were primarily intended to promote technological advances, capital investment for replacing outdated infrastructure and vehicles, and comprehensive, coordinated long-term planning. Over time, the allowable uses of federal aid expanded to include maintenance and operations. Nevertheless, federal assistance has always been mainly for capital investments, which accounted for 82% of the total federal subsidy in 2000 (American Public Transportation Association, 2002).

State and local subsidies as well were essential in rescuing the transit industry from its financial quagmire. Moreover, they have been crucial in offsetting stagnating federal contributions, which fell from 53% of the total subsidy when President Reagan took office in 1980 to only 24% when President Bush left office in 1992. The federal proportion has remained at about one-fourth through 2000 (American Public Transportation Association, 2002). But thanks to rising state and local subsidies, the total subsidy has continued to grow and thus has enabled public transportation to expand its services and offer more attractive fares, thus luring back some lost passengers.

Some studies suggest that transit subsidies

FIGURE 8.2. Through repeated expansions, the Washington, D.C., Metro system has attracted so many riders in the past two decades that it is now the second largest subway system in the United States. Photo courtesy of the Washington Metropolitan Area Transit Authority; Larry Levine, photographer.

have been wasted through inefficiency, excessive wage increases, overstaffing, expensive federal regulations, and overinvestment in high-cost technologies (Pickrell, 1992; Richmond, 2001a, 2001b). Nevertheless, transit subsidies unquestionably have produced substantial benefits. They helped finance an 86% increase in annual vehicle miles of service, almost doubling total transit service supplied (from 2.2 billion in 1975 to 4.1 billion in 2000). Moreover, the quality of public transportation services has risen dramatically. Virtually all rail cars and buses have been replaced by modern, comfortable, air-conditioned vehicles. Old metro systems and suburban rail systems, such as those in New York, Boston, and Philadelphia, have been upgraded through station modernization and by rehabilitation of tracks, tunnels, and signaling systems. In addition, many new rail systems have been built, and existing systems have been extended, amounting to 1,780 miles of new rail routes since 1975, an increase of 42% (American Public Transit Association, 1977, 2000).

Finally, both bus and rail systems have been made more accessible to passengers with physical disabilities. Most buses now have lifts, low floors, or kneeling features for easier boarding. Many rail stations have been built or retrofitted with elevators and ramps. And both bus and rail vehicles offer special seating areas for wheelchair users. Moreover, as mentioned earlier, transit systems now provide extensive van or minibus services for those passengers with such severe disabilities that they need door-to-door transportation.

Government subsidies have not only financed substantial expansion and improvement of services; they have also kept fare increases low. When adjusted to control for inflation, the average passenger fare per unlinked trip rose only 8% over the entire 25-year period between 1975 and 2000, or about 1% every 3 years. Thus, transit fares have increased only slightly faster than the overall inflation rate. Those nationwide averages hide significant variation among different cities, but they suggest that fares have generally remained low in most cities. Indeed, for the transit industry as a whole, passenger fares in 2000 covered less than one-fourth (23%) of combined capital and operating expenses (American Public Transportation Association, 2002).

Equally important, transit fare structures and ticketing procedures have improved. Virtually all large transit systems now offer monthly passes that provide substantial discounts for regular transit riders. Fare technology has improved through the use of smart cards (with computer chips), magnetic-stripe fare cards, and proof-of-payment tickets (self-ticketing). The diversification of fare offerings and more customer-friendly ticketing have surely promoted greater transit usage. That is best seen in New York City, which introduced the discounted MetroCard in 1997, providing a variety of new fare options for all its bus and subway services. That caused ridership to jump 30%, although virtually no new services were added (Pucher, 2002).

For the country as a whole, service expansion and low fares financed through large subsidies succeeded in turning around the decades-long decline in public transport ridership. Total annual passenger trips rose from only 8.0 billion in 1975 to 9.5 billion in 2001, an increase of 36%. That was a considerable accomplishment, given the strong trends promoting ever more auto ownership and use. As discussed by Hanson (Chapter 1, this volume) and Muller (Chapter 3, this volume), rising auto ownership and the continuing decentralization of both residences and employment to low-density suburbs increasingly encourage auto use and inhibit public transportation.

As Figure 8.4 shows, public transit usage has been quite cyclical over the past three decades. Generally, the upswings are due to economic booms (low unemployment), rising gasoline prices, and large funding increases that permit service expansion and low fares. The downswings are generally

FIGURE 8.3. Ridership on suburban rail service boomed during the 1990s, resulting in overcrowded platforms and vehicles on systems such as New Jersey Transit, shown here. Photo courtesy of New Jersey Transit; Michael Rosenthal, photographer.

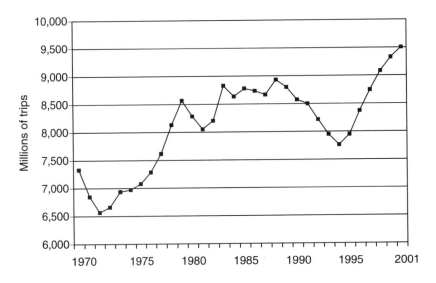

FIGURE 8.4. Long-term trends in public transport ridership in the United States, 1970–2001 (millions of unlinked passenger trips). Source: American Public Transportation Association, *Public Transportation Fact Book*, various years.

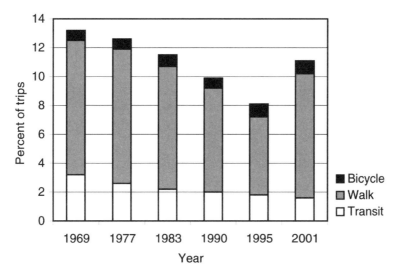

FIGURE 8.5. Trends in public transport, walking, and cycling shares of urban travel in the United States, 1969–2001 (percent of person trips by type of transport, all trip purposes). Source: Pucher and Renne (2003, Table 1), as calculated from the Nationwide Personal Transportation Surveys 1969–1995 and the 2001 National Household Travel Survey. Note: The large jump in walk trips from 1995 to 2001 was due to a significant change in survey methodology that captured previously unreported walk trips. It is not due to an actual increase in walking.

due to recessions (high unemployment), falling gasoline prices, and transit funding cutbacks (or stagnation) that require fare hikes or service cuts. Whatever the ups and downs of public transportation, the long-term trend over the past three decades has been distinctly upward, with total ridership in 2001 at its highest level since 1959, over 40 years ago (Pucher, 2002).

In spite of the recovery of passengers by public transportation over the past 25 years, transit's overall market share of urban travel continues to fall, although at a slower rate than previously. As shown in Table 8.3 and Figure 8.5, public transport's share of total local travel in the United States (all trip purposes) fell steadily from 1969 to 2001: from 3.2% in 1969 to 1.6% in 2001. Those modal split statistics are from the Nationwide Personal Transportation Survey (1969–1995) and the National Household Travel Survey (2001), the only federal surveys of

TABLE 8.3. Trends in Transit, Walk, and Bike Shares of Urban Travel, 1969–2001 (as Percentages of Total Person Trips, All Trip Purposes)

Transport mode	1969	1977	1983	1990	1995	2001
Transit	3.2	2.6	2.2	2.0	1.8	1.6
Walk	9.3	9.3	8.5	7.2	5.4	8.6
Bicycle	0.7	0.7	0.8	0.7	0.9	0.9

Note. The large jump in the modal share of walking from 1995 to 2001 is due to a significant change in survey methodology in 2001 that captured previously unreported walk trips. It is not due to an actual doubling of walk trips from 1995 to 2001. Source: Pucher and Renne (2003, Table 2), as calculated from the 1969–1995 NPTS surveys and the 2001 NHTS survey.

travel for all trip purposes. The decennial census only reports information about the journey to work, which accounts for one-fifth of all urban trips (but for two-fifths of transit trips). Those census data, shown in Figure 8.6, indicate a continuing but slowing decline in transit's share of total work trips: from 12.6% in 1960 to 8.5% in 1970, to 6.2% in 1980, to 5.1% in 1990, to 4.7% in 2000 (U.S. Census Bureau, 2002a; U.S. Department of Transportation, 1993).

It may seem strange that transit's market share fell even during the 1990s, when total transit ridership rose by 36%. The explanation is that total urban travel has risen faster than transit use—due to population growth, increasing per capita incomes, more work trips by women, and continuing decentralization of both housing and jobs to the suburbs, leading to more and longer trips of all kinds.

In short, although public transportation has made impressive progress in expanding and improving its services since 1975—thus attracting 36% more passengers—its share of total urban travel continues to fall: down to 1.6% in 2001. Some might interpret that

low market share as proof of the virtual irrelevance of transit for most Americans. For a number of reasons, that interpretation is not justified.

IMPORTANCE OF PUBLIC TRANSPORTATION

Public transit is far more important than might be suggested by the small percentage of trips it serves in U.S. metropolitan areas. Clearly, federal, state, and local government officials—and their voting constituencies—would not have been willing to invest $23 billion a year in subsidies to public transit if the only benefit were transporting less than 2% of the population. Although economists have called transit subsidies into question, political scientists explain them on the basis of the general public's widespread perception that public transit helps to relieve congestion, save energy, reduce pollution, revitalize cities, provide mobility to the disadvantaged, and ensure basic mobility options for everyone (Altshuler et al., 1979; Dunn, 1997). In short, transit subsidies have been justified on the basis of perceived indi-

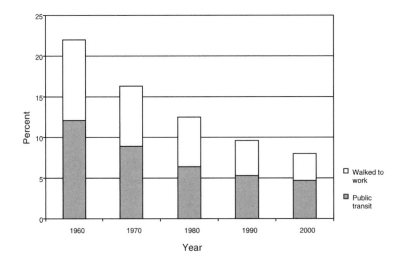

FIGURE 8.6. Trends in public transport and walking shares of work trips in the United States, 1960–2000 (percent of total work trips by type of transport). Sources: Pucher and Renne (2003, Table 1); U.S. Department of Transportation (1993).

rect social and environmental benefits far beyond the direct benefits to riders themselves. One can argue about the actual extent of such indirect benefits, but they have clearly formed the political basis of support for transit.

Transit's Special Role in Large Cities

First of all, public transportation is a crucial travel mode in the largest and densest cities. For example, it accounts for 25% of all work trips in the New York metropolitan area and for 53% of work trips in New York City itself. As shown in Table 8.4, the most important U.S. cities depend on transit to get a significant percentage of their residents to work. Even an extremely auto-dependent city like Los Angeles counts on transit to serve a tenth of all work trips.

TABLE 8.4. Public Transportation's Market Share of Work Trips in Selected Metropolitan Areas, 2000 (Percentage of Total Work Trips)

Metropolitan area	Portion of metropolitan area	
	Central city	Entire metro area
New York, NY	53	25
Washington, DC	33	11
Boston, MA	32	14
San Francisco, CA	31	10
Chicago, IL	26	12
Philadelphia, PA	25	10
Baltimore, MD	22	8
Pittsburgh, PA	21	6
Seattle, WA	18	7
Minneapolis, MN	15	5
Atlanta, GA	15	4
New Orleans, LA	14	6
Los Angeles, CA	10	7

Source: U.S. Bureau of the Census (2002a), 2000 Census, Journey to Work in U.S. Metropolitan Areas, 2002.

Congestion Relief

Transit's contribution to mobility is especially important because it provides the most service at peak travel times in the most congested travel corridors. Indeed, public transportation's unique comparative advantage is that it thrives on high volumes of travel demand concentrated in space and time. Transit's cost per passenger served decreases sharply as travel volume increases, while cost per passenger for the private car (including time cost) increases (Vuchic, 1981).

A metro line, for example, can generally carry up to 36,000 passengers per hour in one direction. In New York City, one subway line exceeds even that volume (39,526), and three lines exceed 30,000. By comparison, a single lane of limited-access highway carries a maximum of 2,500 vehicles per hour. Since most cars contain only the driver during peak hours, the passenger volume rarely exceeds 3,000 per hour per lane, only 1/12 of the metro passenger capacity. In Boston and San Francisco, light rail lines carry 9,600 and 13,100 passengers per hour, respectively, during the rush hour in the peak direction, exceeding the passenger capacity of four or more lanes of highway traffic (Transportation Research Board, 2000). Even bus lanes can carry dramatically higher passenger volumes than normal highway lanes. The single bus lane on I-495 from New Jersey to the Lincoln Tunnel carries 32,600 passengers per hour in the peak direction, roughly five times the combined volume of the three Manhattan-bound car lanes (Levinson & St. Jacques, 1998; Transportation Research Board, 1996; Vuchic, 1981; World Bank, 2000).

Transit takes a load off the roadway network precisely when it is most likely to reach its capacity limit during rush-hour congestion. That benefits not only transit passengers but also car occupants on the less congested roadways. The special contribution of transit in serving peak-hour travel is not reflected in the low aggregate market share percentages.

Of course, not every transit route in every city reduces highway congestion. Clearly, there are some transit services that are so underutilized that they produce little if any congestion relief. For example, buses on some routes run nearly empty in extremely low-density cities, suburbs, and small cities. In contrast, transit vehicles in large, dense cities are usually filled, if not overflowing, during rush hours, thus providing undeniable roadway congestion relief. Provided that public transportation attracts enough riders per vehicle, it inevitably takes some of the traffic load off of parallel roadways or at least permits a higher overall passenger traffic volume in any given corridor.

Energy Savings

Most studies show that public transportation—especially rail transit—is more energy efficient than the private car (Kenworthy, 2002; Lenzen, 1999; Oak Ridge National Laboratory, 2001). As shown in Table 8.5, specific estimates of energy use vary widely because of different methodologies, underlying assumptions, and data sources. Moreover, there is enormous variation due to

different vehicle occupancy rates among cities and within the same modal category. Thus, some single-occupant sport utility vehicles (SUVs) require over 10 times more energy per passenger-kilometer than an energy-efficient subcompact car filled with five passengers. Likewise, a standard-sized bus with one or two passengers uses more than 10 times as much energy as a rush-hour subway in New York City carrying more than 100 passengers per car.

On average, however, transit is usually more energy efficient, especially for the work trip, when transit vehicles are full of passengers and auto occupancy rates are the lowest. In the United States, the average car had only 1.1 occupants for the work trip in 1995. Thus, roughly nine out of every 10 cars had only the driver and no passengers at all on the journey to work. Since the estimates in Table 8.6 are for all times of day and all trip purposes, they underestimate the relative energy efficiency of public transport for the journey to work, the main purpose of transit use.

As shown in Table 8.5, European transit is much more energy-efficient than U.S. public transport, thanks to more advanced technology, higher occupancy rates, and traffic priority measures that allow buses to

TABLE 8.5. Comparisons of Energy Use per Passenger-Kilometer for Different Transport Modes (in Megajoules)

Transport mode	USA		Australia	W. Europe
	ORNL (2001)	Kenworthy (2002)	Lenzen (1999)	Kenworthy (2002)
Car	3.8	3.3	4.4	2.5
Personal truck	4.8	NA	NA	NA
Transit bus	5.1	2.9	2.8	1.2
Light-rail transit	3.3	0.7	2.1	0.7
Metro (heavy rail)	3.3	1.7	2.8	0.5
Suburban rail	3.1	1.4	na	1.0

Note. These estimates are not strictly comparable. The Oak Ridge National Laboratory (ORNL) estimates include rural and intercity car use on highways, which usually achieves higher miles per gallon than in cities. The Lenzen study includes only urban modes, while the Kenworthy study includes only large cities. All figures are for operating costs only; they do not include energy used in capital expenditures.

TABLE 8.6. Modal Shares of Transit Use, Walking, and Cycling by Race/Ethnicity, 2001 (Percentage of Trips in Each Racial/Ethnic Group by Means of Transportation)

Mode of transportation	Race/ethnicity			
	African American	Asian	White	Hispanic
Total transit	5.3	3.2	0.9	2.4
Bus and light rail	4.2	1.8	0.5	2.0
Metro/subway/heavy rail	0.9	1.1	0.3	0.3
Commuter rail	0.2	0.3	0.1	0.1
Total nonmotorized	13.2	12.3	9.6	12.6
Walk	12.6	11.7	8.6	11.8
Bicycle	0.6	0.5	0.9	0.9

Source: Pucher and Renne (2003, Table 13), calculated from the 2001 National Household Travel Survey.

achieve faster speeds. Thus, U.S. transit has the potential to be more energy-efficient than it is now.

Environmental Advantages

Many of the same considerations that complicate intermodal comparisons of energy efficiency also plague comparisons of environmental impacts. As shown in Chapter 13, the environmental effects of transportation are quite varied, including air, water, and ground pollution, ecological disruptions, and global warming. It is difficult to isolate and quantify such environmental impacts, including both short-term and long-term effects.

Most studies find that transit tends to be less polluting on a per-passenger-kilometer basis (Transportation Research Board, 2002; U.S. Environmental Protection Agency, 2001). There are large variations, however, by specific type of pollution. For example, buses as well as light rail systems emit much lower levels of carbon monoxide (CO) and volatile organic compounds (VOCs) than do cars, but higher levels of nitrogen oxides (NOx). Moreover, as with energy comparisons, there are considerable differences by type of transit mode and by specific city. As usual, the higher the vehicle occupancies, the lower is the pollution index per-

passenger-kilometer. Thus transit's relative environmental advantage is greatest in large, dense cities on heavily used routes with crowded vehicles during the rush hour. By contrast, buses with one or two passengers during the evening on lightly used suburban routes are more polluting than most cars.

As with energy use, it is also important to consider the geographic impacts of car use on the environment. Since people travel longer distances in auto-dependent cities than in transit-oriented cities, auto-dependent cities also generate more pollution from personal transportation. Even if the auto were exactly as polluting as transit on a per-kilometer basis, the transit-oriented city would generate less pollution simply because the car-oriented city requires more travel. That long-run spatial component to the environmental impacts of transportation is usually ignored in intermodal comparisons.

Kenworthy (2002) estimates that Western European cities generate less than half as much air pollution per capita from transportation as U.S. cities (98 vs. 265 kilograms per capita). That large difference is due not only to fewer kilometers of travel in European cities (thanks to a more compact land use pattern) but also to much greater use of the less polluting modes of public transportation, walking, and cycling. Air pollution per capita in Canadian cities (179 kilograms per capita) is also less than in the United States, but almost twice the Western European level.

Mobility for the Disadvantaged

Transit plays an important role in ensuring basic mobility needs for many of the poor, minorities, elderly, and disabled who either cannot afford a car or are physically or mentally unable to drive. In 2001 transit accounted for 5.3% of all trips by African Americans and 2.4% of all trips by Hispanic Americans, compared to only 0.9% of trips by whites (see Table 8.6). Perhaps even more striking (but not shown in Table 8.6), African Americans and Hispanic Americans together accounted for 62% of all bus riders, 35% of all metro riders, and 29% of all suburban rail riders (Pucher & Renne, 2003). To put these figures in context, African Americans and Hispanic Americans comprised only 20% of metropolitan households in 2001 (Pucher & Renne, 2003). In auto-oriented cities, transit-dependence among minorities is even more striking. For

example, 80% of bus riders in Los Angeles are persons of color (Duran, 1995).

Similarly, poor Americans are far more dependent on transit than affluent Americans (see Figure 8.7). Public transportation accounted for 5% of all trips by households earning less than $20,000 in 2001, compared to a transit market share of 1.0%–1.5% for households earning over $20,000. Moreover, poor households accounted for 38% of all transit riders in 2001 (far more than their 14% share of the urban population) (see Table 8.7). Indeed, households with incomes less than $20,000 comprised almost half (47%) of all bus riders in 2001 and 20% of all metro riders. By comparison, only 6% of commuter rail riders had incomes below $20,000. Thus, the transit-dependent poor rely primarily on buses to get around, and only secondarily on metro services, mainly in the largest cities with extensive multi-modal transit systems. Overall, however, public transportation clearly plays a special role in serving the transport needs of the disadvantaged.

It is noteworthy that the poor also rely much more heavily than affluent households on walking and bicycling for their ur-

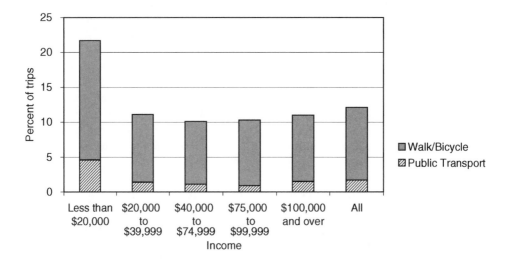

FIGURE 8.7. Differences among income classes in modal shares of public transport, walking, and cycling for urban travel in the United States, 2001 (percent of trips by type of transport, all trip purposes). Source: Pucher and Renne (2003, Table 8 and Figure 2), as calculated from the 2001 National Household Travel Survey.

TABLE 8.7. Income Distribution of Transit Riders, Pedestrians, and Cyclists, 2001 (Percentage Composition of Each Mode's Users by Income Class)

Mode of transportation	Household income					
	Less than $20,000	$20,000 to $39,999	$40,000 to $74,999	$75,000 to $99,999	$100,000 and over	All
Total transit	37.8	19.8	21.0	7.4	14.1	100
Bus and light rail	47.1	21.4	19.0	5.6	6.8	100
Metro/subway/heavy rail	19.7	18.7	25.2	9.1	27.2	100
Commuter rail	6.3	7.0	26.1	19.1	41.6	100
Total nonmotorized	22.7	22.8	27.4	12.8	14.3	100
Walk	23.6	22.6	26.9	12.6	14.2	100
Bicycle	13.5	24.1	32.8	15.0	14.6	100

Source: Pucher and Renne (2003, Table 10), as calculated from the 2001 National Household Travel Survey.

ban travel (see Figure 8.7). While households earning less than $20,000 made 17% of their trips by either walking or cycling in 2001, those nonmotorized modes accounted for only 9–10% of trips by households earning $20,000 or more (Pucher & Renne, 2003). Similarly, Hispanic Americans and African Americans make a higher percentage of their trips by walking and cycling than white Americans (13% vs. 10%) (see Table 8.6). As discussed later in this chapter, walking and cycling are important complementary modes to public transit. Thus policies to improve public transport must be accompanied by measures to promote the safety and convenience of walking and cycling. All three modes—transit, walking, and cycling—are crucial to the mobility needs of the poor and minorities. Chapter 12 takes up these issues in considerable detail.

PRESENT AND FUTURE PROBLEMS AND CHALLENGES

Despite impressive progress over the past 25 years, public transportation in the United States faces enormous challenges. Perhaps the most obvious problem is the need to finance growing operating deficits and in-

vestment costs. While transit subsidies have grown steadily over the past few decades, it is not inevitable that the political support for those subsidies will continue unabated. Thus it will remain crucial for the transit industry to convince the public and its elected representatives that large and growing transit subsidies are justified. In the face of overall budget cutbacks and competing demands for public funds, that political task will be formidable. The geographic concentration of transit use in the United States makes it especially difficult to obtain political support from auto-oriented areas, which include most low-density cities, most suburbs, and all the rural states. In addition, the transit industry could do more to increase the efficiency of its operations and thus reduce costs. For example, transit systems could streamline management, reduce overstaffing, use more part-time labor for peak hours, eliminate routes with few riders, and contract out more services to private firms. In addition, transit might choose simpler, less expensive technologies—such as express bus instead of rail—in those situations where they achieve approximately the same service outcomes (Richmond, 2001b).

Nevertheless, the transit industry has already made considerable progress in controlling costs in recent years. Between 1982 and 1997, hourly transit wages increased 16% less than the wages of government workers in the same metropolitan areas, and 22% less than wages of manufacturing workers in the same areas (Denno & Robins, 2000). Moreover, operating costs per vehicle mile rose more slowly than the overall inflation rate between 1980 and 2000, falling by 4% in real, inflation-adjusted terms (American Public Transit Association, 1992, 2002; U.S. Bureau of the Census, 2002a). That is a dramatic improvement over the decade 1970–1980, when operating costs per vehicle mile tripled, rising 31% faster than the overall inflation rate (Pucher, Markstedt, & Hirschman, 1983).

The most fundamental underlying challenge of public transportation has been—and will continue to be—the need to adapt to increasingly polycentric, dispersed, and usually low-density suburban development. As discussed in Hanson (Chapter 1, this volume) and Muller (Chapter 3, this volume), all indications are that both residences and jobs will become even more decentralized in the coming years and decades. Thus travel will continue to shift away from radial corridors focused on downtown, precisely the travel demands best served by current transit technologies. Almost all growth will be in travel within and between suburbs, between tens of thousands of origins and destinations scattered throughout each metropolitan region.

Transit faces two basic choices. It could ignore the difficult-to-serve suburban market and concentrate only on what it does best by limiting itself to high-volume radial routes and a smaller, more compact service area around a central core. Only the best-used and most efficient routes would be maintained, eliminating much suburban service completely. While that would save subsidy funds, it would destroy the regional systems created over the past few decades, including many lines needed to feed the principal routes. It would also remove the

broad political support that transit needs to continue to obtain state and federal subsidies, as well as metropolitan regionwide dedicated transit taxes, which are the main source of local subsidy funds. From an equity perspective, eliminating most suburban service would reduce the already limited access of the inner city's poor to suburban jobs.

The alternative is to modify transit networks to better serve further decentralizing, polycentric, land use patterns. As discussed later in this chapter, several cities have already established multidestination hub-and-spoke route systems. Their coordinated "timed transfers" at dispersed hubs are specifically designed to promote transit use within and among suburbs, as well as reverse commuting from the downtown to the suburbs. Other such adaptations involve so-called community transit systems that use vans or minibuses for local travel in the suburbs and as feeders to higher capacity transit routes for longer trips.

These approaches would be greatly facilitated if local governments adopted land use policies to encourage clustered development around key transit hubs, thus producing islands of high-density, transit-oriented development (TOD) at suburban centers. Of course, TODs can enhance transit use even within most existing monocentric route networks. The actual experience with TODs is discussed later in this chapter.

The other basic challenge of transit is the difficulty of competing with the omnipresent automobile. Almost all U.S. households (92%) possessed at least one car by 2001, and well over half (59%) had two or more cars. Even among households earning less than $20,000, 48% owned one car, and 25% owned two or more (see Figure 8.8). For most trips, the private car provides higher quality transportation: it is faster, more comfortable, more flexible in timing and routing, and more practical for carrying things. The main exception is for trips in highly congested radial corridors leading to

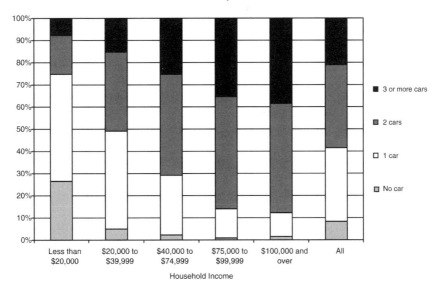

FIGURE 8.8. Auto ownership levels in the United States by income class, 2001 (percent of households in each income class). Source: Pucher and Renne (2003, Figure 1), as calculated from the 2001 National Household Travel Survey.

dense city centers with limited and expensive parking. In spite of the many deleterious social and environmental impacts attributed to cars (see Chapter 14), they remain very popular with private consumers. Thus car use becomes more dominant in U.S. urban transportation each year.

Although transit technology has improved over the past few decades, automobile technology has improved faster. Not only are cars safer now than they were in the 1960s, but they are less polluting, quieter, and include some features hardly dreamed of a few decades ago: advanced entertainment systems, global positioning systems (GPS) for navigation, and elaborate climate control, for example. While a transit ride is more comfortable now than in the 1960s, it can hardly compete with the level of comfort, privacy, convenience, and door-to-door time savings offered by even the simplest cars, especially when one considers the walk to and from the transit stop.

The dual challenges of decentralizing land use and competing with affordable, high-quality private transportation are daunting but not necessarily insuperable. As shown in the following section, European and Canadian transit systems have been quite successful at attracting passengers and maintaining a high modal-split share of urban travel—in spite of high per capita incomes, high and rising auto ownership rates, and decentralization of both jobs and homes. Europe and Canada provide many potential lessons for improving public transportation in the United States.

EUROPEAN AND CANADIAN PUBLIC TRANSPORT

As shown in Figure 8.9, transit's share of urban travel in the United States is much lower than in Canada and Western Europe. The lowest transit market share in Europe (7% in The Netherlands) is still more than three times higher than in the United States. Most other European countries have transit shares at least five times higher than the United States, with some over 10 times higher (e.g., Switzerland at 20%). Similarly, walking and bicycling account for a very

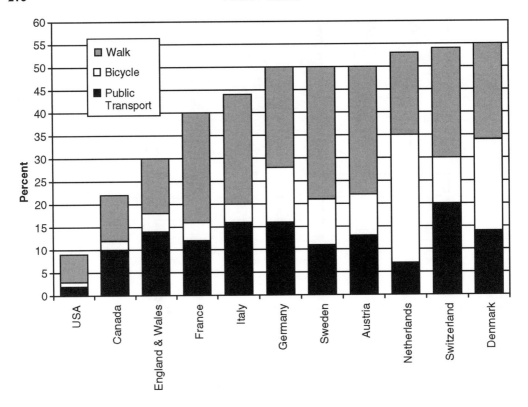

FIGURE 8.9. Variations among countries in modal split shares of public transport, walking, and cycling, 1995 (percent of all urban trips by type of transport, all trip purposes). Source: Pucher (1999, Table 7), based on data provided by each country's national department of transportation or ministry of transport. Note: Although more recent data are available for some countries, all modal split shares shown in this figure are for 1995 in order to ensure comparability among all countries. Due to a change in survey methodology, the 1995 walk share for the United States is considerably lower than that reported by the 2001 National Household Travel Survey (Pucher & Renne, 2003).

low percentage of urban trips in the United States (only 7%) compared to Canada (12%) and most Western European countries (30–40%; see Figure 8.9). It is no coincidence that countries with high transit use also have much walking, since that is the main access mode for transit. Likewise, many European cities depend on bicycling access to their transit services (Pucher, 1997; Pucher & Dijkstra, 2000). In addition, good transit services make auto ownership less necessary, thus enabling some households to be car-free, using transit for long trips and walking or cycling for short trips.

An important question to ask is why public transportation, walking, and cycling are so much more important for personal travel in Canadian and Western European cities. Clearly, it is not the case that transit always provides an inferior level of service used only by the poor and others without any other travel option. Canadians and Western Europeans are as affluent and as technologically advanced as Americans. Yet they willingly choose transit, walking, and cycling for a high percentage of their trips, even though they have auto ownership levels not far below those in the United States.

Differences in Land Use and Urban Development Patterns

There are many reasons for the dramatically higher transit modal-split shares in Canada and Western Europe. Some of these factors are beyond the control of public transport systems themselves. Thus both Canadian and Western European cities are much denser, less decentralized, and less polycentric than U.S. cities (Transportation Research Board, 2001). The most recent international comparison found that large Western European cities had an average density of 55 persons per hectare in 1995, compared to 26 persons per hectare in Canadian cities and only 15 persons per hectare in U.S. cities (Kenworthy, 2002). Thus, on average, even Canadian cities are roughly twice as dense as U.S. cities, and European cities are almost four times as dense. Although not reflected in any quantitative index, Canadian and European cities are also less polycentric than U.S. cities, with central cores that remain vibrant and far more dominant over their metropolitan regions than U.S. downtowns (Cervero, 1998; Transportation Research Board, 2001; Vuchic, 1999). All of these land use differences help explain why Canadian and European transit systems have higher market shares than U.S. systems.

To some extent, the higher European densities are historically based, legacies of development patterns that unfolded over many centuries before the automobile era. That is not the case with Canadian cities, however, which developed at roughly the same time as U.S. cities. For decades, most local and provincial governments in Canada have implemented a range of land use policies encouraging clustered development and higher densities (Cervero, 1998; Newman & Kenworthy, 1999; Pucher & Lefevre, 1996). Moreover, even the most recent suburban development around European and Canadian cities is at much higher densities than those found in the U.S. That is not due to history but to strict land use policies and the much higher price of land in Europe (Downs, 1999; Pucher & Lefevre, 1996). Thus outer suburban densities averaged 39 persons per hectare in Western Europe and 26 persons per hectare in Canada, but only 12 in the United States (Newman & Kenworthy, 1999).

Differences in Taxation of Automobiles and Gasoline

Another important reason for the higher transit market shares in Canada and Western Europe is the range of public policies that both restrict auto use and make it more expensive than in the United States. For example, the price of gasoline is roughly four times higher in Western Europe than in the U.S., and almost all of the price differential is due to much higher gasoline taxes in Europe (see Figure 8.10). Canada also has considerably higher gasoline prices than the United States, due to taxes that are about 50% higher, even though Canada exports petroleum while the United States must import over half its petroleum (International Energy Agency, 2002). In short, higher gasoline prices in Western Europe and Canada are the result of deliberate government tax policies that make auto use much more expensive than it is in the United States.

Likewise, automobile purchase prices are higher in Western Europe and Canada mainly because sales taxes, import fees, and other charges on auto ownership are so low in the United States. While sales taxes, registration fees, and property taxes amount to only about 5–10% of the purchase price of a private car in the United States, they account for 15–50% of the purchase price in most European countries. In a few countries sales taxes, registration fees, and import fees on new cars are so high that they exceed the base price. For example, Denmark charges a 180% sales tax on new cars, which obviously deters auto ownership (Transportation Research Board, 2001).

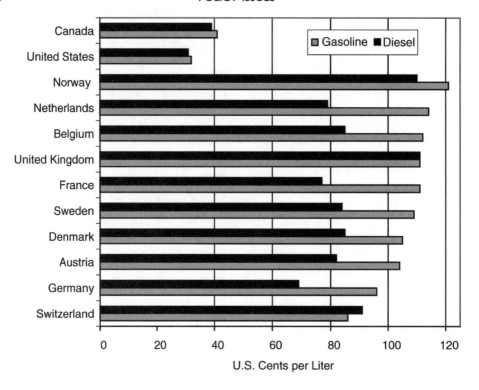

FIGURE 8.10. Variation among countries in after-tax fuel prices, 1998 (in U.S. cents per liter). Source: Transportation Research Board (2001, Figure 3.2, p. 82).

Restrictions on Auto Use That Give Public Transit, Pedestrians, and Cyclists Traffic Priority in Western Europe

In addition to making auto ownership and auto use much more expensive, Canada and especially Western European countries have imposed a wide range of restrictions on auto use. Those comprehensive restrictions on speed, rights-of-way, and routing reduce the mobility advantages of the private car, while dramatically enhancing the speed, convenience, and safety of walking, cycling, and public transport (Cervero, 1998; Pucher & Lefevre, 1996; Transportation Research Board, 2001).

Wide sidewalks, bike paths and lanes, bus lanes, and auto-free zones, for example, generally remove rights-of-way from cars. Complementing those infrastructure provisions, traffic calming limits the speed of motor vehicle traffic in residential neighborhoods, both by law—30 kilometers per hour (19 miles per hour) or less—and through physical barriers. For example, traffic calming forces drivers to slow down through raised intersections and crosswalks, traffic circles, road narrowing, zigzag routes, curves, speed bumps, and artificial dead-ends created by mid-block street closures (Ewing, 1999; Hass-Klau, 1992). That obviously deters car traffic while vastly improving conditions for pedestrians and cyclists, including those headed for transit stops.

Transit vehicles (buses and streetcars) in Europe usually have priority over cars (Transportation Research Board, 2001). The law specifies that cars must yield to buses re-entering the traffic stream after picking up passengers at the curb. Moreover, both buses and streetcars trigger automatic green lights at many intersections, thus speeding up transit travel relative to car travel. In many cities, transit vehicles as well as bicyclists have advance stop lines ahead of cars, and also benefit from advance green lights specifically for them. Together with reserved lanes and auto-free transit-pedestrian malls, all of these traffic priority measures in Europe accelerate transit vehicles, reduce operating costs, attract more riders, and enhance both the safety and the convenience of pedestrians walking to transit stops.

Differences in Parking Availability and Price

The low cost and widespread availability of ample parking also encourages auto use in the United States and deters not only transit use but also walking and cycling (Transportation Research Board, 2001). The 1990 Nationwide Personal Transportation Survey (NPTS) found that over 95% of car trips made in U.S. cities benefit from free parking. Most other parking involves nominal fees, often subsidized by employers (Shoup, 1995, 1997, 1999). U.S. firms usually offer free parking for their customers and employees—a tax-free fringe benefit for employees and a tax-deductible expense for firms. The subsidy entailed in free parking is enormous, estimated by some studies to exceed $1,000 per year per urban vehicle (Litman, 2002; World Resources Institute, 1992). Indeed, the country's leading parking expert provides convincing evidence that free parking entails more subsidy—and more inducement to drive—than would the provision of free gasoline (Shoup, 1997, 1999).

Not only is most parking free in U.S. cities, but there is lots of it, even in downtown areas. While large U.S. cities had 555 parking places per 1,000 central business district (CBD) jobs in 1995, large Canadian cities had only 390 parking places, and large Western European cities had only 261 places, less than half the U.S. parking supply (Kenworthy, 2002).

Virtually all local governments in the United States set high *minimum* parking requirements for new buildings, even in the city center. Such requirements have virtually no valid justification and lead to a large oversupply of parking spaces (Shoup, 1999). Massive parking lots, off-street parking decks, and extensive on-street parking have become trademarks of U.S. cities, claiming vast tracts of valuable urban land. In sharp contrast, Western European and Canadian cities often set strict *maximum* parking limitations for new buildings, or prohibit parking facilities altogether, specifically to reduce auto use in central cities and to encourage transit use, walking, and cycling. In addition, many Western European cities have sharply reduced the supply of on-street parking spaces while raising parking rates, limiting parking to the short term, and reserving some parking for neighborhood residents only (Transportation Research Board, 2001).

Clearly, ample free parking is a strong inducement to drive. One study found that it is also one of the most important factors deterring transit use (Transportation Research Board, 1998b). Most Americans and their politicians appear to be convinced that free parking for their cars is an inalienable right and thus take it for granted. Few realize how much it distorts their transportation choices and travel behavior.

Licensing of Car Drivers

Another policy that discourages car use in Europe, especially among the young, is the very strict and expensive licensing of new drivers. Public schools do not provide free or low-cost driving lessons, as in most U.S. states. Instead, aspiring drivers in Western Europe must purchase mandatory lessons from private firms. At a minimum, that costs $1,000–$1,500 in most countries, and much more if students fail the difficult test and must take additional lessons.

Moreover, the minimum age for obtaining a driver's license is 18 in almost all

Western European countries. Even then, licenses are often provisional for the first few years, subject to revocation (or more driving lessons) if the driver's actual performance is not satisfactory. By comparison, most Americans get their driver's license as soon as they turn 16 years old (Pucher & Lefevre, 1996; Transportation Research Board, 2001).

Although most Europeans eventually get a driver's license, they start their lives as car drivers at least 2 years later and thus continue to walk, cycle, and use public transportation at least through high school and often through their university years. That early experience probably affects their travel choices later in life as well, and also affects how they, as drivers, treat pedestrians and cyclists.

Better, More Extensive, and Cheaper Transit Services in Europe

Most of the factors examined above are beyond the control of transit systems. Of course, transit systems themselves determine, at least to some degree, the extent and quality of the services they provide. Service supply, however, is subject to funding constraints (subsidy availability), government regulations, and provision of rights-of-way by public authorities. In virtually every respect, circumstances in Europe have fostered more and better transit services.

European transit systems have been publicly subsidized and publicly owned roughly five decades longer than U.S. systems (Transportation Research Board, 2001). Some European transit systems were publicly owned as early as the 1890s, and the vast majority of European systems were publicly owned by 1920. Similarly, government subsidies had become an important source of funding for most European transit systems by the 1920s. As a consequence, European transit systems had direct, long-term access to government funds for both operations and capital investment for most of the 20th century.

In contrast, most U.S. transit systems remained privately owned until the late 1960s and early 1970s, roughly the same time that government subsidy programs started up. U.S. transit systems, as explained earlier, were only marginally profitable even in the first half of the 20th century. After World War II, transit systems became increasingly unprofitable. The almost complete lack of government support during the crucial decades from 1920 until the mid-1960s led to repeated fare increases, service cutbacks, and deterioration of rights-of-way, stations, and vehicles. That led to a massive loss of public transport passengers in the United States, while transit use in Europe increased through most of the 20th century.

The generous funding and high quality of public transportation in Western Europe have continued to the present day. With few exceptions, European transit systems offer much higher levels of comfort, convenience, reliability, safety, and speed than U.S. systems. Moreover, European systems generally have a denser and more extensive route network with more frequent service, thus providing more service to a much higher proportion of households than U.S. transit systems. Overall, European cities average about four times as many vehicle hours of transit service per capita as U.S. cities, although there is much variation among cities (Transportation Research Board, 2001, p. 34).

Table 8.8 summarizes the primary strategies European transit systems adopt to provide high-quality service. Measures to increase speed of transit travel are perhaps the most important. Many European systems have extensive, well-integrated rail lines with completely separate rights of way. Even transit vehicles operating in mixed traffic (buses and streetcars) are accelerated by traffic priority measures such as reserved lanes, turn priority, advance stop lines, and automatic green lights (triggered by approaching transit vehicles). European systems have also put great emphasis on providing modern, comfortable, attractive,

TABLE 8.8. Western European Strategies for High-Quality Public Transportation

To improve speed, reliability, and frequency of service:
- Wide spacing between bus stops to increase operating speeds
- Passenger loading platforms and curb extensions to ease bus reentry into traffic streams
- Prepaid tickets and passes to expedite passenger boarding
- Low-floor buses with wide, multiple doorways to speed boarding and alighting
- Transit priority in mixed traffic (e.g., bus lanes, special turning provisions, and priority traffic signals for buses and streetcars)
- Vehicle locator system to facilitate on-time service and provide real-time information to riders
- Extensive light rail, metro, and suburban rail systems with exclusive rights-of-way, thus insulated from roadway congestion delays

To improve comfort, safety, and convenience of service:
- Amenities at transit stops and stations (e.g., shelters with timetables, clocks, telephones; newspaper kiosks, shops)
- Clean, comfortable vehicles and knowledgeable, helpful drivers
- Widespread ticket purchasing places accepting cash, debit cards, and credit cards
- Safe, pleasant, and ubiquitous sidewalks providing easy pedestrian access to stations and secure, well-lit waiting areas
- Extensive bike parking at bus stops and rail stations plus comprehensive network of bike lanes and paths leading to transit stops
- Uniform, well-integrated and simple fare structures coordinated among all transit modes in the metropolitan area
- Wide variety of deeply discounted transit passes tailored to individual rider needs (e.g., students, commuters, elderly, families, tourists)
- Widespread publication of current schedules and posting at every transit stop
- Real-time information at transit stops (about actual arrival times) and onboard digital information screens (about upcoming stops and transfer possibilities)
- Fully integrated service network, with transit stops, schedules, and fares of different transit modes fully coordinated to ensure seamless transfers among modes and routes

Source: Adapted from Transportation Research Board (2001, p. 158) and Pucher and Kurth (1995).

clean, and safe vehicles and stations, often equipped with real-time information about arrivals, departures, and transfer possibilities.

Fare structures and ticket options in Europe are generally more attractive and more innovative than in the United States. Discounts for monthly and annual passes, in particular, are much deeper than those offered by U.S. systems. The result is that over 80% of transit riders in most European cities use monthly, quarterly, semester, or annual passes. In some cities, students and the elderly travel at nominal charges or for free. Moreover, transit systems negotiate with private firms and government agencies to sell deeply discounted tickets to entire groups of their employees. They also work with organizers of conferences, conventions, sports events, concerts, and amusement parks to include a transit ticket as part of the regular price of admission. Such combination tickets are also available from many hotels and tourist agencies (Pucher & Kurth, 1995; Transportation Research Board, 2001).

European systems employ a wide range of techniques to market their high-quality services. Although the specific measures vary from city to city, they always include advertisements in newspapers, on billboards, and on radio and TV. Most cities have networks of booths and mobile vans that distribute information and sell tickets throughout the region. There are also seasonal mass mailings to all households with information on new route schedules and fares, as well as new service offerings. In many cities, transit systems make special presentations in schools to show children how to use transit. Some German transit systems even run annual transit sweepstakes and carnivals to lure potential customers to transit (Pucher & Kurth, 1995).

In short, European transit systems offer a higher quality service at more attractive fares and do a better job at marketing than U.S. transit systems.

FUTURE OF PUBLIC TRANSPORTATION IN THE UNITED STATES

With all the ups and downs of public transit in the United States over the past few decades, it is uncertain what the future holds. The 36% increase in transit usage since 1975 has been impressive. Nevertheless, the widely predicted continuation of metropolitan decentralization and further growth in auto ownership do not bode well for public transportation. In this concluding section of the chapter, we examine a few selected issues relating to alternative transit futures.

Lessons from Europe and Canada

The high taxes and restrictions on auto ownership and use in Europe are probably the most powerful factors encouraging transit use, walking, and cycling. Unfortunately, such measures are so politically unpopular in the United States that they probably cannot be relied on to increase transit use here. In spite of decades of pleas by economists to charge higher prices and taxes for roadway use and parking, very little has changed: almost all roads and parking facilities remain free of user charges. Likewise, auto restraint measures such as auto-free zones, traffic calming, and transit priority measures have made little headway in most U.S. cities, since most car drivers vehemently oppose giving up their current privileges. Stricter land use planning to promote higher densities is probably possible at selected sites in some cities, but not on the comprehensive, coordinated scale found in Europe.

Nevertheless, many of the measures listed in Table 8.8 are indeed feasible for adaptation and implementation in U.S. cities. Certainly it would be possible to better integrate the various modes of public transportation within each metropolitan area. Fare structures, ticketing, routes, stops, and schedules could be as fully coordinated as they are in Europe, and there is really no good excuse for U.S. transit systems not having already done so. Likewise, there is no excuse for not quickly adopting the more modern, more comfortable, and more attractive transit vehicle technology already found in so many European cities. Many U.S. transit systems do not even bother to clearly mark their bus stops, let alone provide attractive shelters and post schedules specifically for that stop, as done in virtually all European cities.

Perhaps an even more important lesson from Europe is the crucial need to facilitate walking and bicycling as essential complements to transit use. These modes have been shamefully ignored in the United States. Walking and cycling in most U.S. cities are not only unpleasant and inconvenient but lethal. Traffic fatality rates per mile walked or cycled in the United States are 10 times higher than in Germany and the Netherlands, and the reason is entirely due to public policy (Pucher & Dijkstra, 2000, 2003). German and Dutch cities provide vastly better facilities for walking and cycling: wide, ubiquitous sidewalks; auto-free zones; bike paths and bike lanes; priority signals at intersections; and highly visible, enforced crosswalks. They also traffic-calm most residential neighborhoods, require rigorous traffic education of both motorists and nonmotorists, and strictly enforce laws protecting pedestrians and cyclists.

Moreover, most European cities have provided exemplary pedestrian and bicyclist access to their transit stops and stations. Safe and convenient bike paths and lanes lead to transit stops with plentiful and often sheltered, secure bike parking. In Munich alone, there are over 50,000 bike racks at rail and bus stops, specifically designed to promote cycling access to transit. There are over a million such bike racks in Germany overall, and the latest German Federal Bike Plan calls for an additional 500,000 in the coming decade (German Ministry of Transport, 2002).

Even more crucial, however, is safe and convenient pedestrian access to transit, since the vast majority of transit riders reach their stops by walking. Just as European cities provide a superior walking environment overall, they make a special effort to coordinate pedestrian facilities with transit stops and stations, since that is a crucial intermodal connection with high volumes of passengers and pedestrians (Pucher & Dijkstra, 2000; Transportation Research Board, 2001).

Through their general neglect of walking and bicycling, U.S. cities not only discourage those modes but transit use as well. Through their specific neglect of intermodal coordination of walking and cycling with transit, they further discourage transit use. As noted later, there are many compelling reasons to promote safe and convenient walking and cycling for their own sake, as they are surely the most energy-efficient, least polluting, least congesting, and cheapest of all travel modes. But they should also be promoted to encourage more transit use, since they are crucial for accessing transit stops.

Promoting Transit, Walking, and Cycling for Improved Public Health

As noted above, public transportation and walking are interdependent and complementary modes. Figure 8.9 shows that countries with high market shares of transit also have high shares of walking, while the United States has very low shares of both transit and walking. Auto-oriented environments, as in most U.S. cities, strongly discourage walking as well as cycling, even for short trips (Pucher & Dijkstra, 2003).

One serious consequence of so little walking and cycling in the United States is insufficient physical exercise on a daily basis and shocking increases in obesity levels over the past two decades, as the United States has become almost completely auto-dependent (Flegal, Carroll, Ogden, & Johnson 2002; Mokdad et al., 1999, 2001). By 2000, 31% of all adult Americans were clinically obese (with body mass index over 30) and 65% were clinically overweight (with body mass index over 25). As noted in the Journal of the American Medical Association, "obesity is a risk factor for many chronic conditions including diabetes, hypertension, hypercholestolemia, stroke, heart disease, certain cancers, and arthritis" (Flegal et al., 2002). Several studies indicate that obesity will soon overtake smoking as the most important cause of premature death in the United States (McGuiness & Foege, 1993; Peeters et al., 2003; Sturm, 2002). Moreover, whether normal-weight, overweight, or obese, physically inactive persons are two to three times more likely to die prematurely (Wei et al., 1999).

In Western Europe, where walking and cycling account for between a third and a half of all urban trips, obesity rates range from 3% in the Netherlands to 9% in Luxembourg, Finland, and Greece. The European average is only 6%, less than a fifth the American level (see Figure 8.11). Average healthy life expectancies in the Netherlands, Denmark, Sweden, and Germany are 2.5 to 4.4 years longer than in the United States, although per-capita health expenditures in those four European countries are only half those in the United States (European Commission, 2001; Organisation for Economic Cooperation and Development [OECD], 2002). One might argue that the inverse correlation between the combined walk/bike/transit share of urban travel and obesity—and its direct correlation with life expectancy—are purely coincidental, but all

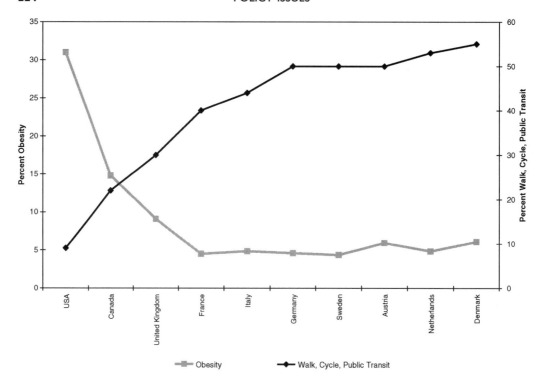

FIGURE 8.11. Variation among countries in obesity rates and modal split shares of urban travel by public transport, walking, and cycling, 1995–1999. Sources: Pucher and Dijkstra (2003); Flegal et al. (2002); European Commission (2002).

the evidence points to a causal link (World Health Organization, 2000). Moreover, the World Health Organization (2000) specifically finds that public transportation is associated with higher levels of physical activity and lower levels of obesity.

Lead editorials in medical journals have strongly encouraged more walking and cycling for daily travel as the cheapest, safest, and most feasible means to increase the physical activity of Americans (Carnall, 2000; Dora, 1999; Koplan & Dietz, 1999). Similarly, the Surgeon General of the United States specifically recommends more walking and cycling for practical, daily travel as an ideal approach to raising physical activity levels (U.S. Department of Health and Human Services, 1996). Even in the sprawling metropolitan areas of the United States, 41% of all trips in 2001 were shorter than 2 miles, and 28% were shorter

than 1 mile (Pucher & Renne, 2003). Bicycling can easily cover distances up to 2 miles and most people can walk up to a mile (Pucher & Dijkstra, 2000).

Publicizing the serious health consequences of physical inactivity and the enormous personal health benefits from more walking and cycling might help to reduce driving. That would also encourage more transit use, since walking and cycling complement transit.

Enhancing Public Transportation through Transit-Oriented Development

Low-density, suburban sprawl is one of the key factors cited by all studies for the long-term rise in automobile use and concomitant decline in public transit's share of urban travel. Recent movements toward so-called smart growth, transit-friendly design, transit joint development, and transit-oriented development all aim at raising development densities at key nodes of the transit network. In different but related ways, they attempt to re-create the sort of compact, transit-friendly land use patterns currently found in most European and Canadian cities and that also prevailed many decades ago in U.S. cities. Although they all include increased transit use as one of their goals, some of them focus on one or more other goals: reduced congestion, air pollution, and energy use; revitalization of urban neighborhoods and a heightened sense of community; greater social and economic diversity; increased traffic safety; and an improved walking and cycling environment.

Transit-oriented development (TOD) is the most comprehensive of these strategies, since it seeks to create "transit villages": socially diverse, compact, and functionally integrated mixed-use neighborhoods within 5- to 10-minute walking distance of key transit stations. Joint development, by comparison, can be part of a TOD but generally involves more isolated, project-by-project arrangements for high-density development at transit stations (Cervero, Ferrell, & Murphy, 2002; Transportation Research Board, 1997).

As noted in several recent surveys, there has been a great deal of enthusiasm about the potential of such transit-friendly developments, but the actual record seems disappointing, with limited results so far (Belzer & Autler, 2002; Cervero et al., 2002). A few TODs have been quite successful, but they account for only a tiny percentage of new metropolitan development. Nevertheless, TODs and related transit-friendly developments are likely to become more widespread in the coming years.

Cervero and colleagues (2002) cite three demographic trends favorable to TODs:

- Aging of the U.S. population, with many elderly households wanting to "downsize" their housing and ensure continued mobility without a car when they can no longer drive.
- Increasing incidence of childless couples, who also need less space, and the continued trend of young professionals, with or without children, locating in denser, more interesting urban environments.
- Increased immigration from countries with traditions of transit use and compact neighborhoods.

The continuing trend toward worsening traffic congestion in virtually all metropolitan areas might also increase the demand for housing within walking distance of rail transit, in particular, which is usually insulated from delays on the roads.

Complementing these demand-side factors supporting TODs, several recent policy shifts may foster TODs from the supply side. Many states and localities have enacted laws to restrict low-density sprawled development at the suburban fringes. By limiting the supply of land available for the usual low-density development, governments are, in effect, forcing developers to consider higher density development such as TODs. Moreover, Fannie Mae is now actively promoting "location efficient" mortgages, preferential home loans for households living within a quarter-mile of a transit stop. Finally, the federal government has recently given priority to funding transit projects specifically integrated with land use developments such as TODs and joint development (Cervero et al., 2002).

The successful implementation of TODs and joint development depends crucially on active cooperation from private developers, transit agencies, and all levels of government. Some communities strongly support TODs because of their potential to revitalize neighborhoods, especially in central cities or older suburbs. Other communities, however, strongly oppose any attempts to raise development densities and block such attempts through restrictive zoning and building codes.

Increasing traffic congestion, the growing need to revitalize older neighborhoods, the potential for local tax revenue generation, and the availability of land near transit stations (sometimes former brownfield sites or underutilized parking lots) may convince local governments to be more supportive of TODs than they have in the past.

Whatever the eventual success of TODs may turn out to be, it seems unlikely that they will ever become the dominant development form. All studies of land use and overall development patterns predict continued decentralization of U.S. cities. At best, TODs will probably not account for any more than 5–10% of total new development, perhaps forming islands of relatively high-density, transit-focused neighborhoods at a few select locations throughout the mostly low-density metropolitan area. The regional transit system would then link these "transit villages" to each other and to the urban core. That might not reshape metropolitan development patterns, but it would at least provide a boost to public transportation. It would also expand housing and travel choices for all Americans by making a less auto-dependent lifestyle a more realistic option than it is now. The vast majority of Americans will probably still choose the traditional single-family detached house in a low-density suburb, but at least they would have the option of shifting to a lifestyle where walking, cycling, and transit use are feasible for most daily travel needs.

Refocusing Transit to Serve Suburban Markets

With most new housing and jobs locating in the suburbs, the growth in travel demand over the coming decades will continue to be fastest in the suburbs, especially for trips within suburbs and from one suburb to another. Most U.S. transit systems are still radially focused on the old CBD and provide few services to meet suburban needs except for the commute to CBD jobs, which has been steadily declining as a percentage of all travel.

Several recent studies suggest a way for transit systems to reconfigure their service networks from an almost exclusive CBD focus to polycentric systems that better serve the suburbs. Thompson (1977) proposed creating route networks that are as multidestinational as the United States's rapidly decentralizing metropolitan areas. Instead of forcing all trips via radial routes to the CBD and then out again to their ultimate destinations, polycentric systems feature many key transit nodes in the suburbs. Those nodes, which could also serve as sites of TODs, would provide timed transfers of suburban routes, enabling quick, guaranteed connections among routes, similar to the hub-and-spoke system of most airline routes. The result would be less circuitous and quicker transit travel within the suburbs, and would also facilitate reverse commuting from the city center to suburban job sites. Such a shift would divert some resources from CBD-focused services, and thus would lose some of those passengers. Nevertheless, the gain of passengers in the rapidly growing suburban market might exceed those losses and result in net gains (Thompson, 1998).

Thompson and Matoff (2003) compared actual trends in passenger and service levels, costs, and various indices of efficiency and effectiveness for nine different transit systems in the United States over the 15-year period between 1983 and 1998. They found that the three systems in the sample that had switched to a polycentric route network (San Diego, Portland [Oregon], and Sacramento) performed better than most of the traditional CBD-focused systems, with larger increases in ridership per capita, more passengers per vehicle, and lower operating costs per passenger mile.

Those findings contradict almost all previous studies, which either estimated or assumed it would be too expensive to provide extensive transit services to the suburbs (e.g., Pushkarev & Zupan, 1977). It remains to be seen whether future studies confirm the potential advantages of polycentric service networks. Because an increasing number of public transportation systems in the U.S. have been shifting to polycentric route networks, there will be a better statistical basis of comparison between the performance of those systems and traditional CBD-oriented systems. At any rate, polycentric systems may offer some hope of capturing part of the booming travel demand in the suburbs and might also help curb the escalating congestion problems on suburban roadways.

For any suburban-based transit routes, it will be essential to include coordinated, flexible feeder services to reach low-density residential areas and to connect them with stops on the main bus and rail routes. Several studies have suggested an integrated system of vans, minibuses, jitneys, and taxis to provide the flexible services needed in most suburbs (Morlok, Bruun, & Vanek, 1997; Transportation Research Board, 1999). These flexible services can be fixed-route circulators, route-deviation paratransit, or fully demand-responsive dial-a-ride systems, but they virtually always utilize smaller, more maneuverable vehicles. So far, the most successful of these flexible suburban services have been minibus and van shuttles to and from suburban rail stations with high-speed links to the CBD and other high-density suburban centers. Rail appears to have enough additional attraction for riders that they are more willing to make the extra transit trip and to transfer in order to benefit from the higher line-haul speed of rail (Thompson & Matoff, 2003; Transportation Research Board, 1999).

Unfortunately, the big disadvantage of most flexible suburban services is that they are much less productive than fixed-route services within or to the central city, generating fewer riders per vehicle and requiring much larger subsidies per passenger trip. They also require more subsidy per passenger than most conventional fixed-route suburban services using full-size buses in higher volume suburban corridors. In many cases, passenger fares on flexible suburban services cover less than one-fifth of operating costs, resulting in operating deficits that exceed $5 per passenger trip (Transportation Research Board, 1999). However useful flexible van and minibus services in low-density suburbs may be, they serve far fewer passengers per subsidy dollar than most fixed-route services.

One possible solution to the problem of providing flexible services to low-density suburban areas might lie in low-cost jitney and shared-ride taxi services (Richmond, 2001a). They are usually run by private firms at a fraction of the cost of public demand-responsive systems and require little if any subsidy. Public transit systems have long opposed deregulated jitney and taxi services, because they have been viewed as a competitive threat. In very-low-density suburban areas, however, traditional transit services, as well as the more flexible minibus and van services noted above, are virtually always unprofitable, often incurring large operating deficits. Thus, low-density suburbs might be an appropriate location for experimenting with such flexible private transit, which could also feed into the main transit routes. Even if routes, schedules, and fares were deregulated, safety standards and insurance coverage would have to be strictly enforced. Of course, it might turn out that low-density areas with high auto ownership would generate so little ridership that even such low-cost jitney and shared-ride taxi services would be unsuccessful. Moreover, if any government subsidy is required, there might be significant labor union opposition to the low wages paid to the non-unionized drivers of jitney and shared-ride taxi services. But it might be worth a try.

Serving the Growing Elderly Population

Several studies suggest a potentially large demand for transit services among the rapidly growing population of Americans over 65. The latest available U.S. census projections estimate a doubling in the number of Americans over age 65 between the years 2000 and 2030, from 34.8 million to 70.3 million. The age group over 85 is projected to increase from 4.3 million in 2000 to 8.9 million in 2030 (U.S. Bureau of the Census, 2002b). The vast majority of those senior citizens will be living in low-density suburbs with little current transit service and no alternatives to the private car for their mobility needs (Transportation Research Board, 1998c). With increasing age, especially beyond 80 years old, it becomes increasingly difficult and dangerous to drive. Senior citizens are faced with few options. Many simply reduce their travel and suffer from lack of mobility. Some become completely dependent on friends or family for transportation. Others continue to drive even when their physical and mental limitations make it unsafe, thus risking their own lives and the lives of others.

Clearly, transit could play an important role in providing the rapidly growing population of senior citizens with mobility options. Already, many senior citizens benefit from dial-a-ride vans and minibuses that provide door-to-door prescheduled, shared-ride services in their local communities. Those demand-responsive paratransit services are not very frequent or extensive, however, and they are poorly integrated with each other and with the main system of fixed-route transit services in the metropolitan area.

Even as far back as the 1970s, many studies recommended the integration and coordination of demand-responsive paratransit services for the elderly, the disabled, and other disadvantaged groups (e.g., Altshuler et al., 1979). Unfortunately, most metropolitan areas still have fragmented services run by scores of different social service agencies funded by a variety of different government programs. That lack of coordination inhibits the frequency, coverage, and quality of service; it also causes very high costs for pro-

viding such service. For the country as a whole, including both central cities and suburbs, demand-responsive dial-a-ride systems cost an average of $2.09 per passenger mile in 2000, compared to $0.59 per passenger mile for full-sized buses, $0.28 for metros, and $0.45 for light rail (see Table 8.1).

Although a few states and counties have begun experimenting with pooling a variety of funding sources to provide coordinated demand-responsive services, they remain a tiny minority. Much more could be achieved with existing resources simply by better coordination (Storen, 1999). Moreover, taxi and jitney services, if deregulated and expanded, might be usefully integrated into such demand-responsive service systems as well. The main obstacles remain institutional, with rivalries, funding conflicts, and legislative barriers among different social service agencies and transportation providers.

Clearly, however, the rapidly growing elderly population represents an important potential market for transit's future. Since 1974, elderly passengers (as well as those with disabilities) have benefited from federally-mandated, off-peak fare discounts of at least 50% off the regular fare. Moreover, fixed-route bus and rail transit services have become more and more accessible to the elderly and passengers with physical disabilities. If carefully integrated with main-line transit services, demand-responsive services could feed into the overall metropolitan transit system and raise ridership on those lines as well.

Serving University Travel Needs

University students and staff comprise one of the most obvious potential markets for public transportation. Especially with their relatively low incomes, flexible schedules, and ability to walk to transit stops, students might be expected to use public transportation much more frequently than most other groups. Most universities, however, make almost no effort to encourage their students to use public transit, let alone coordinate transit services between the campus and surrounding neighborhoods.

Fortunately, that is changing. Over the past decade, at least 35 universities have contracted with their local transit agencies to provide free rides for their students, with university identification cards used as transit passes (Brown, Hess, & Shoup, 2001). Each university negotiates an annual lump-sum contract with the local agency, costing the university an average of only $30 per student per year. The arrangement has been beneficial for both sides and has essentially paid for itself. The new passengers have helped transit systems fill up many empty seats, thereby raising overall ridership while reducing operating costs per passenger. Students benefit from free travel and enhanced accessibility between home and campus. Moreover, universities report reduced congestion and parking problems, thus saving them money on new investments for campus roadways and parking facilities. These so-called "unlimited access" programs (also called "Upass," "ClassPass," and "Super-Ticket") dramatically raised student usage of public transportation, with increases ranging from 71–200% in the first year and by an additional 2–10% in each subsequent year (Brown et al., 2001).

The latest available survey indicates that 825,000 people are now served by such fare-free, university-based transit services, with the potential for many more university customers (Brown et al., 2001). Clearly, such joint partnerships between universities and public transport systems are mutually beneficial and should be actively pursued by the transit industry.

Light-Rail Transit (LRT)

One of the most prominent trends over the past decade has been the proliferation and expansion of light-rail transit (LRT) systems in cities all over the country. Even low-density, auto-oriented cities such as Dallas, Denver, Sacramento, and Los Angeles have turned to LRT. From 1990 to 2000, the amount of LRT service nationwide rose by 118% (from 24.2 million to 52.8 million vehicle miles of service), and travel by LRT rose by 137%, from 571 million to 1,356 million passenger miles (American Public Transportation Association, 2002). In 2000, there were 25 cities with 481 route miles of light rail transit (LRT). An additional 151 route miles of LRT were already under construction in 2000, with another 157 miles of LRT in the design stage. In short, there appears to be a veritable boom in LRT.

Some academic studies have criticized LRT for being too expensive, too inflexible, and not fast enough to attract riders (Kain, 1999; Pickrell, 1992; Richmond, 2001b). Clearly, in some cities LRT has been built, not because it was the most cost-effective technology, but because it had a better image or seemed more fashionable. In some instances, express bus services probably would have carried as many passengers at less cost. Nevertheless, LRT systems usually have lower operating costs per passenger mile than normal bus services ($0.45 vs. $0.59 in 2000). Moreover, the additional capital costs of fully separate rights-of-way for express bus systems (bus rapid transit [BRT]) partly offset the higher capital costs of constructing LRT.

In fact, construction costs for LRT vary widely, depending on the specific system design, tunneling needs, timing, and cost of land acquisition for rights-of-way. For those LRT systems currently in operation or under construction, the construction costs range from $12 million to $120 million per mile, with an average cost of $35 million per mile (U.S. General Accounting Office, 2001). That is much cheaper than the cost of new metro systems. For example, the current 8.5-mile BART extension to the San Francisco Airport costs $175 million per mile, and the new 11.8-mile Tren Urbano in San Juan, Puerto Rico costs $106 million per mile. To put LRT's costs in further perspective, the $14.8 billion in subsidies being spent on the "Big Dig" highway project in downtown Boston would finance 423 miles of new light-rail lines, almost as much as the entire existing mileage of LRT in all U.S. cities. Compared to such notoriously wasteful highway projects, light rail transit doesn't seem cost-ineffective at all ("Big Dig's problems," 2000).

LRT unquestionably has caught the imagination of many people, including politicians and the media. For a variety of reasons, the public generally perceives rail transit as superior to bus transit. Perhaps that is due to LRT's smoother ride, more spacious vehicles, greater permanence and visibility of its rights-of-way and station stops, and insulation from traffic congestion delays. Or perhaps it is simply due to a romantic attachment to rail transportation. Whatever the reasons, it seems legitimate in our democratic political process to honor that public preference in our choices of transportation technologies.

Moreover, the public has responded enthusiastically to new LRT systems by actually using them. Over the past decade, total passenger miles on LRT have increased faster than the rapid growth in vehicle miles of LRT service (137% vs. 118%). As LRT has expanded, it has been used more intensively than before, thus encouraging yet further expansion. The extensive LRT systems in San Diego and Portland (Oregon) have been particularly successful, attracting large increases in riders over the past decade with operating costs per passenger mile less than half the costs for bus services in the same systems. Moreover, it is noteworthy that the expansion in LRT service was accompanied by increased bus usage as well, thus leading to significant growth in total system ridership (Thompson & Matoff, 2003).

Of course, not all LRT systems have been a success. Buffalo's LRT, for example, has been very expensive to build and operate, and has been steadily losing riders since 1990. The country's most wasteful LRT system is probably the River Line from Camden to Trenton in southern New Jersey. Opened in 2004, the line is so expensive and underutilized that passenger fares cover only 7% of operating costs. When amortized capital costs are included, the line requires $31 in taxpayer subsidy per passenger trip (Masnerus, 2004). Fortunately, that is not typical, but it demonstrates the danger of adopting LRT technology in low-density environments where bus transit would be far cheaper and more effective.

Bus Rapid Transit (BRT)

In recent years, the Federal Transit Administration (FTA) has been actively promoting bus rapid transit (BRT) as a cheaper alternative to LRT and other rail transit systems. FTA even employs the slogan "think rail, use buses," explicitly acknowledging the better public image of rail, but hoping that cities will choose buses instead. There are several variations on BRT, but all involve some kind of express bus service. The exclusivity of rights-of-way distinguishes the three main categories of BRT: (1) busways—completely separate roadways for buses only; (2) HOV lanes shared with other high-occupancy vehicles on limited-access highways; and (3) bus lanes on arterial streets within cities. Within these three categories, there is also variation in the extent to which various features are included to enhance the quality of bus travel: (1) traffic signal priority; (2) prepaid or electronic fare passes; (3) low-floor vehicles or platform-level boarding with multidoor entry and exit; (4) increased distance between stops; (5) improved stations and shelters; (6) cleaner, quieter, and more attractive vehicles; and (7) Intelligent Transportation System (ITS) technology to optimize bus scheduling and provide passengers with real-time information (U.S. General Accounting Office, 2001).

Whereas the famous BRT systems in Curitiba, Brazil, and Ottawa, Canada include exclusive busways and incorporate most of the features listed above, only a

ridership since starting serv
The Wilshire–Whittier
carried 40,300 passen
tember 2001. That
previous bus ride
with new tra
third of th
Metro R
investr
vehi

couple of U.S. systems come close to the highest BRT standards. Most existing routes in the U.S. are simply express bus lines running on shared HOV lanes, with little if any signal priority, special station design, advanced ticketing, ITS features, or especially modern, attractive vehicles. Indeed, many transit experts would define BRT more narrowly than the GAO report, excluding almost all currently operating express bus services, even on HOV lanes (e.g., Vuchic, 2002). According to that narrower definition, there are less than 50 route miles of BRT service currently operating in the entire U.S. That may change, however, since the FTA is running a special BRT demonstration program in 10 cities to advance BRT technology in the United States and to convince more cities to adopt it.

The main reason the FTA is now promoting BRT is to save money. The U.S. General Accounting Office (2001) estimates an average capital cost of $13.5 million per mile for the few existing busway BRTs in the United States, which exclude most of the advanced features found in Ottawa and Curitiba. BRT running on HOV lanes is estimated to cost an average of $9 million per mile. BRT running in bus lanes on arterial streets is estimated to cost only $0.7 million per mile. In short, BRT is cheaper than LRT. Moreover, BRT has the potential advantage of greater flexibility to change routing in response to changing travel demands over time. It also can avoid the transfer often required by LRT between the feeder–distributor bus routes and line-haul rail service. Finally, BRT appears to be fairly competitive in speed, although the comparison is misleading, since most BRT lines currently run on freeway HOV lanes without any stops at all for most of the trip.

The new Metro Rapid bus service in Los Angeles does not qualify as true BRT because it does not have any dedicated right of way. Nevertheless, it is probably the most successful express bus service in the United States. Metro Rapid currently runs for 26 route miles in the Wilshire-Whittier Corridor and for 16 route miles in the Ventura Corridor. It operates in mixed traffic but benefits from wide spacing between stops (0.8 miles) and priority traffic signals that turn green for approaching buses. That helps Metro Rapid run on schedule and achieve average speeds 25% faster than regular express buses in Los Angeles. Metro Rapid features easy-to-board low-floor buses fueled by compressed natural gas. The special express buses are bright red for easy identification and run much more frequently than regular buses (every 2–5 minutes in rush hours). Finally, real-time information on bus arrivals is available for passengers at every stop. The two existing Metro Rapid lines have attracted impressive

...ce in June 2000. ...ine, for example, ...gers per day by September ...was a 40% increase over ...ship in the same corridor, ...sit users accounting for a ...e additional riders. The two ...pid lines required a modest capital ...ent of only $225,000 per mile for ...cles, signal systems, and stations. As a result of its success, the Los Angeles MTA is planning 24 additional Metro Rapid lines with 360 route miles throughout the region by 2008.

An important limitation of most express bus service in U.S. cities is that it operates only during the rush hours in the peak direction and thus provides little if any base-level service during the rest of the day. In addition, many express bus systems with separate rights-of-way for the line-haul must operate in congested mixed traffic on downtown streets. Most U.S. cities have been reluctant to set aside bus lanes or busways in the center. Finally, express bus services concentrate on long rush-hour trips between the suburbs and the CBD, with little service for points in-between. LRT generally provides more comprehensive service over most of the day and benefits from permanently separate rights of way.

Nevertheless, BRT and other forms of advanced express bus service (such as Metro Rapid in Los Angeles) are certainly options to consider, especially given their relatively low cost and greater adaptability to decentralizing land use patterns in U.S. cities. Moreover, BRT should be viewed not simply as a cheap alternative to LRT but as a low-cost and highly effective way to upgrade existing bus services throughout the country. Indeed, BRT lines could serve as effective feeders to rail lines as well as operate independently, especially in those cities and those corridors where volumes are not high enough to justify rail transit. It will take at least another 5 years for the 10 BRT demonstration projects of FTA to become fully operational. At that time, a more complete assessment of BRT's costs and benefits will be possible.

OVERALL PROSPECTS FOR PUBLIC TRANSPORTATION IN THE UNITED STATES

The impressive progress of public transit in U.S. cities over the past few decades suggests continued improvement in service quality, technology, and geographic coverage in future years, as well as further growth in passenger levels. Nevertheless, it seems unlikely that transit will be able to increase its overall market share of urban travel, let alone reduce auto ownership and use.

For the most part, that is not the fault of transit. As noted earlier, a wide range of public policies in the United States—from low gasoline taxation and ubiquitous free parking to low-density zoning and subsidized homeownership—make auto use almost irresistible. Moreover, powerful technological, economic, and demographic trends seem to point toward ever greater decentralization of both jobs and housing in the coming years. That will be the most difficult hurdle for transit to overcome.

Various adaptations to the polycentric, low-density suburbs are already being implemented by transit systems in many cities. Together with transit-oriented development at key nodes of bus and rail networks, such adaptations will improve transit's chances of capturing more of that rapidly growing suburban market. The planned expansions of both LRT and BRT systems into the suburbs, and their coordination with other transit routes, will facilitate increased usage of public transportation throughout metropolitan areas.

At the same time, many proven low-cost measures could be taken to improve service quality and fare structure, and thus enhance the overall attractiveness of public transportation. Perhaps the most important step is to better coordinate the existing transit network, completely integrating fares, schedules, and routes. Similarly, even simple improvements like better signage and passenger information, wider spacing between bus stops, transit priority in mixed traffic, and proof-of-purchase ticketing would significantly improve service. The relatively low-cost Metro Rapid lines in Los Angeles are an excellent example of such low-cost but highly effective measures. These less dramatic, cheaper measures may be more important than high-profile capital investments.

One goal of urban transportation policy should be providing Americans with more transportation options than they currently have. Improving public transportation, walking, and bicycling conditions in our cities, and coordinating them better with land use, would contribute to that goal. Currently, the urban transportation system and land use patterns in American cities are almost completely dominated by the automobile, permitting virtually no choice in travel mode. However comfortable, convenient, and popular the private car may be, its complete domination of American cities has serious social and economic consequences that cannot be ignored. It is essential to provide greater balance in our transportation system so that public transportation, walking, and cycling become feasible options for urban travel.

ACKNOWLEDGMENTS

John Pucher would like to thank John Neff, Nigel Wilson, Robert Cervero, Martin Wachs, Donald Shoup, Gregory Thompson, Vukan Vuchic, Eric Bruun, John Renne, Jeffrey Zupan, Thomas Menzies, Susan Hanson, Genevieve Giuliano, Reid Ewing, Jeffrey Kenworthy, John Holtzclaw, Thomas Matoff, and Todd Litman for the helpful suggestions and extensive information they provided for this chapter. Pucher is also indebted to Hyungyong Park, John Renne, and Ralph Buehler for their assistance in gathering information and producing the tables and figures.

REFERENCES

Aaron, H. (1972). *Shelter and subsidies: Who benefits from federal housing policies?* Washington, DC: Brookings Institution.

Altshuler, A., Womack, J., & Pucher, J. (1979). *The urban transportation system: Politics and policy innovation.* Cambridge, MA: MIT Press.

American Public Transit Association. (1977). *Transit fact book.* Washington, DC: Author.

American Public Transit Association. (1981). *Transit fact book.* Washington, DC: Author.

American Public Transit Association. (1992). *Transit fact book.* Washington, DC: Author.

American Public Transportation Association. (2000). *Public transportation fact book.* Washington, DC: Author.

American Public Transportation Association. (2002). *Public transportation fact book.* Washington, DC: Author.

Belzer, D., & Autler, G. (2002). *Transit oriented development: Moving from rhetoric to reality.* Washington, DC: Brookings Institution, Center for Urban and Metropolitan Policy.

Big Dig's problems with accounts and accountability. (2000). *Engineering News Record, 244,* 92.

Brown, J., Hess, D., & Shoup, D. (2001). Unlimited access. *Transportation, 28,* 233–267.

Carnall, D. (2000). Cycling and health promotion. *British Medical Journal, 320,* 888.

Center for Urban Transportation Research. (1998, September). *Public transit in America: Findings from the 1995 Nationwide Personal Transportation Survey.* Tampa: University of South Florida.

Cervero, R. (1998). *Transit metropolis: A global inquiry.* Washington, DC: Island Press.

Cervero, R., Ferrell, C., & Murphy, S. (2002). *Transit oriented development and joint development in the United States: A literature review.* Washington, DC: Transit Cooperative Research Program, Project H-27.

Denno, N., & Robins, M. (2000). *The trend of transit labor costs: 1982–1997.* New Brunswick, NJ: Voorhees Transportation Center.

Dora, C. (1999). A different route to health: Implications of transport policies. *British Medical Journal, 318,* 1686–1689.

Downs, A. (1999). Contrasting strategies for the economic development of metropolitan areas in the United States and Western Europe. In A. Summers, P. Cheshire, & L. Senn (Eds.),

Urban change in the United States and Western Europe: Comparative analysis and policy (pp. 16–32). Washington, DC: Urban Institute Press.

Dunn, J. (1997). *Driving forces: The automobile, its enemies, and the politics of mobility.* Washington, DC: Brookings Institution Press.

Duran, L. (1995). Bus riders union in L.A. *Race, Poverty and the Environment, 6,* 8–9.

European Commission. (2002). *Key data on health 2001.* Luxembourg: European Commission of the European Union.

Ewing, R. (1999). *Traffic calming: State of the practice.* Washington, DC: Institute of Transportation Engineers.

Federal Transit Administration. (2002). *National transit database.* Washington, DC: U.S. Department of Transportation.

Flegal, K., Carroll, M., Ogden, C., & Johnson, C. (2002). Prevalence and trends in obesity among adults, 1999–2000. *Journal of the American Medical Association, 288,* 1723–1727.

German Ministry of Transport. (2002). *Nationaler Radverkehrsplan 2002–2012: Massnahmen zur Foederung des Radverkehrs in Deutschland.* Berlin: Bundesministerium fuer Verkehr, Bau und Wohnungswesen.

Hass-Klau, C. (1992). *Civilized streets: A guide to traffic calming, environment, and transport planning.* Brighton, UK: Brighton Press.

International Energy Agency. (2002). *Energy prices and taxes.* Paris: Organisation for Economic Cooperation and Development.

Kain, J. (1999). The urban transportation problem: A reexamination and update. In J. Gomez-Ibanez, W. Tye, & C. Winston (Eds.), *Essays in transportation economics and policy* (pp. 359–402). Washington, DC: Brookings Institution.

Kenworthy, J. (2002, September). *Energy use in urban transport systems: A global review.* Paper presented at the Third International Biennial Conference on Advances in Energy Studies, Porto Venere, Italy.

Koplan, J., & Dietz, W. (1999). Caloric imbalance and public health policy. *Journal of the American Medical Association, 282,* 1579–1581.

Lenzen, M. (1999). Total requirements of energy and greenhouse gases for Australian transport. *Transportation Research D, 4*(4), 265–290.

Levinson, H., & St. Jacques, K. (1998). Bus lane capacity revisited. *Transportation Research Record, 1618,* 189–199.

Litman, T. (2002). *Transportation cost and benefit analysis: Techniques, estimates, and implications.* Victoria, BC, Canada: Victoria Transportation Policy Institute.

Masnerus, L. (2004, March 13). Light rail, with the emphasis on light. *New York Times,* pp. B1–B4.

McGuiness, J., & Foege, W. (1993). Actual causes of death in the United States. *Journal of the American Medical Association, 270,* 2207–2212.

Mokdad, A., Bowman, B., Ford, E., Vivicor, F., Marks, J., & Koplan, J. (2001). The continuing epidemics of obesity and diabetes in the United States. *Journal of the American Medical Association, 286,* 1195–1200.

Mokdad, A., Serdula, M., Dietz, W., Bowman, B., Marks, J., & Koplan, J. (1999). The spread of the obesity epidemic in the United States, 1991–1998. *Journal of the American Medical Association, 282,* 1519–1522.

Morlok, E., Bruun, E., & Vanek, F. (1997). *The advanced minibus concept: A new ITS-based transit service for low-density markets* (Report prepared for U.S. Department of Transportation, Federal Transit Administration). Philadelphia: University of Pennsylvania.

Newman, P., & Kenworthy, P. (1999). *Sustainability and cities: Overcoming automobile dependence.* Washington, DC: Island Press.

Oak Ridge National Laboratory. (2001). *Transportation energy data book.* Oak Ridge, TN: U.S. Department of Energy, Center for Transportation Analysis.

Organisation for Economic Cooperation and Development (OECD). (2002). *Total expenditures on health per capita, 1960–2000, in US dollars, purchasing power parity.* Washington, DC: Author. Available online at http://www.oecd.org/xls/M0031000/M00031378.xls

Peeters, A., Barendregt, J., Willekens, F., Mackenbach, J., Al Mamum, A., & Bonneux, L. (2003). Obesity in adulthood and its consequences for life expectancy: A life-table analysis. *Annals of Internal Medicine, 138,* 24–32.

Pickrell, D. (1992). A desire named streetcar: Fantasy and fact in rail transit planning. *Journal of the American Planning Association, 58,* 158–176.

Pucher, J. (1995). Urban passenger transport in the United States and Europe: A comparative analysis of public policies. *Transport Reviews, 15,* 211–227.

Pucher, J. (1997). Bicycling boom in Germany: A renaissance engineered by public policy. *Transportation Quarterly, 51*(4), 31–46.

Pucher, J. (1999). Transportation trends, problems, and policies. *Transportation Research A, 33*(7/8), 494–503.

Pucher, J. (2002). Renaissance of public transport in the USA. *Transportation Quarterly, 56*(1), 33–50.

Pucher, J., & Dijkstra, L. (2000). Making walking and cycling safer: Lessons from Europe. *Transportation Quarterly, 54*(3), 25–50.

Pucher, J., & Dijkstra, L. (2003). Promoting safe walking and cycling to improve public health: Lessons from The Netherlands and Germany. *American Journal of Public Health, 93*, 1509–1516.

Pucher, J., & Kurth, S. (1995). Making transit irresistible: Lessons from Europe. *Transportation Quarterly, 49*(1), 117–128.

Pucher, J., & Lefevre, C. (1996) *Urban transport crisis in Europe and North America.* London: Macmillan.

Pucher, J., Markstedt, A., & Hirschman, I. (1983). Impacts of subsidies on the costs of public transport. *Journal of Transport Economics and Policy, 17*, 155–176.

Pucher, J., & Renne, J. (2003). Socioeconomics of urban travel: Evidence from the 2001 NHTS. *Transportation Quarterly, 57*(3), 49–78.

Pushkarev, B., & Zupan, J. (1977). *Public transportation and land use policy.* Bloomington: Indiana University Press.

Richmond, J. (2001a). *The private provision of public transport.* Cambridge, MA: Harvard University Press.

Richmond, J. (2001b). A whole-system approach to evaluating urban transit investments. *Transport Reviews, 21*(2), 141–179.

Salzman, A. (1992). Public transportation in the 20th Century. In G. Gray & L. Hoel (Eds.), *Public transportation* (pp. 24–45). Englewood Cliffs, NJ: Prentice-Hall.

Shoup, D. (1995). An opportunity to reduce minimum parking requirements. *Journal of the American Planning Association, 61*, 14–28.

Shoup, D. (1997). The high cost of free parking. *Journal of Planning Education and Research, 17*, 1–18.

Shoup, D. (1999). The trouble with minimum parking requirements. *Transportation Research A, 33*(7/8), 549–574.

Storen, D. (1999). *Building community transportation systems.* New Brunswick, NJ: Heldrich Center for Workforce Development.

Sturm, R. (2002). The effects of obesity, smoking, and drinking on medical problems an costs: Obesity outranks both smoking and drinking in its deleterious effects on health and health costs. *Health Affairs, 21*, 245–253.

Thompson, G. (1977). Planning considerations for alternative transit route structures. *Journal of the American Institute of Planners, 43*, 158–168.

Thompson, G. (1998). Identifying gainers and losers from transit service change: A method applied to Sacramento. *Journal of Planning Education and Research, 18*, 125–136.

Thompson, G., & Matoff, T. (2003). Keeping up with the Joneses: Radial vs. multidestinational transit in decentralizing regions. *Journal of the American Planning Assocation, 69*, 296–312.

Transportation Research Board. (1996). *Rail transit capacity* (Transit Cooperative Research Program Report No. 13). Washington, DC: National Academy Press.

Transportation Research Board. (1997). *Transit-focused development* (Transit Cooperative Research Program Synthesis No. 20). Washington, DC: National Academy Press.

Transportation Research Board. (1998a). *Consequences of the Interstate Highway System for transit: Summary of findings* (Transit Cooperative Research Program Report No. 42). Washington, DC: National Academy Press.

Transportation Research Board. (1998b). *Strategies to attract auto users to public transportation* (Transit Cooperative Research Program Report No. 40). Washington, DC: National Academy Press.

Transportation Research Board. (1998c). *Transit markets of the future: The challenge of change* (Transit Cooperative Research Program Report No. 28. Washington, DC: National Academy Press.

Transportation Research Board. (1999). *Guidelines for enhancing suburban mobility using public transportation* (Transit Cooperative Research Program Report No. 55). Washington, DC: National Academy Press.

Transportation Research Board. (2000). *Highway capacity manual.* Washington, DC: National Academy Press.

Transportation Research Board. (2001). *Making transit work: Insights from Western Europe, Can-*

ada, and the United States. Washington, DC: National Academy Press.

Transportation Research Board. (2002). *The congestion mitigation and air quality improvement program: Assessing ten years of experience* (TRB Special Report No. 264). Washington, DC: National Academy Press.

U.S. Bureau of the Census. (2002a). *2000 Census of the United States: Journey-to-work in U.S. metropolitan areas.* Washington, DC: U.S. Department of Commerce.

U.S. Bureau of the Census. (2002b). *Projections of total resident population by 5-year age groups and sex, age categories, middle series, 1999–2100.* Washington, DC: U.S. Department of Commerce.

U.S. Department of Health and Human Services. (1996). *Physical activity and health: A report of the surgeon general.* Atlanta: Centers for Disease Control and Prevention.

U.S. Department of Transportation. (1993). *Journal-to-work trends in the United States and its major metropolitan areas.* Washington, DC: Federal Highway Administration.

U.S. Department of Transportation. (2002). *National Transit Database.* Washington, DC: Federal Transit Administration.

U.S. Environmental Protection Agency. (2001). *Our built and natural environments: A technical review of the interactions between land use, transportation, and environmental quality.* Washington, DC: Author.

U.S. General Accounting Office. (2001). *Bus rapid transit shows promise.* Washington, DC: Author.

Vigrass, J. W. (1992). Rail transit. In G. Gray & L. Hoel (Eds.), *Public transportation* (pp. 114–147). Englewood Cliffs, NJ: Prentice-Hall.

Vuchic, V. (1981). *Urban public transportation: Systems and technology.* Englewood Cliffs, NJ: Prentice-Hall.

Vuchic, V. (1992). Urban passenger transportation modes. In G. Gray & L. Hoel (Eds.), *Public transportation* (pp. 79–113). Englewood Cliffs, NJ: Prentice-Hall.

Vuchic, V. (1999). *Transportation for livable cities.* New Brunswick, NJ: Center for Urban Policy Research.

Wei, M., Kampert, J., Barlow, C., Nichaman, M., Gibbons, L., Paffenberger, R., & Blair, S. (1999). Relationship between cardiorespiratory fitness and mortality in normal-weight, overweight, and obese men. *Journal of the American Medical Association, 282,* 1547–1553.

World Bank. (2000). *Urban transit systems: Guidelines for examining options* (World Bank Technical Paper No. 52). Washington, DC: Author.

World Health Organization. (2000). *Transport, environment and health.* Copenhagen: World Health Organization, Regional Office for Europe.

World Resources Institute. (1992). *The going rate: What is really costs to drive.* Washington, DC: Author.

CHAPTER 9

Land Use Impacts of Transportation Investments

Highway and Transit

GENEVIEVE GIULIANO

The role of the transportation system in fostering growth and affecting urban structure has been of great interest to academicians, planners, investors, and politicians. The notion that transportation services and facilities have had an enormous effect on land use seems to be self-evident. Indeed, the evolution of urban form appears to be so clearly linked with transportation technology that Muller formulates his explanation of urban structure in terms of transportation eras (see Muller, Chapter 3, this volume).

Transportation investment decisions are the subject of much debate and lobbying among planners and policymakers, particularly at the local government level. Transportation investments are viewed as growth generators. Public transit investment (usually in rail transit) is promoted as a key to economic revitalization for central cities. Highway facilities (usually expressways) are often promoted as a means to attract economic growth by increasing access to skilled or low-cost labor markets and cheap land in remote areas. Similarly, many see

traffic congestion, the manifestation of transportation system capacity constraint, as a major reason for urban decline and the deterioration of central-city cores. Controls on transportation investments, especially highways, are also viewed as critical to growth management policy objectives. Growth control advocates are thus typically opposed to the construction of new highways for the same reasons that advocates of economic growth endorse them.

The ability of transportation investments to shape or influence urban structure is also a widely held conviction. Today many urban geographers and policymakers are advocating higher density, mixed-use development as a means for solving urban congestion and environmental problems. "Smart growth" proponents see transit-based accessibility as a key element in fostering higher density development patterns (Bernick & Cervero 1997; Newman & Kenworthy, 1998). Smart growth proponents target expressways as a major cause of urban sprawl, meaning low-density, dispersed land use patterns. Among those who

237

prefer low-density development, modal preferences are reversed, but the belief is often the same: transportation investments are *perceived* as having a significant and predictable impact on urban structure.

In this chapter I explore the relationship between transportation and land use: I examine the theoretical basis for expecting transportation to influence land use, and I determine the extent to which such impacts can be documented. It should be noted that the primary focus of this chapter is the impact of transportation on land use, not the impact of land use on travel or transportation. Many researchers are exploring how the spatial structure of metropolitan areas affects travel behavior. They are particularly interested in whether higher density, mixed-used neighborhoods promote more use of public transport and non-motorized modes (see Box 9.1).[1]

The remainder of the chapter begins with a brief description of factors to consider when examining land use impacts of transportation investments. In the second section, the conceptual relationship between transportation and land use is described in the context of accessibility. The third section reviews the major theories that have been developed to explain transportation—land use relationships. The fourth section summarizes the empirical evidence, discusses some specific case studies, and draws conclusions based on these findings.

GENERAL CONSIDERATIONS

Two dimensions are relevant to examining the relationship between land use and transportation: context and dynamics.

Context

Examining the transportation and land use relationship requires that it be placed in some context. First, land use patterns must be distinguished from growth. This is difficult, because most of the land use changes

---

BOX 9.1. Urban Form and Travel

Many urban planners and geographers advocate changes in land use policy to achieve transportation policy objectives. The conventional suburban neighborhood, with access designed around the automobile; shops and offices located in isolated commercial zones; and with walls, cul-de-sacs, and often no sidewalks, making walking or biking difficult for purposeful travel, is seen as a contributing factor to automobile dependence, environmental damage, and, more recently, decline in physical activity. It is argued that higher density, mixed-use neighborhoods built around a grid street pattern promote more walking, biking, and public transit use. Termed "transit-oriented development" or "new urbanism" or "smart growth," proponents advocate more infill development, development around transit stations, and more compact new development.

Would changing land use in these ways lead to significantly different travel outcomes? In one of the most thoughtful and comprehensive studies on this topic, Boarnet and Crane (2001) conclude the following:

- Empirical support for expected travel benefits is inconclusive.
- The extent of the market for higher density, mixed-use neighborhoods is uncertain.
- Land use policy is unlikely to be an effective tool for addressing transportation problems; such problems are better handled directly via pricing or regulation.
- Direct policies targeted at transportation problems might well increase demand for higher density, mixed-use neighborhoods.

we observe in conjunction with transportation improvements tend to be manifestations of economic growth. These manifestations can take many forms, however. Commercial development can occur in the form of garden office buildings or multistory skyscrapers. My emphasis here is on the *form* these land use changes take, rather than on the economic implications of these changes.

Second, the magnitude of the transportation change must be considered. Improvements that represent significant technology change (and hence significant change in service level), such as construction of the electric streetcar systems in the late 19th century, can be expected to have much greater impact than incremental changes, such as hybrid-fuel automobiles. The scale of the investment relative to the existing system is also important. For example, the first 10 miles of expressway constructed within a region should have greater impact than the last 10 miles, all other things being equal, because the relative impact of those first 10 miles will be so much greater.

Third, the geographic scale of analysis must be considered. In a larger metropolitan area, for example, even a very large project in terms of cost, such as a new rail line or expressway segment, might have a limited impact when viewed in the context of the area as a whole. One example is the $2 billion I-105 (Century) Expressway in Los Angeles. The 17-mile segment opened in 1993 and as of 2001 was carrying 242,000 daily trips, less than 1% of the region's daily vehicle trips.[2] This is not to say, however, that significant changes may not be taking place in the local area of new investment. The level or scale of analysis must be defined. Generally, both *micro-* (local) and *macro-* (regional) *impacts* are of interest.

The fourth consideration is the geographical context of the transportation investment. The built environment is very durable; most structures survive 50 years or more. Even in rapidly growing areas, the vast proportion of building stock that will exist 10 or 20 years from now is already built. The cost of reconstruction in developed areas is usually high, because demolition and site preparation must precede new construction. Consequently, changes in land use occur very slowly in already built-up areas. This is a limiting factor, even though utilization of the existing building stock may change much faster. The location of the transportation investment is therefore important. The potential rate of land use change is much lower in developed areas than in undeveloped areas.

Dynamics

It is particularly difficult to examine the land use impacts of transportation because land use and transportation are mutually dependent. A simply illustration is presented in Figure 9.1. The characteristics of the transportation system determine *accessibility*, or the ease of moving from one place to another. Accessibility in turn affects the *location* of activities, or the land use pattern. The location of activities in space, together with the transportation resources connecting them, affects daily *activity patterns*, which in turn result in travel patterns. These *travel patterns*, expressed as flows on the transportation network, affect the transportation system. Land use and transportation are part of the larger urban system: the collection of

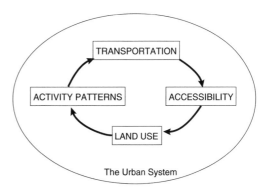

FIGURE 9.1. The transportation–land use relationship.

people, institutions, and infrastructure that together form the urban space economy.

The model in Figure 9.1 shows that a change in land use will affect transportation, just as transportation affects land use. Note that this model does not imply anything about the strength of these relationships. It serves only to illustrate the interdependence of empirically observed land use and transportation changes. It is quite difficult to isolate the impact of transportation on land use (or, conversely, land use on transportation), as it requires examining one half of the figure, holding the other half constant. The land use system is dynamic, however, and what we observe in empirical analysis is the totality of relationships expressed in Figure 9.1.

Not only are land use and transportation patterns interdependent, this interdependency is expressed over long periods of time. Over time, people, institutions, infrastructure, and technology change, making it even more challenging to separate out the role of transportation. As will be explained in a later section of this chapter, the durability of the built environment, the high costs of relocation for both residents and firms, and the long lag times of major transportation projects all make empirical analysis of land use and transportation relationships problematic.

TRANSPORTATION IN THE CONTEXT OF ACCESSIBILITY

The basic concept underlying the relationship between land use and transportation is accessibility. In its broadest context, "accessibility" refers to the ease of movement between places (see Hanson, Chapter 1, this volume). As movement becomes less costly—either in terms of money or time—between any two places, accessibility increases. The propensity for interaction between any two places increases as the cost of movement between them decreases. Consequently, the structure and capacity of the transporta-

tion network affect the level of accessibility within a given area.

Accessibility also includes the concept of *attractiveness*, the opportunities or activities that are located in a given place. Thus the ease of movement between places, as well as the attractiveness of these places as origins or destinations, are expressed in accessibility. More specifically, we can define *accessibility* as the attractiveness of a place as an origin (how easy is it to get *from* there to all other destinations) and as a destination (how easy is it to get *to* there from all other destinations). Note that these two measures are not symmetrical. In the first case, the emphasis is on accessing opportunities located in other places; in the second case, the emphasis is on the opportunities located in that place. It is important to incorporate attractiveness in the concept of accessibility because of the unique function of transportation as the mechanism for spatial interaction. That is, trips are usually a "derived demand," meaning that they are made in order to engage in some other activity, such as working or shopping.

How does a change in the transportation network affect accessibility? I illustrate the effect with a simple network example, shown in Figure 9.2. Each node represents a possible origin/destination, and the numbers near each link represent travel times. I begin with the network on the left side of Figure 9.2 (Network K). I then make an

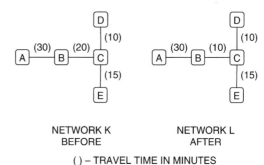

NETWORK K   NETWORK L
BEFORE   AFTER

( ) – TRAVEL TIME IN MINUTES

FIGURE 9.2. Accessibility in a simple network.

TABLE 9.1. Travel Time Accessibility Matrix for Networks Shown in Figure 9.2

Network K							Network L							Change (%)
From/to	A	B	C	D	E	?	From/to	A	B	C	D	E	?	
A	0	30	50	60	65	205	A	0	30	40	50	55	175	−14
B	30	0	20	30	35	115	B	30	0	10	20	25	85	−26
C	50	20	0	10	15	95	C	40	10	0	10	15	75	−21
D	60	30	10	0	25	125	D	50	20	10	0	25	105	−24
E	65	35	15	25	0	140	E	55	25	15	25	0	120	−14

improvement such that travel time between nodes B and C is reduced by one-half, as shown on the right side of Figure 9.2 (Network L).

Network accessibility can be measured by calculating the travel time from each node to every other node and summing over each node, as in Table 9.1 (note that this implicitly assumes that each node is equally attractive). Each row in the matrix corresponds to travel times from one to every other node. The row sums are the accessibility measure for each node. Since I have used travel times in the example, lower numbers mean greater accessibility. Comparing the row sums of the two networks, Table 9.1 shows that the network improvement not only increases accessibility between nodes B and C, it also benefits the entire network. This simple example demonstrates how a network improvement (e.g., a new expressway link between two places) affects both the accessibility of the two directly connected places and the accessibility of the entire network, and as accessibility increases, the level of spatial interaction increases because travel has, in effect, become less costly.

How does this change affect land use? As more interaction occurs, more activities will locate in those places that have become more accessible. In this example, nodes B, D, and C have benefited most from the improvement, and thus the greater land use

changes would be expected to occur at these nodes. This process may be characterized as regional growth: as population and employment increase, their relative location will be affected by the transportation system. In a theoretical world where capital is completely mobile, then, transportation system changes would result in observable shifts in activity location.

The concept of accessibility can also be used to predict changes resulting from deterioration in the transportation system. Suppose that the link between two places has reached capacity and travel speeds have declined. Accordingly, the level of accessibility declines, generating incentives for activities to shift away from these two places—all other things being equal.

The network example above does not explicitly identify the economic benefits generated by reduced travel costs on the B–C link. Reduced transport costs can lead to greater economic productivity. For example, reduced shipping costs might allow a manufacturer to seek out more distant but lower cost input suppliers, hence reducing the price of the finished good. Other firms may gather around this highly accessible link, realizing resulting economies from access to one another (i.e., external economies). Reduced transport costs effectively increase a worker's net wage, allowing her or him to spend more on other goods and services. The result of this process is economic growth.

Shifts in activity location are part of this process.

Is it always that case that transportation investments lead to economic growth? If our sample network was located in a declining city, it is quite likely that the accessibility improvement would not be sufficient to overcome the city's competitive disadvantages that are the underlying cause of its decline. It is also possible that shifts in activity location would occur, but these would simply be a redistribution of economic activity, leaving other parts of the city worse off, as we shall see in some examples later in this chapter.

THEORIES OF LAND USE–TRANSPORTATION INTERACTION

Research on the relationship between land use and transportation has generated numerous theories. This section discusses theoretical expectations regarding transportation and land use changes. Developed by economists or geographers, these theories seek to explain the effect of transportation cost on location choice.

Standard Urban Economic Theory

The basis of the standard model of urban structure is von Thunen's (1826) theory of agricultural land rent and use. This theory was developed to explain the basic structure of cities, meaning land value, population and employment distribution, and commuting patterns. Initially developed by Alonso (1964), Mills (1972), and Muth (1969), the theory focuses on residential location choice. Household location is expressed as a utility-maximizing problem in which choice depends on land rent, commuting cost, and the costs of all other goods and services. The standard theory is based on several simplifying assumptions. In addition to the usual assumption of rational behavior, identical preferences, and perfect information, the following assumptions are also made in the simplest version of the theory:

- The total amount of employment is fixed and located at the center of the city.
- Each household has only one worker, and only work travel is considered.
- Housing is a function of capital and land, and therefore location and lot size are the distinguishing factors.
- Unit transport cost includes both time and monetary costs, and it is constant and uniform in all directions.

Residential location theory gives rise to a city form in which the greatest population density and highest land values are at the center and in which these density and land value gradients decline with distance from the city center. Figure 9.3 shows the land rent curve: the value per unit of land as a function of distance from the city center. According to the standard theory, the land rent curve is nonlinear; its gradient is constantly decreasing in absolute value with distance from the center. The dotted line R_A denotes the nonurban land rent, and the point at which $R(d)$ equals R_A defines the city boundary. The unit cost of housing must decline with distance from the center of the city, since transport (commuting) costs increase with distance. That is, the value of transportation costs savings is reflected in land (and therefore housing) cost. If this were not the case, all households would locate at the center of the city. It therefore follows that more housing will be consumed as distance from the center increases, and therefore population density will also be constantly decreasing. The theory also predicts commuting patterns; the average commute trip length corresponds to the mean distance of total population to the center.[3]

The best location for a given household is the point at which the marginal savings of housing are equal to the marginal cost of transport, or the savings in housing are just offset by the increase in transport cost. If households have identical preferences, they will be indifferent to any given location. If

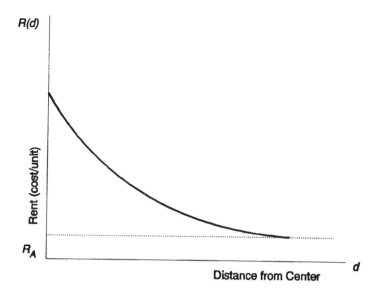

FIGURE 9.3. The relationship between rent and location.

this assumption is relaxed, the particular location of any given household depends on relative preferences between housing and transport. It is generally assumed (and there is some supporting empirical evidence) that housing preferences are the stronger of the two; the higher the household income, the more housing will be consumed, even at the cost of additional commuting. Thus lower-income households would consume less housing and locate closer to the city center, and higher-income households would consume more housing and locate farther from the city center.

What does the theory predict in response to a change in transport cost? If commuting costs are reduced, the theory predicts movement away from the center, or population decentralization. Figure 9.4 illustrates this process. Curve 1 is the land rent gradient before the transportation improvement; Curve 2 is the land rent gradient after the transportation improvement. The decrease in transport cost reduces rent at the city center, because the location advantage of the center declined. Consumers take advantage of lower rents by increasing housing consumption and commuting greater distances. As a

result, the total amount of land consumed increases, and the city boundary extends. If transport cost is increased, the reverse effect will be observed: households will economize on travel by locating closer to the center and consuming less housing.

In a very general way, empirical evidence tends to support the standard theory. Population density does decline with distance from the center, and time series studies document that densities have declined historically in conjunction with transportation system improvements (Anas, Arnott, & Small, 1998). Lower-income households are more likely to live near the city center and to have shorter commutes, while higher-income households are more likely to live in the suburbs and to have longer commutes (Mieskowski & Mills, 1993; see also Hanson, Chapter 1, this volume).

As stated earlier, commuting costs include both time costs and monetary costs. Commuting costs can therefore be reduced either by reducing monetary costs (e.g., reducing the price of fuel or transit fares), or by increasing travel speed (e.g., providing an express transit route or adding a lane to an expressway). Different types of travel cost

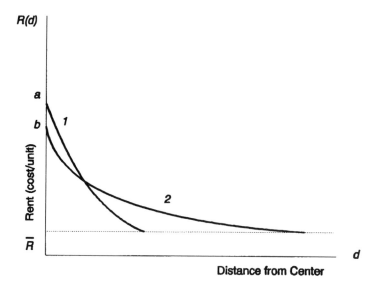

FIGURE 9.4. Response of rent function to a transport cost decline.

reductions are likely to have differing impacts across income groups. A simple price reduction will affect all income groups in the same way: more income would be available for housing, leading to more housing consumption, and hence to more distant location choice.

Travel time reductions tend to have a greater effect on higher-income groups, because the value of time is a function of income.[4] Consequently, the decentralizing effect of a travel time reduction is greater for higher-income households. Suppose that a transportation improvement is accompanied by a price increase—for example, by the construction of a commuter rail system that reduces travel time but charges a higher fare. The fare increase could offset the value of travel time savings for the low-income household, but not for the high-income household. Thus the benefit of this investment would accrue disproportionately to higher-income households. It bears noting that congestion pricing of roads would work the same way: the benefits of travel time savings realized from reduced congestion would be greatest for higher-income households.

Employment Location Theory

In the simplest version of the standard model, employment location is assumed to be fixed in the city center. However, the theory can be applied to employment location (Brueckner, 1978; White, 1988). Another way of expressing this assumption is to say that firms place a higher value on central location than do households. What if we relax the assumption of fixed employment and express it as a function of land rents, commuter costs, and all other costs? It turns out that employment location will always be more concentrated than residence locations. Consider again a reduction in commute cost. Another way of describing the benefit of a commuting cost reduction to an employee is as a net wage increase: reduced commute costs provide more income to spend on other things.

Suppose now that employment location can vary. Households will be indifferent between working far from home at a higher wage and working close to home at a commensurately lower wage, as long as the net wage remains constant (i.e., as long as the wage difference is equivalent to the commute cost difference between the two locations). Then the employer's location decision depends on *total transport costs*: the cost of commuting (as reflected in wage rates) and the cost of transporting goods to market. If it is assumed that the goods export node is at the city center, employers will *centralize* to the extent that commuting costs are less than shipping costs, because the central location minimizes goods transport costs. You can see that applying the standard theory in this way implies that as goods transport costs decline relative to commute costs, employers will *decentralize* and locate closer to their workers.

Finally, suppose that the goods produced have no transportation costs, as, for example, the production of information. If the product can be produced and sold anywhere, where will employers locate? In this extreme case, employment would be distributed identically with the population, so long as there are no scale economies of production, meaning as long as the efficiency of producing the good does not depend on the size of the firm. Again, empirical evidence supports theory in a very general way. Jobs have indeed been decentralizing as transport costs (for both goods and passengers) have declined (Gordon, Richardson, & Yu, 1998).

Other employment location theories deal with certain types of firms or economic activities. It is useful to distinguish between market-sensitive and non-market-sensitive activities. *Market-sensitive activities* (such as sales and services) depend on accessibility to consumers, whereas *non-market-sensitive activities* include industrial or manufacturing activities. Theories dealing with these two categories of economic activities are presented in the following sections.

Central Place Theory

The classical theory of market-sensitive activities is central place theory, formulated by Christaller (1966) and Losch (1954). Cen-

tral place theory gives rise to a hierarchical pattern of market centers. The distribution and size of markets are functions of consumer range and market threshold requirements. "Consumer range" refers to the distance consumers are willing to travel, which varies for different types of goods. "Market threshold" depends on the volume of sales required to market a given good at a profit. The resulting distribution of market centers is that which minimizes total consumer transport cost and fulfills the minimum threshold requirements.

Central place theory predicts larger, more dispersed centers as transport costs decline, because with declining costs consumers are willing to travel greater distances, making it possible for producers to take advantage of the scale economies of larger markets. Increased transport cost would have the reverse effect, leading to smaller, more concentrated centers, all other things being equal. Central place theory has enjoyed renewed interest in recent years, because it predicts a multinucleated land use pattern, consistent with the observed evolution of contemporary large metropolitan areas (see Chapter 3). Central place theory is also consistent with the increase in scale of major retail centers observed over the last several decades.

Industrial Location Theory

Industrial location theory focuses on the transport or shipping costs of product inputs and outputs. Often termed the "classical model" of industrial location, the theory was first formulated by Weber (1928), and later developed by Hoover (1948), Isard (1956), and Moses (1958). It applies primarily to manufacturing activities. Location choice depends on the relative costs of shipping inputs and outputs, as well as economies of scale in production. In contrast to the location theories previously discussed, the focus here is on goods movement rather than passenger (worker) movement. In most model formulations, labor force availability is assumed, although it can be treated as another input shipping cost.

Industrial location models are no longer adequate predictors of industrial location choice. The importance of transport costs in the manufacturing sector has declined as a result of technological shifts and a highly developed goods movement network. In essence, producers have taken advantage of ever decreasing transport and communications costs to restructure economic production. With networked systems, producers can take advantage of specialized or low-cost labor pools and input factors, wherever located. Just-in-time inventory systems further reduce inventory and warehouse costs. At the same time, goods movement transport costs have declined more rapidly than commuting costs, making labor force access a relatively more significant factor in location than market or input access. More recent research suggests that agglomeration economies, economies of scale and scope, linkage patterns, infrastructure availably, and labor force access are more important to industrial location than transport costs (Gramlich, 1994; Ingram, 1998).

The theories discussed in this section are quite different from one another. Each ap-plies only to certain activities; there is no integrated theory that applies to all types of employment. In each case, however, transport costs play a significant role. I turn now to some empirical observations that further complicate attempts to develop appropriate theories for land use–transportation relationships.

Theoretical Simplicity versus Real-World Complexity

The development of theories and models of complex phenomena requires the use of simplifying assumptions. These assumptions are what make the problem manageable. It would be practically impossible, for example, to develop a theory of residential location that takes into account all the factors influencing location choice. (Residential location models have been estimated with 30 or more independent variables, yet we still cannot accurately predict individual choices.) Moreover, such a model would be so complex that the most significant factors might be obscured. In this section I discuss some of the key assumptions associated with the theories described above and examine how relaxation of these assumptions affects our ability to predict land use–transportation relationships.

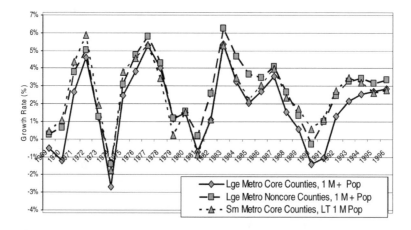

FIGURE 9.5. Average annual private-sector job growth by MSA category, 1969–1997. Source: Regional Economic Information Service data.

The Monocentric City

The standard urban model of location assumes a *monocentric city*: all employment is assumed to be located at the city center, as discussed in the previous section. This assumption is unsupported in contemporary urban areas. Rather, as Muller (Chapter 3, this volume) and others have pointed out, large urban areas today are decentralized and dispersed. To illustrate, Figure 9.5 shows the average annual growth rate of private employment, by county, for all counties in Metropolitan Statistical Areas (MSAs), from 1969 to 1997. Counties containing the central city are designated "core counties"; the figure shows growth for core counties of large MSAs, noncore counties of large MSAs, and core counties of small MSAs. With few exceptions, the slowest growth rate is observed for the core counties of large MSAs.

Although the downtown central business district (CBD) generally remains the location of the greatest employment density and the highest land values within the region, the central city is surrounded by competing subcenters, with much of the employment and population dispersed throughout the metropolitan area. Historical empirical studies of Chicago and Los Angeles show that the monocentric model does an increasingly poor job of describing these metropolitan areas (McMillen, 1996; Small & Song, 1994). Cross-sectional studies of Los Angeles, Chicago, and San Francisco identify multiple subcenters (Cervero & Wu, 1997; Giuliano & Small, 1991; Gordon,

Richardson, & Wong, 1986; McMillen & McDonald, 1998).

The Los Angeles metropolitan area provides an interesting illustration of changing urban form, given its rapid growth over the past decades. Comparing population and employment by county from 1980 to 2000, we find a clear pattern of decentralization, as shown in Table 9.2. Los Angeles County includes the historic core of the region. Orange County emerged as a commuter suburb to Los Angeles in the 1950s. The "outer counties" were functionally separate until the 1970s, when counties to the east of Los Angeles became commuter suburbs to both Los Angeles and Orange County. These shifts are apparent in the population and employment shares, with Orange County maturing as a job center, and the outer counties making big gains in population share.

Using employment data from the same period, but at the level of census tracts, employment centers are identified. In this case, an *employment center* is any cluster of employment with at least 20,000 jobs and a density of at least 20 jobs per acre. There were 10 centers in 1980, 13 in 1990, and 15 in 2000. Figure 9.6 shows the location of the 2000 centers. The map shows county boundaries and the expressway system. The total share of employment in centers remained steady: 17% in 1980 and 16% in 2000, but there is decentralization among the centers. The only center to lose employment in absolute terms was the Los Angeles CBD. In 1980 the Los Angeles CBD accounted for 55% of all jobs in cen-

TABLE 9.2. Los Angeles Region Employment and Population Shares, 1980–2000

	1980		1990		2000	
	Employment	Population	Employment	Population	Employment	Population
L.A. County	73.0	66.6	66.9	63.0	60.0	58.1
Orange County	17.0	17.3	19.0	17.2	20.4	17.4
Outer counties	10.0	16.1	14.2	19.8	19.6	24.5
Region total	5,388,057	11,191,953	6,874,676	14,011,641	7,426,771	16,373,645

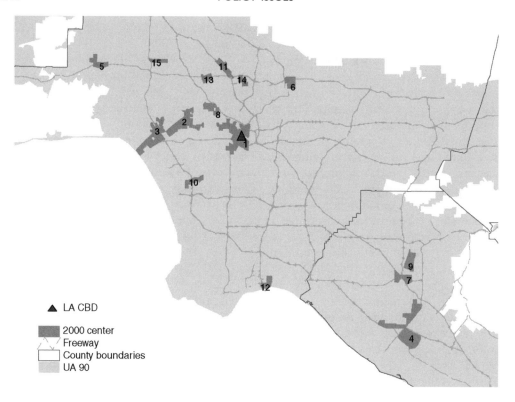

FIGURE 9.6. Subcenters in the Los Angeles region.

ters; by 2000 the share had dropped to 33%. In contrast, no subcenters existed in Orange County in 1980, but by 2000 there were three centers, collectively representing 17% of all jobs in centers. Moreover, the "suburban" subcenters are more spread out. For example, the Los Angeles CBD remains the center with the greatest number of jobs (about 393,000) and the highest job density (68 jobs/acre). The South Coast metro subcenter (No. 4 on the map) has 22 jobs/acre. It is a classic "edge city," concentrated around the intersection of three expressways and spreading west and north along the expressway corridors.

The Housing/Transport Trade-off

As already mentioned, the standard model relies on the assumption that people decide where to live primarily by selecting a trade-off between commuting cost and land cost.

As I noted earlier, given a fixed location of employment, the theory can be used to predict average commute length. It follows that the observed average commute length can be used as a test of the theory. If households do seek to minimize commute cost, observed commute patterns should reveal an average trip length close to that predicted by the model. Numerous studies have conducted such a test. In most cases, the observed average commute length far exceeds that predicted by the standard model, *even when model predictions are based on actual employment and population spatial patterns* rather than the conventional assumption of monocentricity (Giuliano & Small, 1993).

Why is the average commute much longer than the theory predicts? First, imbalances or mismatches between workers and jobs may exist. *Imbalances* occur when the number of workers who can be housed in an area differs substantially from the number of jobs there. *Mismatches* occur when prices or other characteristics make housing unsuitable for workers who hold jobs there. I have stated that studies based on the actual distribution of housing and employment substantially underpredict the actual average commute; hence spatial imbalances likely do not explain observed commute patterns.

The possibility of spatial mismatches is likely. Exclusionary zoning practices, growth controls, rising development costs, and rapid economic growth have been identified as explanations for a shrinking supply of affordable housing (Meck, Retzlaff, & Schwab, 2003). In a search for lower cost housing, workers move to outlying areas far from their jobs, thus incurring long commutes. It is difficult to determine, however, whether such workers are "pushed out" by high housing prices, or "pulled out" by the attraction of affordable single-family dwellings. Two studies that tried to account for such mismatches by introducing some constraint in calculating the model predictions generated estimated average commutes about half as long as actual commutes (Cropper & Gordon, 1991; Giuliano & Small, 1993). For specific population segments there is more conclusive evidence of spatial mismatch. Several studies document both longer commutes and less job access for blacks living in the central city (Ihlandfelt & Sjoquist, 1998), and for low-wage, unskilled workers (Shen, 1998).

A second explanation for longer than expected commutes is related to the job market. Given decentralized employment, households with more than one worker must somehow locate to accommodate multiple job locations. The rise of multiworker households is therefore consistent with longer commutes. Job mobility (i.e., higher rates of job turnover and shorter tenure in any given job) may lead households to locate so as to maximize access to future job opportunities, rather than to a given current job. Finally, the high cost of moving one's residence suggests that households may be willing to trade-off a longer commute in order to avoid a residential move (Crane, 1996).

A third explanation for observed commute patterns is that the cost of transportation has declined more rapidly than the cost of housing. One might argue that the cost of transportation (or, more precisely, the price we pay directly for transportation) is so low that households do not attempt to economize on travel. The real cost of commuting has indeed dropped dramatically: turn-of-the-century commuters paid about 20% of the daily wage (Hershberg, Light, Cox, & Greenfield, 1981), compared to about 7% for commuters today. On the other hand, as per capita income rises, the time cost of commuting should dominate, limiting the willingness of households to incur ever longer commutes. Hence, even in the largest metropolitan areas, we observe a small share of commuters traveling 1 hour or more: 18.4 % for New York, 13.2 % for Chicago, 11.1% for Los Angeles.[5]

Household Preferences

Residential location models are of the form

$$D = f(P_h, P_t, P_g)$$

where D is housing demand, P_h is housing price, P_t is transport price, and P_g is the price of all other goods and services the household consumes. One of the model's key assumptions is identical preferences among households. Preferences are not identical, however, even among households with similar socioeconomic and demographic characteristics. As a result, the demand for housing services (particularly with respect to other goods and services) is not the same for all households. In economic terms, housing demand elasticities are different among households, meaning that relative preferences between transport and housing are not necessarily identical.

The P_g term in the residential location model (representing all other goods and services) also has a location component.

Public goods such as parks, schools, and air quality exhibit a high degree of spatial variation. If households have strong preferences for such goods, location choice will be affected. The quality and availability of public goods are important considerations for many households, such that increased commuting costs might willingly be incurred in order to obtain them.

Residential location theory defines housing in the broadest of terms as the bundle of all housing-related services, so public services and site amenities can be conceptualized as being contained within this housing bundle. The theory also assumes, however, that housing is identical and uniformly available throughout the region, hence public services and amenities are also uniformly available. This assumption is, of course, unsupported in contemporary urban areas.

A further complication arises from the durability of housing stock. As tastes and per capita income change, so does the new housing stock. For example, the average size of a new single-family dwelling has increased from 1,660 square feet in 1973, to 2,320 square feet in 2002.[6] Since cities grow outward, the largest houses, on average, are located at the periphery. Thus urban areas include a wide variety of housing choices differentiated by age, size, general condition, and the like, with older and smaller units closer to the historic center and newer larger units further from the center.

A number of hard-to-quantify factors also affect location choice. Examples include ethnic preferences, racial biases, family loyalty to specific neighborhoods, or preferences for architectural styles. Last, as noted earlier, the financial and psychological costs of moving are so high that a large amount of inertia is associated with location choice.

These observations imply that location choice is much more complicated than the simple trade-offs among housing commute costs and all other goods that location theory posits. On the contrary, work trip cost is just one among many factors that may be considered. It should be no surprise then that the theory is seldom supported in empirical tests.

Economic Structure

Muller (Chapter 3, this volume) and Janelle (Chapter 4, this volume) described the changing structure of the urban space economy. Restructuring of the U.S. manufacturing sector in the 1970s and early 1980s resulted in the loss of large numbers of high-wage, long-term, skilled worker jobs. Job growth in the 1980s was of a different sort, with much of the growth accounted for by lower wage, shorter term, service-sector jobs. Restructuring of the service sector in the late 1980s and early 1990s eliminated many high-wage, long-term professional and middle management jobs. Information and communications technology has made possible far more flexible work arrangements, leading to more contract and temporary work, more self-employment, and a resurgence of small enterprises. Hence the economic expansion of the late 1990s saw increases in both high-wage professional jobs and low-wage service jobs. The recession of the early 2000s has generated another round of restructuring, further increasing the share of "contingent" (contract, part-time, temporary) jobs (see Hanson, Chapter 1, note 2, this volume). Overall, the trend is one of increasing job turnover, which again reduces the relative importance of locating close to any given job.

Restructuring also has implications for employment location. Whereas manufacturing firms typically invest heavily in plant facilities and equipment, making relocation costly, service firms are mobile. Office equipment is relatively inexpensive and easy to move, telephone lines are ubiquitous, and optic fiber is widely available. Information and communications technology allow efficient operation from many different locations. Thus the continued shift to service and information-based activities can be expected to increase the mobility of both firms and workers.

The Supply Side

A number of factors associated with the supply of land in urban areas also affect theoretical expectations. The standard model assumes a perfectly competitive land market, but land markets are affected by many variables. First, land use in the United States is regulated by local jurisdictions. Municipalities influence land use through zoning codes, development requirements and restrictions, and the provision of infrastructure and services.

Tax revenues are a major concern for municipalities. Land uses that contribute positively to the tax base are likely to be encouraged, while uses that add to the public cost burden may be discouraged. Termed the "fiscalization of land use," land use decisions based primarily on revenue considerations have been identified as a leading factor in the growing shortage of affordable housing (Freeman, Shigley, & Fulton, 2003). California municipalities provide some extreme examples. With the property tax accounting for only 6% of local revenue and largely controlled by the state (State of California, 2003), California cities have become increasingly reliant on consumption taxes (e.g., sales tax, hotel tax). Hence high-volume retail sales activities—auto malls, "big box" discount centers—are welcomed. Low-income housing (or in some cases *any* new housing) is not welcomed, however, because future residents must be provided with schools, police, and other public services, while the tax revenue they generate will not be sufficient to cover these added costs (Altshuler, Gomez-Ibañez, & Howitt, 1993). The search for revenue also leads to competition between municipalities for activities viewed as economically beneficial. Competition may range from booster efforts to generous tax reductions or other subsidies, infrastructure provision, and even training programs for prospective workers (Fulton, 1997; Herman, 1994).

Municipalities are also concerned with preserving and enhancing perceived amenities. Affluent municipalities may wish to preserve a suburban or rural atmosphere, protect environmental amenities, or preserve open space. Development fees, growth caps, density restrictions, land dedications, and infrastructure requirements are some examples of the many policy instruments available to municipalities to accomplish these goals.

A second important supply-side factor is the nature of the contemporary land market. The scale of land development has increased over time. Residential development has occurred in gradually larger tract developments, then in planned residential developments, and now in the planned community, which typically includes both residential and commercial uses. A similar process has occurred in the commercial sector, from individual buildings, to office parks, to large multiuse centers. The consequences of these changes are twofold. First, large tracts of open land are a necessary condition for such developments, and large parcels are most likely to be found at the periphery of metropolitan areas—areas with relatively low accessibility. Second, land use changes are determined by fewer decision makers and fewer decisions now than was the case in the past. The role of the large-scale developer is increasingly seen as a key factor in explaining the emergence and growth of suburban centers (Henderson & Mitra, 1996). Since large developments must obtain local government approvals via the environmental review process, local preferences may play a significant role in determining land use development outcomes, further weakening any systematic relationship between land use and transportation.

Statics versus Dynamics

Finally, the various theories discussed previously are static: they assume instantaneous equilibrium across all markets. Metropolitan areas are, of course, always changing, but changes occur at different rates. Employment changes relatively rapidly, as old jobs are eliminated and new ones emerge, and as firms move, or grow, or go out of business. Households change rather quickly as well. But the built environment changes very slowly. A major highway or rail project takes years to build and has an operational life of many decades. Most cities have building stock dating back a century or more. It is therefore likely that metropolitan areas are never in equilibrium, but rather constantly adjusting to the changing population and employment dynamic.

One might ask whether these simple theories are of any use in understanding the land use–transportation relationship. The answer is an unequivocal "yes": cities have decentralized and become less dense as transport costs have declined; more accessible places have higher land value than less accessible places; subcenters in large metropolitan areas are functionally linked; and few workers have 1-hour or more commutes, even in the largest metropolitan areas. One we go beyond these stylized facts, however, we cannot predict whether real-world land use and transportation changes will conform to theoretical expectations.

EMPIRICAL EVIDENCE OF THE LAND USE IMPACTS OF TRANSPORTATION INVESTMENTS

There are two ways to measure the land use impacts of transportation investments. The first is through model simulation, as described in Johnston (Chapter 5, this volume). The new generation of integrated urban planning models allows for the distribution of activities (jobs and residences) to respond to changes in the transportation network. There are many versions of such models under development or in limited use, and it is beyond the scope of this chapter to discuss them.[7] The general concept of an integrated model is shown in Figure 9.7. The model begins with baseline population and employment distributions, as well as the

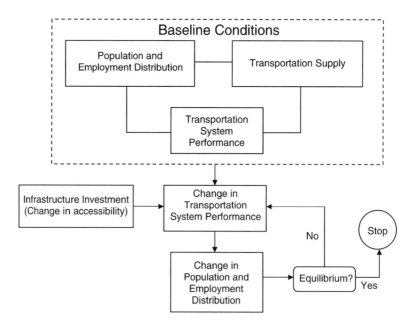

FIGURE 9.7. Transportation–land use modeling framework.

baseline transportation system. The pattern of travel demand results in some equilibrium level of system performance. We then introduce some change in transportation infrastructure—for example, a new rail line. In the short run, travel demand changes in response to the increased transit capacity. These travel demand changes lead to longer run changes in activity location, as described earlier in this chapter. Activity location changes, in turn, result in further changes in travel demand, and therefore further changes in system performance. Through some mechanism (e.g., prices, travel time), an equilibrium solution is obtained that once again balances transportation supply and demand.

Integrated models are the only means for *predicting* the land use impacts of transportation investments, but even the most complex of such models are based on many assumptions and are highly simplified approximations of the real world. Unfortunately, there are few cases of testing these models by "back-casting," or using the model to forecast from a previous time period to the present. Thus we do not know very much about their predictive reliability. Nevertheless, integrated models are useful tools for evaluating alternative scenarios and contributing to our general understanding of transportation and land use relationships.

The second way to measure land use impacts is through empirical study. Empirical documentation has been of interest to researchers for many decades. Studies of specific projects in the United States were conducted as early as 1930, and many more have been performed since. Interestingly, there is little consensus on the conclusions to be drawn from the results of these studies. Results are conflicting for both highway and transit investments. In order to understand why, it is useful to discuss the problems involved in tracing the impacts of transportation changes as well as the context in which transportation investments are made.

Methodological Problems

The ideal research design for establishing causality is the four-way experiment: a "test" group and a "control" group, each monitored before and after the "treatment." Such a design is, of course, impossible in the case of transportation impacts. With enough lead time, a before/after study can be done, but there is no control group with which to compare the before/after results. Researchers are left with two choices: longitudinal studies (comparing conditions before and after a given transportation investment), or cross-sectional studies (comparing conditions in one area with a given transportation investment to another similar area without such an investment). Many researchers have attempted various types of "quasi-experimental design" by, for example, identifying similar communities with and without a rail line or new highway and conducting a longitudinal analysis. Because of the unique characteristics of any given community, the reliability of such comparisons is limited. Other researchers use complex statistical techniques to control for as many contravening factors as possible, but, of course, no study controls for every possible factor. The infeasibility of applying standard research design techniques to the empirical study of transportation impacts explains why a body of consistent results across many studies has yet to be generated. Let me now explain methodological problems in more detail.

Long-Term Dynamics

Two problems plague attempts to identify land use impacts of transportation investments. First, as noted earlier, transportation changes take place in a highly dynamic system. A change in the transportation system is just one of many changes occurring at the same time. Furthermore, changes continue after the investment takes place, making it difficult to attribute observed changes correctly. That is, land use changes that may be observed following a given transportation investment do not necessarily result from that investment.

Second, the interaction of land use and transportation takes place over time. Because of the longevity of capital infrastructure, the market response to changes in transportation may take place only after many years or decades. To make matters even more complicated, it has been argued that land use changes related to the investment may appear *before* the project is actually constructed, because landowners may act in anticipation of higher land values (Boarnet, 1997). The longer the time span over which these effects take place, the more difficult they are to isolate from all the other changes in economic conditions, regional employment, and population demographics that are taking place at the same time.

Longitudinal versus Cross-Sectional Studies

Longitudinal studies focus on one area and study the changes that take place over a given period of time. If lead time is sufficient and data are available, land use changes within the area of impact (however defined) are traced from before the project took place to some time period after its completion. In order to isolate transportation-related impacts, all other factors thought to be relevant are incorporated into the analysis. For example, the regional rates of housing and commercial construction would be considered in the impact assessment. The longitudinal approach, however, may not control for the distributional shifts in land use activities that transportation investments may generate if the area of analysis is limited. The presence of a given transportation investment can attract land use change that might otherwise have located elsewhere in the region. Thus, although significant local changes may have taken place, the net impact on the region may be insignificant. This point is particularly important when decision makers are concerned with the

economic growth impacts of transportation investments.

Cross-sectional studies make comparisons between nearby similar areas as an approximation of the "no project" comparison. Several cross-sectional studies have been performed at the subregional level in order to make the comparison as appropriate as possible. For example, rail transit impact studies are often performed at the corridor level. Land use changes in a rail corridor are compared to those in a similar corridor without rail. This type of analysis has two pitfalls. First, as noted above, seemingly similar corridors may in fact be quite different. Second, if transportation investments cause a shift in activity location because of improvements in relative accessibility, such a comparison will exaggerate the extent of the impact. What is actually being measured is the combined effect of increased accessibility in the corridor with rail and the decreased accessibility in the corridor without rail.

Finally, neither of these approaches allows us to draw any conclusions on the direction of causality. Even when significant land use changes occur in conjunction with transportation investments, there is no way to determine whether the investment *caused* these changes, or whether the changes *created the demand* for the transportation improvements. There is much empirical support for the latter interpretation. The planning and environmental approval process for major projects can span a decade or more, making lags between regional growth and transportation investments quite large.[8] In any event, land use and transportation decisions are so closely tied together that it has been impossible so far to separate their effects.

The Contemporary Context of Urban Highway Investments

The context of urban highway investments can affect the extent of land use impacts. First, any single highway investment is but a part of a much larger urban transportation system. In this sense, most highway invest-ments are marginal. They add some incre-ment of accessibility to the system. In an area that already enjoys a high level of ac-cessibility, we would not expect a new in-vestment to have much of an impact, as in the case of the I-105 Expressway noted ear-lier. In an area with limited accessibility, the same investment should have a much greater impact. Most U.S. urban areas, how-ever, fall into the former category.

Second, the availability of developable land must be a key consideration. As noted earlier, land use change is more likely to oc-cur in the form of new construction, and available land is a prerequisite for new con-struction. All other things being equal, then, the land use impacts of highway invest-ments in areas with vacant land available should be greater than in areas where devel-opment has already taken place. This is not to say that highway investments may not promote redevelopment, but rather that significant change is more likely when de-velopable land is available.

Third, even when land is available, local public policy may not be favorable to devel-opment. Local land use controls are another critical factor in impact assessment. Land use change cannot occur unless local zoning per-mits it. If local opposition to development is strong, highway investments may have little effect, even if market demand exists.

A fourth factor is the state of the regional economy. Little new investment takes place in a stagnant or declining region. Rather, disinvestment occurs. Residential areas de-teriorate, factories close down, and the gen-eral level of economic activity declines. Under these circumstances, highway invest-ment should have little or no impact. If it did, it would most likely be at the expense of other areas within the region.

Fifth, the scale of analysis must be consid-ered. Determining whether highway invest-ments have an impact within their imme-diate vicinity is relatively straightforward. Indeed, any observer of the urban scene can recount changes observed after a new free-way opened or an interchange was con-structed (though the careful observer would also be able to point out interchanges where nothing changed). All that appears to be necessary is a series of before/after studies. But the question in more complicated if the region as a whole is of interest.

Observation of local changes does not support the hypothesis that highway invest-ments generate land use impacts. In order to test such a hypothesis correctly, we must study the region as a whole. Only then can land use changes taking place in the vicinity of a highway improvement be determined to be significantly different from those oc-curring outside that vicinity.

Sixth, and finally, careful impact analysis requires that changes in land use patterns be distinguished from general economic

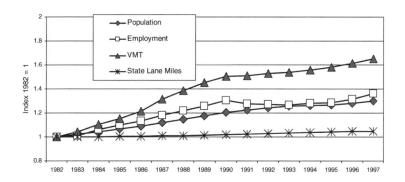

FIGURE 9.8. Indexed growth in California, 1982–1997.

growth. That is, land use change will occur in a growing region with or without new highway investment. This point is illustrated in growing regions throughout the United States. Growth in urban travel far exceeds growth in highway capacity over the last two decades. Despite negligible additional highway investment, new development has been spurred by general economic expansion. California provides a good example. Figure 9.8 shows indexed trends in population, employment, vehicle miles traveled (VMT), and miles of state highway, from 1982 to 1997. While expansion of the state highway system was negligible (4%), population increased by 30%, employment by 36%, and VMT by 65%.

The potential growth effects of highway investments have become a major environmental issue. Many of those concerned with the sustainability of our urban areas argue that provision of more highway capacity generates "induced demand," meaning increased private auto use (independent of any economic growth effect), which creates more demand for low-density development, which in turn increases dependence on the private auto and thus demand for yet more highway capacity (see Box 9.2). Because major public investment decisions are being made on the basis of these claims, it is critical to examine land use–transportation relationships as comprehensively as possible.

BOX 9.2. Induced Demand

Those opposed to new highways argue that adding new highway capacity is poor public policy, because more highway capacity generates more travel. Congestion relief benefits are short term at best, as increased travel demand—demand *induced* by the new capacity—soon results in the same or worse congestion than before, while further increasing auto dependence and its environmental damages. Is it possible that added capacity will generate more travel and hence offset all accessibility benefits?

Consider a congested highway. New lanes are added in each direction. Travel times decline as congestion is reduced. In the short run, travelers using other modes, or taking other routes, or traveling at other times of the day will be attracted to the expanded highway until a new equilibrium across routes, modes, and times is established. The added capacity, by virtue of reducing travel costs, also may attract new or longer trips. Shoppers may choose a bigger mall, farther from home, since the traffic is no longer unbearable. Friends may choose to visit more frequently, hence making new trips. By such choices total travel will indeed increase, and these new or longer trips are induced to the extent that the added capacity reduced travel costs. This is simply supply and demand at work: as the price falls, more of the good is consumed.

In a supply–demand framework, it is obvious that added highway capacity cannot possibly lead to the same or worse conditions, as long as the demand curve remains fixed. If the demand curve shifts to the right (e.g., from population growth), it is possible that congestion will be worse than initial conditions (but not as bad as would have been the case without the added capacity). However, increased travel demand from the added population is not induced.

What if population and jobs were attracted to the area in response to the added capacity? Households and firms residing elsewhere in the region might relocate, redistributing existing economic activity. New households and firms might choose to locate in the area. It is this long-term effect of guiding growth that is at the basis of the highway critics' concerns.

Highway Impacts

The literature on highway impacts is very broad, ranging from national studies of the economic growth effects of highway investment, to regional studies, to highly localized case studies. Research on highway impacts has been influenced by national policy objectives. The first wave of research was stimulated by the highway building boom of the 1950s and 1960s that followed passage of the 1956 Interstate Highway Act. Several studies were commissioned to identify the social and economic benefits of highways and to identify future highway needs. By the mid-1960s, political opposition to highway construction and growing environmental concerns began to shift the national transportation policy focus to public transit. Highway impacts did not become a subject of research again until the mid-1970s, when highways were targeted as promoters of energy consumption and urban sprawl.

By the 1980s, most of the Interstate Highway System had been constructed, and highway building in metropolitan areas seemed to have ended. National policy interest shifted to traffic congestion relief in the context of a fixed highway system, as metropolitan population, employment, and vehicle travel growth overwhelmed transportation system capacity. Public transit continued to gain attention and funding as the best alternative to continued increases in automobile travel.[9]

Another policy issue emerged in the 1980s: the link between economic productivity and highway investment. Researchers demonstrated correlation between the slowdown in economic growth observed since the early 1970s and the decline in public infrastructure investment, spawning a scholarly debate that has continued to the present.[10]

In the 1990s, the growing desire on the part of many urban planners and policymakers to restructure U.S. metropolitan areas into more socially and environmentally desirable form has renewed interest in both highway and transit land use impacts. If indeed highways have played a significant role in the dispersion of urban areas (and if dispersion is causally linked with excess energy consumption, environmental damages, or social inequities), policies restricting further highway expansion might be justified. Hence several studies were funded to reexamine highway impacts. These new studies have the advantage of using geographic information systems (GIS), which has allowed more fine-grained spatial analysis than had been feasible in the past. Moreover, despite the end of the Interstate era, new highways have been built in metropolitan areas, particularly in the high-growth regions of the West and the South.

Strategies for Measuring Impacts

As noted above, empirical studies may be longitudinal or cross-sectional. In some cases pooled cross sections may be used, meaning more than one observation over time of the same cross section. There are three possible ways for measuring impacts: travel outcomes, property values, or changes in the distribution of population and/or employment. Travel outcomes are the weakest test. We might argue that changes in travel are a necessary but not sufficient condition for demonstrating impact. If the new facility has no significant impact on accessibility, it cannot have an impact on land use, because accessibility is the mechanism by which land use changes are affected. Even if travel shifts are observed, they may not reflect enough of a change in accessibility to promote shifts in the location of households or firms.

Measuring changes in property values is a stronger test of land use impacts. Changes in accessibility generated by a new highway or transit investment should be valued in the land market, per economic theory. As land increases in value, capital (structures) will be substituted for land, leading to more intense development. Hence increased land values can be interpreted as a signal for development. Changes in property values can also be used to measure the physical extent of the impact. For example, we expect that the accessibility value gradient is steeper for transit than for highways, since access to transit is mainly by foot, and people are willing to walk only short distances.

The third possibility is to measure changes in land use: changes in employment or population density, commercial building, residential units, and so on. This is the most straightforward measure but also is subject to confounding factors, as discussed above. In order to conduct an analysis based on land use changes, all the other factors that influence land use must be carefully controlled.

Summary of Empirical Results

Early studies of highway impacts tended to validate theoretical expectations (Adkins, 1959; Mohring, 1961). These studies used land values as the measure of impact; results showed significant increases in land value near the highway. It was not clear, however, whether these increases represented a net gain within the region as a whole, or whether they represented a distributional shift, that is, the land value increases near the highway were offset by land value decreases somewhere else. These early highway investments had a dramatic effect on accessibility, and significant impacts on land values were the result.

Highway impact research conducted in the 1970s took place in a very different context. The Interstate Highway System was nearing completion, and expressway systems had been constructed within all the major cities in the United States. The rate of auto ownership had more than doubled, and the shift of both population and employment to the suburbs was well underway (see Muller, Chapter 3, this volume).

One example of this generation of studies is the Beltway Study, conducted for the U.S. Department of Transportation to examine the land use and related impacts of beltways (Payne-Maxie Consultants, 1980). *Beltways* are circumferential expressways that were designed to divert through traffic from congested, inner-city areas. Concerns that beltways promoted central-city decline and suburban growth led to the Beltway Study.

Study results were based on various statistical analyses of 54 U.S. cities (27 with beltways and the remainder without), as well as in-depth case studies of eight beltway cities. The statistical analysis revealed that the land use impacts of beltways have generally been insignificant. Growth effects as manifested by regional population and employment changes during the period of study (1960–1977) were related to the change in manufacturing employment and city age. Growth of the manufacturing sector was associated with growth in total employment and population. The city/age relationship was negative: newer cities grew faster than older cities. The existence of a beltway, that beltway's relative location, and its length had no consistent effect on growth. Selected Beltway Study results are presented in Table 9.3.

Both the statistical analysis and the case studies suggested that land availability and favorable economic conditions were the key factors associated with employment location. Under favorable conditions, these new activities tended to locate near beltways. In cities without beltways, however, the spatial distribution of employment growth (city vs. suburb) was no different. The study authors conclude that land use impacts are largely determined by local market conditions.

It is useful to categorize the recent literature as addressing regional impacts or intraregional impacts. Table 9.4 gives summaries of three regional studies of economic impacts. They were selected on the basis of sound methodology, appropriate data, and representing state-of-the-art research on this topic. All three studies examine factors associated with economic

TABLE 9.3. Selected Beltway Study Results

Measure	Beltway	Other
Growth impacts		
SMSA population growth, 1960–1970	Slight (+)	Change in manufacturing employment +
		City age −
SMSA population growth, 1970–1977	No significant effect	Change in manufacturing employment +
		City age −
Wholesale employment growth, 1972–1977	Beltway distance from CBD +	Change in manufacturing employment +
		City age −
		Arterial mileage −
Distributional impacts		
Central-city population growth, 1960–1970	No significant effect	Change in central-city manufacturing employment +
		City age −
Central-city wholesale employment, 1972–1977	Interchanges per beltway mile −	Change in central-city population +
		Change in central-city manufacturing employment +
Central-city service employment, 1972–1977	Interchanges per beltway mile +	Change in central-city wholesale employment +
	Beltway distance from CBD +	

TABLE 9.4. Summary of Selected Recent Regional Highway Impact Studies

Author	Location	Description	Data and method	Results
Boarnet (1998)	California	Study of economic output to test role of highway capital.	1969–1988 county-level annual data. Production function model, allowing for intercounty spillover effects.	Economic output + for within-county highway capital, − for adjacent county highway capital.
Henry, Barkley, and Bao (1997)	South Carolina, Georgia, North Carolina mixed urban–rural region	Study of urban and rural population and employment growth to determine linkages between rural and urban growth.	1980, 1990 census tract population and employment data. Simultaneous model of regional development.	Coefficient for highway access not significant.
Singletary, Henry, Brooks, and London (1995)	South Carolina	Study of new manufacturing employment to test impact of highway investment.	1980–1989 new manufacturing employment, durable and nondurable; ZIP code level. Regression, accounting for spatial autocorrelation and heterogeneity.	For durable manufacturing, coefficient for highway access generally +. For nondurable manufacturing, coefficient for highway access generally not significant.

growth, including highway infrastructure, and all three consider distributional impacts (spatial spillover). Boarnet (1998) found a strong link between highways and economic growth within counties, as well as strong spillover effects, meaning that some of the within-county growth was at the expense of neighboring counties. Henry and colleagues (1997) found that the strongest predictor for rural population and employment growth was proximity to high-growth urban cores; highways had no significant impact on rate or location of growth. Singletary and colleagues (1995) focused on durable and nondurable manufacturing employment growth in South Carolina and found highway access to be generally significant for durable manufacturing, but generally not significant for nondurable manu-

facturing. The authors conclude that the following factors are important: (1) agglomeration economies; (2) road density and access to I-85, the main connector between Atlanta and Charlotte; (3) location, timing, and type of highway investment. These studies suggest that under the right circumstances highways often do influence patterns of economic growth, but more fundamental regional economic factors are the drivers of economic growth.

Table 9.5 summarizes four intraregional studies. As with the regional studies, all of these utilized appropriate data and methodologies, and hence represent the most reliable studies. Comparing results (last column of table), we observe that the Boarnet and Chalermpong (2001) and Voith (1993) studies show consistently significant and

TABLE 9.5. Summary of Selected Recent Intraregional Highway Impact Studies

Bollinger and Ihlandfeldt (1997)	Atlanta metropolitan area	Study of employment and population changes to test impact of MARTA.	1980, 1990 census tract-level data on employment and population. Simultaneous equation model of employment and population growth as function of proximity to rail and highways.	Coefficient for presence of freeway in tract + for employment growth, not significant for population growth.
Harder and Miller (2000)	Toronto metropolitan area	Study of residential property values to test significance of highway, subway, CBD access.	1995 residential sales data. Hedonic regression, residential sales, accounting for spatial autocorrelation.	Coefficient for presence of highway – Coefficient for presence of subway + Coefficient of access to CBD –
Boarnet and Chalermpong (2001)	Orange County, CA	Study of residential sales to test impact of Orange County toll roads.	1988–1999 residential sales data for two corridors plus control corridor. Hedonic regressions, residential sales, before/after corridor open.	Coefficient for distance from ramp – in after period, both corridors, no difference before/after in control corridor.
Voith (1993)	Philadelphia metropolitan area	Study of residential sales to test significance of highway and commuter rail access to CBD.	1970–1988 residential sales data for Montgomery County. Hedonic regressions, residential sales (individual sales, census tract averages).	Coefficient for presence of station + and increasing in value over period. Coefficient for highway access to CBD – and increasing in value over period.

positive impacts of the highways (a negative coefficient on a distance variable means prices decline with distance from a ramp). In Atlanta, presence of a freeway is associated with employment growth, but not with population growth. In Toronto, presence of a highway nearby (within 2 kilometers) is *negatively associated* with residential property values, although distance to the CBD has the expected negative coefficient, and access to the CBD is in part a function of the highway network. These results are mixed. A closer look at the Orange County study illustrates why highway impacts are so place-specific.

The Orange County, California, Toll Roads

The Orange County toll roads provide a good example of highway construction in the post-Interstate era. The toll roads include 51 centerline miles of limited-access highway.[11] Financed by a combination of developer fees, state and local highway funds, and toll-backed bonds, the toll roads are often considered the model for future highway building in the United States.

Orange County, California, is a rapidly growing part of the Los Angeles metropolitan area. Before 1950 it was the agricultural hinterland of Los Angeles. Much of the land is semiarid and hence was used for livestock production. The flatlands adjacent to Los Angeles attracted postwar housing development, and by 1960 the northern and central portions of the county, made increasingly accessible by new freeways, were bedroom suburbs of Los Angeles. Development continued to spread south, with the large new planned communities of Mission Viejo and Irvine emerging by 1970. Growth in both population and employment has continued unabated; by 2000 the county was home to 2.8 million persons and 1.5 million jobs. Figure 9.9 provides a map of Orange County with the freeway system and toll roads.

Population and employment growth in the 1950s and 1960s was accommodated by

the Interstate Highway System, but by the early 1970s it became clear that continued growth would require more highways. Orange County planners undertook two major highway planning studies, one dealing with the southern coastal area (SEOCCS), the other with the northeastern area (NEO CCS). These studies concluded that new limited-access facilities would be needed in both areas and provided tentative alignments for the proposed facilities. These early studies were the origin of the Orange County Toll Roads.

It is important to understand the context of the toll roads. At the time of the NEOCCS and SEOCCS studies, most of southeast Orange County remained undeveloped and in the hands of major landowners. The largest among them was (and is) the Irvine Company, with holdings of approximately 93,000 acres—nearly one-fifth of the county's total land area—that extended from the Orange County border in the north to the coast in the south. It was clear to the Irvine Company that development of their land was contingent upon providing access (much of the area was open ranchland or hill country with no major roads of any kind). It was also clear that the traditional source for major highway building—federal Interstate Highway System funds—would not be sufficiently available to finance such an ambitious plan. It took years of planning and political deal making to come up with a plan, but in 1986 the Transportation Corridor Agencies (TCAs) were formed (San Joaquin Hills TCA for Route 73 along the coast; Foothill/Eastern TCA for Routes 241 and 261 in the north and east). Landowners (primarily the Irvine Company) agreed to deed rights-of-way, and the county agreed to provide some limited public funding. It was further agreed to establish a fee on all new construction within the corridor areas of impact, with the areas of impact defined by traffic forecasts. But there was still a shortfall in funding, and the final piece of the finance package was tolls to partially cover

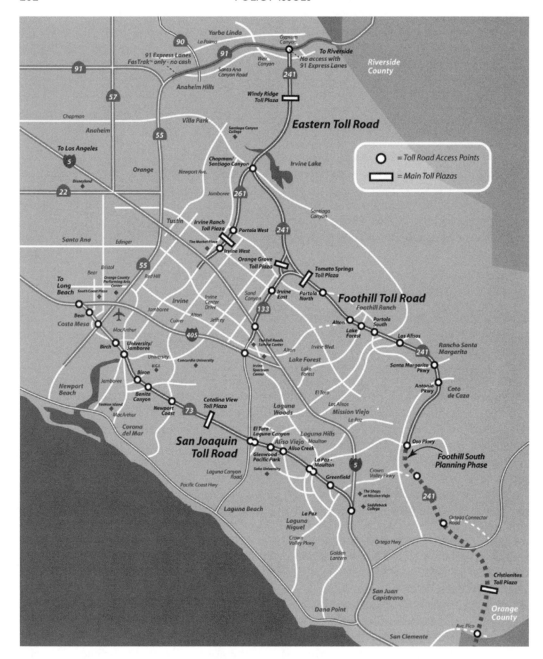

FIGURE 9.9. Orange County, California.

bond payments. Finally, development in the corridor areas was contingent upon construction of the tollways.[12]

Boarnet and Chalermpong (2001) examined the land use impacts of the new toll roads. They investigated residential sales prices from 1988 to 1999 within the Foothill and San Joaquin Hills corridors, before and after construction of major segments of each facility, and also included a control group of residential sales within another freeway corridor that had not undergone any change in capacity over the time period. The authors estimated a series of "hedonic regressions," which estimate the sales price of the residence as a function of the attributes of the residence itself (e.g., number of bedrooms, size of lot), neighborhood attributes, and access to the highway. In this case, highway access is measured as straight-line distance to the nearest interchange. Results for the before/after access coefficients were as expected. For both corridors, the "after" access coefficient is significant and negative (sales prices fall with distance from the nearest interchange), while the coefficients in the control corridor are not significant (Boarnet & Chalermpong, 2001, p. 588).

The authors conducted additional analysis to estimate year-to-year house price indices in each corridor. For the Foothill, house prices closer to an interchange relative to those farther away are consistently higher throughout the period. For the San Joaquin Hills, the same pattern is observed to 1995, after which house prices farther away are higher. The authors argue that the San Joaquin Hills results are less reliable, because of other improvements made on a nearby freeway during the same time period, and because the amenity value of the coast could confound results. They conclude that the toll roads did indeed change accessibility, and the value of the "accessibility premium" is reflected in residential sales prices. The authors further conclude that these accessibility changes also change development and travel patterns.

The evidence from this comprehensive study is quite convincing. It is important to note, however, that the highways were built to allow nearby development in a regulated, supply-constrained housing market. *The development and the highways were planned together.* Indeed, today the San Joaquin Hills corridor is in financial trouble in part because some of the planned development did not occur, leading to lower levels of travel demand. Opposition from local residents has resulted in down-zoning and delays in development. The toll roads did indeed increase the accessibility of south and eastern Orange County, but land use regulation has determined what was built.

Conclusions on Highway Impacts

Several decades of empirical research leads to the following conclusions:

1. Highway investments increase accessibility. The greatest benefits are generated at the site of the new investment. Relative accessibility changes throughout the network.
2. Highway impacts are context-specific. Where there is growing population and employment, and land is available for development or redevelopment, impacts are likely to be significant. When these conditions do not exist, impacts are not likely to be significant.
3. There is little evidence that highway investment is an effective means for promoting *net* economic development. Even when economic growth is documented, it may be a result of redistributing activities from other locations. From a local perspective, of course, this may be the objective.

Transit Impacts

National policy began to focus on public transit in the mid-1960s. Public transit systems in several major cities were in financial trouble, and the federal government was called upon to provide capital for rebuilding the nation's transit industry. The financial needs of big city transit systems, together with a growing public awareness of the adverse environmental and societal impacts of highway building, created a broad consensus for federal support of public transit. Improvement and expansion of public transportation services were justified as the means to reduce use of the private auto, reduce air pollution, save energy, and revitalize central cities. Advocates argued that better transit would improve central-city accessibility, hence attracting development back into cities and halting urban sprawl. The federal government responded with a massive capital grants program that eventually led to the construction of the "new generation" mass transit systems.

These rail systems proved to be very costly public projects. In contrast to highway projects, whose direct costs are funded largely by user revenues (gasoline and vehicle excise taxes) earmarked for highway use, transit capital projects were (and are) funded from general revenues (income and other general taxes). Justifying these projects on the basis of indirect benefits therefore became very important. Land use benefits were of particular interest: increased land values would generate more tax revenue, thus offsetting some of the system costs. Not surprisingly, then, the federal government sponsored impact studies for all three new heavy rail systems: BART in San Francisco, METRO in Washington, D.C., and MARTA in Atlanta.

The new generation rail systems failed to meet expectations. With the possible exception of the Washington METRO, these systems did not achieve predicted ridership levels and failed to attract auto users to transit in significant numbers (Fielding, 1995; Kain, 1999). Most of the systems constructed later have not fared any better (Dunn, 1998; Pickrell, 1992; Rubin, Moore, & Lee, 1999). Despite the disappointing results, public support of mass transit contin-

ues to increase. The 1991 Intermodal Surface Transportation Efficiency Act (ISTEA) made it possible to use federal highway funds for other modes, and the 1998 Transportation Equity Act for the 21st Century (TEA-21) continued these provisions. Metropolitan areas around the country are using state and local funds to supplement federal funds to build or expand rail transit. Transit—particularly rail transit—continues to be viewed as a means for improving air quality, increasing energy efficiency, redeveloping central cities, providing mobility for the transportation-disadvantaged, and creating more sustainable cities. Indeed, transit advocates argue that more investment in transit is warranted: the only way to compete with the private auto is to provide extensive, high-quality, low-fare service (see Pucher, Chapter 8, this volume).

Theoretical Expectations of Rail Transit Impacts

It is useful to examine what land use impacts of rail transit might be expected on the basis of theory. Rail transit systems generate changes in accessibility only in the immediate vicinity of the rail line itself. The construction of a rail transit system should improve accessibility along the rail line corridors and increase the relative advantage of rail corridors compared to areas that are not served by the rail system. All other things being equal, then, activity location should shift toward the rail corridors, and this shift should be reflected in increased land values. In addition, since rail systems are focused on the CBD, the position of the CBD as the most accessible point in the area should be enhanced, leading to an increase in activities and land values in the CBD.

Two additional points follow from theory. First, the potential effect of rail transit is defined by the extent to which accessibility is changed. Because rail service usually replaces existing bus service, its effect on accessibility can actually be quite small. Moreover, the transit system accounts for a very small portion of the entire transportation network. Even a large change in transit accessibility has little effect on the system as a whole. Therefore we should not expect any regional effects, as is the case for highways. Second, a transit improvement is another form of transportation system improvement, and therefore should have the same *decentralizing effect*. To the extent that the transit improvement reduces transport costs, people will use some of the reduced costs to consume more travel. Thus the difference between a highway improvement and a transit improvement should be one of degree: the land value gradient should be steeper for rail than for highway, but any reduction in transport cost will cause the gradient to become flatter, as we saw in Figure 9.4.

Empirical Evidence

There is an extensive literature on the land use impacts of rail transit. As noted earlier, San Francisco's BART, Washington, D.C.'s METRO, and Atlanta's MARTA were the subject of studies shortly after they opened (Dvett et al., 1979; Lerman, Damm, Lerner-Lamm, & Young, 1978; Webber, 1976). Other early studies were conducted for the Philadelphia–Lindenwald commuter line (Allen & Boyce, 1974) and the New York transit system (Pushkarev & Zupan, 1977). The results of these studies showed that there were few significant impacts on land values or development patterns. In the case of the heavy rail systems, it was noted that the first impact studies were too early; it would take decades before the land market

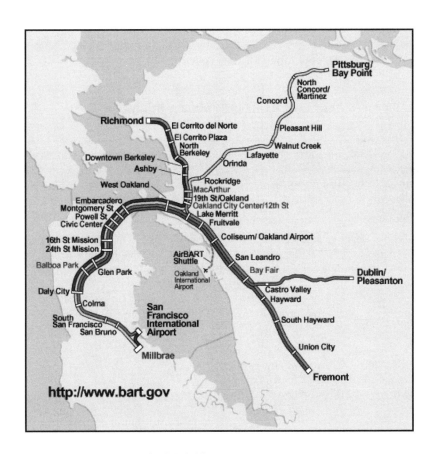

FIGURE 9.10. The BART system.

could respond (e.g., Dvett et al., 1979). Nevertheless, a comparison of rail system impacts led to the following conclusions: (1) a growing local economy is a necessary but not a sufficient condition for land use impacts, and (2) supportive zoning and development policies must be in place (Knight & Trygg, 1977).

It has now been more than 20 years since the opening of BART and the other heavy rail systems. All three have been subject to additional analysis. In addition, there are many recent studies of other rail systems around the country. Here are some results.

The San Francisco Bay Area Rapid Transit (BART) System. The BART system was the first of the "new generation" rail systems to be constructed. The first line began operation in 1972, and the 72-mile system was completed by 1974 (see Figure 9.10). An additional line was constructed along the I-580 corridor and opened in 1997; service to the San Francisco Airport opened in 2003. The BART system was the first to use many new technology elements such as automated fare collection systems and computerized train control. The system was beset with operating problems for several years, and did not achieve its designed service frequency until a decade after its opening.

Cervero and Landis (1997) conducted a "BART @ 20 Update Study" to examine the land use and other impacts of BART 20 years after its opening. The analysis was conducted at several levels, from superdistricts (aggregations of census tracts) to hectare grid cells, and included many different sources of data. Table 9.6 provides selected results from the study. The overall conclusion is that BART impacts have varied dramatically across the system. BART had the greatest impact on downtown San Francisco, greatly increasing labor force access and hence allowing an already vibrant downtown to grow even larger. Favorable zoning and properties available for redevel-

TABLE 9.6. Selected Results of BART@20 Impact Study

Comparison	Results
County business patterns data, 1981–1990 Shift-share analysis of differences in job growth, ZIP codes with BART station vs. ZIP codes without BART station	Job growth in ZIP codes with BART station greater: 57% of all job growth. Result due to job growth in downtown San Francisco; all other BART station ZIP codes had less job growth than non-BART ZIP codes
Employment density, census tract level, 1980, 1990 Data from Census Transportation Planning Package	Employment density increases in San Francisco CBD, Oakland CBD, Concord Line corridor, Fremont Line corridor, but also North Bay, San Mateo SR-101 corridor, Silicon Valley
Office space in various locations, before BART (1962), 1963–1974, 1975–1992	Downtown SF: about 2/3 new San Francisco office space 1963–1974 within ¼ mile of BART station; about ¾ 1975–1992.
Downtown Oakland: less than 10% of new East Bay office space, entire period; most new office space away from BART stations	
Matched-pair comparisons of nine BART stations and similar freeway interchanges	More multifamily housing construction near BART stations, slightly more nonresidential construction near BART stations, 1979–1993

opment were also factors. Between 1975 and 1992, 40 million square feet of new office space was built in San Francisco, about one-half of which was located within one to two blocks of the Embarcadero station.

Across the Bay in Oakland, despite supportive public policies, most of the development that did occur was either public or heavily subsidized. It bears noting that downtown Oakland is the most accessible node in the BART system, so theoretically it should have benefited most. Stations along the Concord and Fremont lines show how local conditions have mediated BART impacts. The Walnut Creek and Pleasant Hill stations have experienced significant commercial and residential development, while other stations have not. Cervero and Landis claim that BART has allowed for more clustered office/commercial development than would have occurred without BART. The Fremont station area has experienced extensive moderate- and high-density housing construction by virtue of ambitious local zoning and municipal support, and there is some evidence that nearby apartment dwellers are paying a rent premium for access to the station. In Berkeley, public opposition to higher-density redevelopment has deterred change around stations, while in Richmond a poor local economy and a high crime rate have discouraged station-area development efforts.

Atlanta's MARTA. In contrast to the San Francisco Bay Area, where the physical geography constrains the spatial extent of development, the Atlanta metropolitan area has been able to expand in all directions. Many critics identify Atlanta as one of the most sprawled metro areas in the United States. Population density of the urbanized area is 1,783 persons per square mile, compared to 6,130 for San Francisco–Oakland and 4,407 for Miami.[13] The metro area has experienced double-digit growth every decade since 1950; between 1990 and 2000 the population increased by nearly 40%, to

4.1 million. As with the other major rail systems, the intent of MARTA was to increase transit use, reduce private vehicle use, revitalize the downtown, and promote growth within the rail corridors.

Bollinger and Ihlandfeldt (1997) conducted a comprehensive study of MARTA impacts. They developed a model of population and employment growth, using census tract-level data from 1980 and 1990. Employment growth is modeled as

$$EMP\Delta_t = f(M_t, EMP_{t-1}, POPA_{t-1}, POPA\Delta_t E)$$

where $EMP\Delta_t$ = change in employment, M = proximity to MARTA, EMP_{t-1} = employment in previous period, $POPA_{t-1}$ = labor force access in previous period, $POPA\Delta$ = change in labor force access, and E = control variables affecting employment change.

A similar model is used for population change, with population change a function of employment access, level of employment in previous period, proximity to MARTA, and other control factors. The authors tested two models, one with a single variable for proximity to a MARTA station, another with stations categorized by type (from "high-intensity urban node" to "neighborhood station"), since early planning for the system identified categories of stations and zoned the surrounding land accordingly. For employment, only one category of station type, mixed-use regional node, was associated with significantly greater employment growth. For population, none of the MARTA measures was associated with population growth. The authors concluded that MARTA had neither a positive nor a negative impact on population and employment growth, but that it did have the effect of altering the composition of employment in some station areas, increasing the share of public employment. Findings are attributed to (1) MARTA's insignificant impact on accessibility, (2) the absence of a significant increase in transit use, and (3) public policy efforts limited to rezoning.

The Miami Metrorail. Miami might be considered a better candidate for transit than Atlanta, with its large share of immigrants, large share of elderly population, and much higher average population density. However, the Miami Metrorail is among the least successful of the new rail systems. Ridership has been extremely low. System planners saw Metrorail as a potential economic development tool, and sought to use it to revitalize economically depressed areas of the city.

Gatzlaff and Smith (1993) used residential sales data to examine the impact of the system on land values. They argued that a change in land value is a precursor to change in land use, so is a good measure of accessibility-related impacts. They conducted a repeated sales analysis, which compares the change in sales price of a given property in successive sales. Repeated sales within station areas (1 square mile) and outside station areas were compared; there was no significant difference between groups. They also conducted hedonic regressions, controlling for before/after the system was approved and for distance from the nearest station. Results were counter to expectations: with one exception, access to a station had either no effect or a negative effect on land values. The authors find their results consistent with the lack of ridership on the Metrorail (implying no impact on accessibility), and with planners' efforts to use Metrorail as an economic development tool, locating the system outside Miami's economic growth corridors.

Rail system impacts have been studied extensively, and it is beyond the scope of this chapter to summarize them all. The results of these studies are as mixed as the three examples discussed here. A comparison of employment growth showed significantly greater employment change around Washington, D.C., METRO station areas than nonstation areas, but the study did not control for other factors that may have affected employment growth (Green & James,

1993). A small but positive effect of proximity to a light rail station on commercial lease rates in San Jose, California, was documented (Weinberger, 2001). This is surprising, given the low ridership of the system. Building quality was not controlled for in the analysis, and since the city rezoned station areas for higher density, the rent effect is more likely due to newer and better quality space than the rail system. Proximity to a subway station has a positive impact on residential land values in Toronto (Harder & Miller, 2000). In Buffalo, New York, the light rail line had no significant effect on land use, employment, or central-city economic development (Banister & Berechman, 2000). Similar results were found for a light rail system in Manchester, U.K. (Forrest, Glen, & Ward, 1996).

Conclusions on Rail Transit Impacts

The past 30 years of research on the land use impacts of rail transit lead to the following conclusions:

1. A necessary but not sufficient condition for land use impacts is a significant impact on accessibility. Rail systems typically have little impact on accessibility; most rail investment replaces existing bus service, and even many route-miles of rail serve only a small share of a city's origins and destinations.
2. Impacts are highly localized and tend to occur in fast-growing, heavily congested core areas.
3. Impacts depend on complementary zoning, parking, and traffic policy, and especially on development subsidies.
4. Rail investment is not sufficient to promote economic development in declining areas.

Conclusions on Land Use Impacts

Empirical evidence on the land use impacts of both highways and transit indicates that transportation investments do not have consistent or predictable impact on land use. The evidence clearly shows that land use change does not necessarily follow transportation investments, even when the dollar value of these investments is large. Nevertheless, land use impacts continue to be a major policy issue in transportation planning. Metropolitan areas around the United States expect to solve congestion and environmental problems, revitalize central cities, reduce urban sprawl, and increase livability through investment in rail transit. Highway expansion is resisted on the grounds that it promotes sprawl and auto dependence. Given the empirical evidence, the reader might ask why. I conclude with some possible explanations.

First, urban planners and local officials are heavily vested in the preservation and revitalization of central cities. Many decades of decentralization has left the central city with a decreasing share of economic activity, but an increasing share of disadvantaged population, and sometimes significant available infrastructure. Rail transit, like sports stadiums and festival marketplaces, is seen as a tool for attracting the middle class back to the city by making commuting to central-city jobs easier. Rail transit has the added advantage of federal and state subsidies—"free" money to a local jurisdiction.

Second, rail transit is often promoted for its economic development potential.[14] Cities are usually willing to subsidize local station area development by assembling and preparing land for development, offering density offsets or reduced parking requirements, low-cost loans, or outright subsidies. Many urban planners believe that there is significant demand for transit-oriented city living among young professionals and older baby boomers, and hence see such efforts as a way of attracting more market-rate development to the city. Some cities seem to be successful in such efforts, for example, Portland and San Diego. Others are not, for example, Oakland or Buffalo. In all cases, outcomes depend more on the underlying conditions of the city (population characteristics, crime, economic potential, aesthetics, competition, etc.) than the presence or absence of a rail system. For example, Seattle and Austin have thriving downtowns without rail transit.

Third, many urban planners believe that public transit investment is one of the few feasible alternatives available for solving transportation and sustainability problems. It is argued that rail transit, with its superior line-haul travel time and higher quality ride, has more potential for attracting discretionary riders out of their cars than bus transit. It is further argued that public transit is an appropriate "second-best" policy. Although charging auto users full marginal costs is the most efficient solution to transportation externalities, it has so far proved politically impossible to do so. If we charged full marginal costs on the auto, transit demand would increase considerably. In the absence of pricing, subsidizing public transit is another way of making it more competitive with the auto.

Fourth, some planners and policymakers are convinced that if transportation capacity increases are restricted to transit, eventually travelers will shift to transit as congestion becomes intolerable. Increased transit mode share will then lead to accessibility-related changes favoring higher density, more compact development. Known as the "field of dreams" or "build it and they will come" belief, the logic holds only in a closed system. That is, if people and firms had no choice to move to another city or region, or to shift travel to other times and places, rising congestion would indeed shift some travelers to transit as the travel time advantage of the auto erodes.

Finally, it has recently been argued that rail transit systems are expanding because their construction is politically acceptable, but highway construction is not (Altshuler & Luberoff, 2003). Rail systems impose a much smaller footprint on urban areas. They require less land; noise and vibration are highly localized. In the dense parts of central cities, rail transit is obviously more appropriate to the urban fabric and to high land values and dense development. Outside such areas, however, rail transit has little likelihood of solving transportation problems.

Metropolitan areas throughout the United States have highly developed transportation systems. Even the largest investments can effect only minor changes in the general level of accessibility within the region. The extent and redundancy of U.S. transportation systems makes them flexible; system users can adapt to existing capacity in many different ways. Even in heavily congested metropolitan areas like Los Angeles or Washington, D.C., developable land with reasonable access remains available. Metropolitan growth is a function of economic conditions—the general state of the economy, labor force quality, the regulatory environment—and increasingly of the preferences of affluent households. The past 20 years of metropolitan growth is illustrative; despite the lack of investment in highways, population and employment decentralization has continued unabated, both in the United States and in Europe.

Broader economic and technological trends suggest that the relationship between transportation and urban form will continue to be uncertain. The continuing shift to an information-based economy is changing the nature of accessibility (see Janelle, Chapter 4, this volume). It is also increasing the locational flexibility of households and firms. Information-intensive activities require relatively little investment in fixed facilities and hence can relocate with ease. For such firms, as long as labor force access is satisfactory and rental price is reasonable, location choice may be based largely on subjective (and hence difficult-to-predict) criteria. A world of flexible work changes the calculus of households, and perhaps makes access to job opportunities more important in location choice than access to any given job. Given the flexibility of the urban transportation system and the many factors that affect the location and travel choices of individuals and firms, it is understandable that the link between transportation and land use in contemporary metropolitan areas is complex, difficult to identify, and almost impossible to predict.

NOTES

1. See Boarnet and Crane (2001) for an excellent review and discussion.
2. Source: *http://www.losangelesalmanac.com/topics/ Transport/tr26.htm.*
3. For formal presentation of the standard theory and explanation of these relationships, see Mills and Hamilton (1993).
4. The value of time is the monetary value an individual places on time savings. For example, you have the choice of taking a shuttle van or a taxi to the airport. The shuttle van fare is lower, but the trip is longer compared to the taxi. Research has shown that the value of time is related to the individual's wage rate. Therefore we observe that college students are more likely to take the shuttle, while business executives will take the taxi.
5. Source: computed by the author from U.S. Census 2000, STF 3 data.
6. Source: *http://www.census.gov/const/C25Ann/ sftotalmedavgsqft.pdf.*
7. See Wegener (1994) for a description of the major operational urban models. See Meyer and Miller (2001), Chapter 6, for a discussion of selected models applied to transportation planning.
8. The I-105 Century Freeway discussed earlier provides an extreme example. The freeway was adopted as part of the State Highway Plan in 1959. Planning and environmental studies began shortly thereafter. The Final Environmental Impact Statement was approved in 1978; subsequent litigation delayed the start of construction until 1982 (Hestermann, DiMento, van Hengel, & Nordenstam, 1993). The facility opened in 1993.
9. Although federal funding for public transit was reduced under the Reagan administration, increased funding from states and local jurisdictions resulted in continued increases in total funding (Dunn, 1998).

10. For reviews, see Gramlich (1994), Boarnet (1997), and Bhatta and Drennan (2003).
11. The completed system is anticipated to be 67 miles. The final 16-mile segment is currently under development.
12. The reader might ask whether there was public opposition to such a massive plan of development. Opposition to the Foothill/Eastern was limited and eventually appeased by various compromises on route location and mitigation. Opposition to the San Joaquin Hills was especially strong, and numerous lawsuits delayed the project.
13. Source: *www.demographia.com/db_uanscan.htm.* The Los Angeles–Long Beach urbanized area has the highest population density in the country, at 7,063 persons per square mile.
14. The economic impact from construction of any public project is a transfer from general taxpayers to the local jurisdiction. It is also temporary.

ACKNOWLEDGMENT

Valuable and greatly appreciated assistance in writing this chapter was provided by graduate students in the School of Policy, Planning, and Development at the University of South California. Patrick Golier provided general assistance. Data and maps for the Los Angeles region were developed by Chen Li and Duan Zhuan under a grant from the Los Angeles County Metropolitan Transportation Authority. All errors and omissions are the responsibility of the author.

REFERENCES

Adkins, W. (1959). Land value impacts of expressways in Dallas, Houston, and San Antonio, Texas. *Highway Research Board Bulletin, 227*, 50–65.

Allen, W., & Boyce, D. (1974). Impact of a high speed rapid transit facility on residential property values. *High Speed Ground Transportation Journal, 8*(2), 53–60.

Alonso, W. (1964). *Location and land use.* Cambridge, MA: Harvard University Press.

Altshuler, A., Gomez-Ibanez, J., & Howitt, A. (1993). *Regulation for revenue: The political economy of land use exactions.* Washington, DC: Brookings Institution.

Altshuler, A., & Luberoff, D. (2003). *Mega-projects: The changing politics of urban public investment.* Washington, DC: Brookings Institution.

Anas, A., Arnott, R., & Small, K. (1998). Urban spatial structure. *Journal of Economic Literature, 36*, 1426–1464.

Banister, D., & Berechman, J. (2000). *Transport investment and economic development.* London: UCL Press.

Bernick, M., & Cervero, R. (1997). *Transit villages in the 21st century.* New York: McGraw-Hill.

Bhatta, S., & Drennan, M. (2003). The economic benefits of public investment in transportation: A review of recent literature. *Journal of Planning Education and Research, 22*, 288–296.

Boarnet, M. (1997). Highways and economic productivity: Interpreting recent evidence. *Journal of Planning Literature, 11*(4), 476–486.

Boarnet, M. (1998). Spillovers and locational effect of public infrastructure. *Journal of Regional Science, 38*(3), 381–400.

Boarnet, M., & Chalermpong, S. (2001). New highways, house prices, and urban development: A case study of toll roads in Orange County, CA. *Housing Policy Debate, 12*(3), 575–605.

Boarnet, M., & Crane, R. (2001). *Travel by design: The influence of urban form on travel.* New York: Oxford University Press.

Bollinger, C., & Ihlandfeldt, K. (1997). The impact of rapid rail transit on economic development: The case of Atlanta's MARTA. *Journal of Urban Economics, 42*, 179–204.

Brueckner, J. (1978). Urban general equilibrium models with non-central production. *Journal of Regional Science, 18*, 203–215.

Cervero, R., & Landis, J. (1997). Twenty years of the Bay Area Rapid Transit System: Land use and development impacts. *Transportation Research A, 31A*(4), 309–333.

Cervero, R., & Wu, K.-L. (1997). Polycentrism, commuting, and residential location in the San Francisco Bay Area. *Environment and Planning A, 29*, 865–886.

Christaller, W. (1966). *Central places in southern Germany* (C. Bushin, Trans.). Englewood Cliffs, NJ: Prentice-Hall.

Crane, R. (1996). The influence of uncertain job location on urban form and the journey to work. *Journal of Urban Economics, 39*(3), 342–358.

Cropper, M., & Gordon, P. (1991). Wasteful commuting: A re-examination. *Journal of Urban Economics, 29*, 1–13.

Dunn, J. A. (1998). *Driving forces: The automobile, its enemies and the politics of mobility.* Washington, DC: Brookings Institute Press.

Dvett, M., Dornbusch, D., Fajans, M., Fackle, C., Gussman, V., & Merchant, J. (1979). *Land use and urban development impacts of BART* (Final Report No. DOT-P-30-79-09, U.S. Department of Transportation and U.S. Department of Housing and Urban Development). San Francisco: Blayney Associates and Dornbusch and Co.

Fielding, G. J. (1995). Transit in American cities. In S. Hanson (Ed.), *Geography of urban transportation* (2nd ed., pp. 287–304). New York: Guilford Press.

Forrest, D., Glen, J., & Ward, R. (1996). The impacts of a light rail system on the structure of house prices: A hedonic longitudinal study. *Journal of Transport Economics and Policy, 30*(1), 15–30.

Freeman, B., Shigley, P., & Fulton, W. (2003). Land use: A California perspective. Available online at http://www.facsnet.org/tools/env_luse/calif1.php3.

Fulton, W. (1997). *The reluctant metropolis: The politics of urban growth in Los Angeles.* Point Arena, CA: Solano Press Books.

Gatzlaff, D., & Smith, M. (1993). The impact of the Miami Metrorail on the value of residences near station locations. *Land Economics, 69*(1), 54–66.

Giuliano, G., & Small, K. (1991). Subcenters in the Los Angeles region. *Journal of Regional Science and Urban Economics, 21*, 163–182.

Giuliano, G., & Small, K. (1993). Is the journey to work explained by urban structure? *Urban Studies, 30*(9), 1485–1500.

Gordon, P., Richardson, H., & Wong, H. (1986). The distribution of population and employment in a polycentric city: The case of Los Angeles. *Environment and Planning A, 18*, 161–173.

Gordon, P., Richardson, H., & Yu, G. (1998). Metropolitan and non-metropolitan employment trends in the U.S.: Recent evidence and implications. *Urban Studies, 35*(7), 1037–1057.

Gramlich, E. (1994). Infrastructure investment: A review essay. *Journal of Economic Literature, 32*(3), 1176–1196.

Green, R., & James, D. (1993). *Rail transit station area development: Small area modeling in Washington DC.* Armonk, NY: M. E. Sharpe.

Harder, M., & Miller, E. (2000). Effects of transportation infrastructure and location on residential real estate values: Application of spatial autoregressive technique. *Transportation Research Record, 1722*, 1–8.

Henderson, V., & Mitra, A. (1996). The new urban landscape: Developers and edge cities. *Regional Science and Urban Economics, 26*, 613–643,

Henry, M., Barkley, D., & Bao, S. (1997). The hinterland's stake in metropolitan growth: Evidence from selected southern regions. *Journal of Regional Science, 37*(3), 479–501.

Herman, T. (1994, August 3). Bidding wars escalate. *Wall Street Journal*, p. A1.

Hershberg, T., Light, D., Cox, H., & Greenfield, R. (1981). The journey to work: An empirical investigation of work, residence and transportation, Philadelphia, 1950 and 1880. In T. Hershman (Ed.), *Philadelphia: Work, space and group experience in the nineteenth century* (pp. 128–173). New York: Oxford University Press.

Hestermann, D., DiMento, J., van Hengel, D., & Nordenstam, B. (1993). Impacts of a consent decree on "the last urban freeway": I-105 in Los Angeles County. *Transportation Research A, 27A*(4), 299–313.

Hoover, E. (1948). *The location of economic activity.* New York: McGraw-Hill.

Ihlandfeldt, K. T., & Sjoquist, D. L. (1998). The spatial mismatch hypothesis: A review of recent studies and their implications for welfare reform. *Housing Policy Debate, 9*(4), 849–892.

Ingram, G. (1998). Patterns of metropolitan development: What have we learned? *Urban Studies, 35*(7), 1019–1035.

Isard, W. (1956). *Location and the space economy.* Cambridge, MA: MIT Press.

Kain, J. (1999). The urban transportation problem: A reexamination and update. In J. Gomez-Ibañez, W. Tye, & C. Winston (Eds.), *Essays in transportation economics and policy* (pp. 359–401). Washington, DC: Brookings Institution Press.

Knight, R., & Trygg, L. (1977). *Land use impacts of rapid transit: Implications of recent experiences* (Final Report No. DOT-TPI-10-77-29, U.S. Department of Transportation). San Francisco: De Leuw Cather and Co.

Lerman, S., Damm, D., Lerner-Lamm, E., &

Young, J. (1978). *The effect of the Washington Metro on urban property values* (Final Report CTS-77-18, U.S. Department of Transportation, Urban Mass Transportation Administration). Cambridge, MA: MIT Center for Transportation Studies.

Losch, A. (1954). *The economics of location* (W. Woglum & W. Stolper, Trans.). New Haven, CT: Yale University Press.

McMillen, D. (1996). One hundred fifty years of land values in Chicago: A nonparametric approach. *Journal of Urban Economics, 40*(1), 100–124.

McMillen, D., & McDonald, J. (1998). Suburban subcenters and employment density in Metropolitan Chicago. *Journal of Urban Economics, 43*, 157–180.

Meck, S., Retzlaff, R., & Schwab, J. (2003). *Regional approaches to affordable housing* (Planning Advisory Service Report No. 513/514). Washington, DC: American Planning Association.

Meyer, M., & Miller, E. (2001). *Urban transportation planning* (2nd ed.). New York: McGraw-Hill.

Mieskowski, P., & Mills, E. S. (1993). The causes of metropolitan suburbanization. *Journal of Economic Perspectives, 7*(3), 135–147.

Mills, E. S. (1972). *Studies in the structure of the urban economy.* Baltimore: Johns Hopkins University Press.

Mills, E.S., & Hamilton, B. (1993). *Urban economics* (5th ed.). New York: HarperCollins.

Mohring, H. (1961). Land values and the measurement of highway benefits. *Journal of Political Economy, 79*, 236–249.

Moses, L. (1958). Location and the theory of production. *Quarterly Journal of Economics, 72*, 259–272.

Muth, R. (1969). *Cities and housing.* Chicago: University of Chicago Press.

Newman, P., & Kenworthy, J. (1998). *Sustainability and cities: Overcoming automobile dependence.* Washington, DC: Island Press.

Payne-Maxie Consultants (1980). *The land use and urban development impacts of beltways* (Final Report No. DOT-OS-90079, U.S. Department of Transportation and Department of Housing and Urban Development). Washington, DC: Author.

Pickrell, D. (1992). A desire named streetcar: Fantasy and fact in rail transit planning. *Journal of the American Planning Association, 58*(2), 158–176.

Pushkarev, B., & Zupan, J. (1977). *Public transportation and land use policy.* Bloomington: Indiana University Press.

Rubin, T. A., Moore, J. E. II, & Lee, S. (1999). Ten myth about US urban rail systems. *Transport Policy 6*(1), 57–73.

Shen, Q. (1998). Location characteristics of inner city neighborhoods and employment accessibility of low-wage workers. *Environment and Planning B, Planning and Design, 25*, 345–365.

Singletary, L., Henry, M., Brooks, K., & London, J. (1995). The impact of highway investment on new manufacturing employment in South Carolina: A small region spatial analysis. *Review of Regional Studies, 25*, 37–55.

Small, K., & Song, S. (1994). Population and employment densities: Structure and change. *Journal of Urban Economics, 36*(3), 292–313.

State of California. (2003). *Cities annual report, FY 2000–2001.* Sacramento, CA: State Controller's Office.

Voith, R. (1993). Changing capitalization of CBD-oriented transportation systems: Evidence from Philadelphia, 1970–1988. *Journal of Urban Economics, 33*, 361–376.

von Thunen, J. (1826). *Der Isolierte Staat in Beziehung ant Landswirtschaft and Nationalekomie.* Hamburg, Germany.

Webber, M. (1976). The BART experience—What have we learned? *Public Interest, 4*, 79–108.

Weber, A. (1928). *Theory of the location industries* (C. J. Freidrich, Trans.). Chicago: University of Chicago Press.

Weinberger, R. (2001). Light rail proximity—Benefit or detriment in the case of Santa Clara County, California? *Transportation Research Record, 1747*, 104–113.

Wegener, M. (1994). Operational models: State of the art. *Journal of the American Planning Association, 60*(1), 17–29.

White, M. (1988). Location choice and commuting behavior in cities with decentralized employment. *Journal of Urban Economics, 24*, 129–152.

CHAPTER 10

Transportation and Energy

DAVID L. GREENE

Energy is essential to transportation. The separation of people and things in space necessitates passenger and freight movements, and energy is required to overcome that separation. Physics defines *work* as the application of force over a distance, and *energy* as the ability to do work. To move any object from one place to another work must be done: force must be applied over distance to accelerate an object from rest, and it must be continually applied to overcome the frictional forces that oppose motion. Thus, transportation requires the use of energy. Energy is measured in several different ways (see Box 10.1).

Over the past century, the world's transportation systems have come to rely on petroleum as a source of energy. While transportation vehicles can, and have been, powered by various fossil fuels (e.g., coal was the fuel of choice for the steam engines that powered ships and railroads in the 19th and early 20th centuries), petroleum has become the overwhelming fuel of choice for modern, internal combustion engine vehicles. Petroleum's key advantages are abundant supplies, low cost, high energy density (in terms of both weight and volume), and

ease of transport, handling, and storage. But transportation's reliance on petroleum brings problems, as well: climate-altering greenhouse gases, air pollution, and, for the United States, serious problems with respect to the reliability and security of vital energy supplies.

As human economies become more complex and specialized, as they become more technologically advanced and wealthier, the demand for mobility both of persons and goods, increases (World Business Council for Sustainable Development, 2001). Based on data collected from developed and developing economies from various times, researchers discovered that, *on average*, human beings spend about 1 hour per day traveling, and about 10–15% of their income on transportation (Schafer, 2000). As incomes grow, people will spend more money on travel. As more is spent on travel, in order to stay within their travel time budgets, travelers will choose faster and faster modes of transport. The implication is that as long as economic growth continues and the technology of travel permits greater speeds, the demand for travel will increase. And unless the energy intensity (energy use per

---

BOX 10.1. Measures of Energy

Energy is most often measured in terms of heat equivalents, such as joules (in the International System of Units or metric system) or British thermal units (BTU) (see Table 10.1). A joule is defined by the force required to move an object, and is equal to 1 kg m²sec⁻². It takes 1,055 joules to equal 1 BTU, which is defined as the amount of heat required to raise the temperature of 1 pound of water by 1°F. A gallon of gasoline contains approximately 121.7 million joules, or 115,400 BTU.[1] At the scale of the entire U.S. economy, so much energy is used that the units commonly used to describe energy use are exajoules (10^{18} joules) or quads (10^{15} BTU). Petroleum fuels are often measured in terms of volume or weight. The most common volumetric measure is the barrel, equal to 42 gallons. U.S. petroleum supply and demand are most often described in terms of millions of barrels per day. One million barrels per day of gasoline is equal to 15.33 billion gallons per year. International data sources often describe petroleum in terms of metric tons of oil. One gigatonne (billion metric tons) of oil is equal to 18.65 million barrels per day.

---

passenger- or ton-mile) of transportation is reduced, transportation energy use will increase, too.

Human beings value mobility highly because it expands the realm of reachable destinations, thereby providing access to greater opportunities. For firms, greater mobility permits increased specialization and facilitates trade. But mobility, especially when fueled by petroleum, also produces unintended and undesirable consequences such as air pollution, traffic congestion, fatalities and injuries in traffic accidents, greenhouse gas emissions, and oil dependence (Greene, 1996). Most of the environmental pollution and all of the oil dependence problems caused by transportation are the result of burning petroleum fuels in internal combustion engines.

Compared with the United States, other countries' transportation systems are far less motorized. Vehicle registrations per thousand persons are shown for several world regions in Figure 10.1. The solid line displays the evolution of U.S. motorization over time, shown on the horizontal axis. Other regions have been plotted on the U.S. curve based on their degree of motorization in the year 2000. Not all countries will follow the path the United States has taken. Higher population and settlement densities, as well as greater investments in non-highway-modes of transport may lead developing economies in the direction of Europe or Japan. Nonetheless, as the world's economies grow, there is clearly enormous room for expansion of the world's motor vehicle fleets, and consequently the demand for en-

TABLE 10.1. Energy Conversion Factors

	Exajoules	Gigatonnes of oil equivalent	Quadrillion BTU	Million barrels of oil per day
Exajoules	1	2.388×10^{-2}	0.948	0.448
Gigatonnes of oil equivalent	41.868	1	39.68	18.76
Quadrillion BTU (quads)	1.055	2.52×10^{-2}	1	0.472
Million barrels of oil per day	2.233	5.33×10^{-2}	2.117	1

Source: Davis and Diegel (2002, Table B-4).

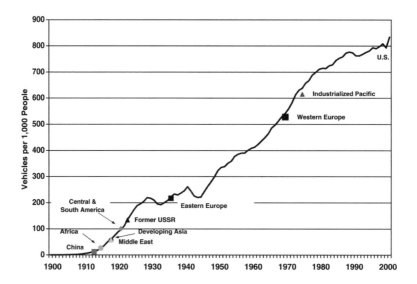

FIGURE 10.1. The motorization of global transport. Source: Davis and Diegel (2003, pp. 3–6).

ergy. It is not yet clear where the energy for a world as mobile as the United States, Europe, or Japan would come from.

PATTERNS AND TRENDS IN TRANSPORTATION ENERGY USE IN THE UNITED STATES

Transportation is a major consumer of energy in the U.S. economy, accounting for 28% of total U.S. energy use in 2001. Transportation's share has been gradually increasing: it stood at 25% in 1975. When it comes to petroleum use, however, transportation is dominant. Transportation vehicles account for over two-thirds of U.S. oil consumption. This share has also been steadily increasing: in 1973, transportation used 52% of the oil consumed in the United States. Since the oil price market upheavals of the 1970s, only the transportation sector's use of petroleum has increased. All other sectors' oil consumption declined except for industry, whose oil consumption has remained constant. As a result, petroleum use has become increasingly concentrated in the economic sector that has shown the least ability to find acceptable alternative sources of energy: transportation.

The highway mode dominates U.S. transportation activity and energy use (see Figure 10.2). Of nearly 5 trillion passenger miles traveled in the United States in 2000 (enough for a trip around the world for 200 million people), almost 90% were traveled on the highway (U.S. Department of Transportation, Bureau of Transportation Statistics, 2002, table 1.31). Nearly all of the remaining

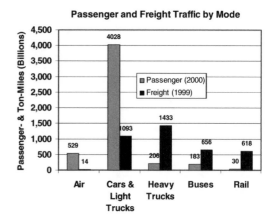

FIGURE 10.2. Distribution of passenger and freight transport by mode. Source: U.S. Department of Transportation, Bureau of Transportation Statistics (2002, tables 1-34 and 1-44).

500 million passenger miles were traveled in aircraft. All types of bus transportation (transit, school, and intercity) account for only 3% of passenger travel, and all mass transport modes combined (excluding air but including bus) comprise only 4% of passenger travel (Figure 10.3). Freight traffic is more evenly distributed across modes. Rail carries the most ton-miles (38%), followed by trucks (29%), water (17%), pipeline (16%), and airfreight (0.4%) (U.S. Department of Transportation, Bureau of Transportation Statistics, 2001, table 1.41).

Energy use by transportation modes reflects the concentration of transportation activity on U.S. highways (see Table 10.2). Highway vehicles account for 75% of total transportation energy use. Air comes in a distant second with a 9% share, waterborne transport with 6%, pipelines with 3%, and rail with 2% (Davis & Diegel, 2002, table 2.5). Energy used by mobile equipment in construction and agriculture comprises another 3% of transportation energy use. Within the highway mode, light–duty vehicles (passenger cars and light trucks) consume three of every four gallons of motor fuel. Energy policy for transportation has focused on passenger cars and light trucks

primarily because they account for 57% of total transportation energy use.

While there are large differences in modal shares, with highway vehicles predominating, in the United States energy intensities vary surprisingly little across modes of passenger transport (see Figure 10.3). *Energy intensity* describes how much energy is required per unit of transportation activity (e.g., per vehicle-mile, passenger-mile, or ton-mile). It may seem odd that mass transit vehicles use, on average, only slightly less energy per passenger mile than automobiles and light trucks and, in fact, the average transit bus is more energy-intensive. In part, this can be attributed to fuel economy regulations enacted over 25 years ago that required a doubling of passenger car fuel economy. In part, it can be attributed to relatively low average load factors for transit vehicles. Cities with exceptionally effective transit systems can achieve energy intensities of 2,000 to 3,000 BTUs per passenger mile (Davis & Diegel, 2002, figures 2.3, 2.4, 2.5), but this is not the norm.

Comparisons of energy intensity among freight modes is hampered by lack of reliable data on tons carried per vehicle mile (load factors) for highway modes. It

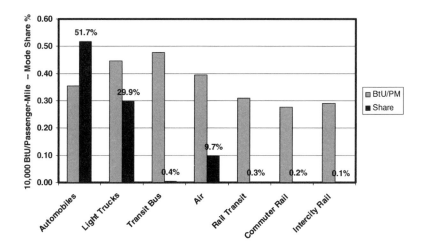

FIGURE 10.3. Average energy intensities and modal shares of passenger travel in the United States, 2000. Sources: Davis and Diegel (2002, pp. 2-15–2-16); U.S. Department of Transportation, Bureau of Transportation Statistics (2002, table 1-31).

TABLE 10.2. Domestic Consumption of Transportation Energy by Mode and Fuel Type, 2000[a] (Trillion BTU)

	Gasoline	Diesel fuel	Liquified petroleum gas	Jet fuel	Residual fuel oil	Natural gas	Electricity	Methanol	Total
Highway	15,952.3	4,742.2	26.1			8.2	0.9	0.0	20,729.7
Light vehicles	15,396.2	299.7	9.6			0.0		0.0	15,705.5
Automobiles	9,031.1[b]	50.6				0.0		0.0	9,081.7
Light trucks[c]	6,338.9[b]	249.1	9.6			0.0		0.0	6,597.6
Motorcycles	26.2								26.2
Buses	10.6	190.9	0.5			8.2	0.9	0.0	211.1
Transit	3.7	88.1	0.5			8.2	0.9	0.0	101.4
Intercity[d]		33.4							33.4
School[d]	6.9	69.4							76.3
Medium–heavy	545.5	4,251.6	16.0					0.0	4,813.1
Off-highway	105.4	838.3							943.7
Construction	23.9	359.1							383.0
Agriculture	81.5	479.2							560.7
Nonhighway	351.6	825.0		2,508.2	1,120.6	664.4	315.4		5,785.2
Air	40.4			2,508.2					2,548.6
General aviation	40.4			134.7					175.1
Domestic air				2,004.0					2,004.0
International air				369.5					369.5
Water	311.2	288.6			1,120.6				1,720.4
Freight		288.6			1,120.6				1,409.2
Recreational	311.2								311.2
Pipeline						664.4	246.5		910.9
Rail		536.4					68.9		605.3
Freight (Class 1)		516.0							516.0
Passenger		20.4					68.9		89.3
Transit							47.2		47.2
Commuter		9.8					16.1		25.9
Intercity[c]		10.6					5.6		16.2
Total	16,409.3	6,405.5	26.1	2,508.2	1,120.6	672.6	315.4	0.0	27,457.7

[a]Civilian consumption only. Totals may not include all possible uses of fuels for transportation (e.g., snowmobiles).
[b]Includes gasohol.
[c]Two-axle, four-tire trucks.
[d]Data for 1999. Data for 2000 are not yet available.
Source: Davis and Deigel (2002, table 2-4).

is clear, however, that airfreight is most energy-intensive, followed by truck. The bulk freight modes (rail, marine, and pipeline) have much lower energy intensities, in the vicinity of 500 BTU per ton-mile, or lower.

A simple equation helps understand how total activity (*A*), modal shares (*S*), energy intensity (*I*), and fuel type shares (*F*) determine total energy use. Energy intensity can be defined in many ways, but is most often defined as energy use per passenger-mile or per ton-mile. The "ASIF" identity is useful for analyzing past trends in transportation energy use, understanding the causes, and also for developing predictions of future energy use (Schipper & Fulton, 2001). Total transportation energy use is the sum across all modes and fuel types of the product of total activity, modal share, modal energy intensity, and modal fuel type share.

$$\text{Total energy} = (\Sigma_m \Sigma_f A S_m \, I_{mf} F_{mf}$$

Where *A* is transportation activity (passenger-miles, ton-miles, or vehicle-miles), *S* is the share of transportation activity carried out by mode *m*, *I* is the energy intensity of mode *m*, and *F* is the share of mode *m*'s activity that is powered by fuel type *f*.

Analyses of trends in U.S. transportation energy use over the past 30 years have shown that the growth of activity has been the most important determinant, followed by decreases in energy intensity. Changes in the distribution of passengers or freight across modes have been much less important, as have changes in fuel types (e.g., Greene & Fan, 1996).

The ASIF equation is also useful because it shows the fundamental ways of reducing transportation's petroleum use and greenhouse gas emissions: (1) decreasing the amount of transportation activity; (2) shifting transportation activity from more energy- (or carbon-) intensive to less energy- (carbon-) intensive modes of travel; (3) reducing the energy- (carbon-) intensity of travel; or (4) switching some transporta-

tion activities to nonpetroleum (lower carbon) fuels that produce less pollution and fewer greenhouse gas emissions.

Late in 1973 the significance of U.S. transportation's need for energy was radically changed by events in the Middle East. In that year, Arab nations belonging to the Organization of Petroleum Exporting Countries (OPEC) declared an oil boycott against the United States in retaliation for its support of Israel in the 1973 October War. Oil prices increased fivefold. Because of government price controls on oil produced in the United States, the boycott resulted in shortages of gasoline and long lines at gas stations across the United States.[2] The gasoline shortages had a powerful effect on Americans used to traveling wherever and whenever they wanted in automobiles powered by cheap and plentiful fuel. Fuel economy standards were imposed on passenger cars and light trucks, freeway speed limits were lowered to 55 miles per hour, and numerous and so far largely futile efforts were made to find substitutes for gasoline.

Prior to the oil market upheavals of 1973–1974 and 1979–1980, energy use on U.S. highways increased in proportion to the growth of vehicle travel. From 1950 to 1975, vehicle miles of travel increased by 190%, from 458 billion to 1,328 billion, at an average annual rate of 4.4% (U.S. Department of Transportation, Federal Highway Administration, Highway Statistics, Table VM-201). Over the same quarter century, energy use by highway vehicles increased from 40.3 billion gallons to 113.5 billion gallons, at an average annual rate of 4.2%. Energy use grew in virtual lock-step with the growth of vehicle travel. After 1975, highway vehicle travel increased by 70%, at an average annual rate of 3.0%, while energy use by highway vehicles has increased by only 50%, at an average annual rate of 1.6% (Davis & Diegel, 2002, table 6.5). Not only did the rate of growth of highway travel slow substantially, but energy use and travel growth were "decoupled" by higher fuel prices and policies that required more energy-efficient vehicles. This decoupling of energy use and activity growth now saves Americans about 60 billion gallons of fuel each year, as shown in Figure 10.4.

While U.S. demand for petroleum has been increasing, U.S. petroleum production has been in gradual decline since the 1970s because of the depletion of domestic oil resources. One of the world's first and most heavily exploited oil regions, the lower 48 states were also one of the first to reach peak production and begin to decline (see Figure 10.5). U.S. oil production peaked in 1970 at 11.3 million barrels per day (mmbd). At that time, imports amounted to 3.1 mmbd, 22% of total oil consumption. Over the next few years U.S. import dependence increased to 47% in 1977, as demand grew and production shrank. Beginning in 1978, a combination of higher oil prices, conservation policies such as fuel economy standards for automobiles and light trucks, and substitution of other fuels for petroleum in electricity generation caused petroleum consumption to decrease by almost 4 mmbd, while domestic oil production increased by only 0.7 mmbd. As a consequence, the lowest level of import dependence in recent history, 29%, was reached in 1985.

But in 1986 world oil prices collapsed,

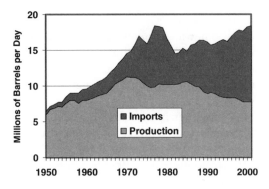

FIGURE 10.5. U.S. petroleum production and net imports, 1950–2001. Source: U.S. Department of Energy, Energy Information Administration (2003, table 5.1).

greatly reducing the economic incentive to use less oil. At the same time, automobile and light truck fuel economy standards, which had increased every year since 1978, were frozen in 1985 and have not been raised substantially since that date. Since 1985, the growth in petroleum demand has been driven almost entirely by transportation energy demand. The result has been strong growth of oil consumption and oil import dependence, which reached an all-time high of 58% in 2001.

TRANSPORTATION ENERGY AND OIL

Almost nothing in the global economy can move without oil. It is this fact that makes oil such a critical resource. Greatly complicating matters is the fact that the world's oil resources are highly geographically concentrated. Two-thirds of the world's proved reserves of crude oil are located in the Middle East region, nearly all of that in the region surrounding the Persian Gulf. Over half of the world's oil reserves can be found in four countries of this region: Saudi Arabia (25%), Iraq (11%), Kuwait (10%), and the United Arab Emirates (10%).

The concentration of critical oil reserves in the Persian Gulf region has created economic, political, and military problems for the United States. The economic costs of

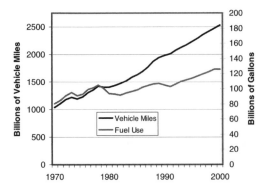

FIGURE 10.4. Decoupling of light-duty vehicle travel and energy use. Source: Davis and Diegel (2003, tables 4.1 and 4.2).

oil dependence to the United States in terms of price shocks, monopolistic oil pricing, and the resulting transfer of U.S. wealth to oil-producing states have been estimated at $7 trillion over the past 30 years (Greene & Tishchishnya, 2000).[3] Oil has greatly complicated U.S. relations with countries in the Middle East. The goal of conquering the oil-rich nations of the Persian Gulf to claim half of the world's proven oil reserves and the enormous wealth they would bring almost certainly motivated Iraq to invade Kuwait in 1990, initiating the Persian Gulf War and conflict with Iraq that has persisted for more than a decade.

The sea-change in the economics and politics of oil began on September 14, 1960, when five oil-producing nations met in Baghdad, Iraq, and formed the Organization of Petroleum Exporting Countries (OPEC).[4] It took 10 years for the members of this new organization to gain control of their national oil reserves from the world's oil companies, and it was not until the fall of 1973 that OPEC exerted a significant influence over world oil markets (Yergin, 1992). In retaliation for the United States's support for Israel in the "October War" of 1973, the Arab members of OPEC declared an oil boycott against the U.S. World oil prices immediately tripled, from $13 to $37 per barrel (in year 2000 dollars; see U.S. Department of Energy, Energy Information Administration, 2002). OPEC's dominance of world oil markets had begun. OPEC held world oil prices at about $35 per barrel until the outbreak of the Iran–Iraq War disrupted world oil supplies. While OPEC did not create the 1979–1980 price shock that nearly doubled the price of oil to $64 per barrel, it capitalized on the increase and attempted to defend the higher price by cutting back on production for the next 5 years. Today, the 11 members of OPEC possess 80% of the world's proved oil reserves (U.S. Department of Energy, Energy Information Administration, 2002) and more than half of all the oil that exists (U.S. Geological Survey, 2000).

As owner of the majority, but not the to-tality, of the world's oil, the market power of OPEC's members is limited by three factors: (1) their share of the world oil market, (2) the ability of non-OPEC producers to increase oil production when prices rise, and (3) the ability of oil users, principally transportation, to reduce demand when prices rise. When OPEC made the decision to defend the high price of oil in 1980, it did so by continuously cutting back on production. Production cuts were necessary to prop up the price of oil because the high price encouraged oil exploration and production outside of OPEC and energy conservation throughout the world. These responses did not occur immediately, because it takes time to find and develop new oil supplies, and to replace older, inefficient transportation vehicles with newer, more energy-efficient technology. As OPEC continued to reduce its oil production from 1980 through 1985, it sacrificed market share, and with the loss of market share it lost market power. Finally, in 1986, OPEC had lost so much of its market power that it was no longer able to control the price of oil. Oil prices collapsed and, with the exception of a brief price spike caused by the Persian Gulf War, remained low until 2000.

With 80% of proved reserves and more than half of the world's ultimate oil resources, OPEC was destined to regain lost market share, as long as the world's growing demand for mobility was dependent on petroleum. In 2000 OPEC, with some assistance from Mexico and Norway, engineered a doubling of the price of oil, reestablishing their dominance of the world oil market.

Thirty years after the first oil price shock of 1973–1974, the U.S. transportation system remains nearly totally dependent on petroleum (see Figure 10.6). The vast majority of nonpetroleum energy used by transportation is natural gas and electricity used by pipelines. The remainder is comprised of alcohols and ethers blended with gasoline (e.g., gasohol) and a tiny amount of electricity used by urban rail systems. Thirty years after the first oil price shocks, alterna-

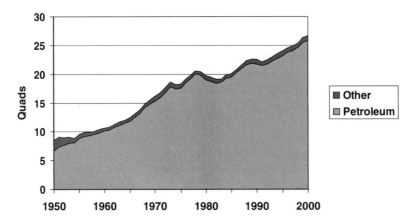

FIGURE 10.6. U.S. transportation energy use, 1950–2000, by energy source. Source: U.S. Department of Energy, Energy Information Administration (2003, table 2.1e).

tive energy sources have had very little impact on transportation energy use.

GREENHOUSE GAS EMISSIONS

Combustion of petroleum for transportation adds significant amounts of greenhouse gases to the earth's atmosphere, contributing to climate change. Greenhouse gases in the atmosphere act like a giant greenhouse, trapping energy in the atmosphere and raising its temperature. Solar radiation passes through the atmosphere or is reflected into space by clouds and dust. Some of the radiation reaching the surface of the earth is also reflected back into space, but much of it is absorbed and reradiated to the atmosphere as infrared (heat) radiation. Greenhouse gases, such as water vapor and carbon dioxide (CO_2), are able to absorb some of the reradiated infrared energy, holding it in the atmosphere and raising its temperature. If there were no greenhouse gases, the Earth's average temperature would be about 59°F (33°C) colder than it is, too cold for life on Earth as we know it.

Since the Industrial Revolution, humans have been adding greenhouse gases to the atmosphere at increasing rates. In the 1860s, human activities added about 100 million metric tons of carbon to the atmosphere

each year.[5] In the year 2000, humans added 6,443 million metric tons of carbon to the atmosphere, and the rate continues to grow (U.S. Department of Energy, Energy Information Administration, 2002, table H1). Approximately half of the CO_2 human beings have added to the atmosphere over the past three centuries remains in the atmosphere; most of the rest has been absorbed by the oceans or terrestrial ecosystems.[6] This accumulation of CO_2 and other human-made greenhouse gases, such as methane, nitrous oxide, tropospheric ozone, and perflourocarbon compounds, is already affecting the climate. A recent panel of the National Academy of Sciences, convened by the president of the United States concluded:

> Greenhouse gases are accumulating in Earth's atmosphere as a result of human activities, causing surface air temperatures and subsurface ocean temperatures to rise. Temperatures are, in fact, rising. The changes observed over the last several decades are likely mostly due to human activities, but we cannot rule out that some significant part of these changes is also a reflection of natural variability. Human-induced warming and associated sea level rises are expected to continue through the 21st century. (National Research Council, 2001)

A large part of the difficulty societies are having in deciding what to do about greenhouse gases and climate change is the result of enormous uncertainty about what the consequences will be. Estimates of the increase in global temperatures that has already occurred, range from 0.7° to 1.5°F (0.4°–0.8°C). Predictions of the rise in temperature by the end of the century show an even greater range: 2.5°–10.4°F (1.4°–5.8°C). The warming is predicted to be greater in the higher latitudes than in the lower latitudes, and larger over land than over sea. Rainfall is predicted to increase in the higher latitudes, but so will evaporation. As noted above, greenhouse gases raise temperatures by increasing the amount of energy in the atmosphere. Increased energy also implies more frequent and more extreme weather events, such as hurricanes and tornados. But little is known for certain about the impacts in any particular country.

Another factor that complicates the problem of deciding what to do is the need for the nations of the world to act together to solve the problem. The United States is by far the largest emitter of greenhouse gases in the world, but even so the United States accounts for only 24% of global carbon dioxide emissions (U.S. Department of Energy, Energy Information Administration, 2002, table H1). The second largest emitter is China, with 12%. The entire continent of Europe accounts for 16%. The only way that greenhouse gas emissions can be effectively controlled is by the coordinated actions of the entire world community.

Within the United States, transportation produces 27% of total greenhouse gas emissions and is the single largest source of CO_2 emissions at 32% of the total (see Figure 10.7) (U.S. Environmental Protection Agency, 2002, tables ES-5 and ES-9). In fact, the U.S. transportation sector produces more CO_2 emissions than any other nation's total CO_2 emissions, except for China. It would be extremely difficult to control global greenhouse gas emissions effectively

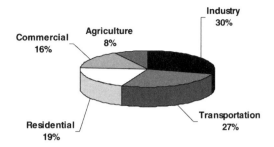

FIGURE 10.7. Transportation's share of U.S. greenhouse gas emissions. Source: U.S. Environmental Protection Agency, Office of Atmospheric Programs (2002, table ES-5).

without curbing transportation's emissions. You may be surprised to learn that your car puts tons of greenhouse gases into the atmosphere each year (see Box 10.2).

CO_2, produced by the use of petroleum fuels in internal combustion engines, accounts for 96% of all of transportation's greenhouse gas emissions (see Figure 10.8). Methane, predominantly a by-product of incomplete combustion of petroleum fuels, accounts for less than 1%. Nitrous oxide, produced in the emissions control systems of modern highway vehicles, comprises 3% of transportation's greenhouse gas emissions, while all other greenhouse gases combined amount to just over 1%.

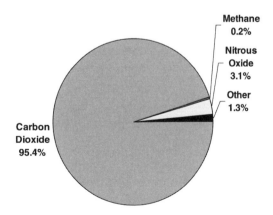

FIGURE 10.8. Transportation emissions by type of greenhouse gas, 2000. Source: U.S. Environmental Protection Agency, Office of Atmospheric Programs (2002, tables ES-5, ES-9, ES-10, and ES-11).

BOX 10.2. Your Car's Greenhouse Gas Emissions

If you own or drive a car, you will probably be amazed at the quantity of CO_2 emissions it produces each year. Each U.S. passenger car and light truck emits well over twice its weight in CO_2 in a single year. A typical passenger car, driven 15,000 miles, puts 8 tons of CO_2 in the atmosphere each year; a typical sports utility vehicle, 12 tons. Over its lifetime, a single passenger car or light truck will emit more than 100 tons of greenhouse gases.

The air may seem weightless, but gases do have weight. Indeed the weight of the CO_2 produced by gasoline-powered vehicles exceeds the weight of the fuel they burn. A gallon of gasoline weights approximately 2.7 kg, of which 2.5 kg are carbon and 0.15 kg are hydrogen. The atomic weight of a carbon atom is 12, while the atomic weight of an oxygen atom is 16. During combustion in a vehicle's engine, each atom of carbon combines with two atoms of oxygen to produce one molecule of CO_2. Of the weight of the resulting molecule, $(16 + 16)/(12 + 16 + 16) \times (100\% = 73\%$ comes from the oxygen, which is taken out of the air. Thus, each gallon of gasoline produces 2.5 kg $\times (44/12) = 9$ kg of CO_2.

China is the second largest producer of CO_2 after the United States. Annual CO_2 emissions in China are 0.56 metric tons per capita (775 million metric tons C / 1,394 million persons = 0.56 metric tons per capita). One U.S. car produces about 15 times that much in a year.

To find the annual greenhouse gas emissions of your car, visit the U.S. Department of Energy's fuel economy information website at *www.fueleconomy.gov.*

Transportation's greenhouse gas emission can be reduced by five types of strategies: (1) reduce the energy intensity of transportation vehicles, so that less carbon is burned per mile of travel; (2) decrease the carbon content per energy content of transportation fuels; (3) shift activity from energy- and carbon-intensive modes to less intensive modes; (4) increase load factors (passenger-miles or ton-miles per vehicle mile; or (5) decrease the amount of transportation required to achieve the same level of accessibility to opportunities.

ALTERNATIVES TO PETROLEUM

Petroleum fuels have remained the dominant source of energy for transportation for four main reasons: (1) they have high energy density, (2) they are liquids at ambient temperatures and pressures, (3) world resources of petroleum are large enough to supply the entire transportation system, and (4) petroleum fuels and internal combustion engines have been able to adapt, through technological advances, to environmental challenges.

Compared to the most promising alternative energy sources for transportation, petroleum fuels have the highest energy density both by weight and volume. This property is key for transportation because increasing either the volume or the weight of fuel that must be carried reduces a vehicle's ability to carry passengers or freight (i.e., payload). Gaseous fuels, like natural gas or hydrogen, even when greatly compressed, have only a fraction of the energy density of gasoline, diesel, or jet fuel. Even compressed to 3,000 pounds per square inch (psi), a natural gas tank would have to be five times the size of a gasoline tank to store the same amount of energy (see Figure 10.9). A tank containing hydrogen compressed to 5,000 psi would have to be 175 gallons in volume to contain as much energy as a 15-gallon gasoline tank. Batteries have the lowest energy density of all, requiring 450–1,000 gallons (depending on the type of battery) to store the energy equivalent of 15 gallons of gasoline.

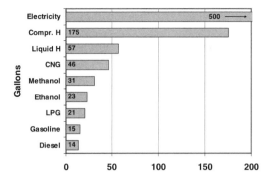

FIGURE 10.9. Energy densities of alternative transportation fuels. Source: Greene and Schafer (2003, table 3).

Because petroleum fuels are liquids under ambient temperatures and pressures, they do not require expensive storage containers capable of withstanding high pressures or very low temperatures. Liquid fuels can be transported and stored easily because they will conform to the shape of any container. That is not to say that petroleum fuels have no storage or handling problems; they are highly flammable, their vapors can be explosive, and they are poisonous if taken internally.

The importance of the fact that world oil resources are large enough to supply energy for the entire world transportation system (at relatively low cost) cannot be overestimated. Liquefied petroleum gases (LPG), largely propane and butane, can be stored as liquids under very modest pressure and have about 70% of the energy density of gasoline. But LPG is a by-product mainly of natural gas production, and nowhere near enough can be produced to satisfy the world's transportation energy demands. Natural gas is about as abundant as petroleum, but has many other competing uses: generating electricity, space heating, and industrial processes. At present, there appears to be enough conventional petroleum to power the world's transportation systems for one or two decades. Beyond that, there are much larger carbon-based resources (in the form of heavy oil, oil and tar sands, shale oil,

and coal) that can be converted to synthetic petroleum, though at higher cost and with greater environmental impacts.

In the 1980s many believed that alternative fuels would be essential if air pollution from transportation vehicles were to be reduced enough for cities to meet national air quality standards. Chemically, petroleum is a hydrocarbon, meaning that it is comprised almost entirely of two elements: hydrogen (H) and carbon (C). Ideally, the combustion of hydrocarbons consists of their rapid (explosive) combination with oxygen (O, from the air) to produce water vapor (H_2O) and carbon dioxide (CO_2). Both of these products are greenhouse gases, but only CO_2 accumulates in the atmosphere and contributes to climate change. In reality, engines produce a variety of other pollutants as a result of incomplete combustion and the oxidation of other components of air (especially nitrogen). These unintended reactions produce carbon monoxide, fine particulates, unburned hydrocarbons, and oxides of nitrogen. In the atmosphere, these pollutants produce smog and cause or exacerbate respiratory problems. Transportation vehicles are a major source of all these pollutants. Natural gas, a much simpler hydrocarbon than gasoline, burns much more cleanly, producing far fewer pollutant emissions. Alcohols, such as ethanol and methanol, also generally burn more cleanly than gasoline or diesel fuel.

Despite the inherently cleaner burning properties of several alternative fuels, advances in emissions control technology and gasoline fuel composition have allowed most of the pollutants created by burning fossil fuels in internal combustion engines to be reduced far more than what was expected from alternative fuels.[7] A modern, properly functioning passenger car produces roughly 1% of the health-threatening pollution emitted by a car built in the 1960s. Still, even these impressive achievements have barely been able to keep pace with the growth of vehicle travel, and many urban areas still fail to meet air quality standards.

In addition, petroleum fuels and internal combustion engines have had much less success in reducing emissions of greenhouse gases, especially CO_2. These continuing environmental challenges, plus the lingering energy security problems caused by dependence on petroleum, create an ongoing interest in finding cleaner, safer, nonpetroleum energy sources for transportation.

In January 2003, President George W. Bush announced a new program, FreedomCAR, to develop hydrogen-powered automobiles and to establish hydrogen as the fuel of the future for U.S. transportation.

A simple chemical reaction between hydrogen and oxygen generates energy, which can be used to power a car producing only water, not exhaust fumes. With a new national commitment, our scientists and engineers will overcome obstacles to taking these cars from laboratory to showroom so that the first car driven by a child born today could be powered by hydrogen, and pollution-free. (State of the Union Address, January 28, 2003)

The program replaced the government's previous research and development partnership, the Partnership for a New Generation of Vehicles (PNGV), a key goal of which was to develop a marketable 80-mile-per-gallon family car by 2004. The PNGV program, which was started in 1993, was abandoned just 2 years before its scheduled completion in 2004. At time of writing it is not clear what impact the aborted program will have on fuel economy, although it has been noted above that the fuel economy of new light-duty vehicles sold in the United States has been declining rather than increasing.

The objective of the new FreedomCAR program is to develop hydrogen-fuel-cell-powered cars and light trucks that will replace internal combustion engine vehicles in approximately one generation. The concept is nothing short of revolutionary. Fuel cells produce electrical energy via the chemical reaction of hydrogen and oxygen. The sole outputs of a hydrogen fuel cell are water and electricity. While there are many fuel cell types, the most promising option for automobile applications is the proton exchange membrane (PEM) fuel cell because of its ability to operate at low temperatures (Burns, McCormick, & Borroni-Bird, 2002). Hydrogen atoms enter the anode side of a PEM cell. Inside, a catalytic coating on the anode splits the hydrogen atoms into an electron and a proton. The PEM membrane permits only the protons to cross over to the cathode. The electrons travel through a circuit to power the car's electric motor or recharge its batteries. They then return to the cathode side of the cell, where they join the protons and oxygen atoms from the air to form water.

In addition to being an inherently clean technology, fuel cells also have the potential to be very energy-efficient. Whereas modern gasoline-powered internal combustion engines can be up to 30% efficient (30% of the chemical energy in the fuel is converted into work at the shaft of the engine), fuel cells can be up to 55% efficient (Burns et al., 2002).

A major hurdle to the success of hydrogen-fuel-cell vehicles will be the lack of an infrastructure for producing the massive volumes of hydrogen required, distributing it across the nation, and delivering it into vehicles. Because hydrogen is an extremely low-density gas, transporting, storing, and delivering hydrogen into vehicles will be much more difficult than distributing petroleum fuels. Unless a major technological breakthrough can be achieved, hydrogen will probably be stored on board vehicles at 5,000 to 10,000 psi, extremely high pressure not used in gaseous fueled vehicles to date.[8] Relatively expensive high-pressure storage tanks will be required onboard hydrogen vehicles and, even so, much less energy will be available to power the vehicle (see Figure 10.7, above). Finally, if fuel cells were mass-produced today, they would probably cost 10 times as much as an internal combustion engine of comparable power.

Fuel-cell vehicles can be run on liquid fuels containing hydrogen, such as alcohols or even gasoline. But these vehicles must have an onboard reformer, a small chemical plant that extracts the hydrogen from liquid fuels and processes it for delivery to the fuel cell. Unfortunately, reformers processing alcohols or hydrocarbon fuels will not be pollution-free and will produce greenhouse gas emissions. Furthermore, if the full fuel cycle emissions of fuel-cell vehicles are considered (the total emissions produced from production of the fuel to the vehicle tailpipe) even fuel-cell vehicles would only be "zero emission" if the hydrogen they used were produced from solar, nuclear, or some other zero-emission source. At present, most hydrogen is produced from natural gas, the lowest cost source, with substantial release of greenhouse gases. Researchers at the Massachusetts Institute of Technology compared the life-cycle greenhouse gas emissions of gasoline-, diesel-, and fuel-cell-powered vehicles (assuming the hydrogen was produced from natural gas) (Weiss, Heywood, Schafer, & Natarahan, 2003). Their results indicate no significant greenhouse gas emissions advantage for fuel cells versus internal combustion engine hybrid vehicles (see Figure 10.10). For the fuel-cell vehicles, nearly all of the life-cycle greenhouse gases are emitted in the production of hydrogen from natural gas. Hydrogen produced from renewable energy, such as solar, wind, or biomass, or from nuclear energy, could have nearly zero life-cycle greenhouse gas emissions.

Clearly, the fuel cell has a long way to go before it can become a successful commercial product. Enormous progress has been achieved in recent years, however, and this is undoubtedly the source of much of the current optimism about the eventual success of fuel cells. But even if fuel-cell research and development is entirely successful, planning and executing a transition from petroleum fuels, with a ubiquitous and efficiently functioning infrastructure, to hydrogen, requiring an entirely new infrastructure, will be a public–private venture of a scale not seen since the construction of the Interstate Highway System.

ACHIEVING PROGRESS: TECHNOLOGICAL CHANGE AND STOCK TURNOVER

Increasing the energy efficiency or fundamentally changing the kinds of fuels used by transportation vehicles takes time. For most purposes, the technological characteristics of transportation vehicles (type and size of engine and its technical efficiency, transmission characteristics, weight, aerodynamic drag, etc.) are fixed when vehicles are

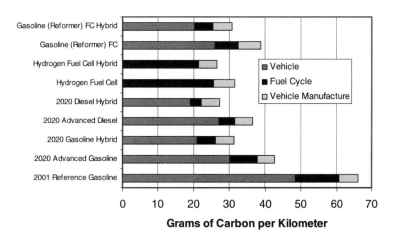

FIGURE 10.10. Life-cycle greenhouse gas emissions of current and advanced vehicle technologies. Source: Weiss, Heywood, Schafer, and Natarahan (2003, table 9).

manufactured. Larger vehicles, such as very heavy trucks, aircraft, and large ships, are to some extent an exception, since it is common to retrofit them with new engines during their normal lifetimes. Still, the original vehicle design limits the kinds of changes that can be made. A complete replacement of an existing stock of transportation vehicles will require from 15 to 30 years.

Significant changes in energy efficiency or fuel types require that old vehicles be replaced by newer ones, and this takes time. The most recent evidence available indicates that 50% of the cars sold in 1990 will survive to be 17 years old or older. This is a substantial increase over 1980 model year cars, which had a median expected lifetime of only 12.5 years. The expected lifetime of a model year 1990 light truck is 15.5 years (Davis & Diegel, 2002, tables 6.9 and 6.10). The expected life of a new heavy truck is 28 years. On the other hand, annual miles of use decrease as vehicles age. Survey data indicate that, on average, a new vehicle owned by a U.S. household will travel 15,600 miles in its first year. By the time that vehicle's age exceeds 10 years, it will be driven only about 9,000 miles (Davis & Diegel, 2002, table 11.12). Because of the greater usage of new vehicles and gradual scrappage of older vehicles, half of all vehicle miles are traveled by vehicles under 7 years old (Davis & Diegel, 2002, tables 6.6 and 6.7). Still, if the fuel economy of new vehicles were immediately increased by 50%, it would take about 15 years for the average fuel economy of all vehicles on the road to approach the same level.

The Corporate Average Fuel Economy (CAFE) standards, passed by Congress in December 1975, provide a clear illustration of how the process of technological change and stock turnover works. In 1974, new passenger cars were averaging about 13.5 miles per gallon (MPG), new light trucks less. The law required each manufacturer to meet or exceed a sales-weighted average fuel economy level of 18 MPG in 1978, in-

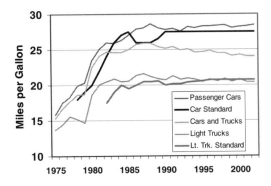

FIGURE 10.11. Passenger car and light truck fuel economy, 1975–2000. Source: Heavenrich and Hellman (2001, table 1).

creasing every year thereafter up to 27.5 MPG in 1985 (see Figure 10.11 and Box 10.3). Standards for light trucks, which at the time comprised less than 20% of total sales of light-duty vehicles, were to be set by the National Highway Traffic Safety Administration (NHTSA). The light truck standard was set at 17.5 MPG for new light trucks sold in 1982, increasing to 20.7 MPG in 1996.[9]

No major manufacturer failed to meet the CAFE standards. From 1975 to 1985, the fuel economy of new passenger cars increased from 15.8 to 28.5 MPG, and new light truck MPG improved from 13.7 to 20.4 MPG. The combined fuel economy of all light-duty vehicles rose from 15.3 in 1975 to a high of 25.9 in 1988, but has since declined to 24 MPG. Fuel economy has decreased in recent years as innovations in the light truck market, such as minivans and sport utility vehicles (SUVs), boosted light trucks' share of the new vehicle market from under 20% in 1975 to about 50% today.

It took approximately 15 years for the total population of passenger cars and light trucks on the road to reach about the same MPG level as new vehicles (see Figure 10.12). The fuel economy estimates cited above and used to monitor manufacturers' compliance with federal regulations are based on laboratory tests conducted

BOX 10.3. Harmonic Mean and MPG

In determining how much energy a fleet of vehicles wil use, low-MPG vehicles count more than high-MPG vehicles. Because of this, the average fuel economy of two vehicles is not the simple arithmetic mean, but rather the harmonic mean.

Consider two vehicles: a passenger car getting 25 miles to the gallon, and an SUV getting 15. Both vehicles are driven 10,000 miles each year. But the average MPG of the two vehicles is not $(25 + 15)/2 = 20$.

In traveling 10,000 miles, the car uses $10,000/25 = 400$ gallons of fuel. The SUV uses $10,000/15 = 667$ gallons, for a total of 1,067 gallons and 20,000 vehicle miles. Dividing miles traveled by gallons gives an average MPG of $20,000/1,067 = 18.74$. This is a different kind of average, called a "harmonic mean." The harmonic mean of two numbers is the inverse of the average of their inverses.

$$\{[(1/25) + (1/15)]/2\}^{-1} = [(0.040 + 0.0667)/2]^{-1} = 1/(0.5335) = 18.74$$

If we measured fuel economy in terms of gallons per mile, or liters per kilometer, as most of the world does, a simple average would be the correct method.

by the Environmental Protection Agency (Heavenrich & Hellman, 2001, table 1). This test is known to overestimate real-world fuel economy by 15%, or more. Estimated on-road fuel economy for new vehicles, passenger cars, and light trucks combined increased from 13.1 to 21.3 MPG over the period from 1975 to 2000. These adjusted new light-duty vehicle MPG estimates are shown in Figure 10.12, along with the estimated average fuel econ-

omy of all cars and light trucks on the road, regardless of age.[10]

Figure 10.12 shows just how long it takes to make significant improvements in the energy efficiency of transportation vehicles. First, it took manufacturers about 8 years to bring new vehicle fuel economy up to the level required by the standards. This is because manufacturers typically redesign 10–15% of the makes and models they sell each year. This schedule allows a relatively constant rate of expenditure of capital and labor on redesign and retooling. Accelerating this process would be very expensive. Second, once new car fuel economy reached the level of the standards, it took another 10–15 years for the new vehicles gradually to replace the majority of used vehicles and raise the total fleet MPG almost to the level of new vehicles.

Significant additional improvements in new car and light truck fuel economy could be made if the trend toward ever bigger and more powerful vehicles could be arrested. A committee of the National Research Council estimated that by 2015, passenger car fuel economy could be increased by 12–27% (depending on size class) and light

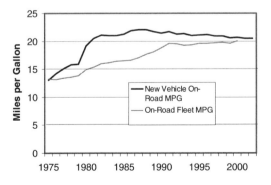

FIGURE 10.12. Fuel economy of new versus the on-road fleet of light-duty vehicles. Sources: Davis and Diegel (2002, table 7.18); Heavenrich and Hellman (2003, table 1).

truck MPG could be raised by 25–42% using proven technologies whose initial cost would be more than paid for by fuel savings over the life of the vehicle (National Research Council, 2002, table 4-2). This study excluded diesel and hybrid technologies, each of which has the potential to increase fuel economy by 30–40%.

Diesel engines ignite an air–fuel mixture by compressing it to much higher pressure than that seen in gasoline engines. Whereas gasoline engines may achieve energy efficiencies of up to 30%, diesels have been constructed with efficiencies exceeding 45%. Modern diesels are greatly improved over the automotive diesels of 20 years ago with respect to performance and quietness; diesel engines have enjoyed enormous success in Europe, where they comprise 40% of new car sales. Manufacturers have been reluctant to introduce diesels in the United States, out of concern that they will not be able to meet more stringent emissions standards that will take effect in 2007. Recent advances in control technology and planned improvements in diesel fuel quality suggest that diesels will be introduced into the U.S. market over the next few years. If they are successful, they could have a major impact on the energy efficiency of light-duty vehicles.

Hybrid vehicles combine an internal combustion engine with an electric motor to maximize fuel economy. Because the electric motor can be used to supplement the internal combustion engine when power is needed for acceleration or hill climbing, a small engine, requiring less fuel, can be installed. Also, the engine can be shut down during deceleration or when the engine would normally be idling, for example, at a stop light. With the large electric motor, the engine can be automatically restarted instantly. To store electrical energy for their motors, hybrids must have additional battery capacity. This also allows the energy normally wasted in braking to be used to drive the electric motor in reverse, to generate electricity which can be stored in the batteries for later use. With all these "tricks"

hybrids achieve 30–40% greater fuel economy over the government's fuel economy test cycle. Hybrid vehicles have their greatest advantage in stop-and-go urban driving and provide little benefit when cruising on the highway. In 2003, three hybrid designs were available, all small passenger cars. Because the extra cost of hybrids cannot at present be paid for by their lifetime fuel savings, hybrids will likely achieve a significant market share only when costs are brought down and manufacturers discover what "extra value" hybrids can provide to consumers.[11]

SOLUTIONS: TECHNOLOGIES, POLICIES, AND INFRASTRUCTURE

The energy challenges facing the United States and the world's transportation systems over the next 50 years are daunting. It will not be easy to find ways to overcome transportation's dependence on petroleum, to reduce its greenhouse gas emissions substantially below present levels, or to make a transition to sustainable energy sources, while at the same time preserving or increasing mobility.

Regulatory Policies

Regulatory standards were a major component of the transportation energy policies instituted in the 1970s and early 1980s in response to the oil price shocks of that era. Passenger car and light truck fuel economy standards were established by the Energy Policy and Conservation Act of 1975, standards that remain in effect today. The same act required the EPA to measure the fuel economy of all light–duty vehicles (less than 8,500 pounds loaded weight) and make that information available to the public. Prior to the act, there was no comprehensive and consistent source of information on fuel economy. There is still none for heavy vehicles. The Alternative Motor Fuels Act of 1989 together with the Energy Policy Act of 1992 set requirements to purchase alternative fuel vehicles for federal and state governments and for the fleets of companies that sold alternative fuels. To date, alternative fuels policies have had a minor impact on transportation energy use. Fuel economy standards have had a far greater impact.

The State of California has had a major impact on vehicle technology, and especially the ability to reduce pollutant emissions. California's "Low Emission Vehicle" program set a timetable for manufacturers to produce and sell "zero emission vehicles" (ZEV) in the state. Initially, the only vehicles qualifying were battery-powered electric vehicles. Over the years the ZEV program has been modified to give partial credit to other cleaner technologies. While battery electric vehicles are not likely to see significant commercial success, California's ZEV program has been a major force behind extremely clean vehicle innovations such as the hybrid vehicle, and the fuel-cell vehicle, as well as ultra-low-emission ("ULEV" and "SULEV") vehicles powered solely by gasoline-fueled internal combustion engines.

Market-Based Policies

During the late 1980s and 1990s, "market-based" mechanisms grew in favor relative to regulations. Rather than setting legal requirements, market-based policies try to use economic incentives to harness the power of the marketplace to make things happen. To date, few market-based approaches other than subsidies have been tried in transportation.

Among subsidies, the most significant is clearly the partial exemption from federal and state highway motor fuel taxes of fuels blended with ethanol. Ethanol, an alcohol predominantly made from corn in the United States (and the intoxicating ingredient in alcoholic beverages), can be readily blended with gasoline in concentrations of up to 10% without requiring changes to vehicles or to the fuel distribution system. Gasoline blended with 10% ethanol pays only 13.0 cents per gallon federal excise tax,

whereas straight gasoline pays 18.3 cents per gallon. This subsidy of 5.3 cents applied to the entire gallon of gasoline is effectively a $0.53 per gallon subsidy for ethanol, since one gallon of ethanol can reduce the tax on 10 gallons of gasoline. In 2001, 1.8 billion gallons of ethanol were blended with gasoline in the United States, up from 0.8 billion gallons in 1985. While 1.8 billion gallons is a small amount relative to annual consumption of 134 billion gallons of gasoline, it is the largest source of nonpetroleum energy in the U.S. transportation sector (excluding electricity and natural gas use by pipelines). Use of ethanol is likely to increase in the future as a result of decisions to ban certain other gasoline additives on the basis of health concerns.

Numerous other market-based strategies have been proposed to boost fuel economy, reduce the growth of vehicle travel, promote ridesharing and mass transit use, and promote low-carbon alternative fuels, but none have been adopted on a national scale. One promising idea is "feebates," a combination of fees levied on low-fuel-economy vehicles and rebates for high-MPG models. In fact, the United States has had a fee system, called the gas-guzzler tax, in effect since 1980. Since 1991, a tax of $1,000 had to be paid by any passenger car achieving less than 22.5 MPG on the federal fuel economy test. The fine increases with decreasing MPG, reaching $7,700 per vehicle for cars getting less than 12.5 MPG. There is no corresponding tax on light trucks, although there are many that achieve less than 22.5 MPG and function primarily as passenger vehicles. One attractive feature of feebates is that they can be designed to pay out as much revenue as they take in, so there is no net flow of funds from the public to the government. At the same time, feebates can provide a clear market signal to manufacturers that improving fuel economy pays (Davis, Levine, Train, & Duleep, 1995).

Other market-base strategies include direct taxes on petroleum or the carbon content of fuels. Such taxes have the advantage of simultaneously encouraging manufacturers to used advanced technologies to improve the fuel economy of all the vehicles they produce, encouraging car buyers to choose more energy-efficient vehicles, and discouraging additional vehicle travel. To date, however, taxes have been neglected because they are perceived to be unpopular.

Research and Development

Society has an interest in research and development of technologies that will protect the environment or enhance national security. Indeed, without the incentives created by environmental regulations or other public policies, businesses are unlikely to invest in developing technologies whose benefits accrue principally to society as a whole. Creating advanced technologies to solve the problems associated with transportation's use of energy (e.g., greenhouse gas emissions, oil dependence, air pollution) therefore requires deliberate and significant governmental action. This action can take the form of direct government investment in research and development or indirect stimulation of technological progress through regulations or other policies that reward the developers of environmentally beneficial and nonpetroleum technologies.

NOTES

1. Because gasoline and other petroleum-based transportation fuels are mixtures of different hydrocarbons in varying proportions, the energy contents of gasoline, diesel, or jet fuel vary slightly from place to place and over time. Energy content is most often specified in terms of "lower heating values," although sometimes "higher heating values" are used. When hydrocarbons are burned, one of the products is water vapor. The higher heating value includes the energy that would be released if the water vapor produced were condensed to liquid water. Such condensation does not occur in internal combustion engines and the extra energy of the higher heating value is generally not available. For this reason, it is more common to use lower heating values when describing the energy content of transportation fuels.

2. Price controls have long been abolished. Supply shocks today raise gasoline prices but do not create shortages.

3. Costs are measured in year 2000 dollars, converted to present value. Unadjusted to account for the present value of past costs, the sum of past costs amounts to $3.4 trillion.

4. Those nations were Iran, Iraq, Kuwait, Saudi Arabia, and Venezuela (see *www.opec.org*).

5. Greenhouse gases are usually measured in equivalent metric tons of CO_2, or of the carbon in CO_2. Since the atomic weights of carbon and oxygen are 12 and 16, respectively, and there are two oxygen atoms per molecule of CO_2, there are $12/(12 + 2*16) = 0.2727$ tons of C per ton of CO_2.

6. Human activities also add substantial amounts of water vapor to the atmosphere. Despite being a potent greenhouse gas, the added water vapor does not affect climate because it does not remain in the atmosphere.

7. The threat of regulations requiring alternative fuels, especially in California, was a key factor in stimulating the advances in emissions control technology and gasoline formulation that permitted petroleum fuels to retain their market dominance.

8. Compressed natural gas vehicles routinely store methane compressed to 3,000 psi.

9. The NHTSA is currently in the process of setting new, higher standards for light trucks. The currently proposed rule would establish a target of 22.2 MPG in 2007.

10. The Environmental Protection Agency (EPA) estimates new vehicle fuel economy, while the average MPG of vehicles on the road is estimated by the Federal Highway Administration (FHWA). The FHWA defines light trucks differently from the EPA, which contributes to differences between the two estimates.

11. Hybrid vehicles have high-voltage electrical systems that could be used to power virtually any household appliance or power tool. They could also be designed to serve as emergency electrical generators. With two power plants, hybrids could be designed to provide a kind of "on-demand" four-wheel drive at little extra cost.

REFERENCES

Burns, L. D., McCormick, J. B., & Borroni-Bird, C. E. (2002). Vehicle of change. *Scientific American, 287,* 64–73.

Davis, S. C., & Diegel, S. W. (2002). *Transportation energy data book* (22nd ed., ORNL-6967). Oak Ridge, TN: Oak Ridge National Laboratory. Available online at www-cta.ornl.gov/data

Davis, S. C., & Diegel, S. W. (2003). *Transportation energy data book* (23rd ed., ORNL-6970). Oak Ridge, TN: Oak Ridge National Laboratory. Available online at www-cta.ornl.gov/data

Davis, W. B., Levine, M. D., Train, K., & Duleep, K. G. (1995). *Effects of feebates on vehicle fuel economy, carbon dioxide emissions, and consumer surplus* (DOE/PO-0031). Washington, DC: U.S. Department of Energy, Office of Policy.

Federal Highway Administration. (2003). *Highway statistics.* Washington, DCP: U.S. Department of Transportation. Available online at www.fhwa.dot.gov

Greene, D. L. (1996). *Transportation and energy.* Washington, DC: Eno Transportation Foundation, Inc.

Greene, D. L., & Fan, Y. (1996). Transportation energy intensity trends, 1972–1992. *Transportation Research Record, 1475,* 10–19.

Greene, D. L., & Schafer, A. (2003). *Reducing greenhouse gas emissions from U.S. transportation.* Arlington, VA: Pew Center on Global Climate Change.

Greene, D. L., & Tishchishnya, N. (2001). The costs of oil dependence: A 2000 update. *Transportation Quarterly, 55,* 11–32.

Heavenrich, R. M., & Hellman, K. H. (2001). *Light-duty automotive technology and fuel economy trends: 1975 through 2001.* Ann Arbor, MI: U.S. Environmental Protection Agency, Office of Transportation and Air Quality, Advanced Technology Division.

National Research Council. (2001). *Climate change science: An analysis of some key questions.* Washington, DC: National Academy Press.

Schafer, A. (2000). Regularities in travel demand: An international perspective. *Journal of Transportation and Statistics, 3,* 1–31.

Schipper, L., & Fulton, L. (2001, January). *Driving a bargain?: Using indicators to keep score on the transport–environment–greenhouse gas linkages.* Paper presented at the 75th annual meeting of the Transportation Research Board, National Research Council, Washington, DC.

U.S. Department of Energy, Energy Information Administration. (2002). *Annual energy review 2001* (DOE/EIA-0384[2001]). Washington, DC: Author. Available online at www.eia.doe.gov

U.S. Department of Energy, Energy Information Administration. (2003). *Annual energy review 2002* (DOE/EIA-0384[2002]). Washington, DC: Author.

U.S. Department of Transportation, Bureau of Transportation Statistics. (2002, December). *National transportation statistics 2002* (BTS02-08). Washington, DC: U.S. Government Printing Office.

U.S. Department of Transportation, Federal Highway Administration. (2001). *Highway statistics 2001* (FHWA-PL-02-008). Washington, DC: Author.

U.S. Environmental Protection Agency, Office of Atmospheric Programs. (2002). *Inventory of U.S. greenhouse gas emissions and sinks: 1990–2000* (EPA 430-R-02-003). Washington, DC: Office of Atmospheric Programs.

U.S. Geological Survey World Energy Assessment Team. (2000). *World petroleum assessment 2000: Description and results.* Denver, CO: USGS Information Services. Available online at http://pubs.usgs.gov/dds/dds-060/

Weiss, M. A., Heywood, J. B., Schafer, A., & Natarahan, V. K. (2003, February). *Comparative assessment of fuel cell cars* (MIT LFEE 2003-001 RP). Cambridge, MA: Massachusetts Institute of Technology, Laboratory for Energy and the Environment.

World Business Council for Sustainable Development. (2001). *Mobility 2001: World mobility at the end of the twentieth century and its sustainability.* Prepared by the Massachusetts Institute of Technology and Charles River Associates. Available online at www.wbcsdmobility.org

Yergin, D. (1992). *The prize: The epic quest for oil, money and power.* New York: Simon & Schuster.

CHAPTER 11

The Geography of Urban Transportation Finance

BRIAN D. TAYLOR

"Follow the money."
—Deep Throat, in Carl Bernstein and
 Bob Woodward's *All the President's Men* (1974)

The geography of urban transportation in U.S. cities is profoundly shaped by the geography of urban transportation finance. Geopolitical struggles over the spatial distribution of public expenditures on transportation have significantly shaped the geography of transportation systems, spatial patterns of travel, and urban form. From where taxes are collected and to where they are distributed are part-and-parcel of political life, but have often been ignored by transportation analysts. In addition to the spatial distribution of public resources, the appropriate level of public spending on transportation has long been hotly debated and varies significantly among both places and transportation modes. Further, urban and transportation economists have long argued that patterns of urban development, economic trade, and travel are strongly influenced by the price of travel, which, in turn, is governed to a large degree by public policies. Put simply, money matters. So to understand the geography of urban transportation, one must first understand the geography of urban transportation finance.

This chapter begins with an overview of the economic geography of transportation finance; we explore economic rationales for public investments in transportation systems, as well as the central role of political geography in determining transportation investments. This is followed by an overview of current surface transportation finance in the United States, with an emphasis on metropolitan areas. Here we examine broad trends in revenues for transportation and patterns of expenditures by both geography and mode. The chapter concludes with thumbnail fiscal histories of the two transportation modes that have most influenced the form of metropolitan areas in the United States: metropolitan freeways and public transit. First we present the story of freeway finance during the middle third of the 20th century as a case study of how the political geography of transportation finance shaped the development of freeway systems, which, in turn, strongly influenced both travel and development in U.S. cities.

Then we look at the fiscal rise, fall, and re-birth of urban public transit systems over the past 100 years, with an emphasis on the political geography of public transit operations and subsidies. So to understand how and why U.S. cities and their transportation systems developed as they have, we must heed Deep Throat's advice to "follow the money."

THE PUBLIC ROLE IN FINANCING URBAN TRANSPORTATION SYSTEMS

Why should the public sector get involved in financing transportation systems? Why not leave it to the private sector? Large portions of urban transportation systems are financed privately: people pay for their own shoes, bicycles, taxi fares, and automobiles; firms pay to move their goods from factory to warehouse to store; retailers pay for parking lots so that shoppers can park for free; advertisers pay transit systems for display space on buses and transit shelters. But in addition to all of this private investment in transportation, governments—local, regional, state, and national—also finance large parts of transportation systems. But just *how* the public sector should collect funds for transportation, *what* systems or modes should receive funding priority, and *where* the funds should be expended are subject to considerable and ongoing political debate. For example:

- Should transportation revenues be collected from user fees (such as bridge tolls, transit fares, or fuel taxes), or with more general instruments of taxation (such as income, sales, or property taxes)?
- Should people pay for transportation systems based on ability to pay? Benefits received? Costs imposed? Should, for example, income-regressive sales tax revenues be used to pay for commuter rail lines that primarily serve higher-income suburban commuters?

- Should streets and highways receive funding priority because they are so heavily used, or should public transit and bicycling receive priority to create more environmentally friendly alternatives to private vehicle use?
- Should transportation taxes and fees collected in one jurisdiction be spent in other places? If so, on what basis should the funds be geographically redistributed?
- Should transportation projects be distributed equally among states and jurisdictions, or should the most needed projects be funded first, regardless of location?

Such questions about public versus private roles, taxes versus user fees, and redistribution of funding among both individuals and places are at the heart of urban transportation finance.

The Political Economy of Transportation Finance

Public finance economists distinguish *public goods* from *private goods* in determining the appropriate level of public involvement. Private goods—like basketball shoes and CD players—are allocated by markets. Public goods, on the other hand, are not typically sold, so markets cannot allocate them. National defense and weather services are examples of public goods. So-called pure public goods share three characteristics: (1) no one can be excluded from consuming them, (2) consumption by one person does not affect consumption by others, and (3) they can only be provided through collective action.

But transportation systems lie on a continuum between pure private goods and pure public goods. At the private end of the spectrum, most public transit systems began as private enterprises funded entirely by passenger fares and the sale of advertising space. People can be excluded from transit vehicles, and consumption of transit capacity does affect others. But except for some specialty services like airport shuttles, the costs of operating comprehensive transit networks when the demand for service is increasingly dispersed spatially and temporally means that public transit service today cannot be financed without collective action. At the other end of the spectrum are sidewalks. Outside of enclosed shopping malls and amusement parks, it is very difficult to exclude people from sidewalks. Sidewalks are only rarely congested, so use of sidewalks does not significantly affect others. And without public regulation, interconnected networks of sidewalks would not exist.

Some observers contend that the public sector has come to play too large a role in the finance and operation of transportation systems (Poole, 2001; Winston & Shirley, 1998). They argue that increasing reliance on private markets to allocate transportation resources would greatly increase economic efficiency. Such arguments led to efforts during the 1980s to increase public contracting with private firms to provide public transit service and during the 1990s to privately finance new toll roads in Southern California. Political debates over the appropriate scope and scale of public involvement in transportation systems are often drawn along ideological lines, especially regarding taxation: conservatives typically favor lower taxes and a larger private-sector role, whereas liberals often favor increased collection of revenues for transportation and substantial public control of transportation systems.

Regardless of ideology, however, transportation public works projects are usually popular with voters and the people whom they elect. Elected officials frequently justify their support for large transportation projects in terms of the jobs they create (Richmond, in press; Wachs, 1995), though usually without regard to whether the jobs were created as a result of the transportation improvement or whether the jobs were simply shifted from some other location

(Boarnet, 1997; Forkenbrock, Pogue, Foster, & Finnegan, 1990). New rail transit projects have been especially popular with elected officials since the 1970s, when public concern with worsening air quality and ambivalence toward the disruptions caused by urban freeway projects began to mount. The 1998 federal surface transportation legislation (Transportation Equity Act for the 21st Century, or "TEA-21") earmarked billions of dollars for 191 new rail transit construction projects (compared to 40 such projects in the similar 1991 legislation) in dozens of sprawling, auto-oriented cities around the United States, including Albuquerque, Austin, Ft. Lauderdale, Galveston, Little Rock, Louisville, Norfolk, Spokane, Stockton, Virginia Beach, and West Palm Beach (Surface Transportation Policy Project, 1998).

When announcing federal funding of new transportation projects to constituents, Congress members often cite the economic benefits such projects will bring to the affected district. Over the past two decades, for example, rail transit projects in Buffalo, Los Angeles, Milwaukee, and elsewhere have been touted by elected officials as tools for job creation and economic revitalization in depressed areas (Boxer, 1998; Dixon, 1997; Kurtz, 1983; Norman, 1996; Richmond, in press). While being debated in Congress, the TEA-21 legislation was repeatedly presented by officials first and foremost as a jobs bill. When finally adopted in the summer of 1998, elected officials around the United States hailed TEA-21 for the jobs it would bring to their districts. In Colorado, for example, TEA-21 earmarked $755 million for light rail and related construction projects along Denver's southeast corridor; news agencies there duly reported that "the influx of spending [on both public transit and highway projects] is expected to provide scores of jobs in Colorado and act as an engine for continued economic growth" (Romano, 1998).

But transportation economists have for decades cautioned policymakers not to confuse the economic effects of expenditures of public funds on transportation facilities and operations with the economic benefits of an improved transportation system brought about by such expenditures. The former—the *expenditure effects*—concern balancing the diminution of economic activity caused by taxation, on the one hand, with stimulation and redistribution of economic activity by tax revenues, on the other. From this perspective, the expenditure effects of a new highway maintenance facility do not directly concern transportation; rather, the expenditure effects of such a project should properly be compared with other possible public expenditures (for schools, health care, etc.) or with not taxing to collect the funds in the first place.

Expenditure effects result directly from the outlay of money, such as when subsidies pay for transit workers' salaries or when public funds pay a construction firm to build a new suburban freeway. These direct payments are then respent in the economy, generating additional economic activity. This expanded effect, known as the *multiplier process*, is the reason for dividing the expenditure effects into direct, indirect, and induced effects in Figure 11.1.

So while local economic activity can be generated by transportation expenditure effects, much—if not most—of this activity may simply be redistributed from other people (taxpayers) and other places that lost out in the geographic competition for subsidy dollars. To transportation economists, policymakers simply miss the point when they focus on the local expenditure effects of transportation expenditure decisions (Forkenbrock et al., 1990; Lewis, 1991).

In contrast, most transportation economists would argue that public expenditures on transportation should properly be judged on their *transportation effects*. In other words, the real economic benefits of public investments in transportation come from improvements to the transportation system. Decreased travel time, increased reliability, reduced emissions, increased safety, and so on—all benefit the economy and society by lowering transportation costs both to users of the transportation system and to society at large. Lower transportation costs facilitate economic transactions and social interactions; they make it cheaper to produce and distribute current goods and services and they make new forms of goods and services

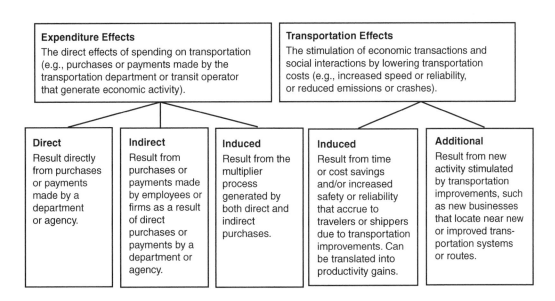

FIGURE 11.1. Economic benefits of transportation expenditures.

possible. Thus transportation improvements "help us do things differently and help us do different things" (Gillen, 1997, p. 11).

In the case of a highway improvement, travelers who accrue time or cost savings from reduced congestion delay realize the transportation effects of highway expenditures. A second source of transportation effects comes from the benefits accruing to others (termed "additional effects" in Figure 11.1). For example, a new high-occupancy vehicle (HOV) lane that is successful in motivating some travelers to carpool may reduce congestion on adjacent unrestricted lanes, thus benefitting all roadway users, not just carpoolers. A new public transit investment may generate new economic activity by attracting new businesses to locate near transit stations. The ability of firms to restructure their logistics and distribution networks to reduce production and distribution costs is another source of economic benefit. As with the expenditure effects, however, the transportation effects of transportation expenditures must be netted against the economic costs of taxation (Lewis, 1991).

But to many policymakers, the transportation effects of public investments can seem rather abstract, arcane, and arbitrary. To a member of the Oregon congressional delegation, a study showing that rail transit investments in New York City yield far greater transportation benefit than those in Portland is not terribly relevant to debates over the equitable geographic distribution of federal transit subsidies. In contrast, the local expenditure effects of transportation facility construction projects in individual congressional districts are clear and unambiguous. A new highway funded with substantial federal funding is a dramatic and highly visible public investment that, during construction, clearly and directly generates employment and economic activity in particular jurisdictions. That much of this economic activity is simply shifted from taxpayers in other jurisdictions is almost beside the point. In a democracy like the United States, comprised primarily of geographically based representatives at the federal, state, and local levels, a stable and popular transportation investment program emerges from a never-ending geopolitical struggle over the collection and distribution of funding (Garrett & Taylor, 1999; Taylor, 1995).

The Geopolitical Equity Imperative

This overriding concern with the geographic equity of transportation funding among states, districts, and jurisdictions ensures a political focus on the expenditure effects of transportation investments, which makes it all but impossible to consider the transportation effects in making transportation funding decisions. Thus from the perspective of most policymakers, it's the transportation economists who simply miss the point by focusing all of their attention on the transportation effects of funding decisions.

So to understand the public finance of urban transportation systems, we must evaluate both the geopolitics of the transportation finance *program* and the effects of this finance program on the deployment and use of the transportation *system*. Table 11.1 offers an overview of how we might simultaneously evaluate the performance of a transportation finance program in each of these realms.

Program performance criteria evaluate how well a finance mechanism meets tests of political acceptability and administrative ease. These questions tend to be prominent in policy debates, and most of them will be familiar even to those new to transportation issues.

System performance criteria, on the other hand, address how finance mechanisms influence the use and performance of the transportation system itself. These criteria acknowledge that finance policies are not just about raising money. Financial instruments also profoundly affect the way transportation services are provided and the way

TABLE 11.1. Program Performance and System Performance Criteria

	System performance	Program performance
Effectiveness	• Optimizes utilization of existing capacity. • Lowers transportation costs and promotes economic development.	• Is politically feasible: has stable political support, is popular with voters, and has little opposition from powerful stakeholders. • Revenues generated meet needs and are stable and predictable.
Efficiency	• Optimizes provision of transportation service for a given level of expenditure.	• Has low administrative and overhead costs relative to the revenue collected.
Equity	• Provides all users with transportation access, regardless of circumstances (age, income, disability, etc). • Is progressive based on ability to pay. • Charges users in proportion to the costs they impose on the system and society.	• Is perceived as treating places and jurisdictions fairly. • Major stakeholders and interest groups perceive they are treated fairly.

Source: Brown et al. (1999).

citizens use them. The fares, fees, tolls, and taxes paid by travelers affect their decisions on where to travel, when to travel, how to travel, and even whether to travel. Use of the transportation system, in turn, greatly influences the maintenance and new capacity "needs" of the system, which affects the finance system. Thus the transportation finance system and the performance of the transportation system are mutually reinforcing.

How both the supply of and demand for transportation are influenced by user costs is neither abstract nor trivial. The issue of truck-weight fees provides an example of how the transportation finance system affects user decisions. The pavement damage caused by heavy trucks increases significantly with the weight per axle. Many people are surprised to learn that a relatively small share of trucks with heavy axle loads do most of the damage to roads (Small, Winston, & Evans, 1989; U.S. Department of Transportation, Federal Highway Administration, 1997a). Yet for decades many states levied truck-weight fees based on the weight of *empty* trucks; and tollways frequently set rates based on the number of axles per vehicle. Both policies encourage

truckers to load heavy weights onto as few axles as possible, thereby *maximizing* damage to roadways. Such truck fee systems increase maintenance and rehabilitation costs in comparison to jurisdictions where fees are assessed in ways to encourage truckers to reduce axle weights. Thus changing the way that fees are levied on trucks can change truckers' behavior, and, in turn, substantially lower maintenance costs without necessarily increasing taxes or revenues.

That transportation finance programs are guided first and foremost by concerns over equity may seem a puzzling, even counterintuitive, assertion. But the way that public officials think of equity in transportation finance is far different from the way that most social scientists, students, or transportation analysts would define the term. Thus "equity" gets defined quite differently by different interests at different times. To paraphrase former Supreme Court Justice Potter Stewart, most of us can't precisely define equity or inequity in transportation finance, but we think that we know it when we see it.

Much of the confusion and debate over equity in transportation finance arises from the competing and contradictory ways that equity is both framed and evaluated (Table 11.2). Debates over equity in transportation finance can arise over the type of equity to be considered—market equity, opportunity equity, or outcome equity—or the appropriate unit of analysis with which to frame the debate—geographic, group, or individual—or both. While "unit of analysis" may seem an abstract concept, it is perhaps the most important consideration in transportation equity.

In general, public finance scholars tend to focus on *individual equity,* advocates and activists are more likely to focus on *group equity,* while elected officials are concerned most with *geographic equity.* This is because representation in the United States is organized geographically into a hierarchy of jurisdictions. And because it is elected officials who oversee the collection and distribution of transportation funds, most debates in transportation finance center, first and foremost, on questions of *geographic equity.*

Geographic equity arises frequently in the context of federal transport policy. For example, the more populous, urbanized states tend to generate more in federal motor fuels tax revenues than they receive in fuel tax-funded federal expenditures, whereas less populous, rural states tend to receive more in federal transportation funding than their motorists generate in federal fuel taxes. This redistribution of federal fuel tax revenues from "donor" states to "donee" states has been hotly debated in Washington for decades and actually delayed the passage of the TEA-21 legislation in 1998.

Supporters of the redistribution of federal transportation funds argue that it enables wealthier states to cross-subsidize poorer states, it supports an interconnected national highway system, and it justifies federal involvement in transportation finance. Critics of the redistribution of federal fuel taxes have countered that the redistribution reflects a rural bias in the federal transportation pro-

TABLE 11.2. Confounding Notions of Equity in Transportation Finance

Unit of analysis	Type of Equity		
	Market equity	Opportunity equity	Outcome equity
Geographic			
States, counties, legislative districts, etc.	Transportation spending in each jurisdiction matches revenue collections in that jurisdiction.	Transportation spending is proportionally equal across jurisdictions.	Spending in each jurisdiction produces equal levels of transportation capacity/ service.
Group			
Modal interests, racial/ethnic groups, etc.	Each group receives transportation spending/ benefits in proportion to taxes paid.	Each group receives a proportionally equal share of transportation resources.	Transportation spending produces equal levels of access or mobility across groups.
Individual			
Residents, voters, travelers, etc.	The prices/taxes paid by individuals for transportation should be proportional to the costs imposed on society.	Transportation spending per person is equal.	Transportation spending equalizes individual levels of access or mobility.

Source: Data from Levy, Meltsner, & Wildavsky (1974).

gram (especially regarding highways), and that research has shown that the federal program actually redistributes funds from poorer states (with less fiscal capacity) to richer states (with more fiscal capacity) (Lem, 1997). Redistribution critics contend further that the national highway system is largely in place and the most significant transportation investment needs are in congested urban areas. However, if all federal fuel tax funds were simply returned to states exactly proportional to their collection, there would be no rationale for a federal fuel tax; it could be eliminated and states would then be free to collect as much as they needed from higher state fuel taxes. Thus some have argued that federal transportation tax collections should be dropped and that each state should be left to make do on its own (Roth, 1998). Such proposals, however, do not support the goal of redistributing funds from the have to the have-not states. In other words, that the current federal surface transportation finance program fails to redistribute funds from states with more fiscal capacity to states with less fiscal capacity is not necessarily an argument against such a goal.

Given the overriding political concern with geographic equity, distortions arise when transportation use or demand does not vary somewhat equally across jurisdictions. Public transit is perhaps the most striking example of this. Transit ridership is concentrated spatially in the largest, most densely developed cities. About one-third of all transit passengers in the United States are in the New York metropolitan area. The 10 largest U.S. transit systems carry over 60% of all riders, while the hundreds of other, smaller systems carry less than 40% of all passengers (Taylor & McCullough, 1998). In the *realpolitik* of public transit finance, however, debates center on how resources are doled out to jurisdictions and the *suppliers* of transit service, with little regard for the enormous spatial variation in the *consumers* of transit service.

For example, the New York Metropolitan Transit Authority (N.Y. MTA) alone carries over 27% of the nation's transit riders each year (American Public Transportation Association, 2003a). During the 6 years between 1995 and 2000, federal capital and operating subsidies combined averaged $0.20 per unlinked passenger trip on N.Y. MTA. In contrast, riders on Chapel Hill Transit in North Carolina, which carries three ten-thousandths (0.03%) of the nation's transit riders, enjoyed federal transit subsidies that averaged $0.97 per trip during the late 1990s (American Public Transportation Association, 2003a, 2003b). Such geographic disparities are not confined to federal transportation finance. In California, the San Francisco Municipal Railway carries nearly half (45%) of all Bay Area transit riders, but receives just 10% of the subsidies allocated through the state Transportation Development Act (TDA). On the other hand, the Santa Clara Valley Transit Authority in the San Jose area carries 11% of all Bay Area transit riders yet receives over one-third of the region's TDA transit subsidies (Metropolitan Transportation Commission, 2003; Taylor, 1991).

The reason for these disparities is quite straightforward: representation in Congress and most state legislatures (with the exception of the U.S. Senate) match the geographic distribution of voters, not urban transit patrons. Geographic equity, therefore, allocates public transit funding "equally" among jurisdictions, regardless of how it is utilized. The centrality of the imperative of geographic equity to transportation policy and planning can hardly be overemphasized. It explains why Texas received $2.7 billion *less* in federal fuel tax revenues between 1956 and 1994 than motorists in Texas paid in federal fuel taxes. In contrast, Hawaii received $2.2 billion *more* than motorists in Hawaii paid in federal fuel taxes; for every $1.00 in federal fuel tax generated in Hawaii, the state received $4.11 in fuel tax-funded appropriations (Poole, 2001). It also explains why new rail transit systems were built in Atlanta, Miami, and many other sprawling Sunbelt cities over the last quarter century, while the long-planned Second Avenue subway in transit-oriented Manhattan has yet to carry a passenger (Lawlor, 1995).

Economists' Solution: Pricing

With all of the attention paid to the politics of geographic equity, public officials frequently fail to consider how a transportation finance program affects the use and performance of the transportation system. Yet the use and finance of the transportation system are tightly intertwined and can not be considered separately (Table 11.1). Fees imposed on users in proportion to the costs those users impose on society are typically the finance mechanisms that will help optimize transportation system performance. User fees make people more aware of the costs of travel (in the form of wear-and-tear on the system, delay imposed on others, environmental damage, etc.). Such information transmitted via prices encourages drivers to shift low-priority trips to less socially costly times of day, routes, modes, or destinations.

Most transportation economists agree that transportation finance programs should, as much as possible, charge users the *marginal* social cost of travel (Murphy & Delucchi, 1998; Small et al., 1989). The term *marginal* refers to the cost of providing for one additional trip, given that others are already using the system at the same time. For example, when a car gets on the freeway, it takes up space that other automobiles can no longer occupy, it imposes some delay on vehicles upstream, and it also causes some amount of pavement damage. If there are very few vehicles already on the freeway, then the cost of providing for that one additional car is very small. On the other hand, if there are many cars already on the freeway, one additional vehicle can slow other cars upstream and increase congestion to a surprising degree. In such cases, the marginal cost of accommodating an addi-

tional car is large. The term *social* refers to the costs that society pays for providing for that one additional vehicle. These social costs result mostly from congestion, pollution, noise, and road wear-and-tear from a trip.

A large body of research shows that most current transportation finance programs do not make users pay the marginal social cost of vehicle use (California Department of Transportation, 1997; Deakin & Harvey, 1995; Forkenbrock & Schweitzer, 1997; National Cooperative Highway Research Program, 1994; Pozdena, 1995; Puget Sound Regional Council, 1997). Yet as the role of the motor fuel tax has declined relative to non-transportation-related instruments like sales taxes and bonds, we are actually moving further away from marginal social cost pricing of transportation (Goldman & Wachs, 2003).

The current transportation finance program generally charges users according to the "average" cost of using the transportation system. That is, when a vehicle gets on the highway and contributes to congestion or an overloaded truck damages a roadway, the "cost" of that delay or road damage is paid by *everyone* experiencing the delay or rough road. The cost is averaged out so that individual users are not made aware of the total costs they are individually imposing on everyone else, and thus have little incentive to alter their travel choices (Pozdena, 1995).

A commonly posed argument against marginal social cost pricing of transportation is that poor people will simply be priced off roads and transit vehicles, leaving free-flowing systems for the wealthy. Such social equity concerns are indeed important, but they ignore the social inequities of our current transportation finance system. The two most common equity principles in public finance are called the "benefit principle" and the "ability-to-pay principle" (Musgrave, 1959). Under the *benefit principle*, equitable taxes are those levied on each individual in proportion to the benefits received by that individual. Within this rubric, charging users according to the incremental social costs they impose on society when using the transportation system is equitable. On the other hand, the *ability-to-pay principle* suggests that a method of finance based solely on benefits received may disproportionately burden the poor. From this perspective, an equitable finance program will treat fairly people who have different abilities to pay, with ability measured primarily by income. Current transportation user fees, like the motor fuels tax and the driver's license fee, fare well under the benefit principle but poorly under the ability-to-pay principle (Chernick & Reschovsky, 1997; Poterba, 1991; Wiese, Rose, & Schluter, 1995). Payment of sales taxes on consumer items are essentially unrelated to use of transportation systems. Yet in recent years many regions around the United States have turned to sales tax increases to pay for transportation improvements (Goldman & Wachs, 2003). Unfortunately, such transportation sales taxes fare poorly under *both* the benefit and the ability-to-pay principles.

Thus, in comparison with our current system of transportation finance, a user fee system based on the principles of marginal cost pricing would clearly increase equity based on the benefit principle and may well increase overall equity under the ability-to-pay principle as well. Deakin and Harvey (1995) found that use of the highway system in congested conditions is positively correlated with income. That is, higher income travelers tend to spend a larger share of their travel time in traffic congestion than do lower income travelers. Thus a shift to a transportation finance system that charges drivers more on congested routes and less elsewhere would fare well under the ability-to-pay principle in comparison to our current finance system.

Given the many potential advantages of a marginal social cost system of transportation pricing, there have been some recent, promising experiments with road pricing in the United States. In 1995, a four-lane facility opened in the median of the State Route 91 freeway through Santa Ana Canyon in metropolitan Los Angeles (see Figure 11.2). The facility, called the "Route 91 Express Lanes," was entirely privately financed and is funded through electronically collected tolls. Use of the 10-mile facility allows toll-paying users to bypass the congestion on the adjacent free lanes and save up to 30 minutes per trip. The tolls vary from a low of $1.00 when the adjacent free lanes are free-flowing to a high (in 2003) of $4.75 during times when the adjacent lanes are most congested. An early evaluation of the facility found that it had increased the throughput of traffic and had decreased congestion in both the toll facility *and* in the adjacent free lanes (Mastako, Rillet, & Sullivan, 1998; Sullivan & El Harake, 1998). Over time, however, growth in demand has increased congestion in the adjacent free

FIGURE 11.2. Congestion pricing experiments, like this one in Southern California, have been effective in improving traffic flow, but many policymakers remain wary of a political backlash against tolls.

lanes, and as a result the peak toll must be increased regularly to keep the Express Lanes traffic free-flowing. With respect to income, higher-income travelers do indeed use the toll lanes more frequently than lower-income travelers, though use of the lanes by all income groups is relatively high. A 1999 survey found that about one-third of the corridor travelers from households with annual incomes below $40,000 used the lanes at least occasionally, compared to about two-thirds of travelers from households with incomes over $100,000 (Sullivan, 2000).

In 1998, underutilized high-occupancy vehicle (carpool) lanes in San Diego, California, were converted to allow single-occupant vehicles to pay a toll to use the carpool lanes and bypass congestion on the adjacent unrestricted lanes. The San Diego facility is unique because the tolls are set in "real time"; they change every few minutes, ranging from $0.50 to $4.00, based on congestion levels. In addition to these facilities, tolls on bridges in the New York City area and in southwestern Florida vary by time of day to encourage travelers to use the bridges at less congested times (U.S. Department of Transportation, Federal Highway Administration, 2002a).

Collectively, these congestion-priced facilities have worked well in practice. Traffic flow is increased, congestion delay is reduced, and the people who voluntarily pay tolls to bypass congestion report high levels of satisfaction. Even the travelers in congested adjacent facilities are generally satisfied with the congestion-priced facilities, because they tend to experience improved traffic flow in the free lanes and have the option to buy out of congestion if they wish to do so (Mastako et al., 1998). Despite the success and local popularity of these congestion-pricing experiments, however, many elected officials remain wary of political backlashes to congestion pricing by drivers angry about paying for something that was formerly "free" (from the drivers' perspective). So while congestion pricing has perhaps the greatest promise to reduce traffic congestion significantly in metropolitan areas, its deployment is likely to continue to be slow and gradual.

Cashing-Out Free Parking

Parking pricing has been proposed as a "second best" approach to marginal cost pricing of private vehicle use. The 1990 National Personal Transportation Survey found that drivers park free for 99% of all vehicle trips (U.S. Department of Transportation, 1994). But, of course, parking is not free to provide. Free parking at a grocery store, for example, is paid indirectly by shoppers, whose groceries are priced to include the cost of the property for, and construction and maintenance of, parking that typically occupies more land than the store. Shoppers who walk to and from a grocery store thus cross-subsidize those who drive and park for free.

Free parking at work is essentially a matching grant for commuting by private vehicle—employers pay the cost of parking at work only if commuters are willing to drive to work. Commuters who do not drive to work do not receive a subsidy. While many commuters would drive even if they had to pay for parking at work, some would choose to carpool, ride public transit, walk, or bike to work instead; such commuters, therefore, drive to work because they park for free. Shoup (1997) has shown that employer-paid parking increases the number of cars driven to work by about 33% when compared with employment sites with driver-paid parking.

Recently, some employers have begun to offer workers the option to "cash-out" free parking by taking the cash equivalent of any parking subsidy offered. Offering commuters the choice between a parking subsidy or its cash equivalent shows commuters that free parking has an opportunity cost, in this case foregone cash. At such work sites, commuters can continue to park free, but the option to receive cash in lieu of free parking also rewards commuters who choose to carpool, ride public transit, walk, or bike to work instead. When given this option, a surprising proportion of employees choose to pocket the cash and commute to work by means other than driving alone (Shoup, 1997).

Finally, public regulations strongly encourage the provision of unpriced parking for drivers. Substantial portions of street and road systems are devoted to curb parking instead of traffic flow. Most curb parking is free, and even where curb spaces are metered they are typically priced well below the market-clearing price (i.e., the price that would ensure at least a few available spots at most times of the day). And because the demand for free or inexpensively priced street parking often exceeds available curb space, local governments typically require all developments (grocery stores, apartment buildings, medical offices, etc.) to provide enough off-street parking to satisfy the peak level of demand for free parking for that land use. Such minimum parking requirements subsidize private vehicle use and can substantially increase development costs—which are borne by everyone, not just parkers. Shoup (1999) has estimated that providing four parking spaces per 1,000 square feet of floor area increases the cost of constructing an office building by between $40 and $100 per square foot of office space. Thus, the high cost of free parking is an important, and often overlooked, part of both transportation finance and pricing.

Given this overview of the theories and rationales for the pricing and public finance of metropolitan transportation systems, we now turn to the current structure of urban transportation finance in the United States.

WHO PAYS FOR WHAT?: THE STRUCTURE OF URBAN TRANSPORTATION FINANCE

Transportation is big business. The movement of people, goods, and information is central to both human culture and economic activity. The U.S. Department of Transportation estimates that over 11% of the U.S. economy relates in some way to transportation (U.S. Department of Transportation, Bureau of Transportation Statistics, 2003). Transportation systems are financed from a labyrinth of public and private sources that defy simple characterization. Individuals and firms pay for private vehicles, insurance, fuel, and fares. Shippers pay fees for the construction of seaports, and passengers pay fees for the development and expansion of airports. Pedestrians pay for shoes, while property owners typically pay for sidewalks. Local governments—cities and counties—use tax money to pay for local streets and roads. And the states and the federal government collect fees and taxes to pay for up to 90% of the costs of freeways, other highways, and many public transit systems. While the structure of transportation finance varies from state to state, and even from county to county, Table 11.3 presents a schematic overview of how five principal metropolitan transportation systems are financed.

Public and private investment in U.S. transportation systems is immense. In 2000, total public (federal, state, and local) expenditures on surface transportation systems exceeded $117 billion. Table 11.4 below breaks down these expenditures by mode and expenditure type. It shows that public transit subsidies exceeded $21 billion, local streets and county roads expenditures exceeded $36 billion, and state and federal highways expenditures were nearly $60 billion (U.S. Department of Transportation, Federal Highway Administration, 2002b). These expenditures amount to $2.48 for every unlinked transit trip in the United States, and to about $0.04 for every vehicle mile of travel in the United States.

The revenue raised for transportation comes from a variety of transportation and nontransportation sources, ranging from vehicle registration fees and weight fees to property taxes and sales taxes. When transportation revenue sources are used to fund nontransportation purposes (like parks or

TABLE 11.3. Schematic Overview of Metropolitan Transportation Finance in the United States

Transport system	Share of metropolitan person trips in 2000	Payments by direct beneficiaries		Payments by others	
		Paid by travelers	Paid by others	Direct subsidies	Indirect subsidies
Pedestrian systems	8.6%	Time, shoes, socks, energy	Sidewalks usually initially funded by developers, and later by property owners via property taxes	Few	Few
Bicycle systems	0.9%	Time, bicycles, energy	Local streets often paid for initially by developers, later by property owners via property taxes	Bikeways, bike paths, bike lanes financed from fuel taxes and other tax revenues for transportation	Roads and highways financed from fuel taxes and other tax revenues for transportation
Public transit systems	3.2%	Time, fares	Local streets on which buses operate often paid for initially by developers, later by property owners via property taxes and transportation impact fees	Transit capital and operations paid from fuel and general taxes. Highways and freeways financed with fuel taxes, license fees.	Minor delay costs borne by other travelers; pollution, noise, and resource-depletion costs borne by all.
Local streets/roads	86.4%	Time, vehicles, fuel, insurance. Some fuel taxes, license fees, etc. for operations, maintenance.	Rights-of-way provided and initial construction often paid by developers. Local streets financed by property owners via property taxes.	Few	Delay costs borne by other travelers; pollution, noise, and resource-depletion costs borne by all.
Metropolitan highways and freeways	86.4%	Time, vehicles, fuel, insurance. Fuel taxes, tolls, license fees, etc. for rights-of-way, construction, operations, maintenance.	Minimal	Some sales taxes and other general taxes for rights-of-way, construction, maintenance	Significant delay costs borne by other travelers; pollution, noise, and resource-depletion costs borne by all.

Mode share source: U.S. Department of Transportation (2001).

TABLE 11.4. Government Expenditures on Surface Transportation, 2000 (in Millions of Dollars)

	Capital funding	% share	Maintenance/ operations funding	% share	Total	% share	Expenditures per transit trip (for transit) or VMT (for streets/highways)
Public transit	$9,270	12.6%	$13,941	30.3%	$23,211	19.4%	$2.48
Local streets/ county roads	$17,814	24.2%	$18,616	40.4%	$36,430	30.5%	$0.03
State/federal highways	$46,405	63.1%	$13,473	29.3%	$59,877	50.1%	$0.04
Total	$73,489	100.0%	$46,029	100.0%	$119,518	100.0%	

Source: Adapted from U.S. Department of Transportation, Federal Highway Administration (2002) and American Public Transportation Association (2003a, 2003b).

schools), or when revenues from general instruments of taxation (like sales or income taxes) are used for transportation purposes, transportation is part of larger programs of public finance. But when revenues from transportation sources are used for transportation purposes, the finance system can be viewed as one of indirect user fees, rather than taxes (Table 11.5). These distinctions are important because highway and, later, public transit interests have successfully lobbied Congress and most states to create special accounts, or trust funds, into which transportation revenues are deposited and out of which transportation programs and projects are funded. These separate accounts have brought stability to the finance of

highways and public transit and for decades reflected a user-fee approach to transportation finance.

Since the 1920s, the principal user fee with which revenues have been raised for the construction and maintenance of U.S. highways (and later public transit systems) has been the motor fuel tax. The motor fuel tax is unique in many respects. Because drivers of motor vehicles impose costs on the transportation network that, to a certain extent, are proportional to their use of fuel, the motor fuels tax has been considered a transportation user fee since its inception in the 1920s. Those who drive more pay more, and those who drive large, fuel-guzzling vehicles pay more. During the Great Depres-

TABLE 11.5. User Fees and General Taxes in Transportation Finance

		Expenditures	
		Transportation purposes	Nontransportation purposes
Revenues	Transportation sources	Transportation user fees • Motor fuel taxes for highways and transit service • Transit fares • Bridge tolls to retire bridge bonds	Transportation taxes for general purposes • Fuel taxes for "deficit reduction" • Parking meter revenue to fund libraries
	Nontransportation sources	General taxes for transportation • Sales taxes dedicated to transportation • General obligation bonds for transportation	General taxes for general purposes • Income taxes for education, health care, and national defense

sion in the 1930s, and then again in the 1990s, the federal government used federal fuel tax revenues for nontransportation purposes, but for the most part fuel tax revenues have been used to finance transportation.

As an instrument of taxation, the motor fuels tax has much to recommend it fiscally, politically, and administratively. First, as motor fuel consumption has soared over the past eight decades, so have tax proceeds. The motor fuel tax is a phenomenal revenue producer, yielding $75.6 billion in the United States in fiscal year 2000 (U.S. Department of Transportation, Federal Highway Administration, 2001). Second, the tax is paid in relatively small increments; the current average levy of $0.377 per gallon ($0.184 federal levy plus an average state levy of $0.193) is relatively hidden in the price of motor fuel (U.S. Department of Transportation, Federal Highway Administration, 2000). This particular feature of the tax has tended to minimize organized public opposition to it, though other taxes collected in even smaller increments (like sales taxes) are eclipsing the fuel tax in popularity. Finally, the tax is, both from the taxpayer's and the government's point of view, easy to administer and collect. The gasoline tax is collected from gasoline distributors rather than directly from retailers or consumers, which serves to minimize the opportunities for gas tax evasion and to reduce the cost of collection to a historical average of only 0.5% of tax proceeds (Brown et al., 1999).

As population, personal travel, and, especially, vehicle use have increased dramatically in recent years, the motor fuel tax has gradually faltered. Three factors have combined to make it difficult for fuel taxes to keep up with expanding needs: increasing vehicle fuel efficiency, the fact that per-gallon fuel tax revenues do not increase with inflation, and increasing transportation program commitments (Brown, 2001; Taylor, 1995).

First, automobile fuel efficiency has increased significantly over the past 30 years, though the growing popularity of pickup trucks and sport utility vehicles (SUVs) has recently slowed fuel efficiency improvements. Table 11.6 shows that for each gallon of fuel consumed, vehicles in 2000 traveled 63% farther than they did in 1970. And since most fuel taxes are levied on a per-gallon basis, fuel tax revenues cannot keep pace with the growth in travel without substantial increases in the per-gallon levy (U.S. Department of Transportation, Bureau of Transportation Statistics, 2002). While improvements in vehicle fuel efficiency have undoubtedly benefitted the environment, they have also substantially reduced fuel tax revenues per mile driven. Plans to promote conversion of the automobile fleet to alternative fuels or electric power (discussed in Chapter 10) further threaten fuel tax revenues. Alternative-fuel and electric-powered vehicles use road in ways similar to traditional gasoline- and diesel-powered vehicles, but they do not produce fuel tax

TABLE 11.6. Trend in Average U.S. Vehicle Fleet Fuel Economy since 1970

	1970	1980	1990	2000
Fleet mileage	13.5 mpg	15.9 mpg	20.2 mpg	22.0 mpg
10-year increase	—	2.4 mpg	4.3 mpg	1.8 mpg
10-year change (%)	—	17.8%	27.0%	8.9%
Increase since 1970	—	2.4 mpg	6.7 mpg	8.5 mpg
Change since 1970 (%)	—	17.8%	49.6%	63.0%

Source: Calculated from U.S. Department of Transportation, Bureau of Transportation Statistics (2002).

revenues (Rufolo & Bertini, 2003; Wachs, 2003).

Second, inflation has diminished the purchasing power of the motor fuel tax. Because the cost of materials used in transportation projects and the cost of land for transportation facilities have risen faster than the general rate of inflation, the buying power of fuel tax revenues has eroded even faster than the inflation rate would suggest. Many taxes, such as sales, property, and income taxes, are able to maintain their productivity in the face of inflation because the tax base rises with inflation. Motor fuel taxes, however, are generally levied on a per-gallon basis, and thus their proceeds do not increase in response to inflation (Taylor, 1995).

To keep pace with rising costs, gas and diesel fuel taxes must be increased periodically by act of a legislature and approval of a governor or the president. Despite public concern over congestion and, to a lesser extent, deteriorating transportation infrastructure, achieving the political consensus necessary to raise fuel taxes has become increasingly difficult. User fee or not, changes to fuel tax levies have been central to many partisan debates over tax increases since the so-called tax revolts of the 1970s. As such, legislators have become increasingly wary of potential voter hostility toward tax increases of any sort, and have been reluctant to accept regular increases in the motor fuels tax levies to keep pace with increasing fuel efficiency and inflation.

Finally, the demands on faltering motor fuel tax revenues have grown as increasing safety standards and environmental safeguards have been added to highway projects, and new types of projects (public transit, bicycle, etc.) have been made eligible for fuel tax funding. This so-called "program creep" has helped to increase the fiscal squeeze on motor fuel tax revenues (Brown et al., 1999).

The result is a widening gap since the 1970s between transportation finance revenues and transportation construction and maintenance needs. Legislators have responded by enacting periodic stopgap revenue-enhancement measures, but no meaningful structural reforms have emerged to stabilize the transportation finance system. Fuel taxes have been occasionally increased since the early 1980s, but the tax increases have failed to keep pace with the combined effects of inflation, increasing vehicle fuel efficiency, and new program responsibilities. Figure 11.3 shows that to return the buying power

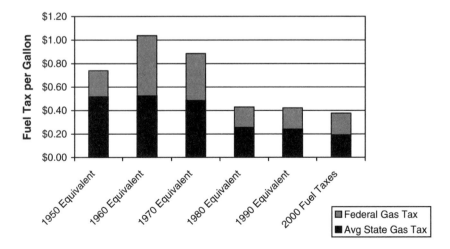

FIGURE 11.3. Changes in per-gallon fuel taxes required in 2000 to restore inflation-adjusted revenues per vehicle mile of travel to level of prior decades. Source: Hill, Taylor, Weinstein, and Wachs (2000).

of the 2000 fuel tax to its 1960 level, the combined state and federal gas tax would have to more than double to over $1.00 per gallon.

Despite their erosion, state and federal motor fuel taxes still account for 57% of all highway revenues (U.S. Department of Transportation, Federal Highway Administration, 2001); in 1960, the figure was 66% (U.S. Department of Transportation, Federal Highway Administration, 1997b). Taken as a whole, transportation-related revenue sources (motor fuel taxes, registration fees, weight fees, tolls, and driver's license fees) accounted for 77% of U.S. revenues for highways in 2000, while non-transportation-related revenue sources (sales taxes, property taxes, etc.) accounted for nearly 23% (U.S. Department of Transportation, Federal Highway Administration, 2001).

Importantly for cities, fuel tax revenues financed most of the extensive metropolitan freeway systems that helped to transform urban America after World War II. How we came to depend on freeways as the backbone of nearly every urban transportation system in the United States and why we choose to bring freeways into the center of most major U.S. cities is substantially explained by the geopolitics of transportation finance. We now turn to this story of fiscal politics and freeways as a case study of the central role of finance in the geography of urban transportation.

CASE STUDY: THE URBAN GEOGRAPHY OF FREEWAY FINANCE

Freeways are the centerpiece of most metropolitan transportation systems, a feature that distinguishes U.S. cities from most others. Though most freeways in the United States were developed as part of the national Interstate System, the most profound effects of freeways have been in cities. Comprising only a tiny fraction (0.5%) of urban street and road mileage, freeways na-

tionally carry over one-third (35%) of all urban vehicle travel. Interestingly, freeways actually play a smaller role outside of cities; about a quarter (25%) of all rural vehicle travel is on freeways (U.S. Department of Transportation, 2001).

Most cities around the globe have some grade-separated, limited-access roadways, but none outside of the United States rely on these roadways for such a significant proportion of metropolitan travel (Jones, 1989). As such, metropolitan freeways in the United States have been embraced by some observers as a foundation of suburban living favored by most Americans and vilified by others as a prime cause of urban decay, urban sprawl, air pollution, and auto dependence. Accordingly, this section examines the development of metropolitan freeways as an important example of the influence of fiscal geopolitics on transportation outcomes.

How did the fiscal politics of freeway finance shape metropolitan freeways and metropolitan areas? The quest for freeway funding led cash-strapped cities to turn to the states and the federal government for help in the 1930s, 1940s, and 1950s. State/federal finance and control of metropolitan freeways, in turn, led to significant changes in the design and routing of freeways. But imposing these new freeways on existing cities was socially disruptive and expensive because the freeways ultimately built in cities were larger, noisier facilities that concentrated traffic and pollution much more than the multimodal expressways envisioned by early urban planners. These more invasive facilities eventually led to a series of "freeway revolts" in the 1960s and 1970s around the United States, whereby citizens and community groups organized to oppose freeways that would displace large numbers of homes or businesses.

The new urban freeways were also very expensive. As discussed above, the eroding motor fuel taxes did not grow in proportion to either travel or rising construction costs, and funding to construct expensive new metropolitan freeways has largely dried up.

While the era of widespread metropolitan freeway development has passed, traffic and congestion on freeways continues to grow (see Figure 11.4).

Early Plans for Freeways

In the 1930s, urban and rural roadway planners were refining plans for hierarchical, interconnected networks of grade-separated, limited-access roadways. However, the scale of the new road networks and the priorities given to different trip types varied substantially between the plans prepared by and for major cities and the plans prepared for rural areas by state highway departments and the federal Bureau of Public Roads. In cities, the planned systems and facilities reflected prevalent concerns with reducing traffic congestion and improving local automobile and transit vehicle circulation; they often included separate lanes for trucks, buses, or streetcars. State and federal highway engineers, on the other hand, were designing new single-mode facilities for high-speed intercity travel and improved safety (Fairbank, 1937; Interregional Highway Committee, 1944; MacDonald, 1954; Purcell, 1940a, 1940b).

So why were high-speed, intercity-style freeways ultimately built in cities? The answer is "money," as the following evidence makes clear:

- In the 1930s, the faltering property tax was abandoned in favor of the gas tax as the main source of urban highway funding. As a result, the opportunity to recapture the positive effects that metropolitan freeway development would have on land values was lost. And this, in turn, prevented highway revenues from keeping pace with increasing metropolitan freeway rights-of-way costs over time.

- In the 1940s, the Interstate System was adopted by Congress and the president, but without funding. In California, the state agreed to raise highway user taxes to begin construction of, among other things, the Los Angeles expressway plan. This model—of state departments of transportation (DOTs) planning and constructing freeways in cities—was later adopted by nearly every other state with the funding of the Interstate Highway System in 1956. The evolution of freeway finance in California, then, is especially relevant to the development of metropolitan freeway systems in the United States.

- Placing state highway departments in charge of all metropolitan expressway development was contrary to the recommen-

FIGURE 11.4. While high costs and lagging revenues have slowed freeway development, traffic on metropolitan freeways continues to grow.

dations of the plan for the Interstate Highway System adopted in 1944, yet this
precedent-setting move helped separate local planning from metropolitan freeway development for the next quarter century.

• In the 1950s, the funding of the Interstate program critically shaped metropolitan
freeway development. First, to secure the
support of urban legislators, the U.S. Bureau
of Public Roads (1955)—again, contrary
to the recommendations of the Interstate
Highway System plan—published routing
plans for every urban Interstate Highway in
the country, which preempted local planning and effectively set the metropolitan
freeway planning process in stone. Second,
to give priority to the Interstate Highway
System, the traditional one-to-one federal/
state matching ratio was changed to nine-
to-one, which encouraged states to put
their resources into Interstate construction
and discouraged them from developing
non-Interstate highways in cities.

The result of these funding actions was to
greatly reduce local control over metropolitan expressway development. The major
planning decisions—the design, routing,
and size of the system—were either specified in advance or delegated to state highway departments (Altshuler, 1965).

Early Urban Road Finance

Until the Great Depression, street and boulevard development in most cities was paid
for jointly by cities and counties with property taxes and bonds; special assessment districts were frequently created to tax the
property owners benefitting from major
boulevard improvements. With the depression came the collapse of the property tax
and urban highway finance. Gas tax revenues, on the other hand, fared quite well
during the Depression. Except for a small
dip in fuel consumption at the onset of the
Depression, fuel consumption and gas tax
revenues increased annually until fuel rationing was adopted during World War II

(U.S. Department of Transportation, Federal Highway Administration, 1997b). So
when property tax and special assessment
funding for streets and roads largely dried
up during the 1930s, cities and urban counties began pushing states for fuel tax revenues to support highway expenditures in
cities (Burch, 1962; Jones, 1989).

In 1944, the federal aid highway program
was substantially revised to allow funding of
"urban extensions" of the (largely rural)
federal highway system (Congressional
Quarterly, 1964; Seely, 1987; U.S. Congressional Budget Office, 1978). In 1947 California agreed to raise motor fuel and related
taxes to help pay for building freeway systems in Los Angeles and other cities, and
placed the state Division of Highways in
charge of all metropolitan expressway development (Collier Burns Highway Act of
1947). So, in separate actions over a 3-year
period, the federal government (in 1944)
and California (in 1947) made substantial
commitments to finance metropolitan freeways. But in each case those commitments
required that cities relinquish control over
expressway development (Brown, 2002;
Jones, 1989).

Evidence of this shift in control was apparent in the earliest designs by the California Division of Highways for freeways in
Los Angeles. The freeways designed by state
highway planners for intercity travel dif-
fered from urban expressways designed by
city planners for intracity traffic in many
important respects. The new freeways were
larger, on much wider rights-of-way, and
designed for much higher vehicle speeds
(60–70 miles per hour instead of 40–50
miles per hour). The Division of Highways
also adopted a practice of uniform design
standards, regardless of location (Taylor,
2000).

Perhaps more significantly, all public transit and joint development components were
eliminated; in most cases freeways became
stand-alone facilities. In an update of the
Los Angeles expressway plan after World
War II, a board of consulting engineers rec-

ommended that rail transit rights-of-way be reserved on five freeways and special provisions for express bus service be included on seven others. The California Division of Highways, however, successfully opposed such provisions for mass transit, which was viewed at the time as a competing mode and an inappropriate recipient of highway funds (Jones, 1989).

Funding the Interstate Highway System

While California was actively building freeways in the first decade after World War II, the national Interstate Highway program received total appropriations of only $50 million in the 10 years after its designation by Congress in 1944 (Congressional Quarterly, 1964). In the mid-1950s, however, funding for the Interstate Highway System was seriously debated by the Eisenhower administration and Congress, culminating in the Highway Revenue Act of 1956, which raised federal motor fuel taxes and fees and created a Highway Trust Fund.

Two aspects of the Interstate funding process have particular importance to the geography of urban transportation: (1) the abandonment of the traditional one-to-one federal/state fund-matching ratio in favor of a nine-to-one ratio that dramatically skewed metropolitan highway investment decisions; and (2) the fixing of urban route locations by the Bureau of Public Roads in consultation with state highway departments prior to the appropriation of funds.

In 1956 the Highway Trust Fund was created and the Interstate Highway System became a national priority. The "national interest" rationale was used to modify the Interstate matching requirement to a nine-to-one federal/state ratio. This shift in the federal/state ratio was so radical that it dramatically altered the planning calculus of state highway departments (Altshuler, 1965). From the states' perspective, building a metropolitan freeway without Interstate funding was now 900% more expensive

than building an interstate freeway; even federal aid urban projects, financed with the traditional one-to-one federal/state match were 400% more expensive than comparable Interstate projects. Since the 10% state contribution bought so much highway, the development of complementary facilities was discouraged, especially as highway revenues began drying up in the late 1960s (Taylor, 1995).

The effects of this nine-to-one match were even more distorted by the mileage limit of the Interstate system. From the original Bureau of Public Roads studies of "a very limited mileage of super-service highways," the Interstates had been planned as a fixed system—fixed in mileage, but not in cost (MacDonald, 1936, p. 19). Beginning in 1960, all Interstate funds were apportioned to states on the basis of each state's estimated cost to complete the system. In concert, the nine-to-one matching ratio and the fixed system length with no cost ceiling had two significant effects on metropolitan freeway planning. First, they encouraged states to design as much capacity as possible—more lanes, more and bigger interchanges—into each mile of Interstate highway. This served to drive up costs and to concentrate very large volumes of traffic on the metropolitan Interstates. From the states' perspective, however, bigger Interstates were still a bargain. And, second, they strongly discouraged states (or cities) from developing comparatively expensive (from the states' perspective) companion facilities or modes to the metropolitan Interstates (Gifford, 1984). In other words, the structure of the Interstate funding program discouraged the kind of multimodal expressways envisioned by the early metropolitan transportation planners to circulate and distribute traffic in cities.

The result, of course, is the dominant role for Interstate freeways in nearly every major metropolitan area in the country—a role, incidentally, that the original architects of the Interstate system emphatically advised against (Interregional Highway Committee,

1944). Nationally, over 60% of the metropolitan freeways and expressways in the country are on the limited 11,500-mile urban Interstate System and 24% of all vehicle miles traveled in cities are on these 11,500 miles of urban Interstates (U.S. Department of Transportation, Federal Highway Administration, 2001).

The second action that determined the routing of metropolitan freeways was the selection of urban route locations prior to the establishment of the Highway Trust Fund in 1956. At the time, federal policy called for the exact routing of the urban freeway routes to be determined by local officials because urban "routes should be located and designed to be an integral part of the entire urban transportation plan" (Clark, 1955, p. A-10). In 1955 Congress failed to enact legislation to finance the Interstate Highway System, in part due to the opinions of many Congress members from urban areas that it was for the most part a rural program. So, in its desire to secure the support from urban Congress members for the Interstate Highway funding proposal, the Bureau of Public Roads (BPR) ignored federal policy regarding local designation of freeway routes. In the months leading up to the 1956 federal highway funding deliberations, the BPR rushed to designate all of the urban interstate routes to show that interstate highways would touch most urban Congress members' districts.

Conclusion

So with the passage of the 1956 Federal Highway Act, local transportation planning and alternative transportation modes were essentially cut out of urban freeway development. Both the nine-to-one matching ratio, which strongly discouraged the development of parallel facilities (arterials, expressways, freeways, and transit), and the preemptive designation of the urban Interstate routes by the Bureau of Public Roads in preparation for the 1956 Congressional deliberations directly contradicted the rec-ommendations of the 1944 plan for the Interstate Highway System. With the funding program in place and control of freeway development in the hands of the state highway departments, the stage was set for the mass production of metropolitan freeways and the transformation of U.S. cities into networks of mostly auto-oriented suburbs linked together by freeways—a transformation determined largely by the geopolitics of transportation finance.

Following three decades of large-scale metropolitan freeway development after World War II, metropolitan freeway funding—and thus construction—has generally languished. Since the 1970s, however, public investment in transit systems has grown dramatically. In some metropolitan areas, expenditures on public transit now exceed expenditures on metropolitan highways. This is largely because many people see new public transit lines and stations as less socially and environmentally invasive than new freeways. Accordingly, we turn now to a case study of the geography of public transit finance.

CASE STUDY: THE GEOGRAPHY OF PUBLIC TRANSIT FINANCE

The story of public transportation in the United States is one of ongoing fiscal crises. These crises, in turn, have shaped the geography of urban public transit. At the turn of the last century, profit-seeking street railways extended lines into the urban fringe, initiating a suburbanization boom in cities around the United States (discussed in Chapter 3). For most of the 20th century, the share of trips made on public transit declined as private vehicles rose to dominate travel in increasingly dispersed metropolitan areas. This continuing loss of market share has severely weakened the financial stability of this formerly for-profit industry, requiring substantial and increasing per-passenger public subsidies. Today, political struggles over the geographic distribution of public funding for transit have created a heavily capitalized industry that tends to emphasize new equipment and politically visible projects over cost-effective operations. As a result, fiscal stability in the transit industry has long proved elusive. Accordingly, this section reviews both the origins of fiscal distress in the transit industry and the influence of subsidy programs on the deployment and performance of public transit systems today.

Private Origins of Fiscal Distress

Many people are surprised to learn that most of the large publicly owned and operated transit systems of today have private, for-profit origins. At the turn of the last century, most large U.S. cities were served by multiple private streetcar operators. As late as the 1950s, urban transit systems were almost exclusively privately owned and operated. By the 1970s, however, the industry had been transformed into almost exclusively heavily subsidized public ownership and operation.

Metropolitan dispersion, increasing competition from automobiles, stormy labor relations, and stringent regulation of fares and service during the first third of the 20th century combined to fiscally squeeze private transit systems. While street railway ridership peaked in 1922, inflation-adjusted investment in street railways had topped out 17 years earlier in 1905. Fare-box recovery[1] began a long period of decline in 1913. By 1916 street railways nationally were in a state of net disinvestment—meaning that transit capital stock was being worn out faster than it was being replaced through reinvestment. Operating expenses were usually easily covered, but any profits were generally being invested elsewhere (Jones, 1985).

During the Great Depression in the 1930s, the erosion of transit ridership, which had begun during the 1920s, accelerated precipitously. Urban public transit went into its first sustained period of patronage decline, and the financial condition of private streetcar companies worsened considerably. Strapped for capital, many streetcar operators chose a short-term strategy of abandoning rail cars and lines in need of expensive maintenance and rehabilitation in favor of buses. In the long run, streetcars on heavily patronized routes were cheaper to operate than buses, but they required much more up-front capital investment. Buses were often more expensive to operate, but they avoided the high costs of extending and maintaining rights-of-way. In many cases, buses replaced streetcars on routes with patronage warranting rail transit; on these routes, the short-term financial strategy of converting to buses put transit operators at a long-term financial disadvantage (Saltzman, 1992).

Corporate Conspiracy or Fiscal Inevitability?

It is within this context of fiscal decline that conspiracy theories about the demise of street railways have arisen. There is perhaps no more enduring story in urban transportation than that of the insidious corporate plot to force Americans into private automobiles by systematically destroying the streetcar systems that a century ago blanketed nearly every U.S. city. In 1939 a corporate coalition, led by General Motors (GM), formed a holding company called National City Lines, which eventually acquired some 100 street railway companies around the United States and converted them to diesel bus operation (General Motors Corporation, 1974). Ten years later GM and its partners were found guilty of violating the Sherman Anti-Trust Act for their involvement in National City Lines, and were slapped with token fines (Plane, 1995). But to what end these corporate giants were conspiring has been the subject of considerable debate.

In 1974 Bradford Snell submitted a report to the U.S. Senate Judiciary Subcommittee on Antitrust and Monopoly that claimed that GM and its allies successfully used National City Lines to destroy viable transit systems and force people into cars (Snell, 1974). While a few scholars have supported Snell's arguments (Yago, 1984; Whitt, 1982; Whitt & Yago, 1985), most of the research on the issue has not been kind to Snell's version of events (Adler, 1991; Bottles, 1987; Brodsly, 1981; Hilton, 1985; Jones, 1985; Richmond, in press; Slater, 1997).

What is clear is that the fiscal decline of street railways began at least a quarter century before the formation of National City Lines, and that the conversion to rubber-tired vehicles was an industry-wide phenomenon in the 1930s and 1940s. In other words, the rate of both bankruptcies and conversion to diesel bus service among the streetcar systems *not* purchased by National City Lines was similar to National City Lines properties. Some scholars believe that the most enduring effect of National City Lines was to accelerate the shift from streetcars to dirtier, noisier diesel buses—and away from cleaner, quieter electric trolley buses on more heavily patronized routes (Jones, 1985; Richmond, in press). But there is simply no evidence that either the nationwide decline of metropolitan street railways or the rise of automobiles during the middle third of the 20th century was caused by a National City Lines corporate conspiracy.

The Fall of Private Transit and the Rise of Public Subsidies

Wartime employment, along with the rationing of oil, rubber, and metal for automobiles, significantly increased transit patronage during World War II. But private investment in transit systems did not match the war-related bump-up in ridership. By the end of the war, the quality of capital stock—vehicles, track, and catenary—was badly deteriorated. When wartime restrictions on automobiles were lifted, transit patronage dropped precipitously and private investment in transit systems dried up completely. In the 10 years after the war, pa-tronage nationwide declined from about 24 billion annual riders to about 10 billion riders, an annual decline of over 7.5% (Figure 11.5).

By the time the Interstate Highway program was funded in 1956, most private transit systems were either bankrupt or on the ropes. At the urging of big-city mayors and other metropolitan interests, the federal government began a program of capital grants for publicly owned transit systems in the early 1960s. As late as 1964, only a handful of transit systems in the largest cities had converted to public ownership, and most of them continued to support all operations out of the fare-box. In contrast, most bankrupt transit systems in smaller cities simply disappeared (Saltzman, 1992; Wachs, 1989).

From a tentative beginning in the early 1960s, federal support of urban public transportation grew rapidly. By the late 1970s, the federal government had become a principal financier of urban public transit. Since total federal appropriations for transit and related programs began in 1961, federal funding of public transit systems has exceeded $150 billion (American Public Transportation Association, 2003b; Li & Taylor, 1998). However, amid growing concerns over taxation, spending, and budget deficits, Congress and recent administrations have focused on capping or reducing federal subsidies to public transit, particularly operations. Beginning in the 1980s, for example, the Reagan administration advocated a transit policy that would increase contracting with private firms for transit service and would gradually withdraw federal operating subsidies while keeping capital subsidies intact (Menzies et al., 2001).

Today, virtually all public transit service in the United States is heavily subsidized. Nationwide, fare-box revenues cover less than 40% of operating costs, and essentially none of the capital costs (such as for vehicles, stations, and facilities). Operating and capital subsidies over and above fare-box revenues and other income (such as income from charter service and advertising) are obtained from local, state, and federal

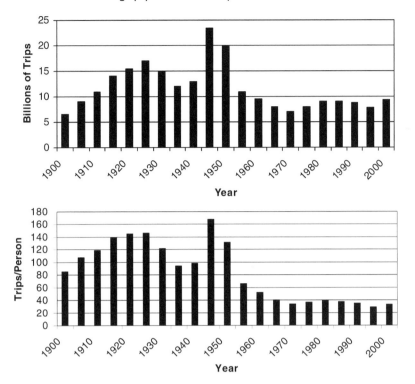

FIGURE 11.5. *Top panel:* Trend in annual transit ridership, 1900–2000. *Bottom panel:* Trend in annual transit ridership per capita, 1900–2000. Source: Adapted from American Public Transportation Authority (2000) and U.S. Bureau of the Census (2003).

sources—and the bill is a big one. Collectively, local,[2] state, and federal transit subsidies totaled $14.4 billion for operations and $10.8 billion for capital in 2001, an average of $2.48 per unlinked passenger trip (American Public Transportation Association, 2003b).

More than one-third (34.5%) of transit operating revenues in 2001 came from local and regional subsidies, and more than one-fifth (21.8%) from state subsidies. The federal government no longer provides operating subsidies to transit operators in large metropolitan areas, so in 2001 federal sources accounted for just 4.8% of transit operating revenues nationwide (American Public Transportation Association, 2003b).

In dramatic contrast to the small federal role in subsidizing transit operations, federal subsidies accounted for over half (50.5%) of all capital revenues in 2001. Local/regional funds from taxes, the sale of bonds, and the

like accounted for 40.1% of capital funding, with state subsidies covering the remaining 9.3% (American Public Transportation Association, 2003b).

The Causes of Transit's Ongoing Fiscal Distress

Despite decades of increasing public financial support, public transit systems in the United States continue to lose metropolitan travel market share. While overall ridership edged up 4.2% during the 1990s, transit's share of metropolitan person travel declined because private vehicle travel increased even faster. Further, inflation-adjusted public expenditures to support this modest growth in transit patronage increased 18.8% during the same period (Figure 11.6) (American Public Transportation Association, 2003a). Why has fiscal stability in the transit industry proven so elusive? Each of the five

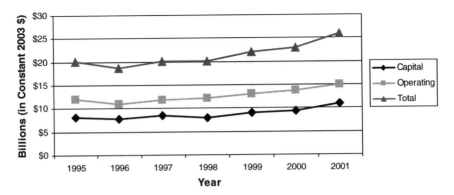

FIGURE 11.6. Trend in inflation-adjusted capital and operating subsidies, 1995–2001. Source: Adapted from American Public Transportation Association (2003b).

factors listed below helps to explain the chronic financial problems faced by U.S. public transit systems.

Loss of Markets and Market Share to Private Vehicles

At the beginning of the 20th century, motor vehicles were a recreational toy for the wealthy. At the dawn of the 21st century, they are a nearly indispensable tool of metropolitan survival. According to the U.S. Census, there were 34 motor vehicles for each 100 households in 1960. By the year 2000 this number had increased to 190 motor vehicles per 100 households (U.S. Department of Transportation, 2001). According to the 2001 National Household Transportation Survey, 86.4% of all metropolitan trips in 2000 were made by private vehicle, compared to just 3.2% on public transit (U.S. Department of Transportation, 2001).

In addition to an overall loss of market share, the types of trips taken on public transit have shifted over time. A hundred years ago, transit carried a large share of all trip types: for work, for shopping, for recreation, and so on. With the arrival of automobiles, recreational trips were among the first to shift from transit to autos. These were followed by shopping trips, personal errands, and most suburb-to-suburb travel. Today, public transit competes effec-

tively with private vehicles in two principal markets: central business district commuters who face parking charges at the destination and people who have limited access to automobiles because of age, income, or disability (Rosenbloom & Fielding, 1998).

Dependence on these two travel markets—downtown commuters and transit-dependents—has serious consequences for the cost of producing transit service. Downtown commuters typically travel inbound on workday mornings and outbound in the afternoons. Such trips are concentrated spatially (to and from large business districts), directionally (inbound in the morning and outbound in the afternoon), and temporally (the early mornings and late afternoons of workdays). In contrast, people with limited access to automobiles who depend on transit tend to be spatially concentrated in lower-income neighborhoods, and their use of transit is less directionally and temporally concentrated than downtown commuters (Rosenbloom & Fielding, 1998).

This spatial, directional, and temporal concentration in the demand for transit service has created serious peaking problems on transit systems. Many systems experience severe crowding on certain lines, especially during the rush hours, but low levels of demand at other locations, in other directions, and at other times. Thus many of the vehicles and drivers needed to satisfy the peak demand for transit service sit idle outside of the peaks. The peaking of transit demand affects costs in at least three ways:

Spatial Peaking. While the demand for transit service tends to be concentrated in and around downtowns and low-income areas, local political officials usually require that public transit operators provide a minimum level of service to all parts of their service areas, regardless of demand. Providing both adequate demand-driven peak service and acceptable policy-driven network coverage is expensive. And despite the spatial concentration of destinations for downtown commuters, and the relative spatial concentration in the location of transit-dependents, average trip lengths for all transit users have been increasing as a result of the ongoing growth and dispersion of metropolitan areas (Federal Transit Administration, 2001). In general, longer trips are more expensive to serve than shorter trips (though transit costs are more closely associated with trip time than trip distance). When trips are long, seats turn over more slowly, and more capacity is needed to accommodate a given number of passengers (Taylor, Garrett, & Iseki, 2000).

Directional Peaking. The demand for transit service tends to be highest heading into downtown and other major activity centers in the morning and away from these centers in the afternoon. The marginal cost of adding service in the peak direction tends to be high because this additional service will be underutilized when the transit vehicle makes its "backhaul" in the off-peak direction. As a result, a large share of transit capacity can go underutilized, even during rush hours. This directional peaking is particularly strong on express bus lines and commuter rail lines that make long express runs into business centers from outlying areas (Vigrass, 1992). Some commuter rail train sets, for example, are limited to just one inbound run each workday morning and one outbound run each workday afternoon. Thus the cost of the labor and equipment needed to provide this service is spread over relatively few passengers (Jones, 1985; Taylor et al., 2000).

Temporal Peaking. Rush-hour transit costs are higher because of the differences in *marginal costs* between peak and off-peak travel. As commute trips have come to form a larger and larger share of overall transit travel, the relative demand for travel during the midday, in the evening, and on the weekend has declined. The result is rush-hour crowding on many transit systems, with underutilized capacity at most other times. During off-peak periods, the demand for service is usually well below maximum system capacity; thus the marginal cost of adding additional service is low because extra vehicles and trained drivers are readily available. During peak periods, however, most of the available vehicle fleet and drivers are already in service; the marginal cost of adding peak service is high because additional service often requires additional vehicles and the hiring and training of additional drivers (Charles River Associates, Inc., 1989; Parody, Lovely, & Hsu, 1990; Saltzman, 1992; Taylor et al., 2000).

Public Regulation of Fares Discourages Efficiency

While the cost of carrying a transit trip can vary dramatically by direction, time, and mode, most transit systems employ flat fares and unlimited ride passes that make these cost variations invisible to the user. One of the earliest and most enduring effects of government regulation of private transit was the flat nickel fare. Not only was the fare generally a nickel, it was flat as well— meaning that the fare did not vary with distance. As cities spread, transit routes did as well, which means that the average fare paid per mile traveled has declined over time. Long distance suburban-to-central-city riders reap significant benefits from flat fare policies, but the flat fare has become such a fixture of public transit systems, it was—and is—very difficult to overcome politically (Jones, 1985; Koski, 1992; Wachs, 1989).

After decades of decline, inflation-adjusted fares edged up 5% during the 1990s, to an average of just under $1.00 per trip (American Public Transportation Association, 2003b). And the recent development of "smart cards" and other fare media have made it much easier to implement discounted off-peak pricing and distance-based transit fares (Fleishman, Shaw, Joshi, Freeze, & Oram, 1996). But in practice, systems tend toward wider varieties of unlimited ride passes (daily, weekly, monthly, etc.) and are actually moving away from transfer charges and distanced-based pricing. Between 1990 and 2001, the number of transit systems with transfer charges decreased from 29 to 20%, and those with some form of distance-based fares decreased from 39 to 32%. In 2001, just 7% of U.S. transit operators reported having some sort of peak-period surcharge (or, put another way, off-peak discounts) (American Public Transportation Association, 2003b).

Work Rules and Peaking

Low wages and long hours were common in the early streetcar era, so street railway workers were among the first to organize a national union. On many of the oldest and largest transit systems, labor structure and work rules in the transit industry were established very early. For example, two-person crews were a holdover from horsecar days when the teamster had to devote his full attention to the horse team. Increased automation allowed a single driver/operator in the later electric cars, but the switch to one-person crews was bitterly resisted. In some cases, transit companies switched lines to bus operation for no reason other than to eliminate two-person crews (Black, 1995; Jones, 1985; Saltzman, 1992).

Today, work rules on many larger transit systems stipulate full-time employment for a large share of the workforce. Such rules are increasingly at odds with the temporally uneven demand for transit service. If the drivers who provide extra service during the peak periods are only needed part-time but are required by labor agreements to be paid full-time, the costs of providing peak-period service increase even more. Research on the cost of labor allocation in public transit suggests that restrictive work rules do far more to increase transit costs than rising wages and benefits (Gray, 1992; McCullough, Taylor, & Wachs, 1998; Menzies et al., 2001; Saltzman, 1992)

Pressure to Expand with Cities

Rapidly expanding systems of street railways were enormously popular investments around the turn of the last century. Investors often would purchase undeveloped land on the urban fringe, subdivide the land for sale as homes or businesses, and then extend existing streetcar lines into the subdivided land to increase accessibility, and hence land values for sale. In addition to real estate development, many transit companies also retailed electricity, thus serving as the earliest electric utility companies (Jones, 1985; Saltzman, 1992).

Increasing auto ownership, rising incomes, and easy credit for home buyers have combined to expand home and lot sizes, as well as driving. These trends have increased the political pressure to extend transit lines to these newer, lower-density suburbs while maintaining flat fares. Today most new transit systems are formed in lower-density suburban areas. These systems, while often able to produce transit service at much lower costs than older, larger, central-city transit systems, typically have very low levels of ridership per unit of service. Yet the political pressure to extend unproductive service into outlying, low-density areas continues to increase (Kirby, 1992; Taylor & McCullough, 1998).

Chronic Overcapitalization

A century ago, even many conservatively managed streetcar companies suffered from overcapitalization. In addition to the speculative extensions of service as part of land development schemes described above, private transit systems often competed with one another to win the right from cities to be the exclusive operators of service along the busiest commercial corridors. To attract paying customers from the most heavily patronized routes, some systems built competing service along nearby parallel streets. To head off such competing service, many transit systems built otherwise unwarranted parallel service to prevent competitors from developing competing routes. In either case, the industry suffered from overexpansion, overcapitalization, and excess investor confidence in its early years (Jones, 1985; Saltzman, 1992).

In the more recent era of public transit subsidies, the transit industry has also suffered from overcapitalization, but for reasons related to the politics of transit finance discussed earlier in this chapter. From the start, federal subsidy of public transit systems has been strongly oriented toward capital (e.g., vehicles and facilities) and away from operations (e.g., wages, fuel, and maintenance). While federal dollars could be used to support operating costs, beginning in 1974 federal assistance was limited to 50% of operating costs, and in 1998 the operating assistance program was eliminated for large cities. In contrast, federal grants for transit facilities and equipment can cover up to 80% of costs, requiring only a 20% match in funds from state and local governments (Kirby, 1992; Taylor & Garrett, 1998). Many observers have argued that the political emphasis on capital expenditures has created inefficiencies in transit operations by encouraging systems to replace labor with capital even when it is uneconomic to do so. Most analysts agree that transit subsidies have stemmed ridership losses, but at the cost of declining productivity (Gomez-Ibanez, 1996; Pickrell, 1986; Wachs, 1989).

Measuring the Effects of Peaking on Transit Costs

Many transportation managers in the private sector might be surprised to learn that their public-sector counterparts often have limited information on the costs of providing public transit service. Airlines and private shipping companies, for example, develop highly sophisticated models to estimate how the cost of carrying passengers or freight varies by season, day of the week, time of day, direction, and mode.

In contrast, public transit managers typically have only rudimentary information linking budgetary inputs to service outputs. One might argue that, as publically subsidized services, transit systems need not be as concerned with such fine-grained cost-estimation detail as profit-driven private businesses. But the broad social policy objectives of public transit do not obviate the need for good cost information to guide managers, transit policy boards, and funding agencies. For example, most policy boards adopt fare structures without a clear understanding of how the cost of service varies from passenger to passenger or trip to trip. Similarly, in making decisions on adding or deleting peak-period or off-peak service, transit managers and boards often may have limited or incomplete information regarding the cost or savings from such changes.

In addition to the variation in transit costs owing to peaking discussed above, the cost of various modes of transit service—buses, trains, vans, and so on—also varies significantly. Yet the techniques employed by most public transit operators to estimate these costs do not account for this variability, nor are they structured to distinguish the estimation of overall costs from those at the margin.

Data from a recent study of costs at the Los Angeles Metropolitan Transporation Authority (LA MTA) illustrates this point. A typical model used to estimate the costs of producing transit service was compared to a more sophisticated model that also takes into consideration (1) annualized vehicle and nonvehicle capital costs, (2) the higher seating capacity of streetcars (now called "light-rail cars"), and (3) variance in costs by time of day. The findings of this comparison are summarized in Figure 11.7.

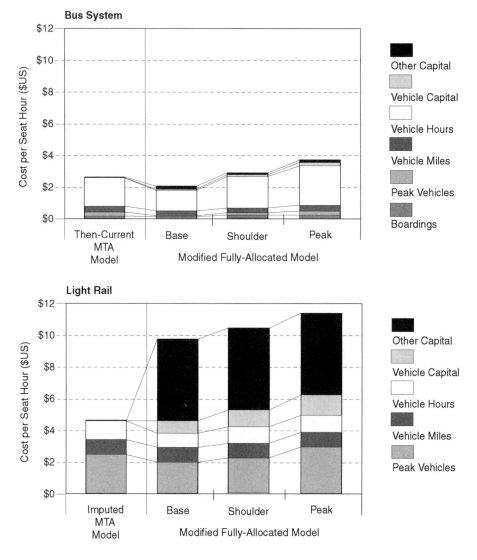

FIGURE 11.7. Comparison of estimated bus system and light-rail costs using the LA MTA Model and a Fully-Allocated Model. Source: Taylor, Garrett, and Iseki (2000).

This comparison suggests that transit operators tend to overestimate the costs of providing off-peak service and underestimate the costs of peak service. In addition, by ignoring the (mostly federally subsidized) capital costs of rail lines, transit operators tend to substantially underestimate the costs of rail service. Why the significant differences in capital costs between buses and rail? Buses operate on streets and highways paid largely by others: property owners (via property taxes) for local streets and private vehicle operators (via motor fuels taxes) for highways. For the light-rail line in this example, the cost of right-of-way, track, catenary, and stations (which were annualized over the life of the investment) are financed almost exclusively with transit subsidies.

Since nearly three-quarters of all transit operating revenues nationwide are locally generated (from the fare box and from local and regional subsidies), while over half of the capital funding for transit capital comes from an outside source (the federal government), transit operators tend to view capital costs as "cheaper" than operating costs, which encourages overcapitalization.

The Geopolitics of Transit Subsidies

The geopolitical equity imperative in transportation finance described early in this chapter has shaped public transit service in two important ways. First (discussed in the previous section) is the tendency to subsidize voters and jurisdictions somewhat equally, but transit passengers rather unequally. This is because transit passengers tend to be concentrated in low-income neighborhoods and in the downtowns of the oldest and largest cities. More than one-third of all transit passengers in the United States are in the New York metropolitan area, and over 60% of all transit passengers are carried on the 10 largest U.S. transit systems (Taylor & McCullough, 1998). Second is that transit policy, and in particular federal transit policy, has strongly favored capital expenditures over operating expenditures.

Why the emphasis on capital subsidies? There are at least three reasons. First, federal capital assistance was initially conceived as a short-term infusion of capital to help restore a deteriorating and depleted industry. Proponents of federal transit subsidies argued that capital subsidies would enable transit to compete more effectively with the private automobile and address social goals that could not be readily achieved via the private market, such as income redistribution, the reduction of air pollution, and energy savings. Operating subsidies, on the other hand, were seen as an ongoing commitment to support the day-to-day operations of local transit systems. Critics of federal involvement in transit subsidies feared that operating support would remove incentives to control costs and lead to a cycle of dependence on ever increasing operating subsidies (Black, 1995; Li & Taylor, 1998; Wachs, 1995; Weiner, 1992).

Second, the long-standing preference for capital subsidies over operating subsidies entails more than just concerns over cost control and dependence. It is perhaps equally important that capital grants have more "political caché" than operating grants. Major transit projects—especially new rail transit projects, funded largely with federal capital subsidies—have proven popular with local voters, local officials, and local members of Congress. New rail transit lines, busways, and transit vehicle maintenance facilities are high-profile public investments that fulfill political promises and reward places (see Figure 11.8). Ribbon cuttings at new rail transit stations attract media attention; more frequent service on a local bus line resulting from increased federal operating support generally does not.

Third, concerns over reducing federal expenditures have also worked against federal operating subsidies. The National Mass Transportation Assistance Act of 1974 provided federal assistance for transit operations for the first time. Between 1975 and 1980, federal operating subsidies escalated from $300 million to $1.1 billion per year (Li & Taylor, 1998). The rapid increase in federal expenditures was cause for concern in the Reagan administration, Congress, and among analysts in the early 1980s.

FIGURE 11.8. New streetcar services, like this one in Los Angeles, have proven politically popular despite high construction costs.

Many critics complained that federal operating subsidies permitted a rapid escalation of costs that disproportionally benefitted transit operators and their employees, instead of providing more, better, or cheaper service for transit riders (Li & Taylor, 1998; Pickrell, 1986). As a result, there has been strong pressure since the late 1970s to cap or curtail federal operating support, such that in 1998 federal operating subsidies for transit were eliminated in metropolitan areas with more than 200,000 population (Garrett & Taylor, 1999; Price Waterhouse LLP, Multisystems Inc., & Mundle & Associates, 1998).

The Waxing Emphasis on Capital Subsidies

In contrast to the recent phase-out of operating support for public transit, federal support for transit capital expenditures—vehicles, equipment, stations, and the like—has increased. The 1998 federal surface transportation legislation (TEA-21) dramatically increased federal subsidy of transit capital projects specifically earmarked by legislators in the bill (Garrett & Taylor, 1999). Many argue that the emphasis on capital expenditures in the federal program has created little incentive for transit operators to seek more efficient, less capital-intensive services. For example, Pickrell (1992) reviewed the costs and performance of rail transit projects in eight U.S. cities built with substantial federal capital subsidy over the past two decades and found that many of the projects cost much more than initial estimations and attracted far fewer riders than projected. He concluded that the emphasis in the federal transit program on capital expenditures, particularly the small local matching requirement for many of the new rail transit projects, discourages local officials from carefully considering less costly alternatives to large new transit projects.

Nationwide, buses carry over half again (52.9%) as many passengers as all rail transit modes combined, and most rail transit passengers are in the New York metropolitan area. But for years transit capital subsidies have been skewed toward rail—mostly to new lines and systems outside of the densely developed, transit-friendly New York area. For example, while more than two-thirds (71.1%) of all operating subsidies supported bus service in 2001, more than two-thirds (68.4%) of all capital subsidies went to rail. As a result, total revenue vehicle miles of bus service nationwide increased just 17.3% between 1991 and 2001, while total rail transit service increased 22.0% during the same period (Federal Transit Administration, 2001). These patterns are summarized in Figure 11.9, which presents the share of passenger trips, capital revenues, and operating revenues during the 10-year period from 1992 to 2001.

Discretionary federal funding for transit also favors capital-intensive programs. Since 1998, rail modernization and new starts have received 80% of the funding, while bus capital projects have received the remaining 20%. For example, of the $3 billion in budget appropriations for the 2003 fiscal year, the Federal Transit Administration allocated $2.4 billion to the fixed guideway rail modernization and new rail starts programs but just $607 million to bus-related projects (Federal Transit Administration, 2003).

In addition to federal funding mechanisms, the states also contribute significantly to transit finance, frequently with strong geographic restrictions on the allocation of transit subsidies. The State of California, for example, subsidizes transit but does not allow funds from sales taxes collected in one county to be expended in another county, and within counties (with one exception) state law distributes transit funds based on the service area population only, not its ridership. Since larger, more densely populated areas have a higher percentage of transit riders, the transit subsidy allocation rules favor smaller, more suburban areas with low levels of transit ridership. As a result, transit subsidies tend to be distributed somewhat

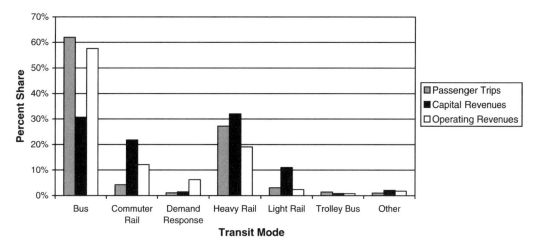

FIGURE 11.9. U.S. passenger trips, capital and operating revenues by mode, 1992–2001. Source: Adapted from American Public Transportation Association (2003a, 2003b).

equally among voters, but quite unequally among transit patrons (Taylor, 1991).

Recap: The Geography of Transit Subsidies

The geopolitics of public transit finance are at odds with the geography of transit ridership. Transit patronage today is spatially concentrated on the largest transit systems operating in the centers of the oldest U.S. cities. Further, transit patrons are disproportionately low-income bus riders (especially outside of New York City) (Giuliano, Hu, & Lee, 2001). The geopolitics of transit finance, however, make it all but impossible to spatially concentrate transit subsidies and service commensurate with transit use. Instead, transit funds are spread across jurisdictions and expended disproportionately on politically visible capital projects.

The combined effect of federal and state transit subsidy policies is that many growing southern and western cities around the United States—like Miami, Portland (Oregon), and San Jose—have built new, heavily subsidized rail transit systems in recent years, while most rail transit passengers remain concentrated in the New York area. Furthermore, within metropolitan areas

there is strong political pressure to spread transit subsidies around the region, which has significantly increased lightly patronized suburban transit services. The popularity of transit subsidy programs suggest that they are performing well politically. But by the measures of transportation system efficiency, effectiveness, and equity outlined in Table 11.1, they emphasize both suburban and high-cost services that provide most public transit patrons (primarily CBD-bound commuters and people without cars) with disappointingly little "bang for the buck."

CONCLUSION: LOOKING TO THE FUTURE

Given this overview of the geography of urban transportation finance, what trends do we see as we look to the future? First, and perhaps foremost, the geopolitics of transportation finance are firmly established at all levels of government and are likely to remain so for the foreseeable future. This means that public investment in transportation systems will continue to be determined more by political struggles over the spatial distribution of resources than by other factors such as transportation system use, costs imposed, or ability to pay.

There have been a few notable challenges to the prevailing focus on big-ticket transportation capital projects. "Freeway revolts" by citizens and environmental advocacy groups in cities around the United States during the 1960s and 1970s halted some high-profile freeway projects in cities like Boston, New Orleans, Reno, and San Francisco. These social movements contributed to a weakening of political support for funding metropolitan freeway development and a shift toward increased investment in public transit systems (Taylor, 1995). More recently, transit passenger advocacy organizations have challenged fare increases and service cuts in Atlanta, Detroit, New York, Philadelphia, and, most notably, Los Angeles. In 1994, a coalition of civil rights and advocacy organizations filed a federal civil rights lawsuit against the LA MTA over a proposed fare increase. The plaintiffs alleged that the LA MTA was increasing fares and short-shrifting investments in the local bus system (which was disproportionately patronized by the poor and by racial-ethnic minorities), while pouring resources into new rail lines and commuter-oriented services (which disproportionately benefitted higher-income, nonminority riders). While the LA MTA denied the allegations, the suit was settled with a consent decree in 1996, which, among other things, limited future fare increases and mandated improvements to local bus service for 10 years (Brown, 1998; Grengs, 2002).

Despite these high-profile challenges to the status quo, however, the emphasis on transportation capital investments over spending on operations and maintenance also shows little sign of abating. While the overcapitalization of transportation systems is clearly economically inefficient, it is politically popular and likely to remain so. This means that transportation investments will continue to emphasize projects over programs, lines and modes over integrated systems, and expenditures over savings.

Finally, the trend away from user fees (like the motor fuels tax and distance-based transit fares) is also likely to continue, at least in the short term. This means that funding for transportation will come increasingly from general public finance instruments like general obligation bonds and sales taxes. This trend is popular with elected officials and voters, but it is likely to reduce both the economic efficiency and the social equity of transportation systems. Economic efficiency is diminished because bonds and sales taxes separate the prices travelers pay from the costs they impose on society. Social equity is reduced because the relative burden of transportation finance is shifted toward lower-income people (who pay proportionally more in sales taxes) and less extensive users of transportation systems.

Over the longer term, however, it is possible that both the introduction of new propulsion technologies and electronic payment media will begin to turn the tide back toward user fees. New propulsion technologies, like gas–electric hybrid engines and fuel cells, will further erode the ability of motor fuel taxes to keep pace with the growth in vehicle travel. And smart card payment media that can remotely deduct payments as travelers pass through fare gates or under payment gantries on roads are likely to become widespread. In concert, these developments could usher in a new era of easy-to-pay and inexpensive-to-collect user charges to replace the fuel, sales, and other taxes we currently use to finance transportation. Such payment systems could allow marginal-cost pricing of roads, transit services, and so on to reflect the costs individual travelers impose on transportation systems, on other users, and on society. This, in turn, would encourage travelers to use transportation systems in places and at times when excess capacity exists, and would signal travelers to be more judicious in their consumption of roads and transit seats in places and at times of peak demand. Such changes would increase both economic efficiency and social equity of transportation finance, but widespread implementation of such user charge systems remains a decade or more away.

In this chapter we have "followed the money" to argue that the political geography of transportation finance has enormous influence on the deployment of urban transportation systems, travel in cities, and patterns of metropolitan development. We reviewed (1) the economic and political factors guiding public investments in transportation systems, (2) a schematic overview of urban transportation finance, (3) the role of fiscal politics in shaping metropolitan freeway development, and (4) the role of finance in shaping urban public transit systems. Collectively, these reviews have shown that the geography of urban transportation is to a large degree determined by the geography of urban transportation finance.

NOTE

1. Defined as the percentage of operating costs covered by fare revenues.

2. In some cases, data for local subsidies had to be estimated. Details available from author.

ACKNOWLEDGMENT

Norman Wong of the UCLA Institute of Transportation Studies assisted with the preparation of this chapter, including data collection, figure and table construction, and photography. I am very grateful for Mr. Wong's able research assistance.

REFERENCES

Adler, S. (1991). The transformation of the Pacific Electric Railway: Bradford Snell, Roger Rabbit, and the politics of transportation in Los Angeles. *Urban Affairs Quarterly, 27*, 51–86.

Altshuler, A. (1965). *Locating the intercity freeway.* Indianapolis: Published for the ICP by Bobbs-Merrill.

American Public Transportation Association. (2003a). *Public transportation ridership statistics.* Available online at http://www.apta.com/research/stats/ridershp

American Public Transportation Association. (2003b). *Transit statistics.* Available online at http://www.apta.com/research/stats

American Public Transportation Authority. (2000). *2000 transit fact book.* Washington, DC: Author.

Bernstein, C., & Woodward, B. (1974). *All the president's men.* New York: Simon & Schuster.

Black, A. (1995). *The history of urban transit: Urban mass transportation planning.* New York: McGraw-Hill.

Boarnet, M. G. (1997). Highways and economic productivity: Interpreting recent evidence. *Journal of Planning Literature, 11*, 476–486.

Bottles, S. L. (1987). *Los Angeles and the automobile: The making of a modern city.* Berkeley and Los Angeles: University of California Press.

Boxer, B. (1998, July 14). *Boxer wins funding for Sacramento South Rail Extension: Senate Appropriations Committee approves $23.48 million for project.* Press Release, Senate Office of Barbara Boxer (D-CA).

Brodsly, D. (1981). *L.A. Freeway: An appreciative essay.* Berkeley and Los Angeles: University of California Press.

Brown, J. (1998). Race, class, gender, and public transportation planning: Lessons from the Bus Riders Union lawsuit. *Critical Planning, 5*, 3–20.

Brown, J. (2001). Reconsider the gas tax: Paying for what you get. *Access, 19*, 10–15.

Brown, J. (2002). Statewide transportation planning: Lessons from California. *Transportation Quarterly, 56*, 51–62.

Brown, J., DiFrancia, M., Hill, M. C., Law, P., Olson, J., Taylor, B. D., Wachs, M., & Weinstein, A. (1999). *The future of California highway finance.* Berkeley: California Policy Research Center.

Burch, P. H. (1962). *Highway revenue and expenditure policy in the United States.* New Brunswick, NJ: Rutgers University Press.

California Department of Transportation. (1997). *Transportation financing—Vehicle miles traveled (VMT) measurement and assessment.* Sacramento: California Department of Transportation, Transportation Planning Program.

Charles River Associates Incorporated. (1989). *Transit deficits: Peak and off-peak comparisons* (CRA Report No. 784.30C). Washington, DC: U.S. Department of Transportation.

Chernick, H. A., & Reschovsky, A. (1997). Who pays the gasoline tax? *National Tax Journal, 50*, 233–259.

Clark, A. C. (1955, June 9). *Criteria for selection of additional Interstate System routes at urban areas.*

Circular memorandum to Division and District Engineers, U.S. Bureau of Public Roads, Department of Commerce.

Collier-Burns Highway Act of 1947. (1947). Sacramento, CA: State Printing Office.

Congressional Quarterly. (1964). *Congress and the nation: A review of government and politics* (Vol. 1). Washington, DC: Congressional Quarterly Service.

Deakin, E., & Harvey, G. (1995). *Transportation pricing strategies for California: An assessment of congestion, emissions, energy and equity impacts* (Draft Final Report). Sacramento: California Air Resources Board.

Dixon, J. (1997, January 27). Response by Congressman Dixon to "Perspective on Metro Rail: Red Line Should Follow Wilshire." *Los Angeles Times*, p. B7.

Fairbank, H. S. (1937). Objects and methods of the State-wide Highway Planning Surveys. *American Highways, 16.*

Federal Transit Administration. (2001). *National transit summaries and trends.* Available online at http://www.ntdprogram.com/NTD/ NTST.nsf/NTST/2001/$File/01NTST.pdf

Federal Transit Administration. (2003). Department of Transportation Federal Transit Administration FY 2001 thru FY 2003 budget. Available at http://www.fta.dot.gov/office/ program/2003/budget.html

Fleishman, D., Shaw, N., Joshi, A., Freeze, R., & Oram, R. (1996). *Fare policies, structures, and technologies* (TCRP Report No. 10). Washington, DC: Transportation Research Board.

Forkenbrock, D. J., Pogue, T. F., Foster, N. S. J., & Finnegan, D. J. (1990). *Road investment to foster local economic development.* Iowa City: Midwest Transportation Center, University of Iowa.

Forkenbrock, D. J., & Schweitzer, L. A. (1997). Intelligent transportation systems and highway finance in the 21st century. In *Transportation finance for the 21st century, TRB Conference Proceedings 15* (pp. 73–82). Iowa City: University of Iowa, Public Policy Center.

Garrett, M., & Taylor, B. D. (1999). Reconsidering social equity in public transit. *Berkeley Planning Journal, 13,* 6–27.

General Motors Corporation. (1974). The truth about *American ground transport*: A reply by General Motors. In *Hearings before the Subcommittee on Antitrust and Monopoly of the Committee on the Judiciary* (Appendix to Part A, pp. A107–A127). Washington, DC: U.S. Senate.

Gifford, J. L. (1984). The innovation of the Interstate Highway System. *Transportation Research A, 18A,* 319–332.

Gillen, D. (1997, December). *Evaluating economic benefits of the transportation system.* Paper presented at the conference on Transportation and the Economy: The Transportation/Land Use/Air Quality Connection, Lake Arrowhead, CA.

Giuliano, G., Hu, H., & Lee, K. (2001). *The role of public transit in the mobility of low income households, final report.* Los Angeles: Metrans Transportation Center.

Goldman, T., & Wachs, M. (2003). A quiet revolution in transportation finance: The rise of local option transportation taxes. *Transportation Quarterly, 57,* 19–32.

Gomez-Ibanez, J. A. (1996). Big-city transit, ridership, deficits, and politics. *Journal of the American Planning Association, 62,* 30–50.

Gray, G. E. (1992). Perceptions of public transportation. In G. E. Gray & L. A. Hoel (Eds.), *Public transportation* (2nd ed., pp. 617–633). Englewood Cliffs, NJ: Prentice-Hall.

Grengs, J. (2002). Community-based planning as a source of political change: The transit equity movement of Los Angeles' Bus Riders Union. *Journal of the American Planning Association, 68*(2), 165–178.

Hill, M. C., Taylor, B. D., Weinstein, A., & Wachs, M. (2000). Assessing the need for highways. *Transportation Quarterly, 54,* 93–103.

Hilton, G. W. (1985). The rise and fall of monopolized transit. In C. A. Lave (Ed.), *Urban transit: The private challenge to public transportation* (pp. 31–48). Cambridge, MA: Ballinger.

Interregional Highway Committee. (1944). *Interregional highways* (House Document 379, 78th Congress, 2nd Session). Washington, DC: U.S. Government Printing Office.

Jones, D. W. (1985). *Urban public transit: Economic and political history.* Englewood Cliffs, NJ: Prentice-Hall.

Jones, D. W. (1989). *California's freeway era in historical perspective.* Berkeley: University of California, Institute of Transportation Studies.

Kirby, R. F. (1992). Financing public transportation. In G. E. Gray & L. A. Hoel (Eds.), *Public transportation* (2nd ed., pp. 445–460). Englewood Cliffs, NJ: Prentice-Hall.

Koski, R. W. (1992). Bus transit. In G. E. Gray &

L. A. Hoel (Eds.), *Public transportation* (2nd ed., pp. 148–176). Englewood Cliffs, NJ: Prentice-Hall.

Kurtz, H. (1983, April 18). Buffalo: Great dreams, but greater needs. *Washington Post*, p. A4.

Lawlor, M. J. (1995). Federal urban mass transportation funding and the case of the Second Avenue Subway. *Transportation Quarterly, 49*, 43–54.

Lem, L. L. (1997). Dividing the federal pie. *Access, 10*, 10–14.

Levy, F., Meltsner, A. J., & Wildavsky, A. (1974). *Urban outcomes: Schools, streets, and libraries.* Berkeley: University of California Press.

Lewis, D. (1991). *Primer on transportation, productivity, and economic development* (National Cooperative Highway Research Program Report No. 342). Washington, DC: Transportation Research Board.

Li, J., & Taylor, B. D. (1998). Outlay rates and the politics of capital versus operating subsidies in federal transit finance. *Transportation Research Record, 1618*, 78–86.

MacDonald, T. H. (1936). *Roads we should have.* Address to the annual meeting of the Councillors of the American Automobile Association, Washington, DC.

MacDonald, T. H. (1954). *The engineers' relation to highway transportation* (Sixth Salzberg Memorial Lecture). College Station: Texas A&M University.

Mastako, K. A., Rillet, L. R., & Sullivan, E. C. (1998). Commuter behavior on California State Route 91 after introducing variable-toll express lanes. *Transportation Research Record, 1649*, 47–54.

McCullough, W. S., Taylor, B. D., & Wachs, M. (1998). Transit service contracting and cost efficiency. *Transportation Research Record, 1618*, 69–77.

Metropolitan Transportation Commission. (2003). *2002 annual report.* Available online at http://www.mtc.ca.gov/publications/AnnualReport-02/MTC_02_Annual_Report.pdf

Menzies, T. R. Jr., Baker, J. B., Gilbert, G., Grande, S. A., Marsella, C. W. Jr., McLary, J. J., Pettus, C. L., Piras, P., Sclar, E. D., Tauss, R., Taylor, B. D., Teal, R., & Wilson, N. H. M. (2001). *Contracting for bus and demand-responsive transit services: A survey of U.S. practice and experience* (Transportation Research Board Special Report No. 258). Washington, DC: National Academy Press.

Murphy, J. J., & Delucchi, M. A. (1998). Review of the literature on the social cost of motor vehicle use in the United States. *Journal of Transportation and Statistics, 1*, 15–42.

Musgrave, R. A. (1959). *The theory of public finance.* New York: McGraw-Hill.

National Cooperative Highway Research Program. (1994). *Alternatives to the motor fuel tax for financing surface transportation improvements* (Draft Summary Report, NCHRP 20–24[7]). Washington, DC: Transportation Research Board.

Norman, J. (1996, February 5). Field of dreams' thinking criticized: Rail not enough for jobs, broad strategy needed for rapid transit to boost poor area, report says. *Milwaukee Journal Sentinel*, p. 2.

Parody, T. E., Lovely, M. E., & Hsu, D. S. (1990). Net cost of peak and off-peak transit trips taken nationwide by mode. *Transportation Research Record, 1266*, 139–145.

Pickrell, D. H. (1986). *Federal operating assistance for urban mass transit: Assessing a decade of experience* (Transportation Research Record No. 1078). Washington, DC: Transportation Research Board.

Pickrell, D. H. (1992). A desire named streetcar: Fantasy and fact in rail transit planning. *Journal of the American Planning Association, 58*, 158–176.

Plane, D. A. (1995). Urban transportation: Policy alternatives. In S. Hanson (Ed.), *The geography of urban transportation* (2nd ed., pp. 445–447). New York: Guilford Press.

Poole, R. W. Jr. (2001). *Commercializing highways: A "road -utility" paradigm for the 21st century.* Los Angeles: Reason Public Policy Institute.

Poterba, J. M. (1991). Is the gasoline tax regressive? *Tax Policy and the Economy, 5*, 145–164.

Pozdena, R. J. (1995). *Where the rubber meets the road: Reforming California's roadway system.* Los Angeles: Reason Foundation.

Price Waterhouse LLP, Multisystems Inc., & Mundle & Associates. (1998). *Funding strategies for public transportation* (TCRP Report No. 31). Washington, DC: Transportation Research Board. Vols. 1 and 2 available online at http://nationalacademies.org/trb/publications/tcrp/tcrp_rpt_31-1-a.pdf and http://nationalacademies.org/trb/publications/tcrp/tcrp_rpt_31-1-b.pdf

Puget Sound Regional Council. (1997). *The effects of the current transportation finance structure*

(Expert Review Draft, Paper 2 of a Series on Transportation Financing). Seattle: Author.

Purcell, C. H. (1940a, May). *California highway program requires more federal aid for projects within cities—Part I.* Sacramento: California Highways and Public Works.

Purcell, C. H. (1940b, June). *California highway program requires more federal aid for projects within cities—Part II.* Sacramento: California Highways and Public Works.

Richmond, J. (in press). *Transport of delight: The mythical conception of rail transit in Los Angeles.* Akron, OH: University of Akron Press.

Romano, M. (1998, June 2). State highway spending soars: In five years Colorado building costs to nearly $1 billion for 28 big-ticket projects. *Denver Rocky Mountain News,* p. 5A.

Rosenbloom, S., & Fielding, G. J. (1998). *Transit markets of the future* (TCRP Report No. 28). Washington, DC: Transportation Research Board.

Roth, G. (1998). *Roads in a market economy.* London: Avebury Press.

Rufolo, A. M., & Bertini, R. L. (2003). Designing alternatives to state motor fuel taxes. *Transportation Quarterly, 57,* 33–46.

Saltzman, A. (1992). Public transportation in the 20th century. In G. E. Gray & L. A. Hoel (Eds.), *Public transportation* (2nd ed., pp. 24–45). Englewood Cliffs, NJ: Prentice-Hall.

Seely, B. E. (1987). *Building the American highway system: Engineers as policy makers.* Philadelphia: Temple University Press.

Shoup, D. (1997). Evaluating the effects of cashing out employer-paid parking: Eight case studies. *Transport Policy, 4*(4), 201–216.

Shoup, D. (1999). The trouble with minimum parking requirements. *Transportation Research Part A: Policy and Practice, 33A*(7/8), 549–574.

Slater, C. (1997). General Motors and the demise of streetcars. *Transportation Quarterly, 51,* 45–66.

Small, K. A., Winston, C., & Evans, C. (1989). *Road work: A new highway pricing and investment policy.* Washington D.C.: Brookings Institution Press.

Snell, B. C. (1974). American ground transport: A proposal for restructuring the automobile, truck, bus, and rail industries. *Hearings before the Subcommittee on Antitrust and Monopoly of the Committee on the Judiciary* (Appendix to Part A, p. A1–A103). Washington, DC: U.S. Senate.

Sullivan, E. (2000). *Continuation study to evaluate the impacts of the SR-91 value-priced express lanes: Final report.* Sacramento: California Department of Transportation.

Sullivan, E. C., & El Harake, J. (1998). The CA Route 91 toll lanes—Impacts and other observations. *Transportation Research Board Record, 1649,* 47–54.

Surface Transportation Policy Project. (1998). *TEA-21 users guide: Making the most of the new transportation bill.* Washington, DC: Author.

Taylor, B. D. (1991). Unjust equity: An examination of California's transportation development act. *Transportation Research Record, 1297,* 85–92.

Taylor, B. D. (1995). Program performance versus system performance: An explanation for the ineffectiveness of performance-based transit subsidy programs. *Transportation Research Record: Public Transit, 1496,* 43–51.

Taylor, B. D. (2000). When finance leads planning: Urban planning, highway planning, and metropolitan freeways. *Journal of Planning Education and Research, 20*(2), 196–214.

Taylor, B. D., & Garrett, M. (1998, October). *Equity planning in the 90s: A case study of the Los Angeles MTA.* Paper presented at the annual conference of the Association of Collegiate Schools of Planning, Pasadena, CA.

Taylor, B. D., Garrett, M., & Iseki, H. (2000). Measuring cost variability in the provision of transit service. *Journal of the Transportation Research Board, 1735,* 101–112.

Taylor, B. D., & McCullough, W. S. (1998). Lost riders. *Access, 13,* 26–31.

U.S. Bureau of the Census. (2003). *Population clock.* Available online at http://www.census.gov

U.S. Bureau of Public Roads. (1955). *General location of national system of Interstate Highways: Including all additional routes at urban areas designated in September 1955.* Washington, DC: U.S. Department of Commerce.

U.S. Congressional Budget Office. (1978). *Highway assistance programs: A historical perspective* (Background Paper, Congress of the United States). Washington, DC: U.S. Government Printing Office.

U.S. Department of Transportation. (1994). *1990 National Personal Transportation Surveys.* Available on line at http://ntps.ornl.gov/npts/1990/index.html

U.S. Department of Transportation. (1998).

TEA-21: Moving Americans into the 21st century. Washington, DC: U.S. Department of Transportation. Available online at http://www.fhwa.dot.gov/tea21/index.htm

U.S. Department of Transportation. (2001). *2001 National Household Travel Survey*. Available online at http://nhts.ornl.gov/

U.S. Department of Transportation, Bureau of Transportation Statistics. (2002, July). *National transportation statistics 2001* (BTS02-06). Washington, DC: U.S. Government Printing Office, July 2002. Available online at http://www.bts.gov/publications/national_transportation_statistics/2001/

U.S. Department of Transportation, Bureau of Transportation Statistics. (2003). *Pocket guide to transportation 2003* (BTS03-01). Washington, DC: Author. Available online at http://www.bts.gov/publications/pocket_guide_to_transportation/2003/index.html

U.S. Department of Transportation, Federal Highway Administration. (1997a). *Federal highway cost allocation study final report*. Washington, DC: Author.

U.S. Department of Transportation, Federal Highway Administration. (1997b). *Highway statistics summary to 1995*. Available online at http://www.fhwa.dot.gov/ohim/summary95

U.S. Department of Transportation, Federal Highway Administration. (2000). *Highway statistics 2000*. Available online at http://www.fhwa.dot.gov/ohim/hs00

U.S. Department of Transportation, Federal Highway Administration. (2001). *Highway statistics 2001*. Available online at http://www.fhwa.dot.gov/ohim/hs01

U.S. Department of Transportation, Federal Highway Administration. (2002a). *Highway statistics 2001*. Available online at http://www.fhwa.dot.gov/ohim/hs01/hf2.htm

U.S. Department of Transportation, Federal Highway Administration. (2002b). *Value pricing pilot program* (FHWA-PL-99-014 HPTS/3-99[5M]E). Available online at http://www.fhwa.dot.gov/policy/vppp.htm

Vigrass, W. J. (1992). Rail transit. In G. E. Gray & L. A. Hoel (Eds.), *Public transportation* (2nd ed., pp. 114–147). Englewood Cliffs, NJ: Prentice-Hall.

Wachs, M. (1989). U.S. transit subsidy policy: In need of reform. *Science, 244*, 1545–1549.

Wachs, M. (1995). The political context of transportation policy. In S. Hanson (Ed.), *The geography of urban transportation* (2nd ed., pp. 269–285). New York: Guilford Press.

Wachs, M. (2003). Commentary: A dozen reasons to raise the gas tax. *Public Works Management and Policy, 7*, 235–242.

Weiner, E. (1992). History of urban transportation planning. In G. E. Gray & L. A. Hoel (Eds.), *Public transportation* (2nd ed., pp. 46–76). Englewood Cliffs, NJ: Prentice-Hall.

Whitt, J. A. (1982). *Urban elites and mass transportation: The dialectics of power*. Princeton, NJ: Princeton University Press.

Whitt, J. A., & Yago, G. (1985). Corporate strategies and the decline of transit in U.S. cities. *Urban Affairs Quarterly, 21*, 37–65.

Wiese, A., Rose, R., & Schluter, G. (1995). Motor-fuel taxes and household welfare: An applied general equilibrium analysis. *Land Economics, 71*, 229–249.

Winston, C. M., & Shirley, C. (1998). *Alternate route: Toward efficient urban transportation*. Washington, DC: Brookings Institution Press.

Yago, G. (1984). *The decline of transit*. New York: Cambridge University Press.

CHAPTER 12

Social and Environmental Justice Issues in Urban Transportation

DEVAJYOTI DEKA

Transportation-related expenses constitute a significant portion of household income and an especially large proportion of the incomes of low-income households. Despite these expenditures, some segments of society lack adequate access to employment, health care, shopping, and recreation. Although government subsidizes components of the urban transportation system, there is an increasing concern that the subsidies are not reaching those in need. Recent years have also seen a growing recognition that certain segments of society are disproportionately affected by transportation-related pollution. Because of these concerns, a study of the social and environmental justice issues in urban transportation is important.

As the concerns about social and environmental justice in urban transportation pertain primarily to the poor and minorities, namely, African Americans and Hispanics, this chapter focuses on these groups. Data are analyzed separately for income, racial, and ethnic groups, and intergroup comparisons are made whenever necessary. The first half of the chapter deals with social justice, while the second half deals with

environmental justice. I begin with a general description of the notion of social justice and then take up some specific social and environmental justice issues in subsequent sections of the chapter. I argue that social and environmental justice have been largely overlooked by the urban transportation planning process, even though a segment of the urban transportation literature has dealt with related issues for a long time. Because many existing urban transportation problems were created by the urban planning and transportation planning processes themselves, those involved in these processes should consider the long-term consequences of a plan in terms of both efficiency and justice, keeping in mind that the United States is a pluralistic society where different groups have differing needs and interests. Although the focus here is on the United States, the concepts and principles apply in other national contexts as well.

VIEWS ON SOCIAL JUSTICE

To most people, "social justice" means fairness in the distribution of goods, services, rights, and opportunities. Fairness is a value-laden term, however. For example, one group may believe that affirmative action is just and fair, while another group feels that it is not. Similarly, some people may think that it is fair to subsidize rail transit, but others may disagree.

Although there is no single, generally accepted standard of measurement for social justice, its various doctrines help us understand its main features or concerns. According to Barr (1998), there are three broad sets of views on social justice: the libertarian, the liberal, and the collectivist views.

Libertarians generally prefer the endowment-based criterion of social justice, which favors a reward system that lacks public intervention in the redistribution of private endowment. According to Barr, there are two libertarian schools of thought, namely, the natural rights libertarians and the empirical (or, new rights) libertarians. The natural rights libertarians believe that individuals have absolute rights to their earnings and inheritance, and hence government should not interfere under any circumstances. The empirical libertarians are also opposed to government intervention in private matters, but their opposition is based on the belief that government intervention reduces social welfare.

Liberals are of two types, namely, utilitarians and Rawlsians. Utilitarians believe that goods, services, rights, freedom, and political power should be distributed among individuals in a manner that maximizes the total utility for society. They therefore espouse the idea of the greatest benefit to the most people. Rawlsians, on the other hand, believe that social justice is useful for the survival of social institutions. According to Rawls (2001), the two main principles of social justice are the greatest benefit to the least advantaged and equal basic liberties and opportunities for all. Rawlsians assume that individuals make decisions under a veil of ignorance (i.e., ignorance about their relative position in society), and because of their risk-averse nature they adopt the principle of greatest benefit to the least advan-

taged (because they themselves may end up disadvantaged at some point).

To the *collectivists*, equality, fraternity, and freedom are the main principles of social justice (Barr, 1998). Of these, equality is perhaps the most central and the most controversial principle. To many collectivists, equality of *opportunity* is not enough—they promulgate the need for equality of *outcome*. Yet even the collectivists do not necessarily espouse the idea of complete equality of well-being. When needs for basic goods—such as food, clothing, and shelter—are met, some collectivists believe, individuals should be rewarded on the basis of their ability.

Although these popular doctrines of social justice appear to be simple, it is difficult for many reasons to translate them objectively into public policies. First, no doctrine is perfect. If the libertarian doctrine fails to consider problems associated with market imperfections, externalities, and public goods, the collectivist view is unrealistic because it ignores self-interest. A collectivist may argue that automobile dependence is undesirable because it incurs social costs, but an individual, perhaps the collectivist himself, may still make all his trips by automobile out of self-interest. This is a serious problem because even the most socially concerned individual is likely to be driven by self-interest.

The second reason lies in the ambiguities within each doctrine. Even though people generally understand the nature of a doctrine, no one has a detailed list of objectives that are met by a specific doctrine. Moreover, like a move in a game of chess, the consequence of a public policy also has consequences, and it is virtually impossible to comprehend all of the eventual consequences of a public policy. Third, disagreements often exist among members of a society about the suitability of a doctrine to their particular situation. For example, one group may benefit more from a policy that is influenced by the libertarian doctrine, whereas another group may benefit more from a policy influenced by the collectivist doctrine. When such conflicting interests exist, naturally one group opposes one doctrine and the other group opposes the other doctrine. That makes decision making complex.

The fourth difficulty in translating doctrines into policies is that objective categorization of members of a society is problematic for the purpose of redistribution. There are no generally accepted standards to determine who deserves benefits from a social welfare policy. Even though the poverty rate is often a benchmark, its appropriateness is questionable because it is so simplistic. Although race and ethnicity are often important considerations for public policies, not all minorities are underprivileged or deserving of welfare benefits. Finally, technical difficulties abound in measuring, aggregating, and comparing utility among individuals. Although people generally understand the concept of utility, translating the concept to public policy objectives is difficult, if not impossible, because of the problems associated with interpersonal comparisons of utility.

In theory, a society may choose a doctrine of social justice anywhere between the extremes of a pure endowment-based reward system and total collectivism. In practice, however, there is no country in the world today where a purely endowment-based reward system prevails. In the United States, the most common approach seems to be the basic-needs or the merit-goods approach, where society tries to ensure that everyone has access to at least the bare minimum of goods and services. Yet history shows that there is no general agreement among people regarding the constituents of basic goods and the length of time for which a household should be entitled to supplemental benefits.

Even though it is recognized in today's society that the extraction of resources from the better-off is necessary for providing basic goods to the worse-off, society is also concerned that there is a limit to such extraction. When the extraction of resources from the better-off goes beyond a particular point, there is a loss of efficiency (Musgrave & Musgrave, 1989). Thus efficiency and social justice may be at odds in certain circumstances, although a conflict between the two is not necessary.

SOCIAL JUSTICE IN URBAN TRANSPORTATION PLANNING

In the 1960s and the 1970s, social justice within cities received serious attention (Davidoff, 1965; Harvey, 1973). In subsequent years, *equity planning*, which emphasizes redistributive policies in favor of the least powerful and their participation in the planning process, emerged as a popular and well-defined planning approach (Krumholz & Clavel, 1994). In urban transportation planning, however, little has been done to include objectives pertaining to social justice. Even though scholars in urban transportation have often shown concern about the problems of the transportation-disadvantaged (Altshuler, 1979; Kain & Meyer, 1970; Meyer & Gómez-Ibáñez, 1981; Pucher, 1981), the standard transportation planning process, described in Chapter 5, rarely considers the disadvantaged.

The neglect of social justice issues in urban transportation planning reflects the highly technical nature of the profession. Among all branches of urban planning, transportation planning perhaps adopts rational planning principles to the greatest extent. Transportation planners also make use of a substantial amount of quantitative data, which makes it easier to adopt a rational planning approach. They often seek to evaluate alternative scenarios of a plan in an objective manner. The objectives of a plan, including saving of human life from accidents, are set in dollar figures. In contrast to the objectives and methods of conventional transportation planning, the issues in social justice are highly subjective, which may account for the failure of conventional transportation planning to incorporate social justice into the planning process.

To a great extent, today's transportation planning has its roots in the discipline of traffic engineering. Although efficiency and equity are not always at odds, traffic engineers have traditionally placed more emphasis on the former whenever there has been a conflict between the two. In traffic engineering, efficiency and effectiveness are perhaps the most important criteria for evaluation. "Efficiency," in this context, means maximizing tangible benefits for given costs. "Effectiveness," on the other hand, is a measure of degree of accomplishment of goals: the higher the degree, the greater the effectiveness of a program or policy. Typical textbooks in traffic engineering and operational transportation planning (e.g., Garber & Hoel, 1999; Khisty & Lall, 1998; Papacostas & Prevedouros, 1993; Wright & Ashford, 1997), from which professional traffic engineers and transportation planners learn the skills of the trade, have little to do with equitable distribution of transportation infrastructure and services.

The dependence of the transportation planning process on computer models to forecast traffic volumes and capacity requirements has hindered efforts to increase social justice. Urban transportation planning models predict future growth patterns on the basis of existing patterns; such forecasts, which are the basis for transportation investments, do not aim to disrupt existing patterns of inequity. When a computer model generates and distributes trips, splits trips among modes, and assigns trips to transportation networks on the basis of existing land use distribution and travel behavior, it reinforces the status quo without any consideration of redistribution or change. If workers who live in predominantly low-income and minority areas have short commutes at present, computer models will predict a short commute for them in the future as well, regardless of their true commuting needs.

Although some efforts have been made recently in large metropolitan areas to include equity objectives in the transportation planning process, it is yet to be known how operational these efforts will be, or whether they will bear any fruits. (To note how equity can be incorporated into the planning process, see Johnston, Chapter 5, this volume.)

In the 1960s, the McCone Commission and the Kerner Commission were set up to study the causes of urban riots. While the Kerner Commission reported discrimination in many spheres of life, the McCone Commission specifically stated that the lack of adequate transportation services restricted the mobility and accessibility of inner-city residents, thereby adding to their social problems. About the same time, public transportation began to be viewed as a solution to urban social problems, and heavy subsidization of transit began in the name of justice or fairness. But despite large government subsidies, little has been accomplished over the years in terms of equitable distribution of transportation services.

Until the early 1990s, legislation to protect or aid the poor, minorities, and other transportation-disadvantaged was rare. Since 1990, however, several pieces of legislation have passed, including the Americans with Disabilities Act of 1990, the Intermodal Surface Transportation Efficiency Act of 1991, and the Transportation Equity Act of 1998; these were supplemented by a presidential Executive Order, entitled "Federal Action to Address Environmental Justice in Minority Populations and Low-Income Populations," in 1994. Although much is yet to be known about their long-term impact, these acts have the potential to encourage the incorporation of social justice into operational urban transportation planning process.

SOCIAL JUSTICE IN ACCESS TO ACTIVITIES

A transportation system provides access to many activities, including work, health care, shopping, and recreation. Concerns about justice or fairness exist regarding access to all these activities. Take the case of health care. Although clinics and hospitals are often located close to inner cities, members of minority and low-income populations generally have less access to health care than do others (Ginzberg, 1991; Todd, Seekins, Krichbaum, & Harvey, 1991). For them, structural barriers, lack of access to proper transportation modes, and distance are the common transportation-related impediments to health care facilities (Malgren, Martin, & Nicola, 1996). Studies have concluded that a clear disparity exists in health care access among races and ethnic groups even when the economic differences between the groups are controlled for (Council on Ethical and Judicial Affairs, 1990; De Lew & Weinick, 2000). Some authors even claim that it is not disparity, but demonstrable discrimination, that reduces access to health care for minorities (Council on Ethical and Judicial Affairs, 1990; Sullivan, 1991).

Although people generally make far fewer trips to health care facilities than to work locations or to grocery stores, good access to health care is essential. Many people needing frequent health care services are incapable of driving to hospitals or clinics because of their age or ailment. Even if trips to other activities can be made by bus or train, trips to health care facilities often require special provisions, depending on the nature of the ailments people have. Despite the seriousness of the problems associated with lack of access to health care facilities, transportation planners attached little importance to planning for health care access.

As in the case of access to health care, concerns about social justice exist regarding people's access to recreational activities. Because low-income and minority populations have limited mobility, their access to suburban, rural, and out-of-town recreational activities is low. Differences in income, knowledge base, and/or tastes mean that low-income individuals tend to attend different types of recreational facilities than higher-income individuals (Floyd, Shinew,

McGuire, & Noe, 1994). Museums and art galleries are of little value to most low-income persons because of their knowledge base and entrance fees. Parks are important, but many urban parks located in or close to low-income and minority neighborhoods are inadequate and in disrepair.

From the social justice point of view, access to stores is perhaps more important than access to recreational activities. This is particularly the case for grocery stores because grocery shopping is a necessity for every household. Because of the disappearance of supermarkets from central-city neighborhoods and a low level of mobility for inner-city residents, there is a concern that the urban poor and minorities are "captive shoppers," and therefore pay higher prices for their purchases (Caplovitz, 1963). A recent empirical study (Chung & Myers, 1999) observed that people living in poor neighborhoods pay more for their groceries owing to the absence of chain grocery stores in close proximity. In another study, Finke, Chern, and Fox (1997) found that low-income, urban African Americans pay a higher price for food products than similarly endowed whites. The lack of local stores that carry fresh foods may have serious health-related consequences for inner-city residents.

For those concerned with social justice, access to work is more important than access to any other activity because most people depend on work for their livelihood. Access to jobs for low-income and minority populations is particularly important because of their continued clustering in inner-city neighborhoods and the dispersal of jobs to suburban areas. These two phenomena led to the *spatial mismatch hypothesis.*

According to Holzer (1991), "spatial mismatch" exists when decentralization of employers to suburban areas leads to unemployment among African Americans living in inner cities. One of the concerns of spatial mismatch is that discrimination in suburban housing markets prevents many minority families from moving to the suburbs even though many new jobs are located there. The spatial mismatch hypothesis proposes that the suburbanization of jobs and limited residential choice for African Americans create a surplus of workers in inner-city neighborhoods, resulting in unemployment, lower wages, and longer commutes for these workers (Ihlanfeldt & Sjoquist, 1998). The hypothesis is thus about suburbanization of jobs and its labor-market impact on central-city African Americans. Although the spatial mismatch literature focuses mainly on African Americans, a mismatch can also affect other poor people without adequate mobility.

The spatial mismatch hypothesis became popular in the 1960s as a result of a seminal study by Kain (1968). In the aftermath of the "welfare-to-work" legislation, formally known as the Personal Responsibility and Work Opportunity Reconciliation Act of 1996 (PRWORA), the significance of spatial mismatch increased enormously. The PRWORA introduced the Temporary Assistance for Needy Families program (TANF), which mandated that a welfare recipient would be eligible for benefits for only 5 years in an entire lifetime and would never receive benefits for more than 2 consecutive years. Under these mandates, job accessibility has become more important to welfare recipients than ever.

There are three potential solutions to spatial mismatch, namely, locating appropriate jobs in central cities, dispersing low-skilled minorities to job-growth areas in the suburbs, and making suitable transportation connections between inner cities and suburban job-growth areas. Of these solutions, the residential dispersal strategy seems to receive the least support from local politicians and planners despite being favored by many academics (Danielson, 1976; Downs, 1973; Kain, 1992; Kain & Persky, 1969). A reason for the lack of popularity of the residential dispersal strategy among politicians is that suburban residents, fearing a decrease in their property values, do not wish to have a diversity of people or housing types in their

neighborhoods. Because metropolitan areas are fragmented into many jurisdictions, little can be done to promote residential dispersal without first introducing drastic measures like metropolitan consolidation.

Attracting jobs to inner cities is equally problematic. Although many jobs are located close to inner cities, they are mainly white-collar jobs in the central business districts (CBDs). These jobs do not fit the skill levels of inner-city residents. Traditionally, low-skill jobs in the retail and manufacturing sectors provided employment to inner-city workers. With changes in technology, manufacturing jobs relocated to outlying areas. Similarly, with the suburbanization of the middle class, retail jobs also moved to the suburbs. To attract these jobs back to central cities is a Herculean task, for the dispersal in the first place was a market-driven reaction to changes in manufacturing and transportation technology. Efforts have been made under different government programs to attract jobs back to the central cities, but such efforts have had little success.

The failure of both the residential dispersal strategy and the job attraction strategy has placed the burden on transportation, especially public transportation, to solve the job accessibility problem of inner-city residents. Following the McCone Commission Report in the 1960s, the federal government sponsored several demonstration projects to provide public transit connection between inner cities and suburban job-growth areas (Altshuler, 1979). As a result of the welfare-to-work legislation, similar enthusiasm is evident again in linking suburban job-growth areas and inner cities by public transit.

The spatial mismatch hypothesis has three premises (Ihlanfeldt & Sjoquist, 1998). First, the demand for labor has shifted from central cities to suburban locations. Second, racial discrimination in housing and mortgage markets has prevented African Americans from moving to the suburbs. Third, owing to discrimination in the labor market, poor information about distant jobs, and limited public transportation services between their neighborhoods and job-growth areas, central-city African Americans have failed to find suitable employment. Three social justice issues are important here. They are discrimination in the labor market, discrimination in the housing market, and discrimination or disparate treatment by a transportation system, including highways and transit.

A large body of literature shows evidence of housing and labor market discrimination. Despite landmark court cases like *Shelley v. Kraemer* in 1948 and the Fair-Housing Act of 1968, discriminatory practices in the housing market continue to impede suburbanization of minorities, especially African Americans (Bullard & Feagin, 1991; Darden, 1995; Feagin, 1999; Orfield, 1974; Yinger, 1998). Similarly, several studies document evidence of discrimination against minorities in the labor market (Fix & Austin Turner, 1999; Kirschenman & Neckerman, 1991; Nord & Ting, 1994; Turner Meiklejohn, 2000).

Although disparity in access to activities is evident among different races and income groups, it is difficult for several reasons to prove that discrimination exists in an urban transportation system (Steinberg, 2000). First, an urban transportation system does not exist independently of the activities or facilities it connects. Sometimes disparate treatment may occur at the activities (e.g., suburban housing or labor market), and a transportation problem (e.g., lack of transit service) may merely be an accentuating factor. Second, an urban transportation system contains multiple components, of which some are owned by private individuals and others are owned by government. Third, it is virtually impossible to determine precisely who pays how much for the use of transportation infrastructure and services. Fourth, the components of an urban transportation system are heavily subsidized by different levels of government, and some components are cross-subsidized by others. Fifth, many of the costs and benefits from

an urban transportation system are indirect and difficult to measure. Finally, as Steinberg (2000) points out, interpretation of discrimination laws, such as Title VI of the Civil Rights Act, is fairly complex. Because of these difficulties, researchers have generally addressed the justice-related issues in urban transportation in a piecemeal manner, and they have avoided discussions of some components altogether.

DISTRIBUTION OF TRANSPORTATION COSTS AND BENEFITS

Urban transportation analysts often examine whether people pay a fair share of the costs of the transportation infrastructure and services they use. Although relative costs and benefits for payer and payee are important in the contexts of both highways and transit, transit has received far greater attention than highways in this regard.

In the two sections below, I discuss some issues relating to distribution of costs and benefits in urban transportation, taking the case of both highways and transit. In the first section, I examine the household expenditures for transportation of different socioeconomic groups. In the second section, I discuss how the government distributes its expenditure on transportation and how this distribution affects different socioeconomic groups.

The Cost of Transportation to a Household

The private cost of transportation is not borne identically by different segments of society. Since the poor earn less, they spend a greater proportion of their income on consumption of goods and services, including transportation. More serious, however, than relative expenditures on transportation by the rich and the poor is the difference in the rate of change in transportation expenditure between the two groups. Table 12.1 shows how expenditure on transportation

TABLE 12.1. Percent Increase in Household Expenditure on Transportation at Current Prices

Category of household	Percent increase
Household income (1992–2000)[a]	
Up to $4,999	41.8%
$5,000–$9,999	57.0%
$10,000–$14,999	52.9%
$15,000–$19,999	36.5%
$20,000–$29.999	17.1%
$30,000–$39.999	25.4%
$40,000–$49,999	22.5%
$50,000–$69,999	19.0%
$70,000 and above	16.8%
Race (1990–2000)[a]	
Black	62.5%
White and other	43.9%
Ethnicity (1994–2000)[a]	
Hispanic	45.5%
Non-Hispanic	24.3%

[a]Since data are available for different time periods for income, race, and ethnicity, increases are calculated for the longest time period allowed by data availability.
Source: Bureau of Labor Statistics (2002).

has been increasing over time for different socioeconomic groups. For those concerned with social justice in urban transportation, these figures are worrisome. Because the increases shown in the table are based on current dollars rather than constant dollars, the magnitude of these increases does not tell much. A comparison of these increases across groups, however, shows that the rate of increase in transportation expenditure for low-income groups, African Americans, and Hispanics is significantly higher than it is for others. An increase in transportation expenditure for low-income and minority households may be an indication, in a relative sense, of reduced consumption of other goods and services. Because, as explained in Chapter 1, travel is generally seen as a derived demand—that is, one does not usually receive any benefit from travel per se—an increase

in transportation expenditure for these groups cannot be viewed as a welcome change.

One reason for the increasing expenditure on transportation by low-income and minority households is their increasing dependence on the automobile. While affluent and white households became dependent on the automobile several decades ago, poor and minority households have adapted to the automobile more recently. Although an automobile provides them with a greater mobility than conventional public transit, ownership of an automobile also puts a burden on their meager household budgets.

The Cost of Transportation to the Public

The government plays a significant role in urban transportation. To some extent, urban transportation problems are due to underpricing and government subsidization of automobile use (Meyer & Gómez-Ibáñez, 1981). Although the highway sector recovers a significant portion of its expenditure from its own sources, such as fuel taxes, vehicle taxes, and tolls, these sources contribute only about 64% of the total funds available for highways (estimated from Federal Highway Administration, 2000, Table HF-10A). The remaining 36% of highway expenditure come from general fund appropriations, property taxes and assessments, investment income, bonds, and other taxes and fees.

If subsidizing highways is an impetus to drive, subsidization of new housing makes it lucrative to seek suburban housing. As a result of subsidized roads and subsidized housing, many activities dispersed from the central cities to the suburbs in the post-World War II period. Suburbanization of activities created mobility problems for those without access to an automobile, and this made transit subsidization a necessity. Transit subsidization gathered momentum in the 1960s and 1970s, and by the late 1990s annual capital and operating subsidies for transit from the three levels of government amounted to $17.7 billion (Federal Transit Administration, 1997), or 69% of total transit expenditures for the year.

Transit is expected to provide several social benefits, such as mobility to the disadvantaged, reduction in air pollution, and reduction in congestion. Because of the expected role of transit in providing social benefits, researchers have often been curious to find out who pays transit's costs and who benefits from the transit expenditure. Some have tried to answer this question on the basis of a rich–poor dichotomy (e.g., Pucher, 1981, 1983), while others have tried to answer it on the basis of a city–suburb dichotomy (e.g., Hodge, 1988). In their efforts to answer the question, researchers have touched upon such critical issues as rail versus bus transit, flat versus distance-based fares, central-city versus suburban services, and peak versus off-peak service.

Empirical studies suggest that because of their relatively small tax contributions, the poor contribute far less to transit subsidies than do the affluent (Pucher, 1981, 1983). When progressive taxes like an income tax are used to subsidize transit, the poor contribute proportionally less, but when regressive taxes like a sales tax are used for subsidization, the poor contribute the same proportion of their consumption expenditure as the rich, but a greater proportion of their income. Because the affluent contribute more to transit subsidies by paying higher income taxes, and since the poor constitute a larger proportion of transit users than the affluent, it appears that the *affluent population* subsidizes *poor users*. A comparison of the contributions of affluent users and poor users to fare-box revenue, however, shows that the poor cross-subsidize the rich for reasons described later in this section.

Studies that have used the city–suburb dichotomy to determine who pays and who benefits from transit draw similar conclusions (Hodge, 1988). Because of its higher tax payments, the suburban population contributes more to transit subsidies than does the central-city population, but central-city users cross-subsidize suburban users by paying a greater proportion of costs per trip.

Poor and central-city users can subsidize affluent and suburban users in many ways. First, this cross-subsidization can take place through the transit fare structure. Central-city residents and the poor make short trips, whereas affluent and suburban users make long trips. Because transit agencies often charge a flat fare and almost never charge a fare exactly proportional to distance, those who make short trips subsidize those who make long trips. The significance of not charging a distance-based fare is enormous from the social justice point of view. For a particular transit system, Wachs (1989) noted that a user who travels 1 mile pays more than twice the true cost of the trip, whereas a user who travels 20 miles pays only 10% of the cost.

Another reason for poor and central-city users subsidizing affluent and suburban users is that contribution to fare-box revenue is higher in central-city services than in suburban services (Cervero, 1990). The low income of the population, the higher density, and the mixed land uses in the city contribute to the popularity of transit in central cities. The higher household incomes, the lower density, and the homogenous land uses of the suburbs all mean that transit is less popular in the suburbs, and hence fare-box collection is also low.

The relatively higher cost of transit provision in peak periods (see Taylor, Chapter 11, this volume) also leads to cross-subsidization of the affluent by the poor. Capital costs for peak-period service are higher than costs for off-peak service because transit systems are designed to accommodate peak volumes. Because the people who travel only in the off-peak period could do without the capacity built for peak periods, they subsidize those who travel during peak periods. Even operating costs for peak-period service are higher because labor hired to serve peak demand cannot be efficiently used in the off-peak periods (Wachs, 1989). Since the poor make proportionally more trips in off-peak periods than do the affluent, poor users subsidize affluent ones.

Figures 12.1 and 12.2 show the arrival time at work and the departure time from work for workers belonging to different socioeconomic groups. In order to avoid casual trips in and out of work, only home-to-work and work-to-home trips are included. The figures show the peaking pattern of commute trips by several groups of workers. It is evident from the figures that commute trips by low-income and African American workers peak far less than do trips by workers as a whole. The reason for less peaking of trips by low-income and African American workers is that many of them are part-time workers who stay at work for only 4 or 5 hours a day, and many commute to and from work at odd hours. A significant proportion of low-income and

minority workers leave work between 9:00 P.M. and 1:00 A.M.—a time when transit services are either totally unavailable or barely available in most cities. It is also noticeable that the commute trips by high-income workers peak more than the trips made by workers as a whole. Under these circumstances, concentrating transit service in peak periods helps the affluent more than the poor.

Table 12.2 shows the median income of transit users in peak and off-peak periods. For all three transit modes shown, the median income of users is higher in peak than in off-peak periods. When the higher median income of peak-period users is considered in conjunction with the higher cost of service provision in peak periods, the provision of a greater level of transit service in peak periods does not appear to be fair.

In any discussion of justice in the provision of transit service, perhaps the most contentious issue is the relative subsidization of bus and rail transit. Table 12.3 shows

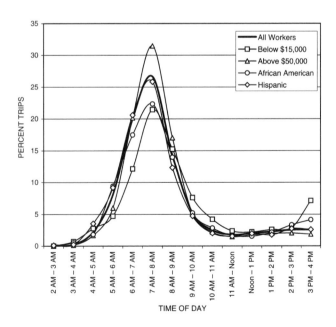

FIGURE 12.1. Arrival time at work for workers belonging to different socioeconomic groups. Source: Federal Highway Administration (1995).

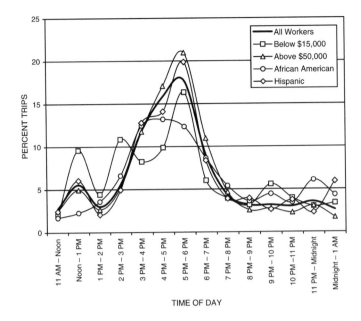

FIGURE 12.2. Departure time from work for workers belonging to different socioeconomic groups. Source: Federal Highway Administration (1995).

the median income of users of different modes, separately for all trips and for commuting trips. Mode-specific median household income was obtained by restricting the trips made by a particular mode and subsequently sorting the household incomes of the persons making the trips. The data presented in Table 12.3 show that bus riders have a lower median income than commuter rail and subway/elevated rail users for both all trips and commuting trips. In fact, commuter rail users have a higher median income than automobile users for both types of trips, while subway/elevated rail users have identical median income as automobile users for commuting trips.

All of transit's annual capital funds of $7.6 billion in 1997 came from the three levels of government: federal, state, and local (Federal Transit Administration, 1997). In contrast, of transit's $17.9 billion operating expenses that year, close to 40% came from fare-box revenue, 56% came from government sources, and 4% came from other sources.

Because rail is capital-intensive and bus is labor-intensive, a greater emphasis on capital subsidies favors rail, and thus affluent us-

TABLE 12.2. Median Household Income of Peak Period and Off-Peak Period Trip Makers

Category of trip	A.M. and P.M. peak periods (trips starting between 6–8 A.M. and 4–6 P.M.)	Off-peak period (trips starting between 8 A.M. and 4 P.M.)
Bus	$25,000–$29,999	$15,000–$19,999
Commuter train	$55,000–$59,999	$30,000–$34,999
Subway/elevated rail	$35,000–$39,999	$25,000–$29,999

Source: Estimated from Federal Highway Administration (1995, Trip File).

TABLE 12.3. Median Household Income by Mode of Travel

Category of trip	All trips	Commuting trips
Automobile	$35,000–$39,999	$35,000–$39,999
Van	$45,000–$49,999	$45,000–$49,999
SUV	$50,000–$54,999	$50,000–$54,999
Pickup truck	$35,000–$39,999	$35,000–$39,999
Bus	$15,000–$19,999	$25,000–$29,999
Commuter rail	$40,000–$44,999	$55,000–$59,999
Subway/elevated rail	$30,000–$34,999	$35,000–$39,999
Taxicab	$35,000–$39,999	$35,000–$39,999
Bicycle	$30,000–$34,999	$30,000–$34,999
Walk	$30,000–$34,999	$30,000–$34,999

Source: Estimated from Federal Highway Administration (1995, Trip File).

ers, whereas a greater emphasis on operating subsidies favors bus, and hence poor users. It is evident from Table 12.4 that operating expense per passenger mile is higher for bus than for rail. Operating expense per trip is also fairly high for bus. However, capital funds per trip and per passenger mile are higher for the rail modes than for bus. When only bus and rail modes are considered, bus receives only 31% of the capital funds, although it carries more than 60% of the trips made by transit.

A reduction of capital subsidies and an increase in operating subsidies would be beneficial to the poor. Because the federal government pays a far greater proportion of the capital subsidies than of the operating subsidies (54.2% vs. 3.4%), and since federal

TABLE 12.4. Performance Indicators for Transit Modes

	Bus	Heavy rail	Light rail	Commuter rail
Operating expense in millions	$9,421.9 (60.2%)	$3,473.7 (22.2%)	$471.4 (3.0%)	$2,274.7 (14.5%)
Capital funding in millions	$2,227.9 (30.7%)	$2,346.1 (32.3%)	$873.2 (12.0%)	$1,817.4 (25.0%)
Annual unlinked trips in millions	4,602.0 (60.2%)	2,429.5 (31.8%)	259.4 (3.4%)	357.2 (4.7%)
Annual passenger miles in millions	17,509.2 (45.3%)	12,056.1 (31.2%)	1,023.7 (2.7%)	8,037.5 (20.8%)
Operating expense per trip	$2.05	$1.43	$1.82	$6.37
Capital funding per trip	$0.48	$0.97	$3.37	$5.09
Operating expense per passenger mile	$0.54	$0.29	$0.46	$0.28
Capital funding per passenger mile	$0.13	$0.19	$0.85	$0.23

Source: Federal Transit Administration (1997).

revenues are collected more progressively than state and local revenues, in a particular way capital subsidies from the federal government can also help redistribution from the affluent population to poor users. The downside of federal subsidies, however, is that it leads to expansion of transit service to suburban areas—areas where the affluent live (Anderson, 1983). Moreover, because a substantially greater share of federal capital subsidies goes to rail transit, the benefits from capital subsidization of transit predominantly go to high- and moderate-income users.

Suggestions to make transit more favorable to the poor include implementing distance-based fares, shifting subsidies from rail to bus, providing off-peak discounts and discount fares for the poor, and improved service in low-income areas (Pucher, 1983). Federal capital subsidization policies, local political interests, and the predilection of urban planners, however, all support the continued favoring of rail transit at the expense of bus transit. As an increasing number of taxpayers now live in the suburbs, the political influence of suburbanites in transit decision making has also increased (Wachs, 1989). The outlook among transit officials today is such that suburban populations must be provided a substantial level of service, even if such service remains underutilized, because these populations provide the lion's share of transit subsidies. Providing underutilized service to suburban populations is stultifying, however, as it creates further deficits for transit.

For decades, scholars have considered subsidizing poor people's personal vehicles (Kain & Meyer, 1970; Meyer & Gómez-Ibáñez, 1981). Claims are made even today that the annual cost of leasing most automobiles is less than the annual cost per commuter for many light rail systems, and, for some of these systems, the annual cost per commuter is greater than the annual household income of the lowest income quintile of the population (The Public Purpose, 2002a, 2002b).

A case for subsidization of poor people's vehicles can be made on the basis of other characteristics of transit as well. First, there is evidence that subsidization can make production of transit service inefficient (Anderson, 1983; Cervero, 1984; Gómez-Ibáñez, 1982). When production of service becomes inefficient, distribution of service may also suffer. Second, even when transit subsidies provide mobility to low-income and minority populations, as they did during the demonstration projects of the 1960s, there is no guarantee that those who benefit from subsidies will continue using transit once they gain access to an automobile (Altshuler, 1979). Even if transit is successful in its mission to provide job access to the poor, to continue to do so, it will constantly have to find new poor. Third, as only a small proportion of the poor use transit, transit subsidies can benefit only a small group of poor (Gómez-Ibáñez, 1982; Pucher, 1981). On the other hand, since more than three-quarters of poverty households own at least one vehicle (Lave & Crepeau, 1994), subsidization of vehicles would reach a much larger proportion of the poor. Fourth, since suburban jobs are located in a more dispersed manner than are central-city jobs, it is difficult to access them by traditional fixed-route transit. Paratransit is more suitable under those circumstances, but the expensive nature of this service increases the relative attractiveness of subsidization of vehicles for the poor. Finally, as shown in Figures 12.1 and 12.2, a fairly large proportion of the poor and minority workers commute at odd hours. Because traditional transit service at those hours is either absent or infrequent, access to an automobile is almost a necessity for these workers.

Although many arguments can be made in favor of subsidizing poor people's automobile use, automobile subsidization will have its own problems. First, subsidization will add to traffic congestion and air pollution. Second, raising gas and vehicle taxes to subsidize automobile use will cause a public outcry. Besides, gas taxes and vehicle taxes are already regressive; raising them to help the poor may make the situation worse. One could argue that roads should be priced and the generated revenue should be used to subsidize poor people's vehicle use, but one of the most common fears about implementing road pricing is that it will disproportionately harm the poor. Yet subsidizing poor people's automobile use is not at all absurd. In fact, in places like Ventura County, California, former welfare recipients are sold retired city cars at subsidized rates under the Temporary Assistance for Needy Families program.

ENVIRONMENTAL JUSTICE IN PUBLIC TRANSPORTATION

Environmental justice picked up momentum in urban transportation in the mid-1990s. On February 11, 1994, Presidential Executive Order No. 12898, known as the "Federal Action to Address Environmental Justice in Minority Populations and Low-Income Populations," was signed. Although this order was not specifically meant for transportation, it had a significant bearing on transportation policymaking in the subsequent years. Following up on the Executive Order, in May 1997 the U.S. Department of Transportation issued an order (DOT Order No. 5610.2), titled "Department of Transportation Actions to Address Environmental Justice in Minority Populations and Low-Income Populations." In December 1998, the Federal Highway Administration issued its own order (FHWA Order No. 6640.23), titled "FHWA Actions to Address Environmental Justice in Minority Populations and Low-Income Populations" (Federal Highway Administration & Federal Transit Administration, 2000).

Although environmental justice became an issue in urban transportation in the 1990s, the environmental justice movement is said to have begun in the late 1970s or early 1980s (Cole & Foster, 2001; Dobson, 1998). In the initial stages of the movement,

its concerns were mainly about disposal of toxic wastes in predominantly minority areas. Grassroots activists have always played a dominant role in the movement. These activists often come from minority communities that have suffered from environmental degradation. Traditional environmental planners or professionals have practically nothing to do with the environmental justice movement. In a way, the term "environmental" in the environmental justice movement is an irony, for its proponents are often at odds with traditional environmental planners and professionals.

Mainstream, or traditional, environmentalism started at the beginning of the 20th century under the auspices of the likes of Theodore Roosevelt, whose primary concerns were about preservation of the natural environment (Cole & Foster, 2001). In the post–World War II period, traditional environmentalism gathered momentum and reached its peak in the 1960s and the 1970s. Environmental planning emerged as a profession around that time, but its goals were the same as those of traditional environmental enthusiasts (Collin, Beatley, & Harris, 1995). The primary concern of environmental planning was, and to a great extent still is, protection of the natural environment from polluting industries. With an increase in society's concern about pollution, industries have been forced to pay more attention to the collection and disposal of their wastes. Industries have complied, but many minority communities believe that wastes are being disproportionately located in their communities.

Although a small number of urban planners, such as advocacy planners and equity planners, have always been concerned about the plight of minority and low-income communities, the urban planning profession has generally adopted the principles of traditional environmentalists rather than being concerned with environmental justice issues. To a great extent, mainstream urban planning practices like growth control and growth management are based on traditional environmental principles, such as preservation of farmland and natural habitat. Obviously, these practices can be detrimental to the interests of minorities and the poor insofar as they are used to protect the status quo in suburban communities. In general, traditional environmental groups like the Sierra Club, the Friends of the Earth, and the Wilderness Society have alienated themselves from minority interests (Westra & Wenz, 1995). For example, in a 1971 national membership survey of the Sierra Club, 58% of the members voted against the club's involvement in the conservation problems of minorities and urban poor. Despite lacking support from traditional environmentalists for a protracted period, the scope of environmental justice has increased significantly to include issues that can hardly be called "environmental" in the traditional sense.

The legal basis for implementing environmental justice in transportation is provided by Title VI of the Civil Rights Act of 1964, the National Environmental Policy Act of 1969 (NEPA), and administrative orders by the U.S. Department of Transportation (USDOT) and its components (Steinberg, 2000). Title VI of the Civil Rights Act prohibits discrimination on the basis of race, color, or national origin. Under NEPA, environmental justice is enforced in transportation projects through a review of environmental impact statements.

Although environmental justice was born in a sphere only peripherally related to transportation, today its scope within the transportation sector alone is enormous. This scope is evident from the circumstances in which activists have challenged transportation plans and projects. In Atlanta, civil rights groups filed a lawsuit complaining that the region's long-range transportation plan favored the predominantly white and affluent parts of the region (Steinberg, 2000). In Los Angeles, a class action lawsuit was filed by the Bus Riders Union against the Metropolitan Transportation Authority, claiming that the practice of subsidizing rail

transit took away resources from bus transit and thus adversely affected the predominantly minority and low-income bus riders. Despite the authority's continued resistance to a 1996 court order requiring it to buy 248 additional buses, its appeal finally ended in March 2002, when the U.S. Supreme Court declined to review the case (*Los Angeles Times*, March 19, 2002). Although the successful Los Angeles class action lawsuit is the most prominent of all environmental justice cases in transportation, citizens' groups have raised their voices in the name of environmental justice in many other cities. In Washington, D.C., activists successfully challenged the routing of a segment of the Metro's Green Line that goes through a predominantly minority area. In New York City, environmental justice activists contested the use of polluting transit vehicles in a minority area. In Oakland, California, citizens' groups challenged the efforts to rebuild an earthquake-damaged freeway in a minority area (Bullard & Johnson, 1997).

Despite the widening scope of environmental justice in transportation, ambiguities in terminology and difficulties in implementing regulations continue to exist. According to Steinberg (2000), in implementing environmental justice, questions arise in simple matters like protecting low-income populations under Title VI, separating "disparity" from "discrimination," interpreting terms like "mitigation" and "exposure," and identifying control groups for the purpose of comparison. Although the environmental justice movement has been successful in influencing urban transportation policies, serious efforts are needed to clarify ambiguities and aid implementation of policies.

According to Whitelegg, the "disbenefits [from transportation] are not evenly distributed throughout society, and it is often the most disadvantaged groups and those least able to be heard who bear the brunt of the disbenefits and who pay directly through their health and their quality of life for other people's mobility" (1997, p. 139). In the two sections below, I discuss some of the transportation-related urban environmental problems, or disbenefits, that can disproportionately affect minority and low-income communities.

IMPACT OF FREEWAYS ON LOW-INCOME AND MINORITY COMMUNITIES

The highway era began immediately after World War II. Under the Interstate Highway program, initiated by the 1956 Federal-Aid Highway Act, more than 41,000 miles of highways have been built. As described in Chapter 11, mainly for political reasons, a substantial portion of the Interstate Highway System was built within urban areas (Lewis, 1997). When the decision to build the Interstate system was made, Americans were mainly happy and enthusiastic about the construction of new highways. Apparently the media complained, but it was not about the environmental problems such highways would create; the complaints were about the pace of construction being too slow. By the mid-1960s, however, protests against freeway construction became common. In 1965, people challenged the connection of Vieux Carré Expressway to the Interstate system in New Orleans. Citizens' groups protested the demolition of low-income and minority housing to build expressways in New York City during the same year. In 1966, beltway construction through minority areas in Washington, D.C., was contested. During the same year, protesters in Memphis, Tennessee, claimed that Interstate-40 was intentionally aligned through minority areas without consideration of other options. These are only a few examples of early protests leveled against highway construction in minority and low-income areas; protests against freeways became common in most urban areas over time. One of the most illustrative cases is that against Interstate-105 in Los Angeles, which underwent 30 years of planning and review because of citizens' concerns. Despite its prolonged planning and review

process, Interstate-105 is still blamed for blighting entire communities.

The people who benefited the most from freeway construction were suburban landowners. Another group to benefit from freeways are the workers who commute to work in downtown locations using automobiles. Suburban residents belong to upper- or middle-income groups, and their automobile ownership rates are generally high. High speeds on urban freeways permit automobile-owning households to live in distant neighborhoods, where they can take advantage of inexpensive residential land. Minority and low-income households are typically located in central cities, and their automobile ownership rates are lower than those of predominantly white and affluent suburbanites. Lower automobile ownership rate means less use of roads, including freeways. Besides, freeways are not meant for short trips—the kind of trips central-city residents typically make. As most jobs were traditionally located within central cities, minority and low-income populations did not have a reason to use urban freeways for commuting purposes. The disproportionately greater use of urban freeways by white suburbanites led minority protesters to complain about "white man's road through black men's bedrooms" (Lewis, 1997, p. 197).

As land value was an important consideration in selection of freeway alignments, freeways often went through poor neighborhoods. Because minorities predominantly live in poor urban neighborhoods, they were seriously affected by the intrusion of freeways. Freeway construction led to the taking of properties in their neighborhoods and evictions. Grade-separated freeways split their neighborhoods and disrupted community life, creating isolated pockets of space—areas where people would rarely go because of fear of crime. The degradation of the physical quality of space in these areas, together with the enhanced accessibility that the new freeways provided, led to the dispersal of well-to-do households and employment activities to suburban areas. This led to further degradation of cities, causing a vicious circle of activity decentralization and degradation of cities. Figure 12.3 provides a conceptual description of the effect of freeway construction on cities.

IMPACT OF POLLUTION ON LOW-INCOME AND MINORITY COMMUNITIES

Street maps of metropolitan areas show that road density is higher in central locations— places where the poor and minorities predominantly live—than in the suburbs. Higher road density means higher traffic density, and higher traffic density means more accidents and greater exposure to pollution. People living in such conditions complain about speed, noise, vibration, dirt, fumes, and traffic accidents (Appleyard, 1981).

As described by Bae (Chapter 13, this volume), vehicular emissions are responsible for various cardiovascular and respiratory illnesses, including cancer. Whitelegg (1997) describes several empirical studies conducted in different parts of the world where traffic-related air pollution was found to have a significant adverse effect on human health. Some of these studies found that the adverse health effects of emissions were felt the most by those living along congested streets—streets resembling those in U.S. central cities. Particulate matters, classified as PM-10, are a major health concern for those living by and walking along roads with heavy traffic. Of particular concern is road dust, which accounts for over 40% of all anthropogenic and biogenic PM-10 emissions. While many other types of pollution showed a decline, dust from roads has increased substantially over time (Bureau of Transportation Statistics, 1996).

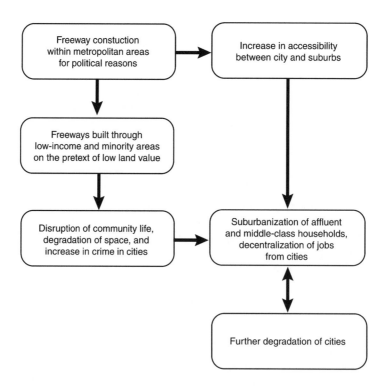

FIGURE 12.3. A conceptual description of the effect of freeway construction on cities.

Low-income and minority neighborhoods are affected by traffic-related noise also. Freeway traffic can be a constant source of noise in these neighborhoods throughout the day and night. Trucks, transit buses, commuter and freight trains, and even car alarms contribute to noise pollution in these areas. Although people generally perceive traffic-generated noise as only an annoyance that disrupts speech and sleep, constant exposure to such noise can actually damage people's physiological and mental health (Whitelegg, 1997).

While discussing the impact of transportation-related pollution, one is bound to wonder who generates the pollution and who bears its effects. The poor, for example, own older vehicles, known to generate more pollution than newer makes. On the other hand, the poor make fewer and shorter vehicular trips than the affluent, thus complicating the issue. In the following sections, data are analyzed to address this issue.

One determinant of air pollution is vehicle age. Analysis of data from the 1995 NPTS reveals that low-income households usually own more old and used vehicles than do higher-income households (Table 12.5). Although the difference is not as significant as that between low-income and high-income households, black and Hispanic households also use more old and used vehicles than white and non-Hispanic households. Since older vehicles consume more fuel and pollute more because of poor technology, from the data presented in Table 12.5, one could argue that low-income and minority households contribute more to air pollution than the rest. Such an argument would be biased against the poor and minorities, however, because they perhaps pollute more only on a per-vehicle basis.

Table 12.6 shows the mean number of vehicles per household, the ratio of trips by personally operated vehicles (POV) to total trips, and the mean distance of POV trips for different income groups, races, and ethnic groups. Each of the variables indicates

TABLE 12.5. Mean Age of Vehicles and Proportion of Vehicles Purchased New by Income, Race, and Ethnicity

Category of household	Mean age of vehicles (years)	Proportion of vehicles purchased new
Household income		
Up to $9,999	10.2	22.6%
$10,000–$19,999	8.6	31.9%
$20,000–$29,999	7.6	38.9%
$30,000–$39,999	6.9	43.7%
$40,000–$49,999	6.6	48.4%
$50,000–$59,999	6.0	54.6%
$60,000–$69,999	5.9	56.8%
$70,000–$79,999	5.5	61.1%
$80,000 and above	4.8	70.3%
Race		
Black	7.0	42.1%
White	6.8	48.1%
Ethnicity		
Hispanic	7.2	41.9%
Non-Hispanic	6.8	47.8%

Note. Pre-1970 vehicle models are omitted because of possible antique value, and post-1994 models are considered as new, or 0 years old.
Source: Estimated from Federal Highway Administration (1995, Trip File).

that higher-income households are more apt to own and use vehicles than are poorer households. All three indicators also show that white households and non-Hispanic households are more likely to own and use vehicles than are black households and Hispanic households, respectively. Thus, in spite of owning a greater proportion of older vehicles, the poor and minorities may add less to overall pollution than their counterparts because their use of personal vehicles is less extensive.

To understand who pollutes and who is affected by transportation-related pollution, one should also consider fuel consumption of vehicles. All else being equal, vehicles that consume a larger amount of fuel not only accelerate the depletion of fossil fuel,

TABLE 12.6. Comparison of Vehicle Ownership, Proportion of POV Trips, and Mean POV Trip Distance by Income, Race, and Ethnicity

Category of household	Mean number of vehicles per household	Ratio of POV trips to all trips	Mean distance of all POV trips (miles)
Household Income			
Up to $9,999	0.88	73.2%	8.47
$10,000–$19,999	1.31	81.2%	8.73
$20,000–$29,999	1.65	86.8%	10.09
$30,000–$39,999	1.88	88.2%	10.24
$40,000–$49,999	2.06	87.9%	9.99
$50,000–$59,999	2.19	89.7%	10.85
$60,000–$69,999	2.26	88.8%	10.43
$70,000–$79,999	2.34	89.0%	11.54
$80,000 and above	2.39	89.4%	11.28
Race			
Black	1.26	75.6%	9.61
White	1.87	88.5%	10.24
Ethnicity			
Hispanic	1.60	81.8%	9.90
Non-Hispanic	1.80	86.9%	10.14

Source: Estimated from Federal Highway Administration (1995, Vehicle File, Household File, and Trip File).

TABLE 12.7. Price and Fuel Consumption of Selected Vehicle Types

Vehicle type	Price range of new car (year 2000)	Fuel consumption in city environment (mpg)
Small car		
Honda Civic	$12,000–18,000	30
Toyota Corolla	$12,000–22,000	28
Medium car		
Honda Accord	$16,000–25,000	20
Toyota Camry	$17,500–25,500	19
Van		
Dodge Grand Caravan	$18,000–32,000	18
Toyota Sienna	$21,500–30,000	18
Pickup truck		
Ford Ranger	$12,000–25,000	17
Toyota Tacoma	$13,500–25,500	17
Sport utility vehicle (SUV)		
Ford Explorer	$21,000–38,000	15
Jeep Grand Cherokee	$26,000–35,000	15

Source: American Automobile Association (2000).

but they also generate a greater amount of air pollution. Table 12.7, which shows the price range and fuel-consumption level of some popular new cars, vans, pickup trucks, and sport utility vehicles (SUVs), indicates that barring the case of pickup trucks, there is an inverse relationship between the price of vehicles and fuel efficiency—the higher the price, the lower the fuel efficiency. Since the income of consumers and the price of consumed commodities are positively associated, the users of expensive vehicles are likely to be relatively affluent. As expensive vehicles also consume more fuel, we can infer that the affluent consume more fuel and add more to air pollution than do the poor.

To comprehend the relationship between income and type of vehicle used, recall Table 12.3, which showed the median household income of users of different modes from the 1995 NPTS. The table shows that automobile and pickup truck users have lower median household incomes than do SUV users and van users. Among the POV modes considered, SUV users have the highest median income. Table 12.7 shows that some of the most popular SUVs are also the least fuel-efficient.

To summarize, low-income populations contribute more to air pollution through the use of older vehicles, whereas affluent populations contribute more to air pollution through the use of less-fuel-efficient new vehicles such as SUVs and vans. Low-income households own fewer vehicles, however, and they make fewer and shorter POV trips. When these factors are taken into account, the poor seem to generate less air pollution than the affluent. In other words, the data on income, price of vehicles, and fuel consumption indicate that the poor pollute less than the affluent because they depend less on vehicular travel. As African Americans and Hispanics have lower average incomes than their white counterparts, it is possible that what is true for the poor is also true for minorities.

SAFETY AS AN ENVIRONMENTAL JUSTICE ISSUE

In the year 2000, motor vehicle accidents resulted in almost 42,000 fatalities and 5.3 million nonfatal injuries in the United States, costing the nation more than $230 billion (National Highway Traffic Safety Administration, 2000a). Although private insurance companies paid 50% and government sources paid 9% of these costs, individual crash victims had to pay as much as 26% of the costs on their own. (Other parties, such as uninvolved motorists, charities, and health care providers, paid the remainder.) Available data suggest that the poor and minorities bear a greater burden of the personal costs of certain types of road accidents, particularly those involving pedestrians and vehicles.

A roadway accident involving a vehicle imposes an external cost on its occupants as well as on nonoccupants, the latter being pedestrians, bicyclists, and persons in other vehicles. Studies show that two-thirds of the external costs from a passenger car accident are imposed on nonoccupants, while only one-third of the costs are borne by the occupants (Persson & Ödegaard, 1995). Since pedestrians and bicyclists are highly vulnerable to serious injuries and fatalities in a vehicular accident, it is likely that they bear a substantial portion of the external costs imposed by a passenger car accident. The external cost imposed on them by a larger vehicle, such as a truck or an SUV, is likely to be even higher.

As was shown in Table 12.6, vehicle ownership and use are lower for low-income and minority households than for other groups. Among the poor and minorities is a substantial proportion of vehicleless households, for whom walking and bicycling to activity sites is a necessity rather than a choice. Data provided in Table 12.3 show that the median income of pedestrians and bicyclists is lower than users of all modes except bus. Walking and bicycling are fairly risky, as they expose an individual to vehicular traffic. In 2000, a total of 4,739 pedestrians were killed in traffic accidents in the United States, comprising 11% of all traffic fatalities during the year (National Highway Traffic Safety Administration, 2000b). Since pedestrian trips constitute only about 7% of all trips and these trips are the shortest among trips by all modes, there is little doubt that pedestrians are over-represented in accident fatalities.

While it can only be inferred from available data that low-income persons are more likely than the affluent to be victims of pedestrian–vehicular roadway accidents, there is direct evidence that minority individuals are overrepresented among the victims of such accidents. According to the U.S. Census for 2000, whites constitute 75.1% of the country's total population, while African Americans and Hispanics constitute 12.3% and 12.5%, respectively. Among pedestrian–vehicular accident victims, however, whites constitute only 45%, while African Americans and Hispanics constitute 37% and 14%, respectively (Stutts & Hunter, 1999). Similarly, whites constitute only 51% of the victims of road accidents involving bicycles and vehicles, whereas African Americans and Hispanics constitute 30% and 15%, respectively.

Another safety concern in urban transportation is crime against transit passengers. Passengers become victims of crime not merely in transit vehicles or at stations/stops, but also while walking between transit stations/stops and their actual origins and destinations (Deka, 2002; Levine & Wachs, 1986). Low-income and minority populations live predominantly in the central locations of metropolitan areas, where crime rates are usually high. As a result, minorities and low-income persons are also more susceptible to crime. In their study of Los Angeles, Levine and Wachs (1986) found that Hispanic transit users were more concerned about crime than were members of other ethnic groups. In my own study on the relationship between crime and transit

use in Florida, I also observed a greater concern about crime among Hispanic respondents than among non-Hispanic respondents (Deka, 2002). In addition, I found that fear of crime in one's own neighborhood is higher among low-income respondents than among high-income respondents, and among urban residents than among suburban residents. In spite of higher crime rates in their neighborhoods, however, many poor and minority individuals continue to use transit because they lack alternatives.

CONCLUDING REMARKS

Many concerns about social and environmental justice in urban transportation are genuine. Although empirical data and the existing literature give many reasons to be concerned, the standard urban transportation planning process has generally ignored these concerns. The concerns about social and environmental justice in urban transportation are mainly subjective or normative. Converting them to operational measures is necessary if they are to be universally accepted as transportation planning objectives. Although awareness of these concerns may help advance the cause of the poor and minorities, significant changes in the distribution of transportation infrastructure and services—and the way they are funded—cannot be expected without incorporating these concerns into the rational transportation planning process.

Perhaps the most important question facing those concerned with social and environmental justice in urban transportation is whether mobility and accessibility are basic needs. Today's society generally acknowledges that the needs for food, clothing, and shelter are basic needs and agrees that it should provide supplemental income to support these needs. Society is less clear, however, in determining whether needs for mobility and accessibility are basic needs.

Transportation professionals should also ask themselves about the relative importance of efficiency and justice—both social and environmental. While efficiency is certainly desirable, it does not guarantee a fair distribution. In the search for efficiency, transportation planning can cause long- and short-term externalities that hurt the poor and minorities. It would be helpful if the transportation planning profession were to take such externalities into account and include objectives pertaining to social and environmental justice in the planning process.

REFERENCES

Altshuler, A. J. (1979). *The urban transportation system: Politics and policy innovation.* Cambridge, MA: MIT Press.

American Automobile Association. (2000). *Year 2000 model reviews: New car and truck buying guide.* Heathrow, FL: Author.

Anderson, S. C. (1983). The effect of government ownership and subsidy on performance: Evidence from the bus transit industry. *Transportation Research A, 17A,* 191–200.

Appleyard, D. (1981). *Livable streets.* Berkeley and Los Angeles: University of California Press.

Barr, N. (1998). *The economics of the welfare state* (3rd ed.). Stanford, CA: Stanford University Press.

Bullard, R. J., & Feagin, J. R. (1991). Racism and the city. In M. Gottdiener & C. Pickvance (Eds.), *Urban life in transition* (pp. 55–76). Newbury Park, CA: Sage.

Bullard, R. D., & Johnson, G. S. (1997). Just transportation. In R. D. Bullard & G. S. Johnson (Eds.), *Just transportation* (pp. 7–21). Gabriola Island, BC, Canada: New Society Publishers.

Bureau of Labor Statistics. (2002). *Consumer Expenditure Survey.* Washington, DC: U.S. Department of Labor. Available online at http://www.bls.gov/cex

Bureau of Transportation Statistics. (1996). *Transportation statistics annual report 1996: Transportation and environment.* Washington, DC: U.S. Department of Transportation.

Caplovitz, D. (1963). *The poor pay more: Consumer practices of low-income families.* New York: Free Press.

Cervero, R. (1984). Cost and performance impacts of transit subsidy programs. *Transportation Research A, 18A,* 407–413.

Cervero, R. (1990). Profiling profitable bus routes. *Transportation Quarterly, 44,* 183–201.

Chung, C., & Myers, S. L. Jr. (1999). Do the poor pay more for food?: An analysis of grocery store availability and food price disparities. *Journal of Consumer Affairs, 33,* 276–296.

Cole, L., & Foster, S. R. (2001). *From the ground up.* New York: New York University Press.

Collin, R. W., Beatley, T., & Harris, W. (1995). Environmental racism: A challenge to community development. *Journal of Black Studies, 25,* 354–376.

Council on Ethical and Judicial Affairs. (1990). Black–white disparities in health care. *Journal of the American Medical Association, 263,* 2344–2346.

Danielson, M. N. (1976). The politics of exclusionary zoning in suburbia. *Political Science Quarterly, 91,* 1–18.

Darden, J. T. (1995). Black residential segregation since 1948 *Shelley V. Kraemer* decision. *Journal of Black Studies, 25,* 680–691.

Davidoff, P. (1965). Advocacy and pluralism in planning. *Journal of the American Institute of Planners, 31,* 596–615.

Deka, D. (2002). *An assessment of the relationship between crime, security, and transit use in Florida using GIS and econometric approaches.* Tallahassee: Florida Department of Transportation, Office of Public Transportation.

De Lew, N., & Weinick, R. (2000). An overview: Eliminating racial, ethnic, and SES disparities in health care. *Health Care Financing Review, 21,* 1–7.

Dobson, A. (1998). *Justice and the environment.* Oxford, UK: Oxford University Press.

Downs, A. (1973). *Opening up the suburbs.* New Haven, CT: Yale University Press.

Feagin, J. R. (1999). Excluding blacks and others from housing: The foundation of white racisms. *Cityscape, 4,* 79–91.

Federal Highway Administration. (1995). *Nationwide Personal Transportation Survey data files* [CD ROM: FHWA-PL-97-034]. Washington, DC: U.S. Department of Transportation.

Federal Highway Administration. (2000). *Highways statistics 2000.* Washington, DC: U.S. Department of Transportation, Federal Highway Administration.

Federal Highway Administration and Federal Transit Administration. (2000). *Transportation and environmental justice case studies.* Washington, DC: U.S. Department of Transportation.

Federal Transit Administration. (1997). *1997 national transit summaries and trends.* Washington, DC: U.S. Department of Transportation, Federal Transit Administration.

Finke, M., Chern W., & Fox, J. (1997). Do the urban poor pay more for food?: Issues in measurement. *Advancing the Consumer Interest, 9,* 13–17.

Fix, M., & Austin Turner, M. (1999). Testing for discrimination: The case for a national report card. *Civil Rights Journal, 4,* 49–56.

Floyd, M. F., Shinew, K., McGuire, F., & Noe, F. (1994). Race, class, and leisure activity preferences: Marginality and ethnicity revisitied. *Journal of Leisure Research, 26,* 158–173.

Garber, N. J., & Hoel, L. A. (1999). *Traffic and highway engineering* (rev. 2nd ed.). Pacific Grove, CA: Brooks/Cole.

Ginzberg, E. (1991). Access to health care for Hispanics. *Journal of the American Medical Association, 265,* 238–241.

Gómez-Ibáñez, J. A. (1982). Assessing the arguments for mass transportation operating subsidies. In H. S. Levinson & R. A. Weant (Eds.), *Urban transportation: Perspectives and prospects* (pp. 126–132). Westport, CT: Eno Foundation.

Harvey, D. (1973). *Social justice and the city.* London: Edward Arnold.

Hodge, D. C. (1988). Fiscal equity in urban mass transit systems: Geographic analysis. *Annals of the Association of American Geographers, 78,* 288–306.

Holzer, H. J. (1991). The spatial mismatch hypothesis: What has the evidence shown? *Urban Studies, 28,* 105–122.

Ihlanfeldt, K. R., & Sjoquist, D. L. (1998). The spatial mismatch hypothesis: A review of recent studies and their implications for welfare reform. *Housing Policy Debate, 9,* 849–892.

Kain, J. F. (1968). Housing segregation, negro employment, and metropolitan decentralization. *Quarterly Journal of Economics, 82,* 175–197.

Kain, J. F. (1992). Spatial mismatch hypothesis: Three decades later. *Housing Policy Debate, 3,* 371–460.

Kain, J. F., & Meyer, J. (1970). Transportation and poverty. *Public Interest, 18,* 75–87.

Kain, J. F., & Persky, J. (1969). Alternatives to the gilded ghetto. *Public Interest, 18,* 77–91.

Khisty, C. J., & Lall, B. K. (1998). *Transportation engineering: An introduction* (2nd ed.). Upper Saddle River, NJ: Prentice-Hall.

Kirschenman, J., & Neckerman, K. (1991). We would love to hire them, but . . . : The meaning of race for employers. In C. Jencks & P. E. Peterson (Eds.), *The urban underclass* (pp. 203–232). Washington, DC: Brookings Institution Press.

Krumholz, N., & Clavel, P. (1994). *Reinventing cities.* Philadelphia: Temple University Press.

Lave, C., & Crepeau, R. (1994). Travel by households without vehicles. *Travel Modes Special Report, 1,* 1–47. Washington, DC: U.S. Department of Transportation, Federal Highway Administration.

Levine, N., & Wachs, M. (1986). Bus crime in Los Angeles II—Victims and public impact. *Transportation Research A, 20A,* 285–293.

Lewis, T. (1997). *Divided highways: Building the Interstate Highways, transforming American life.* Harmondsworth, UK: Penguin Books.

Los Angeles Times. (2002, March 19). Final MTA appeal of bus accord fails. California section, p. 1.

Malgren, J. A., Martin, M., & Nicola, R. (1996). Health care access of poverty-level older adults in subsidized public housing. *Public Health Reports, 111,* 260–263.

Meyer, J. R., & Gómez-Ibáñez, J. A. (1981). *Autos, transit, and cities.* Cambridge, MA: Harvard University Press.

Musgrave, R. A., & Musgrave, P. B. (1989). *Public finance in theory and practice* (5th ed.). New York: McGraw-Hill.

National Highway Traffic Safety Administration. (2000a). *The economic impact of motor vehicle crashes 2000.* Washington, DC: U.S. Department of Transportation.

National Highway Traffic Safety Administration. (2000b). *Traffic safety facts 2000: Pedestrians.* Washington, DC: U.S. Department of Transportation.

Nord, S., & Ting, Y. (1994). Discrimination and the unemployment durations of white and black males. *Applied Economics, 26,* 969–979.

Orfield, G. (1974). Federal policy, local power, and metropolitan segregation. *Political Science Quarterly, 89,* 777–802.

Papacostas, C. S., & Prevedouros, P. D. (1993). *Transportation engineering and planning* (2nd ed.). Englewood Cliffs, NJ: Prentice-Hall.

Persson, U., & Ödegaard, K. (1995). External cost of road traffic accidents: An international comparison. *Journal of Transportation Economics and Policy, 29,* 291–304.

Pucher, J. (1981). Equity in transit finance: Distribution of transit subsidy benefits and costs among income classes. *Journal of the American Planning Association, 47,* 387–407.

Pucher, J. (1983). Who benefits from transit subsidies?: Recent evidence from six metropolitan areas. *Transportation Research A, 17A,* 39–50.

Rawls, J. (2001). *Justice as fairness: A restatement.* Cambridge, MA: Harvard University Press.

Steinberg, M. (2000). Making sense of environmental justice. *Forum for Applied Research and Public Policy, 15,* 82–89.

Stutts, J. C., & Hunter, W. W. (1999). *Injuries to pedestrians and bicyclists: An analysis based on hospital emergency department data.* McLean, VA: U.S. Department of Transportation, Federal Highway Administration, Office of Safety Research Development.

Sullivan, L. W. (1991). Effects of discrimination and racism on access to health care. *Journal of the American Medical Association, 266,* 26–74.

The Public Purpose. (2002a). Comparison of light rail cost per trip and the cost of leasing or buying a new car. In *Urban Transportation Fact Book.* Belleville, IL: Wendell Cox Consultancy. Available online at http://www.publicpurpose.com/ut-newcar.htm.

The Public Purpose. (2002b). U.S. public transport new starts (rail) cost per new compared to monthly 5th quinti In *Urban Transportation Fact Book.* IL: Wendell Cox Consultancy. Availab line at http://www.publicpurpose.co railpoor.htm

Todd, J. S., Seekins, S., Krichbaum, J., Harvey, L. (1991). Health access America Strengthening the U.S. health care system. *Journal of the American Medical Association, 265,* 2503–2306.

Turner Meiklejohn, S. (2000). *Wages, race, skills and space: Lessons from employers in Detroit's auto industry.* New York: Garland.

Wachs, M. (1989). U.S. transit subsidy policy: In Need of Reform. *Science, 244,* 1545–1549.

Westra, L., & Wenz, P. S. (1995). Introduction. In L. Westra & P. S. Wenz (Eds.), *Faces of environmental racism* (pp. xv–xxiii). Boston: Rowman & Littlefield.

Whitelegg, J. (1997). *Critical mass: Transportation, environment, and society in the twenty-first century.* London: Pluto Press.

Wright, P. H., & Ashford, N. J. (1997). *Transportation engineering, planning, and design* (4th ed.). New York: Wiley.

Yinger, J. (1998). Housing discrimination is still worth worrying about. *Housing Policy Debate, 9,* 893–927.

...portation and
...e Environment

CHANG-HEE CHRISTINE BAE

The issues linking transportation and the environment are multiple and complex. I give priority to automobile-related air pollution, reflecting its emphasis in the transportation-environment literature. Many other important environmental impacts in addition to air emissions are important, however, such as noise, urban runoff, impacts on water quality, and consequences for the natural habitats of animals. In addition to automobiles, ships, pleasure boats, airplanes, off-road vehicles, construction equipment, lawnmowers, snowmobiles, and agricultural equipment (not discussed here because of their negligible urban impacts) also have nonnegligible environmental consequences.

Societal concerns with these problems are motivated by their impacts on health, noise, visibility, aesthetics, and natural environment. There are many strategies to mitigate these problems, but the major choice is between *pricing* approaches (i.e., use of incentives/disincentives) and *regulatory* approaches. Other relevant issues include the equity consequences of environmental impacts, global warming, and the measurement of environmental externalities. Overall, I strike a moderately optimistic note. The environmental impacts of transportation are substantial, but many of them are remediable via technological change and changes in household behavior if subject to appropriate price incentives and disincentives.

TRANSPORTATION AND AIR POLLUTION

Table 13.1 lists the U.S. Environmental Protection Agency (EPA) standards for the criteria pollutants (carbon monoxide [CO], nitrogen dioxide [NO_2], ground-level ozone [O_3], sulfur dioxide [SO_2], particulate matter [PM_{10} and now $PM_{2.5}$], and lead [Pb]) controlled by the Clean Air Act of 1970 (revised in 1977 and again in 1990) and some of the symptoms associated with excessive exposure. The Environmental Protection Agency has two types of standards. One is a primary standard to protect public health; the other is a secondary standard to protect public welfare. However, the allowed maximum concentration levels for primary and secondary standards are the same, except CO (there is no secondary

TABLE 13.1. Clean Air Act Criteria Pollutants and Their Attributes

Air pollutants	Chemical description	Properties	CAA criteria	Pollutant sources	Impact on humans	Impact on nature	Impact on atmosphere	Trends and regional concentration
Carbon monoxide	CO	Odorless, colorless, poisonous gas	8-hr: 9 ppm ($10\ \mu g/m^3$) 1-hr: 35 ppm ($40\ mg/m^3$) Should not be exceeded more than once a year	Incomplete internal combustion; fuel combustion; biomass burning; wildfires; industry; woodstoves; cooking, cigarette smoke; indoor heating	Headache; dizziness; nausea; cardiovascular system (e.g., angina pectoris); heart rate (heart attack); respiratory; Ischemia-related leg cramps; impaired vision/ cognitive ability; brain damage and death	Plant respiration	(Nature produces more CO, but regarded as not harmful.)	Highest concentration in the Northeast, the West, and Fairbanks, AK; higher level of concentration in urban than in rural and suburban areas; significant improvement from catalytic converters (1975) and oxyfuel introduction (1992)
Nitrogen dioxide	NO_2	Reddish brown, highly reactive gas contributing to ground-level ozone, PM, haze, and acid rain formation	Annua avg: 0.053 ppm ($100\ \mu g/m^3$)	High-temperature fuel combustion in automobiles, power plants, furnaces, turbines, household heaters/gas stoves	Lung infection; children's bronchitis; eye and nose irritation	Various impacts on plants (e.g., growth, injuries, deaths, etc.); fish kills plant diversity via increased nitrogen inputs to wetland/ terrestrial/aquatic ecosystems (eutrophication)	Reduces visibility; causes acid rain; reduces ozone shields; ozone precursor (form photochemical smog: NOx + HC)	High concentration in coastal areas of Southern California and New England. Increasing in the north central states. Highest concentration in urban areas; very low concentration in rural areas.
Particulate matter	PM_{10}, $PM_{2.5}$	Dust; soot; dust cloud/haze in group; liquid droplets in the air; some particles are invisible; contributing factors include sulfur dioxide, nitrogen dioxide, VOC in the atmosphere	PM_{10} 24-hr avg: 150 $\mu g/m^3$ Annual avg: 50 $\mu g/m^3$ over 3 years $PM_{2.5}$ 24-hr avg: 65 $\mu g/m^3$ Annual avg: 15 $\mu g/m^3$	Fuel combustion; waste incineration; automobile emissions; diesel trucks; power plants, residential fireplaces and wood stoves, industrial processes. Fugitive and nonfugitive dust; driving unpaved roads. Sulfates, nitrates, organic carbon contributions to $PM_{2.5}$	Serious impact on human health; Carcinogen (lung and stomach cancer); cardiovascular system; reduces lung function; lung inflammation; asthma, bronchitis, cardiac arrhythmias; heart attacks. Children, elderly, and people with lung and heart disease are more susceptible.	Deposits on soil, plants, water, leaves, etc. Affects nutrient balance and acidity; corrodes leaf surface; affects plant metabolism	Reduces visibility; Affects vegetation and ecosystem; damages paint; erosion of building materials; precursor to acid rain	Higher concentration in Southern California, eastern U.S. ($PM_{2.5}$), western and the north central states (PM_{10}); urban concentrations are much higher than rural

(continued)

TABLE 13.1. (continued)

Air pollutants	Chemical description	Properties	CAA criteria	Pollutant sources	Impact on humans	Impact on nature	Impact on atmosphere	Trends and regional concentration
Sulfur dioxide	SO_2	Orange; pungent odor	24-hr: 0.14 ppm (365 $\mu g/m^3$) Annual avg: 0.03 ppm 180 $\mu g/m^3$ 3-hr: 0.50 ppm (1,300 $\mu g/m^3$)*	Mainly from fuel combustion, power stations; petroleum refineries; metal smelting; industrial processes	Respiratory illness including bronchitis; temporary breathing impairment of asthmatic children; wheezing; chest tightness; shortness of breath; cardiovascular disease; toxic synergic effects of aerosol; mortality. Children, elderly, and people with lung and heart disease are most susceptible.	Acidification of soils, lakes, and streams; plant with high physiological activity; damages forests	Acid rain; damages plants and marble and limestone buildings); reduces visibility; fog	Higher concentration in the eastern states; urban and suburban concentrations
Ozone (ground level)	O_3	Colorless; pungent odor	1-hr: 0.12 ppm (a day/year) 8-hr: 0.08 ppm (156 $\mu g/m^3$)	Fuel combustion; Reaction of VOC and NO under heat and sunlight	Respiratory illness (lung endema, vulnerable to people with asthma, emphysema, and chronic bronchitis) coughing, sneezing, pulmonary congestion, chest pain.	Reduces agricultural crop/commercial forest yields; susceptible to disease, pests, environmental stress; damage to plant foliage. Reduces aesthetics of trees.	Forms urban smog; contributes to the oxidation efficiency of the atmosphere.	Higher concentration in Southern California, the Gulf Coast, the Northeast and the north central area. Improvement in the West and Northwest, but declining in the Southeast, Southwest, and north central regions in 1990s because of favorable wind conditions. Increasing in National Parks
Lead	Pb		3 mo avg: 1.5 $\mu g/m^3$	Automobile exhaust; Antiknock additives; old dried paint; chemical plants, refineries, industry; aviation fuels	Affects brain, kidney, liver function, and nervous system; Affects children/fetus IQ, learning ability; seizure; anemia; blood pressure/heart disease; death	Affects plants near the source	Can be traced over distance, e.g., Antarctica	Successful cleanup from mobile sources because of the introduction of unleaded gasoline (1975). Most lead concentrations now are related to stationary sources

Note. Measurement units of pollution levels: Each monitoring station measures a criteria pollutant. mg/m³, milligrams per cubic meter; ppm, parts per million; μg/m³, micrograms per cubic meter; m³, cubic meter; PM_{10}, particles with diameter less than 10 micrometers (about one-seventh of the diameter of a human hair); $PM_{2.5}$, particles with diameter less than 2.5 micrometers; *, secondary standard. Sources: Godish (1997); TRB (1997); U.S. EPA *Air Trends* website (*http://www.epa.gov/airtrends*).

standard for CO) and SO_2. Mobile sources[1] account for sizeable proportions of total emissions for several of these pollutants. The details of the standards and their rationale are quite technical; explaining them is not a major priority here (see EPA Green Book, Part 50 for details). A more important point is that if metropolitan areas do not meet all federal clean air standards, they are designated "nonattainment areas" and subject to sanctions—for example, withdrawal of federal highway funds. In fact, as of 2000, there were 127 metropolitan areas that were "nonattainment," most of them for one pollutant, none for more than three. Out of 166 cases, 68 were for PM_{10} and 57 for O_3. As in the case of ozone standards in Los Angeles, metro areas can obtain an extension of the attainment date. Unlike Western Europe, the United States does not have controls on carbon dioxide (CO_2) emissions, although California plans to introduce such controls. Recently, there have been 27 states that have completed *voluntary* state action plans to reduce greenhouse gases as of September 2003. However, many of the proposed measures are dealing with stationary sources (e.g., cleaner and more efficient energy sources, building types, etc.) rather than mobile sources (e.g., alternative fuel fleets, modal shift from trucks to trains, etc.) (EPA Global Warming-Actions, 2004). This reflects the lower degree of concern

with global warming issues in U.S. policymaking compared with abroad.

Table 13.2 presents national emission trends over the past 60-plus years. Most pollution emissions peaked around 1970, suggesting that federal Clean Air Acts have made a substantial difference. The major success story has been lead, where emissions have declined by about 98% over the past 30 years, largely as a result of the switch to unleaded gasoline. CO and VOC emissions have also declined sharply over the last 30 years, but NO_2 emissions have remained stable for more than two decades. Because VOC emissions interact as precursors of ozone, the decline in VOC emissions has had a beneficial effect. PM_{10} emissions have continued to increase.[2] The California numbers are much more impressive, with air pollution declines across the board. This is the result of a particularly strong regulatory regime, although it is arguable that a more efficient approach might have emphasized market incentives/disincentives. Some of the data for the United States as a whole are presented visually in Figure 13.1.

Table 13.3 shows that in the United States transportation is the major source of CO (about four-fifths of the total) and accounts for approximately one-half of ozone precursors (NO_2 and VOCs). It is not a major source of PM (perhaps the most dangerous pollutant), although transport-related

TABLE 13.2. National Emission Trends, 1940–1999 (1,000 tons)

Year	CO	NO_2	VOC	SO_2	PM_{10}	$PM_{2.5}$	Lead	Ammonia
1940	93.6	7.4	17.2	20.0	16.0	NA	NA	NA
1950	102.6	10.1	20.9	22.4	17.1	NA	NA	NA
1960	109.7	14.1	24.5	22.2	15.6	NA	NA	NA
1970	129.4	20.9	31.0	31.2	13.0	NA	220.9	NA
1980	117.4	24.4	26.3	25.9	7.1	NA	74.2	NA
1990	98.5	24.0	20.9	23.7	30.0	8.0	5.0	4.3
1999	89.5	24.5	17.9	18.9	34.7	8.4	4.0	5.0

Note. The PM_{10} trends are misleading because data on major sources (e.g., fugitive dust, agriculture, and wind erosion) were not available before 1985. Source: U.S. Environmental Protection Agency, National Air Pollutant Emission Trends Website, available online at *www.epa.gov/ttn/chief/trends.*

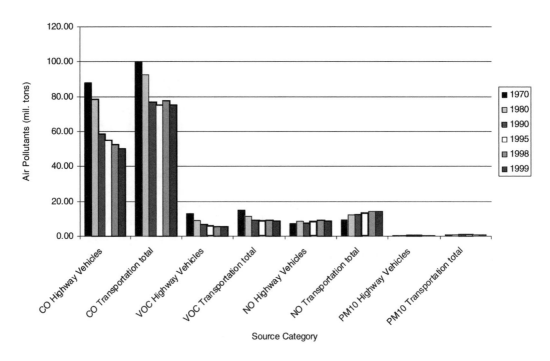

FIGURE 13.1. Trends in U.S. transportation-related pollution, 1970–1999. Source: U.S. Department of Energy, *Transportation Energy Databook: Edition 2001*—Table 4.2, Total National Emissions of Carbon Monoxide, 1970–99; Table 4.4, Total National Emissions of Nitrogen Oxides, 1970–99; Table 4.6, Total National Emissions of Volatile Organic Compounds, 1970–99; Table 4.8, Total National Emissions of Particulate Matter (PM-10), 1970–99; Table 4.10, Total National Emissions of Particulate Matter (PM-2.5), 1990–99.

TABLE 13.3. Transportation's Share in Pollutant Emissions, 1999 (Millions of Short Tons, Percentage)

Sector	CO	NO$_x$	VOC	PM$_{10}$	PM$_{2.5}$	SO$_2$	NH$_3$
Highway vehicles	49.99	8.59	5.30	0.3	0.23	0.36	0.26
	55.9%	35.1%	29.6%	0.8%	2.7%	1.9%	5.2%
Aircraft	1.00	0.16	0.18	0.04	0.03	0.01	0.00
	1.1%	0.7%	1.0%	0.1%	0.3%	0.1%	0.1%
Railroads	0.12	0.95	0.05	0.03	0.03	0.11	0.00
	0.1%	3.9%	0.3%	0.1%	0.4%	0.6%	0.0%
Vessels	0.14	1.00	0.04	0.04	0.04	0.27	0.00
	0.2%	4.1%	0.2%	0.1%	0.5%	1.4%	0.0%
Other off-highway	18.71	3.17	2.19	0.35	0.31	0.54	0.00
	20.9%	13.0%	12.2%	1.0%	3.7%	2.9%	0.1%
Transportation total	70.3	13.05	7.79	0.72	0.64	1.30	0.27
	78.6%	53.4%	43.5%	2.1%	7.6%	6.9%	5.4%
Nontransportation sources	19.15	11.42	10.13	34.03	7.75	17.57	4.7
	21.41%	46.71%	56.53%	97.96%	92.48%	93.11%	94.76%
Total of all sources	89.45	24.45	17.92	34.74	8.38	18.87	4.96
	100.0%	100.0%	100.0%	100.0%	100.0%	100.0%	100.0%

Source: U.S. Environmental Protection Agency, National Air Pollutant Emission Trends, available online at *www.epa.gov/ttn/chief/trends* (additional resources: *www.epa.gov/oar/oaqps*), Table 4.1., Total National Emissions of the Criteria Air Pollutants by Sector, 1999.

activities such as road construction and fugitive dust on roads (both paved and unpaved) are listed under nontransportation sources (if these were included, transportation's share of PM_{10} would rise from 2% to about 50% of the total). Transportation used to be, but no longer is, a major source of lead emissions.

National emission trends mask differences among metropolitan areas. Table 13.4 lists the year-2000 levels for the major transport-related pollutants in the 10 largest metropolitan areas. As expected, Los Angeles appears to be the worst on most indicators, although the main problem is not the counties closer to the ocean (Los Angeles and Orange) but the peripheral inland counties (the so-called Inland Empire) of Riverside and San Bernardino. On the whole, however, the variations are relatively modest among the metropolitan areas. The final column of Table 13.4 (the number of days when the more generalized Air Quality Index (AQI) exceeded 100) again shows that Riverside–San Bernardino is the main problem area in Southern California. The reason is that regional topography is a major contributing factor to air pollution, for example, the

San Gabriel and San Bernardino Mountains in the Los Angeles Basin.

Another point is that there are significant year-to-year variations for climatic or other reasons; for example, Atlanta's ozone concentration was 0.08 particles per million in 2000, but three times that level the year before.

IS AUTO-RELATED AIR POLLUTION A TRANSITIONAL PROBLEM?

Of all the environmental impacts of transportation, air pollution may have received a little too much attention. This argument is based on the thesis that best practice technology is either in place or imminent to reduce emissions from any vehicle to minimum levels (e.g., the California SULEV [super low-emission vehicles], gasoline-electric "hybrids," and fuel-cell vehicles in the pipeline; see Bae, 2001). If this is the case, then automobile-related air pollution may be a transitional problem, although the transition period is far from short. As automobile fleets turn over, emissions will continue to fall, especially if market disincentives (e.g., emission fees) are introduced to

TABLE 13.4. Transportation-Related Pollutants in the 10 Largest PMSAs, 2000

PMSAs	CO ($\mu g/m^3$)	NO_x (ppm)	O_3 (1-hr, ppm)	PM_{10} (8-hr, $\mu g/m^3$)	No. of days AQI 100
Los Angeles–Long Beach	10	0.044	0.11	46	30
New York	4	0.038	0.09	23	16
Chicago	4	0.032	0.08	35	15
Philadelphia	4	0.028	0.10	29	22
Washington, DC	5	0.023	0.09	24	22
Detroit	5	0.024	0.08	43	16
Houston	4	0.015	0.08	32	28
Atlanta	3	0.023	0.11	36	8
Dallas	2	0.014	0.10	29	14
Riverside–San Bernardino	4	0.038	0.12	59	94

Source: U.S. EPA (2000), National Air Pollutant and Emission Trends Report, Table A-15.

wean consumers away from the less environmentally friendly vehicle types.

It is common practice to refer to the high percentages of emissions resulting from automobiles. Although focusing on auto emissions is reasonable, it can be somewhat misleading. First, despite a huge increase in annual per capita vehicle miles traveled (VMT), from around 3,000 miles to almost 10,000 miles over the past 50 years, emissions have declined sharply as pointed out above (especially for CO and VOC, but not for NO_2) because of emission regulations and the updating of the automobile fleet. Second, the Clean Air Acts restrict their purview to six criteria pollutants, ignoring other more dangerous pollutants such as toxic gases and compounds that are covered by other legislation but are more likely to be associated with stationary sources. Third (as shown in Table 13.3), transportation's direct contribution to PM is relatively small (2.1% of PM_{10} and 7.6% of $PM_{2.5}$). However, some scientists and health experts have recently demonstrated the dangers of some types of PM on human health, mainly asthma attacks and cancer risks. PM is probably the most dangerous criteria pollut-

ant. Its impact would be much less if the emissions from diesel trucks could be significantly reduced (the phenomenon of "hot spots" is usually associated with stationary sources, but there are a few transport-related PM examples such as ports; see Box 13.1). On the other hand, as pointed out above, transportation does account for the majority share of CO and NO_2 and a high share of VOC emissions. Fourth, technological changes are dramatically reducing the emissions from new vehicles. Whether these reductions are the result of autonomous innovations by the automobile manufacturers or because of technology-forcing governmental mandates is a topic of substantial debate.[3] In the end, the reasons do not matter much; the innovations are either here or in the pipeline (Bae, 2001). The probable "end game" technology, hydrogen-powered fuel cells, is only at the concept vehicle stage, and there are some technical problems (e.g., on-vehicle storage and the fuel supply infrastructure) to be resolved, but the obstacles are not insuperable and other fuel-cell options (e.g., using methane) are possible in the short run (Box 13.2).

BOX 13.1. "Hot Spots"

A "hot spot" is a local neighborhood where pollution levels are much higher than in other parts of a metropolitan region. In terms of air pollution, they are usually related to stationary sources, such as one or more petroleum refineries or petrochemical plants. However, some mobile source pollutants (especially CO and PM) have a very local effect. Although traffic is more or less ubiquitous, there are some locations (e.g., close to freeway interchanges or bus depots) that qualify as "hot spots."

One example is Mira Loma in Riverside County, California. With a population of 17,600, it was originally developed as a rural community with clean air. More recently, however, a decision was made to promote it as a warehouse hub, and as a result property tax revenues tripled between 1997 and 2002. There are now more than 70 warehouses and distribution centers located there, occupying space equivalent to 434 football fields and employing more than one-fifth of Mira Loma's labor force. Traffic counts at the intersection of Etiwanda Avenue and Van Buren Boulevard indicate 400 diesel trucks per hour. As a result, Mira Loma has the worst PM_{10} readings in Southern California, 40% higher than the federal standard, whereas Los Angeles is 20% below it. The lung function of Mira Loma's children is 10% below the regional average.

BOX 13.2. Technology versus Land Use Controls to Reduce Auto-Related Pollution

An interesting aspect of the attempts to reduce auto-related air pollution is the debate about the relative importance of changes in land use (e.g., making cities more compact) or technological change to reduce tailpipe emissions. Planners and geographers are usually more favorably inclined to the view that altering spatial configurations can achieve wonders. However, the impact of this approach is problematic. Changing the metropolitan landscape is a very slow process. Higher densities mean more air-pollution health damages per unit area, unless automobile driving is dramatically reduced, and that appears not to be the case.

On the other hand, technology is making great strides. Electric vehicle projects continue to flounder because of lack of progress on perfecting batteries that give the vehicles range. But there is significant progress in other vehicle types, especially in California: gasoline–electric hybrids with their capacity to recharge batteries via regenerative braking; the SULEVs (super ultra-low emission vehicles), gasoline-powered cars with close to zero emissions; and potentially zero-emission fuel-cell vehicles soon to be tested under actual road conditions.

These developments suggest that the bottleneck is not one of technology but of production and marketing. Can enough of these low-polluting vehicles be produced? Perhaps not; hybrid capacity is currently less than 55,000 vehicles a year, trivial in the context of 15 million car sales each year. Can an incentives/disincentives structure be created that will encourage consumers to buy or lease enough of them? Of course, even if the answers to both questions is yes, there will be a long transition period, because the automobile fleet turns over slowly (the average age of the car on the road is 8.9 years, up from 5.6 years in 1970). Furthermore, although there has been some progress with diesel technology (e.g., soot traps), this is potentially much more dangerous, and trucks have an even longer life than cars (17.6% of trucks are more than 15 years old).

The range of low-emission vehicles will be broadened significantly in the next few years. A significant problem remains: consumer acceptance. The appetite for SUVs (sport utility vehicles) and trucks remains strong (currently, SUVs and light trucks account for 48.1% of new sales). There has been a dramatic increase (about 430%) in small truck registrations over the past 30 years. Because increased average fuel efficiency (from 10.0 to 17.1 miles per gallon) more than offset increasing VMT, the rise in truck fuel consumption (329%) was somewhat more moderate. A particular problem with the appetite for trucks is that they are significantly more polluting than cars. Figure 13.2 shows that the typical new light truck emits about 40% more CO_2 and NO_2 than the average new car, and about 30% more HC (VOC) and PM.

BACKGROUND ISSUES

Sustainable Transportation

Definitions of the concept of sustainable transportation vary, but most examples allude to reduced automobile dependence, more public transit, and more reliance on nonmotorized modes (walking and bicycling). Originally, the term was used in the context of saving nonrenewable resources (especially petroleum), although technological trends suggest that we will not need all the petroleum reserves available for automobiles. More recently, the concept has become synonymous with travel modes that maintain environmental quality—hence, the emphasis on public transit and nonmotorized modes.

An important issue in transportation-environment discussions is the feasibility of

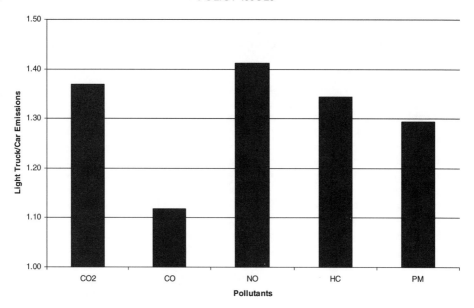

FIGURE 13.2. Index of light truck to car emissions, new 2001 vehicles (autos = 1.00). Source: U.S. Department of Energy, *Transportation Energy Databook: Edition 2001*—Table 4.14, Pollution from a Typical New Car and Light Truck, 2001 Model Year (based on pounds of pollutant per 15,000 miles of travel).

the dominant goal of many public policy-makers and some transportation planners: reducing automobile dependence, primarily but not solely for environmental reasons (e.g., Newman & Kenworthy, 1999). A key question is how many resources should be invested in alternative modes. Although many countries have more balanced modal splits than is the case in the United States, they also have rapidly increasing automobile ownership. In the United States, even if public transit ridership could be doubled, the environmental impacts of automobile use would be little affected because the transit ridership share is so low and continues to decline (4.7% of work trips in 2000 compared with 5.3% in 1990 and 12.1% in 1960; for all trips, the 2001 transit share was only 1.6%; see Pucher & Renne, 2003). However, some places (New York, Chicago, and many other major world cities) and some people (with particular types of workplace and residential location and life-style choices) rely upon public transit. The nonmotorized modes, walking and bicy-

cling, have also declined as a commuting mode to 3.3% in 2000, although the National Household Travel Survey (NHTS) data show some increase in walking for all trips to 8.6% in 2001 (perhaps an artifact of improvements in the survey instrument) (Pucher & Renne, 2003).

Is it reasonable, from a financial standpoint, to think of shifting a large proportion of trips from auto to transit? In most cases, providing the public transit services with the frequency, reliability, geographical coverage, and speeds that might have a significant impact upon automobile dependence is not affordable, regardless of the degree of cross-subsidization from auto user fees to public transit. A recent study of Atlanta (Bertaud & Richardson, 2004) explores the "counterfactual planning" example (Bae & Jun, 2003) of the extent of the MARTA rail system in Atlanta that would be necessary to achieve the public transit ridership share of Barcelona, Spain. In Atlanta, only 4.5% of all trips are made via public transit, while in metropolitan Barcelona 30% of all trips are made on public transit. MARTA has 46 miles of rail and 38 stations. As a result of Barcelona's compactness and Atlanta's extreme dispersion, Bertaud and Richardson estimate that MARTA would need 2,125 miles of rail and 2,800 train stations.

Public transit (when measured on a product life-cycle basis, e.g., taking account of the production of public transit vehicles) is not particularly environmentally friendly, unless it is based on alternative-fuel vehicles. On the other hand, the automobile is becoming much more sustainable, once the transitional problems associated with fleet turnover are overcome (almost all new cars with the exception of some SUVs and trucks are low polluters; as an example, the recent smog emission test of my 1999 car showed that its emissions were 2% of the maximum emissions permitted under California's tough smog control laws). Although congestion and travel time valuations remain significant issues, they can be addressed (in part, at least) by congestion pricing and other policy instruments. Of course, there are other problems in addition to the emissions issue, such as the disposal of batteries, old engine oil, old vehicles, and the risk of oil spills.

For many, the desire to own a car as soon as they can afford it seems irresistible; "automobility" is almost as important as home ownership. If this argument is valid, the key to a successful transportation–environment strategy, at least in the United States, is to devise effective means to accommodate the automobile while reducing its negative environmental consequences. This recognizes that automobile use has adverse environmental impacts but does not accept the proposition that these problems are intractable. In most Western countries automobile ownership is rising faster than in the United States despite much higher transit use in these countries. For example, the unweighted average annual rate of growth in automobile registrations, 1970–1999, was 4.0% in the United Kingdom, France, Germany, and Japan, compared with

1.4% in the United States. These differentials exist despite sharply higher gasoline prices compared with the United States (2.5–3.5 times higher), although of course other countries are playing catch-up in automobile ownership rates.

Outside the large cities and ignoring the case of intercity travel, the European experience is even more compelling than that of the United States. Automobility is becoming ubiquitous, despite pro-transit policies, high fuel prices, and congestion levels that would not be tolerated in the United States (Nivola, 1999).

Compact Cities

The literature on the transportation–land use nexus has emphasized the possibility of reducing the adverse environmental impacts of transportation via changing land uses. This chapter does not engage the sprawl and compact city debate in detail (land use issues are discussed in Giuliano, Chapter 9, this volume). However, the benefits from a land use approach are probably modest. First, the settlement pattern is largely determined, so changes in land use are marginal, although there is some debate about how large that margin may be (e.g., in newly developed suburban areas, revitalized core areas, and infill development). Second, travel behavior may be more susceptible to policy interventions than are land use preferences. Third, contrary to common belief, any visit to European or Asian cities confirms that compact cities with a substantial reliance on automobiles (the increasingly dominant case) aggravate rather than alleviate the health-related environmental impacts from automobiles. This is because for all those who can manage their lives in such an environment without automobiles there are many others who will not. Fourth, land use changes take longer to implement and may even be more costly than transportation changes; automobile fleets change slowly (the expected lifetime of a 1990 model is 16 years), but still much more quickly than does the housing stock. Fifth, there is as yet no convincing evidence that higher densities reduce automobile dependence or even (at least significantly) result in lower VMTs (Boarnet & Crane, 2002). Greenwald (2003) found that higher densities resulted in more walking, but had no influence on automobile use.

Nevertheless, many urban planners and social scientists believe that physical planning policy options can solve many social problems. Among the most popular notions is the idea that compact urban structures can reduce automobile dependence and yield significant environmental benefits. With the possible exception of living in the world's largest cities, this notion is problematic. Currently, without major restrictions on personal mobility, it would be difficult to provide sufficient public transit services, especially at the intraregional level. There are at least three reasons: (1) decision making for transportation planning is typically regional while land use decisions are primarily local; (2) NIMBYism (Not-In-My-Backyard) is widespread in the United States, as many elected officials and community groups are strongly opposed to transit stops or rail lines in their neighborhoods, some cities (e.g., Arlington, Texas) even having no public transportation services whatsoever; and (3) the costs of extensive public transportation networks are prohibitively high, especially rail services (ridership levels never reach ex ante projections and fare-box revenues do not cover operating costs).

In addition, the environmental gains are not substantial. Although new technologies are also transforming public transit vehicles, the pace of innovation and cost constraints make progress slower than in the case of private automobiles. A diesel-powered bus is many times more polluting than 50 best-practice automobiles, and natural gas and other alternative-fuel buses are up to 25% more expensive than conventional buses. Technology may not be a panacea, but sometimes urban planners are too quick to

underestimate the rate of technical innovations and their potential.

The impact of living in a compact neighborhood on *out-of-neighborhood* automobile trips is uncertain and probably small. Moreover, if such a lifestyle results in more shorter trips, emissions could get worse because more than three-fifths of emissions are connected with starting and stopping (the "cold start" and "hot soak" problem, i.e., the emissions associated with starting and turning off the engine). There are technological solutions, such as prewarmed catalytic converters, but none are in production and compliance could be a major problem even if they were available.

A more important consideration is the difficulty of effecting major changes in the built environment. In the United States, urban settlement patterns are more or less given. There can be small changes at the microlevel (e.g., infill projects, New Urbanist settlements), but the macrogeographical distribution remains more or less the same. As an example, consider Fairfield Village outside Portland, Oregon, a little-known model of New Urbanism. It is a small community of 600 rather compact (and quite attractive) dwellings of different types (single-family homes on small lots, townhomes with accessory units, live-and-work row houses, condominiums, and apartments). However, it is accessible only via the freeway. Another Portland, Oregon, example is Orenco Station, which has housing but also has light-rail access as its primary amenity; Orenco Station has been promoted as a blueprint for transit-oriented development (Bae, 2002). About 20% of its residents occasionally use the MAX light-rail service to downtown; this percentage is not higher because even if you are traveling to downtown Portland (about 15 miles away) it takes twice as long by MAX as by private automobile. On the other hand, this share is large by national public transit use standards, and the rail line has been the major marketing instrument at the Orenco Station sales office.

Unless further neighborhood design features are incorporated, compactness in itself is not a solution. Evidence in the United States suggests that the elasticity of automobile ownership with respect to population density is about −0.10 (Ingram & Liu, 1999), indicating that a doubling of population density results in only a 10% decline in auto use. The number of cars per unit area significantly increases. Compact cities would, as a result, be more congested and more polluted, and the aggregate negative human health impacts (especially to pedestrians) would be inevitably worse. This problem could be minimized by establishing an aggressive pedestrian-only section in the city center combined with peripherally located parking facilities, but this approach has made little headway in the United States, although it has been adopted extensively in some parts of Europe. Also, this approach works much better in small- or medium-sized towns.

The contrast between the sprawling United States and compact Europe (and, to a lesser extent, compact Asia) is narrowing rather than widening, despite major differences in public policies (Nivola, 1999). In France, for example, lifestyle preferences favor a quasi-rural life close to a big city (Prud'homme & Nicot, 2003), not only in Paris but also in other large cities (such as Lyon, Marseilles, and Bordeaux). Because public transit services outside the large urban cores are limited (except for intercity rail service), this lifestyle can be accommodated only via the private automobile.

The case for public transit has been skewed geographically. Newman and Kenworthy (1999), and many others, have largely used statistical data from the world's largest cities. There are major economies of scale in public transportation, so it is possible to live easily in a large city like New York, London, Paris, Madrid, Tokyo, Seoul, Hong Kong, Sydney, or Rio de Janeiro without a car, with minor adjustments in lifestyle. But away from a large city, managing without a car requires drastic restrictions in lifestyles,

such as fulfilling all goals within walking distance or organizing longer trips to fit sporadic public transit schedules. U.S. planners often refer reverentially to Europe with its compact cities and its pro-transit policies. Yet in the small towns of many European countries (such as France, Spain, and Italy) buses are rare, usually have few passengers, and schedules are infrequent (typically five buses per day and none in the evening). The obvious conclusion is that a life based on public transit is barely viable outside the largest cities, unless one's daily activities can be confined within the boundaries of a walkable small town or village. Also, public transit is not free from negative environmental impacts. Furthermore, despite high gasoline prices ($55 per fill-up on a compact car), severe traffic congestion, and high parking fees ($2.60 per hour, even in some small towns), Europeans continue to drive. The obvious reason is that mobility is a prized asset.

Nonmotorized Transportation

A more recently emphasized dimension to the public health discussion (that ties in closely with the compact city issue) is the need to emphasize nonmotorized modes (bicycling and walking). Although some research dates back to the 1970s, Frank and Pivo (1994) were among the first to revive the issue; Southworth (1997), Moudon, Hess, Snyder, and Stanilov (1997), Pucher, Komanoff, and Schimek (1999), and Audirac (1999) are other recent examples. Also, Khan and Vanderburg (2001) have published an annotated bibliography about healthy cities that refers to many papers on this issue.

Many Americans bicycle and walk for leisure. They have a wide choice of routes (sometimes aided by the provision of bikeways by state and local agencies). Such activity can have favorable health effects, unless it occurs in high-pollution zones. Commuting is a different matter because bicycling (and to a lesser extent walking) en route to work often involves exposure to noxious automobile fumes. Bicycling is popular among students who live near campus and for some other specific groups, but it takes dedication and commitment to make it viable as a commuting mode unless the work trip length is short (and perhaps suburban).

This is a United States–specific statement based on U.S. land uses. The answer may be different in some European cities: Groningen (The Netherlands), Cambridge (United Kingdom), and Freiburg (Germany) are merely three of many examples (Beatley, 2000). In the United States, the nonmotorization modal share for commuting is modest, as mentioned above, 3%, according to the 2000 Census; in contrast, the shares in The Netherlands and Denmark are 30% and 20%, respectively (Pucher et al., 1999).

The Centers for Disease Control (a federal agency) and the Robert Woods Johnson Foundation have recently begun to sponsor research into changing land uses as a stimulus to walking, bicycling, and better health. The key hypothesis is that compactness would result in more walking. The concept is somewhat problematic. There are several criticisms of this approach. First, leisure walking is more pleasurable and more common in suburban neighborhoods; walking in compact centrally located communities may be less satisfying because of congestion and pollution. Second, privatized solutions such as treadmills and stationary bicycles in the home and enrollment in health clubs may be more effective in maintaining physical health (working out on an inclined treadmill uses on average three times more calories per hour than walking outside, and up to seven times more with extreme use). Third, the existing built environment is inflexible. The most physically active living communities are in new suburban areas. There is no academic consensus on this issue yet. Researchers are only now testing the hypotheses, and there is no clear consistency in empirical results.

TRANSPORTATION-RELATED ENVIRONMENTAL IMPACTS

Health Impacts

Environmental impacts of transportation have major consequences for human health. Most of the benefits from cleaner air accrue to health benefits; these weigh heavily in cost–benefit analyses of control measures. Quantifying these health benefits is beyond the scope of this chapter; it requires deep medical and epidemiological knowledge, although social scientists can be helpful in monetizing the results given to them by the health specialists. Any review of the less technical literature reveals that assessments of health damages varies from study to study and changes over time. We know that people with pulmonary problems such as asthma and chronic bronchitis suffer from exposure to high levels of ozone, as do children (even healthy children, not only those with asthma) and the elderly. We also now know that PM (especially very small particles) is the most dangerous pollutant of all, and this finding requires a major redirection of our control efforts, which has now begun. Box 13.3 summarizes some recent research results about PM levels close to freeways. Although very high, they decline rapidly with distance.

A major health problem is diesel particulates. This is important because the diesel share in fuel consumption rose from 8.8% in 1975 to 19.9% in 1999. Recent medical research has revealed that these are the most dangerous of air pollutants, especially in terms of carcinogenic effects, far more dangerous than ozone (where the damage is largely related to lung function, asthma, etc.) and more general PM. Yet the trucking lobby in the United States has, until recently, been successful in resisting regulation. The new diesel regulations will not be implemented until 2007 (although some states are going ahead with their own regulations sooner), and there is anecdotal evidence that trucking companies are buying up the cheaper and better performing high-

BOX 13.3. Freeway Proximity

It is widely known that living or working close to freeways exposes people to higher levels of air pollution (and noise). Now the pollution impacts have been quantified. Recent research suggests that close proximity exposes individuals to 30 times the risk compared with more neutral locations (Zhu, Hinds, Kim, & Sioutas, 2002). Moreover, whereas standard discussions refer to CO and BC (black carbon), this research especially focused on nanoparticles (tiny particulate matter <0.1 micrometers; the EPA is starting to monitor $PM_{2.5}$, but up until recently only PM_{10} was regulated). These nanoparticles can result in neurological changes, pulmonary malfunctions, and cardiovascular problems. Based on their research on two Los Angeles freeways, the 405 (San Diego Freeway) and the 710 (Long Beach Freeway), the analysts found that pollution levels had declined by 70% at a distance of 330 feet.

reduce the rate of global warming. Fuel consumption is more efficient with diesel, and weaker performance characteristics are mitigated by the proliferation of turbo-diesel engines; these factors induce many Europeans to buy diesel cars, which account for almost one-half of the fleets in some countries. Although European diesel vehicles are cleaner than those in the United States, total emissions are substantial as a result of compactness and narrow streets where four- and five-story buildings trap pollutants in narrow corridors. There is a potentially important trade-off here, but the topic has been little researched, perhaps because the research into the health impacts of diesel fuel is so recent. The preferential taxation for diesel fuel may be an example of wrong-headed policies based on incomplete or out-of-date information, but this must await the outcome of a detailed cost–

polluting trucks in anticipation of the new rules (note that the expected lifetime of 1990 heavy trucks is 29 years). Off-road diesel vehicles are also an important environmental problem (see Box 13.4).

One important development in Southern California is that the Air Quality Management District (the agency in charge of pollution control) has imposed a regulation requiring government agencies to substitute alternative fuel for diesel vehicles. A challenge by the Engine Manufacturers Association and the Western States Petroleum Association was rejected by the U.S. Ninth Circuit Court of Appeals in October 2002.

In Europe, the situation is potentially even worse. Because diesel fuel has lower CO_2 emissions than gasoline, many European countries tax diesel fuel at a significantly lower rate than gasoline in order to

BOX 13.4. Off-Road Diesel Vehicles

Whereas diesel particulates from cars and other on-road vehicles have declined from 6.6 million tons in 1980 to 2.4 million tons in 2001, the particulates from off-road vehicles such as bulldozers and tractors have increased from 0.9 to 1.6 million tons over the same period. The United States has proposed new regulations to cut pollution (primarily soot and NO_2) from these sources by 90–95% by 2030, with the possibility of avoiding 9,600 premature deaths and 8,300 hospitalizations per year. California, always ahead in these matters, already imposes the sulfur standard in diesel fuel of 500 parts per million that will be the federal standard from 2007 to 2010. The proposed regulations make sense in cost–benefit terms, with an annual cost of about $1.5 billion that is tiny compared to an estimated $80 billion savings in mortality and medical costs.

benefit analysis and perhaps better data (the U.S. EPA announced in September 2002 that diesel exhaust may account for one-quarter of all soot particles and declared it a proven carcinogen).

Noise Impacts

Noise is another problem, especially within urban areas and particularly from trucks and motorcycles, but mitigation is more difficult here. Technology has some solutions, but they are imperfect, and noise ordinances are outdated. At the federal level, the Noise Control Act was passed in 1972, requiring a noise analysis of federal projects, but transportation problems have intensified considerably since then. There is a trade-off between noise and traffic speeds. For example, as congestion increases along highways, noise impacts in surrounding neighborhoods decline. Of course, a more important noise problem is from airplanes. However, this has been addressed in most locations via alternative or complementary mitigation measures, most often heavily subsidized insulation (double- or triple-pane windows) and noise barriers. These problems can usually be solved because one side of the negotiations (the airport authority and the airlines) is relatively few in number. Purchase of houses on the flight path or subsidies for insulation are feasible. Also, new airports (such as Kansai, near Osaka, Japan; Lamtiur, near Hong Kong; Incheon International Airport, near Seoul, Korea; and Denver International Airport) are more likely to be built at remote locations with no nearby residential development yet. With older airports near densely populated areas, night and early morning restrictions on flying hours are common.

Transaction costs become very high, on the other hand, in the case of automobile noise because both drivers and noise victims are numerous, and the polluters are difficult to identify. It becomes too difficult for the parties to negotiate a solution. This ex-plains why government agencies often intervene and construct noise barriers beside freeways close to residential developments (considered valuable if they reduce noise levels by at least 7 dBA [decibels]; in terms of cost, as an example, Washington State allows a noise mitigation cost of $21,500 per household for a noise level of 70 dBA). This approach is rarely feasible for arterial roads. In these cases, market forces attempt to deal with the problem either via mitigation (inside insulation) or by compensating discounts (lower house prices and rents on or near major roads). What helps in this situation is that individual responses to different dBA levels vary widely so that buyers or renters can usually be found for noise-impacted residences, albeit at a lower price. Manufacturers of transportation equipment also have a role to play: both airplanes and automobiles are much quieter now than they were a few decades ago.

Aesthetic Impacts

Aesthetic concerns are also related to transportation infrastructure. It is difficult to design roads, bridges, and rail infrastructure that satisfy these concerns. When roads are built in rural areas, landscaping designs that mask the roads can help, but in urban areas the price of land limits the scope of what is feasible to enhance the visual appearance of roads. There are several ways to improve streetscapes, for example, landscaping on medians and/or sidewalks, street furniture (ornamental lamp posts with hanging flags or baskets, benches, visual arts, etc.), arches, and so on. Rail is more complicated. Elevated rail is the cheapest approach but also the most aesthetically displeasing; this explains why Chicago's "El" is the only major elevated rail system in the United States, although Seattle has a monorail project from Ballard to West Seattle in the planning stage.[4] The most extensive new elevated rail system is Bangkok's "Skytrain," which is an exemplar of aesthetic ruin at the street level because of its massive four-story concrete pillars (Jenks, 2003). Surface rail is not aesthetically pleasing and can be dangerous without grade separation from roads; many pedestrian and vehicle accidents occur in Los Angeles on the suburban rail Metrolink and the light-rail Blue Line. Subways are the most satisfactory from an aesthetic point of view, but the higher costs are close to being prohibitive.

On the whole, however, aesthetic impacts are much less important than health impacts. As an example, the visibility costs of Los Angeles's air pollution were valued at $164 million per year compared with partial estimates of health costs that ranged up to $11 billion per year (Bae, 1994).

Mitigation Strategies

How should society deal with these problems? One possibility is "laissez-faire," that is, doing nothing. This is hardly attractive given the scale of these negative externalities. Another approach is negotiation between the polluters and the polluted, which can work in some cases, such as the airport noise issue, where the number of parties on one side is relatively small. Negotiation regarding auto-related pollution is unmanageable because of the high transaction costs, and most of us are both polluters and the polluted. Thus we are left with choosing between price incentives/disincentives or regulation. Fortunately, these approaches are quite effective. The major concern is the efficiency costs of these alternatives. In general, pricing works best, but even regulations are better than nothing.

Pricing versus Regulation

There are many policy instruments to mitigate the adverse environmental impacts associated with transportation. Most of them can be classified as either regulatory controls and standards (e.g., compulsory automobile emissions equipment, storm water runoff standards) or price incentives/disincentives (e.g., emission fees, parking fees, transit passes). The issue of pricing versus regulation as environmental policy tools is controversial. Pricing has at least two major advantages: it offers consumers choices, and prices are easier to adjust over time. Regulation, on the other hand, requires new legislation, revised ordinances, executive actions, or other bureaucratic interventions that are difficult to change.

Economists tend to exaggerate the need for optimal pricing rules (i.e., whereby prices are equated with marginal social costs), but the real world is a second-best world, and the search for the optimal price may be fruitless. Emission fees, or registration fees related to emissions rather than to vehicle price, are an example of a second-best policy option that has yet to be tried. Prices can be adjusted easily to affect behavior in desired ways and/or to raise revenues to pay for environmental amelioration; there is nothing wrong with this in a world of political realities. Of course, the impacts of price instruments can be exaggerated. Australian data suggest that a 40% reduction in bus fares would reduce CO_2 emissions by only 1.26%, while a carbon tax of U.S. 5 cents per pound would reduce them by 2.39% (Hensher, 2003). Another study, of the Dortmund region in Germany, predicted some reduction in CO_2 emissions per capita because the improved energy efficiency of automobiles more than offset rising auto ownership and increases in VMT. A goal of a 25% reduction could not, however, be achieved even with the help of alternative price mechanisms: higher gasoline prices, increased core-city parking fees, or reduced bus fares. Making driving more expensive via higher gasoline prices came closest to achieving the goal (Wegener, 2000). Occasionally, minor programs of price incentives can have a nonnegligible effect (see Box 13.5). This compares with unfavorable outcomes from mandated regulations (see Box 13.6).

BOX 13.5. Price Incentives and Regulation: The Lawnmower Exchange Program and MTBE Restrictions

Sometimes, environmental quality improvement is the accumulation of small steps. One recent innovation in Southern California was a lawnmower exchange program (lawnmowers are classified as a "mobile source," in effect, an off-road vehicle). The program permitted exchanging a working gas lawnmower plus $150 for a $449 cordless electric mower. The program eliminated 2,000 gas lawnmowers accounting for 10 tons of smog per year. In a few cases, however, there may be both regulation and price effects. An example is the MTBE (methyl tertiary butyl ether) regulations introduced in California in 2003. Here, environmental goals conflict. MTBE is a gasoline additive with beneficial effects on air quality, but there are risks of leakage from gasoline station storage tanks into groundwater supplies; hence, the regulation forbidding its use is being introduced for health reasons. But there will also be a price effect. Other additives are more expensive, perhaps adding 5 cents to the cost of a gallon of gas. The higher cost may marginally influence behavior (e.g., driving less or buying a more fuel-efficient car), but this would be a by-product of the policy, unrelated to its main objective.

BOX 13.6. Trip Reduction Programs

A simple regulatory idea to reduce auto-related emissions is to introduce measures aimed at reducing trips. Trip reduction, however, is much more difficult than appears at first sight. Consider two examples adopted by the Air Quality Management District in Southern California, the agency leader in this regard.

Regulation XV mandated ride-sharing plans aimed at achieving reductions in peak hour congestion and air pollution for firms with more than 100 workers. The goal was to increase the ratio of the numbers of morning commuters to vehicles (AVR: average vehicle ratio) via measures such as transit passes, van pools, carpooling incentives, walking and bicycling, telecommuting, and receiving cash in return for giving up parking privileges. It worked reasonably well in its first year, increasing the mean AVR by 2.7% (from 1.213 to 1.246), but the rate of improvement declined once those with the most flexibility to adopt alternatives to solo driving had adjusted their commuting behavior.

Another measure is the promotion of compressed work weeks, replacing the traditional 5 days of 8 hours with a 4-40 schedule, usually not working on Fridays. It has made major inroads in the public sector (where it is very popular with employees), but it has been less popular in the private sector, where only about 11% of employers are subject to trip reduction regulations. Unfortunately, its impact in reducing emissions is negligible. Work trips account for less than one-fifth of total trips. When workers have the day off, they do not stay at home but take other trips. Although the VMT may be reduced, because these nonwork trips are relatively short, the high emissions associated with starting and stopping offset any VMT savings. In the words of a converted original proponent of the program, "People drive like crazy on their days off."

OTHER ISSUES

Marine, Airport-Related, and Leisure Vehicle Pollution

In considering the transportation-environment link, it is important to consider modes other than the automobile. Ships are major polluters, not only via oil spills but also in ports. Oil spills receive considerable attention, but ports-related air pollution has been more or less ignored, perhaps because of the economic benefits of global trading and the enforcement difficulties with foreign-owned ships. Ships in the twin Ports of Los Angeles and Long Beach and related trucking activities are the largest source of air pollution in the Los Angeles area (Box 13.7). Although the aviation issue has focused primarily on noise, there are substantial environmental problems associated with airport-related ground transportation, both for freight and passengers. All terrain vehicles (ATVs) and snowmobiles have adverse environmental impacts in deserts, national parks, and other environmentally sensitive lands, and the key question regarding their use relates to the choice of policy options: prohibition, restricted areas, or pricing. Another problem not discussed in detail here is the transfer of nonindigenous species (especially pests) by air or ocean freight and vessels from one region or continent to another.

Urban Runoff

If air pollution is, on the whole, a transitional problem, other environmental consequences of automobile transportation deserve more attention than they have received. An important example is the urban runoff problem associated with the paving of roads, but also with related urban development; here, transportation and land use become closely intertwined. Rain or snow on impervious surfaces results in the runoff of highly contaminated water (including trash, bacteria, and toxic compounds) into the ocean, lakes, rivers, and streams.[5] The U.S. EPA has highlighted this problem by articulating new goals for storm water runoff, namely, that such runoff should attain the quality of drinking water. This would require tertiary-level treatment by reverse osmosis. To give an example of the potential costs of these regulations, in semiarid Los Angeles County (where it rains on only 9% of the days, with substantial rain [i.e., 0.5 inches] on only 6% of the rainy days; data based on 70-year averages), the esti-

---

BOX 13.7. Ports: Pollution Ignored?

Los Angeles still remains the pollution capital of the United States (despite Houston's intermittent appearance as No. 1), and we now know that the twin ports of Los Angeles and Long Beach are the largest source of pollution in the region, accounting for 14% of the total. A major dilemma is the trade-off between economic development and the environment, given that the ports are a major growth engine. This port pollution is also a major environmental justice issue because the pollution largely affects low-income and minority communities such as Wilmington, Harbor City, and San Pedro.

No regulations restrict emissions from international ships in U.S. ports, despite the fact that an idling ship can emit 1 ton of NO_2 per day. Port-related truck traffic generates up to 100 pounds of PM each day.

There are signs of progress, however. The lease of a new terminal to the China Shipping Holding Co. has been accompanied by a legal settlement that includes docking ships plugged into an electric grid rather than keeping their engines running and a $10 million fund for retrofitting trucks.

mated cost of providing facilities for 100% tertiary treatment could be as high as $198 billion (about $42,000 per household; see Gordon, Kuprenas, Lee, Moore, & Richardson, 2003)! Obviously, this will never happen; a major compromise will be necessary. These costs will vary widely among metropolitan areas, depending upon rainfall levels, the topography of watersheds, land availability, and other factors, but the total cost will be very high. Cost–benefit calculations would be helpful, but the opportunity costs of these sums in terms of medical care, education, childcare, community development, and other worthwhile societal goals are obviously significant.

Runoff problems are more severe in rainy regions such as the Pacific Northwest and the South, but they are far from negligi-

ble even in relatively arid regions such as Southern California because of underinvestment in storm water systems in these locations. There can be critical trade-offs, however. Box 13.8 describes an example. In the Pacific Northwest, certain species of salmon are protected under the Endangered Species Act, and their habitat in urban rivers and streams can be damaged by the increased urban runoff associated with the more compact urban development (including roads) promoted by growth management legislation.

Disruption of Natural Habitats

Another quite different issue is that roads, especially in environmentally sensitive areas, can disrupt the migratory pathways of wild-

---

**BOX 13.8. Salmon, the Endangered Species Act,
and the Washington State Growth Management Act**

In the Pacific Northwest in 1999 six types of salmon, including the Chinook, were listed under the Endangered Species Act. Transportation barriers (debris and broken branches blocking culverts, faster stream turbidity, and higher water temperature from urban runoff) hinder their upstream swimming. A survey found 4,463 sites where highways crossed "fish-bearing" streams. Because this number was too large to be dealt with given the resources available ($9.3 million, 1999–2001), a priority index was developed to reduce the number for treatment to 500. Methods included removing barriers (McDonal Creek, WA), reducing water temperature via vegetation and installing fish ladders (Bear Creek, WA), replacing concrete archways (Rasmussen Creek, WA), and controlling the depth and velocity of water via rock replacement (Canyon Creek, OR).

   An interesting aspect of the potential conflicts between competing environmental goals in the Washington case is the trade-off between the Endangered Species Act and the Growth Management Act (GMA; Washington's statewide growth management legislation). Preserving salmon habitats in urban streams requires an avoidance of impervious urban paving, including road construction, given that roadways account for 20–30% of the land used for urban development. The GMA, however, dictates compact urban development. On the one hand, the GMA mandates development within the "urban growth area" (or urban growth boundary); on the other hand, the pressures for compactness result in more people per unit of pavement. On balance, the area of impervious pavement within urban growth boundaries is probably more than in the absence of the GMA, so there is a contradiction between the two laws. Recently, Washington State passed a supplemental ordinance to protect critical areas. A compromise might be to accept the loss of some urban salmon habitats combined with a much more rigorous preservation of others, but this option meets with disapproval from strict environmentalists and conflicts with the Endangered Species Act.

BOX 13.9. Wetlands Banking

One of the primary expressions of antiroad sentiment is related to wetlands loss, for example, swamps, floodplains, lake fringes, and estuaries. Perhaps a million acres of U.S. wetlands have been lost to highways in the past 40 years. Because wetlands are rich in soil nutrients, have an important storm-water-retention function, absorb water pollutants, and provide quality habitat for birds and fish, their ecological value is high. As a result, the potential conflicts between environmentalists and transportation engineers are severe. The problem is more prominent in the Northwest and the South, regions subject to high annual rainfall, complex watersheds, and sometimes endangered wildlife. Dredging and filling associated with urban development results in increased runoff and permanent wetlands loss.

Wetlands banking is an appealing policy instrument to preserve wetlands. A well-known example is the case of San Francisco Airport, where an agreement was made to expand the existing airport rather than building a new one. Expansion involved the loss of up to 90 million cubic feet of mud, but the Airport Authority agreed to "bank" the adjacent wetlands in return. Some states, for example, Washington, have passed legislation that imposes a strict principle of "no wetlands loss." There are three ways to compensate for wetlands loss: (1) on-site or off-site preservation or development; (2) in-lieu fees; and/or (3) mitigation credits prior to granting a development permit.

---

life and/or disturb wetlands (see Box 13.9). Roadkill is a major problem with respect to some species; highway deaths have almost eliminated the ocelot (an endangered wild cat) and have been especially damaging to slow-moving animals such as salamanders and turtles. This is not an insignificant problem. In addition to the thousands of animals killed, more than 200 motorists are killed each year and thousands are injured in animal–vehicle collisions, with $200 million of insurance reimbursements. Forest carnivores are especially at risk because of their small populations, low reproduction rates, and extensive home ranges. Interestingly, landscape ecologists report that the ecological impact of "road avoidance, especially due to traffic noise" are greater than that of roadkill (see Forman and Alexander, 1998, for details on other roadside ecological impacts). While not discounting the importance of this problem, it is amenable to mitigation measures (e.g., grass-covered bridges over roads [see Box 13.10]) at a tolerable but not negligible cost; also funds are available under the Transportation Equity Act for the 21st Century (TEA-21).

Global Warming

The global warming issue remains somewhat controversial (National Research Council, 2002). There is probably now a consensus that global temperatures are rising and that this has all kinds of climatic consequences. There is still a debate about the rates of change, depending on the time period chosen, and on the relative contributions of natural and man-made sources. Also, although many of the predicted impacts are clearly adverse (e.g., coastal flooding because of rising ocean levels, more skin cancers), not all are (e.g., some areas will experience increased agricultural productivity). The interesting point, as noted earlier, is that the United States does not control auto-related CO_2 emissions, despite an opportunity to include this in the 1990 Clean Air Act. CO_2 emissions from motor gasoline are 60.3% of the national total, and 98% are from petroleum fuels; on the other hand, only 0.5% of lead emissions (a criteria pollutant) are associated with motor vehicles.

California was the first state to address this issue, requiring automobile manufac-

---

BOX 13.10. Wildlife Overpasses and Underpasses

To prevent highways from disrupting the movements and migration of wildlife, it is possible to build highway overpasses and underpasses. Banff National Park in Canada has two of the few animal overpass structures in the world. Traffic records indicate that more than 20,000 animals use these structures daily during the summertime peak in busy sections of the Trans-Canada Highway (Clevenger, 1997). The area is home to many large mammals, such as deer, elk, moose, wolves, and bears. When park authorities decided to widen the highway, they placed an 8-foot fence along the road to protect the animals from running across it and at the same time built 22 underpasses and two overpasses to protect wildlife passage, especially for deer and elk. Parks Canada reports that 43,725 wildlife crossings have been monitored from 1996 to 2002 (Government of Canada, 2002).

Other examples include a tortoise underpass in the Mojave Desert in San Bernardino County, California; badger tunnels in The Netherlands; salamander tunnels in Massachusetts; goat underpasses in Glacier National Park near Mt. Rainier, Washington; amphibian-reptile walls and tunnels in Florida; and an overpass over I-75 in the Marjorie Harris Carr Cross-Florida Greenway in Marion County intended to serve humans by day and animals by night.

Is this more of a rural than an urban issue? Perhaps. But metropolitan areas consist of whole counties, and include many locations that are not primarily urban. Moreover, geographers have for a long time subscribed to the concept of a "functional urban region," typically defined in terms of locations within 2 hours driving distance of an urban core that includes recreational areas that can be accessed daily. Hence, recreational areas that are accessible to metropolitan residents are important to maintaining urban quality of life.

Source: *http://www.fhwa.dot.gov/environment/wildlife crossings.*

---

turers to reduce CO_2 emissions by 2005 with rules taking effect in 2009. However, in August 2003 the U.S. EPA issued a ruling that does not require the reduction of greenhouse gas emissions (CO_2, hydrofluorocarbons, and some other emissions) on the grounds that they are not pollutants under the Clean Air Act. The EPA position is that any change would require legislative action by Congress. Whether this will happen probably depends on if, and when, the Democrats take over Congress. In any event, the State of California and environmental groups are likely to sue.

Europeans have been much more aggressive in this area, but the resulting priority given to diesel fuel (because of its lower CO_2 emissions) now seems to be problematic. Another relevant point is that much of the growth in transport-related energy consumption over the past 30 years is linked to

the growth in light and heavy trucks; energy consumption by autos has remained flat (see Figure 13.3).

Social Equity and Environmental Justice

The equity aspect of environmental exposure is an important social issue. The environmental justice literature contains many examples of poor and minority people being exposed to hazardous wastes, toxic releases, and other environmental hazards (e.g., Tiefenbacher & Hagelman, 1999; Saad, Pastor, Boer, & Snyder, 1999). The debate hinges around which came first, the people or the facilities? Regardless of the answer to this question, most of the pollution sources in these instances are stationary. Major exceptions include cutting new highways through low-income neighborhoods and transportation that is closely connected to a stationary source (such as trucking associated with a port or a major distribution yard).

On the other hand, some measures to reduce mobile source pollution may have a pro-equity impact. For example, consider the case of automobile-related air pollution. Although, as pointed out above, pollution decays with distance from freeways, serious attempts to improve mobile source pollution have to tackle the problem regionwide, mainly because ozone is widely diffused and may not be correlated with local traffic levels. Such measures to improve *regionwide* air quality tend to benefit the poor and minorities because the rich have previously purchased the better air quality sites (Bae, 1997a, 1997b).

For example, in the Los Angeles metropolitan region, the well-off tend to live on the Westside and near the ocean, where the ocean breezes blow the pollutants inland and where there are few major stationary pollution sources. The poor and the lower middle class, on the other hand, live inland

(even as far east as San Bernardino) in areas that are both "smog traps" and exposed to higher levels of stationary source pollution. The geography of pollution is very complex, but antismog strategies have to apply to the region as a whole and cannot be targeted to specific sites. Hence, air quality improves throughout the region, but it improves by much more in the poor, previously polluted, neighborhoods.

However, it is not all good news. The costs of emission equipment are more or less fixed from vehicle to vehicle so the costs of additional emission controls tend to have a regressive impact (i.e., their cost accounts for a higher proportion of a poor household's income). If emission fees were to be introduced (an interesting idea that has not been given much attention), their distributional impact is unclear. In the used car market, the poor would probably pay more because they own older cars; when they enter the new car market, however, they may pay less because they generally purchase less expensive, less powerful cars.

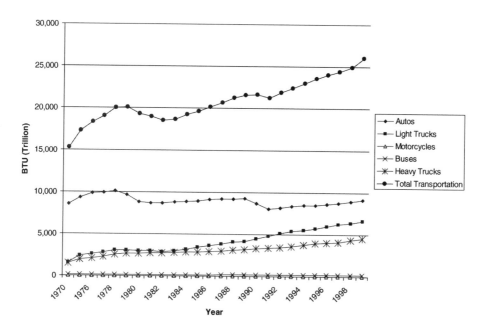

FIGURE 13.3. U.S. energy consumption by transportation mode, 1970–1999. Source: U.S. Department of Energy, *Transportation Energy Databook: Edition 2001*—Table 2.6, Transportation Energy Use by Mode, 1998–99.

Valuing the Environment

One of society's most important questions is how much to pay for the prevention and mitigation of environmental impacts. Answering this question is difficult, not merely because costs are unknown, but primarily because of the ubiquity of unpriced environmental externalities. The issues are even more complicated when we take account of the risks and uncertainties associated with long-term environmental impacts, such as global warming.

Two major aspects need to be addressed: equity concerns and the valuation of environmental "intangibles." Of course, the standard cost–benefit approach, with its emphasis on the net present value of benefits minus costs, is neutral with respect to equity goals. A benefit of $286 million to Bill Gates or the sultan of Brunei is treated the same as a benefit of $1 to each U.S. resident. Some techniques have been available for 30 years or more to address equity impacts, for example, by defining all the stakeholders and attempting to calculate the costs and benefits for each group (e.g., the Net Welfare Impact measure suggested by Bae, 1997a, 1977b), but many of these attempts are in the test stage compared with the straightforward net present value estimation approach.

Pricing environmental "intangibles" is the other problem, and it is not insignificant. The standard technique is the "contingent valuation" (CV) approach, based on hypothetical questions to individuals (or households) about what they would be willing to pay for a given environmental "good" or the dollar amount they would be willing to accept to tolerate an environmental "bad." The CV approach has been roundly criticized, and rightly so (Lindsey, 1994; Sugden, 1998). For example, the "willingness to pay" for environmental quality tends to be about only one-third of the "willingness to accept" environmental deterioration.

Yet it would be premature to dismiss the CV method, especially as there are few good alternatives. One modification, if an expensive one in terms of research budgets, is to expand the survey instrument to supplement the hypothetical willingness-to-pay questions with real-world behavioral correlates. For example, if an "environmentalist" is willing to pay much more for environmental quality, and environmentalists are $x\%$ of the population, then their mean estimate should have a weighting of $x\%$ in an aggregate societal estimate. These weighted societal estimates would be much more accurate than individual mean or median estimates. This illustrates the principle of modifying CV estimates via reference to real-world rather than solely hypothetical behavior.

CONCLUSIONS

This chapter has explored several key issues relating to the environmental impacts of transportation. The major point focuses on the link between automobile use and air pollution and on how this link is tied to the concept of sustainable transportation. The chapter offers a more cautiously optimistic perspective than that of many geographers or urban planners.

First, the automobile is a fact of life, almost a necessity in some places, and therefore it should not be treated as an evil to be eliminated at any cost. Globally, its use is increasing at a rapid rate and cannot be stopped, almost regardless of the price disincentives adopted.

Second, if accepted, this thesis leads to the conclusion that the way to proceed is to accommodate the automobile by minimizing its adverse environmental consequences. The societal consequences of ignoring mitigation of automobile pollution in favor of promoting other transportation modes are very risky.

Third—and here are the grounds for optimism—technology is generating the solutions that will eventually make the air pollution impacts of the automobile minimal (although other environmental consequences may persist, if in an attenuated form). SULEVS, hybrids, and eventually fuel-cell vehicles will become best-practice technology.

Fourth, there is a long transition period, so the remedies are not immediately imminent. Price disincentives may be necessary to push consumer tastes toward more fuel-efficient and environmentally friendly vehicles. Also, the diesel truck problem will haunt society for a long time.

Fifth, the argument is neutral about public transit as an approach for mitigating environmental impacts, although its impact may be minimal in view of the fact that only 1.6% of national daily trips are by public transit (National Household Travel Survey, 2001). Public transit options are valuable, especially to transport people in large cities and to meet the needs of the transportation-disadvantaged. Also, as a decreasing long-run marginal cost industry, public transit should not be required to pay for itself; hence, subsidies may be socially optimal. The technical reason is that the efficient marginal cost fare is below average cost. Buses and paratransit options are more appealing than rail, on the grounds of cost and flexibility. "New rail" cities have paradoxically experienced a decline in public transit ridership because the fall in bus ridership has exceeded the number of the new rail riders (many of those who now use rail previously used the bus).

Sixth, the automobile–air pollution link is only part of the transportation-environment story. Automobiles create other environmental problems, including storm water runoff associated with paved roads and urban development, damage to wildlife habitats, excessive noise, and aesthetic concerns. Solutions are technologically feasible in most cases, but they are often expensive.

Seventh, other transportation modes (e.g., ships, airplanes, snowmobiles, ATVs, and water-related recreational vehicles) have nonnegligible environmental impacts that have received relatively limited attention. Where these problems have been addressed,

the typical approaches have emphasized regulation more than pricing strategies.

Eighth, more sophisticated valuations of environmental quality are needed. Clearly, the prescription of some environmental groups in favor of a pristine environment at any price is impractical in a world of scarce resources. On the other hand, and increasingly so, society values environmental quality highly, so obtaining more precise valuations is important to determine how much to invest in mitigating the environmental consequences of transportation. Mitigation may be a sounder approach than restrictions and constraints on individual mobility.

Finally, it is necessary to establish a stronger relationship between air pollutants and human health impacts based on science, and to collect better air pollution data about micro environments.

NOTES

1. The major sources of air pollution emissions are classified into two types. One is stationary sources that are the emissions caused by factories, buildings, and other facilities. The other type is mobile sources, that is, emissions from on-road and off-road vehicles. This chapter deals with the latter type of emissions.
2. The long-term trends in PM_{10} are misleading because data on some key sources (e.g., fugitive dust, agriculture, and wind erosion) were not available before 1985.
3. The ZEV (zero emission vehicle) rules were eventually nullified in federal court and abandoned by the California Air Resources Board (CARB) in April 2003; CARB shifted its priorities from electric vehicles to hybrids, demanding an annual production of 125,000 hybrids by 2010. Nevertheless, the concept of a ZEV vehicle is retained by requiring some fuel-cell vehicles by 2008.
4. Seattle has some experience with elevated monorail with a short system from Seattle Center (near the Space Needle) to downtown.
5. Many urban areas have interconnected stormwater systems. Drains on pavements in coastal cities often carry the sign: "No Dumping, Drains to Ocean."

REFERENCES

Audirac, I. (1999). Stated preference for pedestrian proximity: An assessment of new urbanist sense of community. *Journal of Planning Education and Research, 19,* 53–66.

Bae, C,-H. C. (1993). Air quality and travel behavior: Untying the knot. *Journal of the American Planning Association, 59,* 65–75.

Bae, C,-H. C. (1994). *The income distribution impacts of air quality controls: A Los Angeles case study.* Unpublished doctoral dissertation, University of Southern California, School of Urban and Regional Planning.

Bae, C,-H. C. (1997a). The equity impacts of Los Angeles' air quality policies. *Environment and Planning A, 29,* 1563–1584.

Bae, C,-H. C. (1997b). How Los Angeles' air quality policies benefit minorities. *Journal of Environmental Planning and Management, 40,* 235–260.

Bae, C,-H. C. (2001). Technology vs. land use: Alternative strategies to reduce auto-related air pollution. *Planning and Markets, 4,* 37–49.

Bae, C.-H. C. (2002). Orenco Station, Portland, Oregon: A successful transit oriented development experiment? *Transportation Quarterly, 56,* 9–15.

Bae, C.-H. C., & Jun, M.-J. (2003). Counterfactual planning: What if there had been no greenbelt in Seoul? *Journal of Planning Education and Research, 22*(4), 374–383.

Beatley, T. (2000). *Green urbanism: Learning from European cities.* Washington, DC: Island Press.

Bertaud, A., & Richardson, H. W. (2004). Transit and density: Atlanta, the United States and Western Europe. In H. W. Richardson & C.-H. C. Bae (Eds.), *Urban sprawl in Western Europe and the United States* (pp. 293–310). London: Ashgate.

Boarnet, M., & Crane, R. (2002). *Travel by design.* New York: Oxford University Press.

Brindle, R. (1994). Lies, damned lies and "automobile dependence." *Australasian Transport Research Forum, 19,* 117–131.

Clevenger, A. P. (1997). *Highway effects on wildlife: A research, monitoring and adaptive mitigation study.* Report prepared for Parks Canada, Banff, Alta. Available online at http://www.hsctch-twinning.ca/assets/pdf/PR3.pdf

Forman, R. T. T., & Alexander, L. E. (1998). Roads and their major ecological effects. *Annual Review of Ecology and Systematics, 29,* 207–231. Available online at http://arjournals.annualreviews.org/doi/full/10.1146/annurev.ecolsys.29.1.207

Frank, L. D., & Pivo, G. (1994). Impacts of mixed use and density on utilization of three modes of travel: Single-occupant vehicle, transit, and walking. *Transportation Research Record, 1466,* 44–52.

Gordon, P., Kuprenas, J., Lee, J. J., Moore, J. E. II, & Richardson, H. W. (2003). *Financial and economic impact study of storm water treatment—Los Angeles County.* Los Angeles: University of Southern California, School of Policy, Planning and Development.

Government of Canada. (2002). *The G8 Kananaskis environmental legacy website. Existing science—The effects of highways on wildlife population.* Available online at http://www.g8legacy.gc.ca/english/assessment_dead_mans_flats/existing_science.html

Greenwald, M. J. (2003). The road less traveled. *Journal of Planning Education and Research, 23*(1), 39–57.

Guensler, R., & Sperling, D. (1994). Congestion pricing and motor vehicle emissions: An initial review. In *Curbing gridlock: Peak-period fees to relieve traffic congestion* (Transportation Research Board Special Report) (Vol. 2, pp. 356–379). Washington DC: National Academy Press.

Hensher, D. A. (2003). Integrated transport models for environmental assessment. In D. A. Hensher & K. J. Button (Eds.), *Handbook of transport and the environment* (pp. 787–804). Amsterdam: Elsevier.

Ingram, G., & Liu, Z. (1999). Determinants of motorization and road provision. In J. A. Gomez-Ibanez, W. B. Tye, & C. Whinston (Eds.), *Essays in transportation and economics and policy* (pp. 325–356). Washington, DC: The Brookings Institution Press.

Jenks, M. (2003). Above and below the Line: Globalisation and Urban Form in Bangkok. *Annals of Regional Science, 22*(3), 547–557.

Khan, N., & Vanderburg, W. H. (2001). *Healthy cities: An annotated bibliography.* London: Scarecrow Press.

Lindsey, G. (1994). Planning and contingent valuation: Some observations from a survey of homeowners and environmentalists. *Journal of Planning Education and Research, 14,* 19–28.

Moudon, A. V., Hess, P. M., Snyder, M. C., & Stanilov, K. (1997). Effects of site design on pedestrian travel in mixed-use, medium density environments. *Transportation Research Record, 1578,* 48–55.

National Research Council. (2002). *Climate change science: An analysis of some key questions.* Washington, DC: National Academy of Sciences.

Newman, P., & Kenworthy, J. (1989). Gasoline consumption and cities: A comparison of U.S. cities with a global survey. *Journal of the American Planning Association, 55,* 24–37.

Newman, P., & Kenworthy, J. (1999). *Sustainability and cities: Overcoming automobile dependence.* Washington, DC: Island Press.

Nivola, P. S. (1999). *Laws of the landscape: How policies shape cities in Europe and America.* Washington, DC: Brookings Institution Press.

Oakridge National Laboratory. (2001). *Transportation energy data book* (Contract No. DE-AC05-00OR22725). Oak Ridge, TN: Author. Available online at http://www.cta.ornl. gov/data

Pucher, J., & Renne, J. L. (2003). Socioeconomics of urban travel: Evidence from the 2001 NHTS. *Transportation Quarterly, 57*(3), 49–78.

Pucher, J., Komanoff, C., & Schimek, P. (1999). Bicyclying renaissance in North America?: Recent trends and alternative policies to promote bicycling. *Transportation Research A, 33,* 625–654.

Prud'homme, R., & Nicot, B.-H. (in press). Urban sprawl in France in recent decades. In C.-H. C. Bae & H. W. Richardson (Eds.), *Urban sprawl in Western Europe and the United States.* London: Ashgate.

Saad, J. L., Pastor, M. Jr., Boer, J. T., & Snyder, L. D. (1999). Every breath you take: The demography of toxic air releases in Southern California. *Economic Development Quarterly, 13,* 107–123.

Southworth, M. (1997). Walkable suburbs?: An evaluation of neotraditional communities at the urban edge. *Journal of the American Planning Association, 63,* 28–44.

Sugden, R. (1998). Anomalies and biases in the contingent valuation method. In D. Banister (Ed.), *Transportation policy and the environment* (pp. 298–319). New York: Routledge.

Tiefenbacher, J. P., & Hagelman, R. R. III. (1999). Environmental equity in urban Texas: Race, income, and patterns of acute and chronic toxic air releases in metropolitan counties. *Urban Geography, 19,* 516–533.

Transportation Research Board. (1997). *Toward a sustainable future: Addressing the long-term effects of motor vehicle transportation on climate and ecology.* Washington, DC: National Academy Press.

U. S. Environmental Protection Agency. (2000). *Green book: Part 50—National primary and secondary ambient air quality standards.* Available online at http://www.epa.gov/oar/oaqps/greenbk/40cfr50.html#Sec.%2050.1

U. S. Environmental Protection Agency. (2003). *Global Warming—Actions.* Available online at http://yosemite.epa.gov/oar/globalwarming.nsf/content/ActionsStateActionsPlans.html#Mitigation

U. S. Environmental Protection Agency. (2003). *National air pollutant emission trends.* Available online at www.epa.gov/ttn/chief/trends [Also see www. epa.gov/oar/oaqps]

Wegener, M. (2000). *Overview of the IRPUD model.* Dortmund, Germany: Institut fur Raumplanung.

World Bank. (1996). *Sustainable transport: Priority for policy reform* [Online]. Washington, DC: Author. Available online at: http://www-wds.worldbank. org/servlet/WDSContentServer/WDSP/IB/1996/05/000009265_3961214175350/Rendered/PDF/multi_age.pdf

Zhu, Y., Hinds, W. C., Kim, S., & Sioutas, C. (2002). Concentration and size distribution of ultrafine particles near a major highway. *Journal of Air and Waste Management Association, 52,* 1032–1042.

CHAPTER 14

Managing the Auto

GENEVIEVE GIULIANO
with SUSAN HANSON

The previous 13 chapters have described the many and complex dimensions of the contemporary urban transportation system in the United States. We have seen consistent historical trends over many decades, especially rising private vehicle ownership and use accompanied by declining market share for public transit and non-motorized modes. These trends have gone hand-in-hand with rising per capita income, the decentralization of metropolitan areas, and changes in economic structure that make households and firms ever more mobile. Furthermore, these fundamental trends are observed throughout the Western world.

Nevertheless, the United States is an extreme case when we look at travel patterns around the world. The United States has the highest rate of private vehicle ownership, the highest level of daily miles traveled, and the lowest rates of trip making by modes other than the auto. The transit mode share in Canada is far larger than that of the United States, despite comparable per capita income and geography. Countries of northern Europe also have comparable per capita income, yet private vehicle ownership is much lower, and vehicle miles trav-

eled (VMT) per capita is on average half that of the United States (Transportation Research Board, 2001). These differences are explained by a combination of history, culture, and policy. We have seen how the contemporary U.S. situation developed from decisions and actions taken by many actors, including households, government, and business.

The private vehicle-oriented system—some would say the private vehicle-dependent system—that has evolved in the United States provides unprecedented mobility to those people who have access to an auto, and that mobility has created many benefits for both individuals and society. But the U.S. system also has generated many problems. Those who do not have access to a private vehicle have suffered declining mobility as other modes have become less-effective substitutes. The transport sector consumes enormous amounts of energy, accounts for more than 40,000 accidental deaths each year, affects human health and the well-being of the environment, and is bound up with urban sprawl. Understandably, the current urban transportation system in the United States has been the target of a great deal of criticism (see, e.g., Kay, 1997; Nadis & MacKenzie, 1993; Whitelegg, 1993).

What can be done to change things? It is important to recognize that the urban transportation system we now experience is the result of choices made by individuals and institutions; it is not the outcome of some relentless natural or necessary process. As several authors in this book have noted, the transportation planning process itself has contributed significantly to the current situation of urban transportation in the United States. The traditional four-step modeling process, for example, emphasized highway building over other approaches to providing accessibility. Through the ongoing decisions and actions of a variety of actors (including planners, firms, citizens, and government), the transportation system is constantly changing. How would *you* like to see it changed? What new policies might bring about the kinds of changes *you* would like to see? How can citizens and activist groups help to bring about positive change in the urban transportation system?

Because the automobile is usually identified as *the* source of contemporary transportation-related problems, it is the focus of our thoughts in this chapter on potentials for change. We consider various approaches to managing the auto. Our focus is on policy: what governments at all levels can do, or how citizens can influence what governments do. We acknowledge at the outset that policies aimed at getting people out of their cars are unlikely to be broadly successful, simply because the mobility cars provide is of great value to most people. Nevertheless, change is both possible and necessary.

The remainder of this chapter discusses the context of transportation policy decision making, strategies for making positive change, and examples of how change is accomplished. Our "Top Eight" strategies are based on their potential effectiveness as well as their implementation feasibility. We close with some success stories of positive change.

THE POLICY CONTEXT

Developing effective strategies to solve the problems generated by our contemporary urban transportation system requires an understanding of context, of what works and what doesn't. We make four observations related to context.

There Is No Magic Bullet

Transportation policy history is littered with failures. Many of these failures are the result of thinking that there is a "magic bullet"—a single, painless solution (meaning one that is low in cost and requires no big change in individual behavior) to the urban transportation problem. In the 1950s, urban leaders saw the problem as congestion and the solution as the circumferential beltway. The Interstate Highway Program offered 90/10 funding (whereby the federal government provided 90% of the funds, the states 10%), making highway building a bargain from the perspective of states and local leaders. In the 1970s, and again in the 1990s, rail transit was seen as the answer to an expanding list of problems. Again, the provision of 80/20 (federal/local) capital funding made rail transit attractive to local leaders. Decision makers believed that better and faster transit would surely draw people out of their cars. In the 1980s, many claimed that Transportation System Management (TSM) was the answer. TSM proponents argued that major highway building was over (highways are too costly, disruptive, and environmentally damaging); we would simply use existing capacity more efficiently. Needless to say, increasing efficiency was no match for increasing VMT. In the 1990s, electric vehicles were going to solve the vehicle emissions problem. In the Los Angeles region, numerous imagined magic bullets have appeared in regional transportation plans over the past decade. These include investments in rail transit, increases in telecommuting, mag-lev commuter trains, and shifting population and employment growth to the core of the region, all at a massive scale.

No single policy can solve the urban transportation problem. Highways, transit, system management, and a host of other strategies all have a role to play. As with many areas of public policy, many strategies, each contributing incrementally, may collectively result in significant change. Effective approaches are based on a suite of complementary and mutually reinforcing measures and policies. We will describe a range of possible approaches that need to be considered as a whole so that the synergies among them can be realized.

Effective Solutions Are Tailored to Specific Places and Circumstances

Many authors in this book have emphasized the distinctiveness of place. Each urban area combines resources, governance structures, and geography in a different way. Communities and neighborhoods within urban areas are distinctive as well. Furthermore, transportation policy decision making is increasingly local (see Johnston, Chapter 5, this volume, on the transportation planning process, and Taylor, Chapter 11, this volume, on transportation finance). It is therefore not surprising that what works in one place does not necessarily work well in another, or that a given policy will require different implementation strategies in different places to be effective. In addition, the synergies among a suite of policies will come together and work differently in different places.

Many examples illustrate that what works in one place may not work in another. An obvious example is rail transit. New York City is utterly dependent upon its extensive subway system, yet the rail systems in Miami or Sacramento or Atlanta have no significant impact on accessibility or travel patterns. Deep discounts on monthly transit passes seem to be an obvious strategy for increasing transit ridership. For people with extremely low income, however, it is im-

possible to accumulate the funds to buy the monthly pass, so the discount is effectively unavailable.

Vehicle emissions provide a good example of the need for solutions that match local conditions. The particular contribution of emissions by the transport sector depends on industry mix, weather, air currents, and the state of the local economy. Each air basin therefore must develop its own plan for improving air quality. Similarly, the environmental impacts of roads, transit lines, or bicycle paths depend on the local area. Imposing uniform rules on water quality or runoff restrictions results in less efficient and more costly mitigation.

This is not to say that we cannot generalize about strategies for solving transportation problems or that lessons learned in one community are not transferable to another. From the perspective of the transit customer, for example, lower fares and higher service quality are always better. Parking charges provide a second example. Extensive evidence shows that hefty parking charges promote carpooling and transit use wherever they are imposed (Shoup, 1997); increases in the use of these modes depend, however, on local circumstances. Our point about uniqueness and generalization is simply that analysts and policymakers need to be sensitive to local differences when thinking about creating or adopting various policy options.

Change Is Gradual and Incremental

Policymaking in the United States is typically a slow and incremental process, punctuated by occasional major shifts. Even major pieces of federal legislation, such as the Clean Air Act (CAA) and the National Environmental Policy Act (NEPA) in 1970 or the Intermodal Surface Transportation Efficiency Act (ISTEA) in 1991, were preceded by several years of effort.[1] As participation in the policy process has broadened to include more stakeholders and interest groups and as decision making has devolved to lower levels of government, change has become even more gradual and some would say more difficult. Wachs (Chapter 6, this volume) notes the great variety of stakeholders in transportation and observes that there is often little consensus on how transportation problems should be solved. As these groups increase in number and narrow in perspective, it becomes more difficult to achieve consensus on any given issue. The downside of this rich complexity of stakeholders and participants is paralysis. We have many examples of transportation projects that are mired in lawsuits for decades and of policy proposals that are abandoned for lack of consensus. While some of these projects are deservedly stalled, others are worthy initiatives that would benefit society as a whole.

Furthermore, if we accept that for all its imperfections, the policy process in the United States is fundamentally democratic, it follows that policy change is really a function of citizen values and preferences. Environmental policy provides some examples. It was not until there was widespread public outrage at the state of many of the nation's rivers and lakes that regulations to restore water quality were passed and enforced. A decade ago, urban runoff was not even a topic of public policy discussion; now state departments of transportation (DOTs) around the country are changing the ways roads are designed so as to capture and treat runoff before it drains into nearby waterways.

Fuel economy standards (Corporate Average Fuel Efficiency, or CAFE) are another example. While the establishment of CAFE in 1975 represented a major policy change (this time the motivating factor was the energy crisis of 1973), the standards were introduced gradually, and the legislation was modified in response to U.S. automakers' claims that the standards could not be met as quickly as the legislation required (Nivola & Crandell, 1995). More recently, the debate is about extending standards to light trucks (see Greene, Chapter 10, this vol-

ume), and as of this writing, sport utility vehicles (SUVs) remain exempt from the tougher car standards.

Beware of Unintended Consequences

It is, of course, impossible to know in advance how people will react to changing circumstances, so there is always some uncertainty when new policies are implemented. In some cases, however, policies have unintended consequences that create new problems. In California, the South Coast Air Quality Management District (SCAQMD) established Regulation XV in 1988, which required all employers with 100 or more employees to develop and implement a plan to reduce commute trips. The regulation was quite complex but essentially required that each employer file a rideshare plan to show how the employer would achieve the target reduction in commute trips (targets were set based on employer size and location). An employer could be fined if the SCAQMD did not approve the plan but not if the employer had an approved plan and then did not reach the target reductions. The regulation was imposed first on the largest employers and phased in over a period of years.[2]

Employers became increasingly resistant to the regulation, as SCAQMD exercised its power and imposed large fines on some employers and as it became apparent that the employer was placed in the position of having to impose restrictions on employees or incur significant costs to offer large subsidies to persuade workers to rideshare or use transit. Employers argued that the regulation placed them at a disadvantage in the labor market (small employers were exempt) and added to their costs, placing them at a disadvantage in the national marketplace. In addition, commuting in effect became another labor relations issue. Eventually employer organizations went to the state legislature with their concerns, and in 1996 SB 836 was passed, prohibiting mandatory rideshare programs. Not only did

Regulation XV fail to achieve its trip reduction goals, it soured employers on the concept of ridesharing, and most existing programs were abandoned shortly after the 1996 legislation was passed.

Unintended consequences are sometimes the result of conflicting policies. The emergence of the SUV is a good example. The CAFE standards provided a loophole for supplying the U.S. market with large vehicles via the lower fuel efficiency standards for light trucks. U.S. travelers have no incentive to economize on fuel when its real price (i.e., price adjusted for inflation) is dropping. When a consumer considers the extra cost of fuel compared to the comfort, power, passenger capacity, and perceived safety of a larger vehicle, the larger vehicle becomes more attractive to him or her as the price of fuel declines. The price of fuel is itself the result of a policy decision to keep fuel taxes as low as possible. Most other countries see the fuel tax as an important lever to discourage both fuel consumption and private vehicle use.

THE CONTEXT FOR CHANGE

Many books have been written on how we might improve the U.S. urban transportation system or on how we might move toward a more sustainable transportation system. Before we discuss specific strategies for achieving these goals, we focus in this section on the context within which strategies for change are conceived.

Rules-Based versus Market-Based Policies

We organize our discussion in terms of two broad categories of policy: rules-based policies and market-based policies. *Rules-based policies* are those that mandate specific outcomes. They may be applied at any level of government and to either firms or individuals. Examples include Regulation XV (a regional policy applied to firms), CAFE (a national policy applied to auto manufactur-

ers), and seat belt use laws (a national policy applied to individuals). *Market-based policies* use price incentives or disincentives to promote desired outcomes. They also may be applied at any level of government and to either firms or individuals. Examples include city parking taxes (a local policy applied to individuals), vehicle registration fees (a state policy applied to vehicle owners), or off-peak discounts on transit fares. Less obvious policies are various forms of subsidies, such as transit capital subsidies or the support of the local road system by property taxes rather than by direct user charges.

The two policy approaches have contrasting advantages and disadvantages. First, because a rules-based policy has a specific standard or requirement that must be met, some believe that the outcome is more predictable than that of a market-based policy. For example, emissions standards on new vehicles require manufacturers to produce vehicles that meet the standard, and the standard can be enforced via random testing. If we were to impose instead an "emissions tax" on vehicles—for example, a charge per mile—individuals would respond by some combination of driving less and purchasing cleaner vehicles. In theory, we can set the tax to generate the same outcome as the regulation, but in practice people's response to the tax is much harder to predict.

Second, many claim that rules-based policies are more politically acceptable. Regulation appears to impose the same burden on everyone (e.g., all auto manufacturers face the same emission standards), whereas pricing generates equity concerns (e.g., an emissions tax would fall disproportionately on the poor because the poor drive the oldest and dirtiest cars). Of course, those who are regulated have different capabilities and resources for complying with the regulation, and regulation has its own inequities (e.g., why should drivers in Kansas pay the extra cost of emissions controls?). Moreover, pricing inequities can be addressed via targeted subsidies or reductions in other taxes. Rules-based policies are more politically palatable also because costs are hidden. New car buyers do not know how much they are paying for state-of-the-art emissions control systems, but an emissions tax would be an obvious "extra cost" of using a car. As the vast majority of voters are also car owners, it is not surprising that direct taxes on car use are not popular among elected officials.

In addition, industry may prefer rules-based policies, as regulatory regimes are more easily "captured." Some observers, for example, argue that the U.S. auto industry much prefers the emissions regulatory regime to an emissions tax because it protects them from competition. With an emissions tax, auto producers would have to compete to produce the cleanest car at the lowest cost, whereas today they need only meet the federal standard and pass the associated costs on to the consumer. The auto industry's incentive is therefore to minimize the burden of the regulation but not to eliminate it.

The major advantage of pricing policy is that it is more efficient: the policy objective—less congestion, less pollution, or less energy consumption—can be achieved at lower cost because those who are creating the problem have choices on how to respond. To continue with the emissions discussion, an emissions tax would impose a fee-per-mile based on the vehicle's emissions rate. Hence SUV drivers would pay more per mile than hybrid-vehicle drivers, and drivers of new cars would pay less than would drivers of old cars. As noted above, drivers would have the choice of driving fewer miles, driving a cleaner car, or some combination of both. Such a policy would generate the following: (1) old cars would be retired more quickly from the fleet; (2) demand for cleaner cars would increase; (3) industry would compete to produce cleaner cars; (4) demand for car use would decline. Hence we could reduce vehicle emissions much faster and more efficiently with an emissions tax than by continuing the current regulatory approach.

The Role of Government in Transportation Supply

It is difficult to overstate the role of government in the urban transportation system, even in an era of devolution and privatization. In addition to the role of policy in influencing the decisions of individuals and firms through prices and rules, the government is the supplier of transportation infrastructure. Federal, state, and most local streets and roads are collectively funded and maintained. Even when roads are built by private entities (as in planned communities), they are approved by public agencies and must comply with government standards. Public transit is almost entirely a public enterprise, funded by taxes and built and operated by public agencies. Bicycle and pedestrian facilities are also typically publicly funded and maintained.

The transportation systems operating in U.S. metropolitan areas are the result of decades of public decisions made by federal, state, and local agencies. Differences in transportation infrastructure investment explain some of the major differences among metro areas. Houston has one of the nation's highest rates of carpooling and one of the lowest rates of transit use, in part because of its extensive investment in high-occupancy vehicle (HOV) lanes and lack of investment in a high-quality public transit service. The extreme auto orientation of cities like San Diego and Atlanta is explained in part by freeway investments made before these cities entered their phase of rapid growth. The lack of sidewalks in the suburbs of Tallahassee, Phoenix, and Dallas is the result of local government decisions not to require them. Portland, Oregon, is conducting a major experiment in using transportation infrastructure to try to change growth patterns by making large investments in public transit and no investment in new highways. Thus transportation infrastructure decisions are a major element in the public policy toolbox.

EIGHT STRATEGIES FOR BETTER URBAN TRANSPORTATION AND MORE LIVABLE COMMUNITIES

We noted earlier that there is an extensive literature on improving urban transportation. Here we focus on the strategies that in our judgment are most likely to be both effective and feasible to implement; we provide our "Top Eight" choices for policies that can make a real difference. These policies pay attention to specific market segments and places, address the source of the problem, acknowledge financial costs and equity concerns, and at least recognize the political challenge of fostering change.

1. *Selectively* Implement Pricing Strategies to Reduce Problems of Auto Use

The fundamental problem of the U.S. urban transportation system is that private vehicle users do not pay the full costs of vehicle use. We automobile travelers do not fully pay for the health and environmental damage resulting from auto emissions, for the additional congestion we cause, or for the many other social and environmental problems discussed in Chapters 12 (Deka, this volume, on equity) and 13 (Bae, this volume, on environmental impacts); in addition, we pay only indirectly for parking and other services. This causes two problems. First, we use private vehicles more than we should from the perspective of society as a whole. Second, all other competitors to private vehicle travel are placed at a disadvantage. In economic terms, the private vehicle is underpriced.

As long as the private vehicle remains underpriced, it will be very difficult to develop viable alternatives. This is clear from recent history. We have made large investments in public transit and HOV lanes, yet transit market share remains flat and carpooling continues to decline. Why? The low price and high convenience of private vehicle travel makes it the preferred mode of travel for most people most of the time. We engage in wasteful public policy when we invest in alternative modes while doing nothing to correct for the underpricing of the private vehicle. Therefore, despite the difficulties involved, our top-priority strategy is to "level the playing field" by implementing pricing strategies.

Any effort to broadly implement a significant increase in the price of private vehicle travel in the United States is bound to fail. For example, the last increase in the federal per-gallon fuel tax took place in 1993, in the amount of 4.3 cents per gallon, raising the federal total to 18.4 cents. Efforts in the past decade to raise the tax have met with failure, despite growing concerns about the consequences of declining gas tax revenues for investment and maintenance of the transportation system. Since the fuel tax is charged per gallon, increased fuel efficiency has resulted in proportionately less tax revenue per VMT. Since the tax is not indexed to inflation, each dollar of tax revenue has less buying power as other prices increase.

The combination of rising congestion, limited public funding, and public resistance to new roads is, however, creating new opportunities. We now have many examples of pricing projects around the U.S. In addition to the Orange County toll roads discussed by Giuliano (Chapter 9, this volume), two expressways in California have tolls that vary by time of day. The purpose of variable tolls is to reduce congestion by discouraging some trips during peak hours. Variable pricing also exists on bridges in Florida and on the bridges and tunnels that cross the Hudson River to connect New York City and New Jersey. Toll roads exist in Colorado, Virginia, Arizona, Texas, and Florida, as well as California. Various types of pricing programs are being planned in Minneapolis, New York City, New Jersey, Florida, and California. London's success with its downtown pricing scheme will eventually be imitated in other heavily congested downtowns (see Box 14.1).

Parking policy can be used as another form of pricing. Shoup and Willson (1992) have pointed out that the subsidy represented by free parking at the workplace is greater than the price of fuel for the commute. That is, free parking is a more powerful incentive to drive alone than would be paying for an employee's gasoline! Shoup and others have shown that charging employees for parking, *even when combined with a "transportation allowance" of the same amount,* leads to significant reductions in drive-alone commuting. When offered a choice of free parking or the equivalent amount of cash, some employees will choose the cash and change commute mode, because the additional cash is worth more to them than the free parking. Other parking strategies include limiting the number of parking spaces permitted for new development (i.e., setting a maximum rather than a minimum number, as is now the case), parking taxes, or special reduced parking requirements in areas with high transit accessibility.

A discussion of pricing requires a discussion of equity. What about the low-income traveler, who is already spending a disproportionate share of his or her income on transportation? There are three responses. First, pricing policies generate revenue, and this revenue can be used to provide subsidies to low-income households (imagine a "lifeline transportation allowance"), to reduce other taxes that fall disproportionately on the poor, or to subsidize alternative modes. Second, pricing will increase demand for alternative modes, thereby increasing public transit revenues and enabling expanded transit services. Third, when pric-

---

BOX 14.1. London's Congestion Pricing Scheme

In February 2003 London became the first Western city to use pricing to reduce downtown traffic congestion. London has long been known for its congestion. In recent years, congestion had reached such a state (average travel speed was down to 10 miles per hour) that residents and businesses were demanding relief. Mayor Kenneth Livingstone proposed a cordon pricing scheme: charge private vehicles £5 (about $8) for entry into an 8-square-mile area of downtown between 7:00 A.M. and 6:30 P.M. on weekdays. In order to make the plan politically acceptable, many exemptions were included, notably a 90% fee reduction for those living within the cordoned area and a total exemption for taxis.

Early results of the program are promising. Traffic within the cordon area declined by about 20%, resulting in increased average speeds of about 15%. Traffic also declined in areas just outside the cordon zone, suggesting that the new fee did not simply shift traffic from one area to another. Perhaps the biggest beneficiary of the scheme is public transit. Reduced traffic on the roads greatly increased reliability on the bus system, and public transit is carrying about 20,000 more daily trips. In addition, net revenues from the fee (fee revenue less costs associated with operation the pricing system) are reserved for public transit improvements.

What made implementation of the London congestion fee possible? We suggest the following. First, there was a widespread perception that congestion was a serious problem. Second, a rather small share of people traveling into downtown London traveled by car; the public transit mode share for commuters is 75%. Very high parking charges were already a deterrent for most commuters. Third, key stakeholder groups that could have defeated the proposal were neutralized by exemptions to the fee. Fourth, the fee provided a new source of funding for badly needed public transport improvements. Finally, Mayor Livingstone was an adept entrepreneurial leader.

ing is used to reduce congestion, public transit users receive additional benefits because transit travel times are reduced, as illustrated in the current case of London (see Box 14.1).

2. *Selectively* Increase and Improve Public Transit Service

Many advocates of sustainable transportation are convinced that investment in public transit—particularly rail transit—is the key to solving transportation problems and building more livable communities. Such advocates argue that if we stop building roads, limit suburban development, build extensive rail transit networks, and promote higher density, mixed-use development around the rail network, we can transform the urban landscape and reduce auto-dependence. Unfortunately, this is a tall order that is unlikely to be fulfilled; transit cannot and should not be viewed as the "magic bullet" that will solve urban problems. Today's metropolitan areas are a mix of higher-density clusters of activities and broad expanses of moderate- to low-density development. Jobs and the activities they represent are dispersed largely in parallel with the population. Traditional central cities, the prime market for fixed-route transit, make up an ever-smaller proportion of metropolitan areas (see Muller, Chapter 3, this volume). It is simply not financially feasible to build extensive rail systems in places that not only do not have the density and clustering of activity to support it but are unlikely ever to have it, even under the most optimistic scenarios.

Rather, we should focus on making public transit as attractive as possible in places where transit is competitive. There are many possibilities, as Pucher (Chapter 8, this volume) describes. Extensive research tells us that people want high-quality service and low fares. High-quality service means frequent, reliable, and safe service to and from a wide variety of origins and destinations. Frequent service reduces wait and transfer

times, and therefore travel time. Buses with wide doors or multiple doors reduce the time required to board and off-load passengers, reducing travel time. Smart fare cards reduce fare payment transaction time. The various "bus rapid transit" services are good examples of how higher-quality service can increase transit use. Perhaps the best example is Los Angeles's Rapid Bus Wilshire line, where ridership has increased nearly 40% in the Wilshire and Ventura corridors since Rapid Bus was introduced in June 2000.

Reliable and safe public transit is fundamental. If buses or trains do not keep to their schedule or if buses are so crowded that they don't stop, the transit user has more uncertainty about how long the trip will take. The traveler must then build in some extra time to account for the possibility of the bus being late, adding an even greater time penalty to using transit. People with a choice will avoid the uncertainty and inconvenience. New information technology provides transit managers and sometimes users with the tools to ensure service reliability by real-time tracking of vehicles and wireless communications with drivers and dispatchers. Similarly, few of us are willing to put ourselves in personal danger. If passenger safety cannot be guaranteed, the transit system will be avoided. Perhaps the best example here is the New York subway system, which has enjoyed a large increase in ridership since it reduced crime and cleaned up the system (and changed the fare structure).

High-quality transit also means a transparent route network that is easy to understand and negotiate. In our judgment, part of the attractiveness of the San Francisco BART or the Washington METRO is due to the small number of routes, coded by color, and the availability of clear information about the system at every station. Transit systems should be organized so that the effort required to use the system is as small as possible. Strategies for large systems would include naming routes based on main streets or final destinations, using color coding for major routes, and providing route information at every stop and interactive maps at major stops.

As Pucher (Chapter 8, this volume) notes, we can borrow many practices from Canada and Europe to make transit more attractive. Among them are various bundled fare strategies. Transit fares should be available in the form of day or several-day passes. Transit passes can be bundled with hotel reservations and air tickets. Group fares can be negotiated, so that greatly reduced fares can be offered. The best U.S. example is the university pass program, where the university negotiates a fixed price per student per year with the transit provider, and students get unlimited-use transit passes. Typically the program is funded via student fees. Student pass programs have resulted in significantly greater transit use (Brown, Hess, & Shoup, 2003).

Pucher (Chapter 8, this volume) also noted that one of the greatest challenges facing U.S. public transit is how to provide service at a reasonable cost outside traditional central cities. Fixed-route service in low- or moderate-density areas is both ineffective (not enough origins and destinations can be served) and expensive (not enough demand). Demand-responsive services, where the traveler arranges for a specific trip, are even more costly. While technology should make it possible to develop more efficient demand-responsive services, the problem is that such services are very labor-intensive (one driver for a few passengers, plus dispatchers).

It is clear, however, that some form of publicly provided transportation must be available; many transportation-disadvantaged households live in low- or moderate-density areas. Some old solutions, rarely implemented because of institutional barriers, are worth revisiting: jitneys, shared-ride taxis, and user-side subsidy programs. Jitney service operates along a major street, picking up and dropping off passengers on demand, rather than at predesignated stops. Jitneys already operate in a few U.S. cities (notably New York, Los Angeles, and Mi-

ami) even though they are illegal. Shared-ride taxis serve two or more different trips at the same time and are legal only in a few places around the United States. Such services are opposed by transit operators and taxi companies, who argue that they take away business. If they provide better service at a competitive price, we believe that they should be able to operate legally. Nonprofit organizations that seek to offer transportation services to specific groups (e.g., the elderly) face another obstacle, namely insurance. Some form of public/private partnership, where liability can be shared but limited, would open more possibilities for special neighborhood services.

User-side subsidies have the added advantage of giving people choices. Unlike a transit pass program, user-side subsidies can be used on public transit or taxi service. We could imagine a "transportation voucher" that would be good for anything, including gas for a neighbor's or a relative's car. The major barriers to such programs are cost and perceptions. Because the transportation-disadvantaged have so much unmet demand, user-side subsidies would be widely used and hence quite costly to provide. Also, user-side subsidies are perceived as a form of welfare.

3. Make Walking and Biking Safer

Walking and biking account for a miniscule amount of urban travel in the United States. In 1995, only 5.5% of urban trips were made on foot, less than 1% were made by bike, and the nonmotorized portion of distance traveled was half of 1% (Pucher & Dijkstra, 2000). These puny figures compare with a nonmotorized mode share of more than 12% in Canada, 34% in Germany, and 45% in the Netherlands. Reliance on motorized modes need not be a hallmark of a developed economy. Davis, California, where 22% of trips are made by bicycle, proves that even in an American city people can forgo the car for a substantial proportion of their travel.

One reason for the tiny mode share of walking and biking in the United States is that using these modes can be treacherous. Each year about 6,000 pedestrians and cyclists are killed in the United States and an additional 85,000–90,000 are injured by a moving vehicle (National Safety Council, 2001). The fatality rate for pedestrians and cyclists in the United States is about three times higher per trip than is that for occupants of cars and light trucks,[3] and more than 30 times higher for pedestrians (more than 10 times higher for cyclists) per kilometer traveled. Comparable figures for pedestrians and cyclists in Germany and the Netherlands are dramatically lower—less than one-quarter of the fatality rates for the United States per kilometer traveled and less than one-thirteenth the U.S. fatality rates per trip (Pucher & Dijkstra, 2000).

How can we make walking and biking safer in the United States? Loukaitou-Sideris (2003) points out that sources of danger to pedestrians and cyclists are not limited to motorized vehicles. Human sources of danger include heavy traffic, reckless drivers, and criminals; nonhuman sources of danger include unattended dogs (who can inflict bites) and poor roadway infrastructure, which can lead to injuries from falls as well as collisions between motorized and nonmotorized modes. But making walking and biking safer will require above all changing the ways that motorized and nonmotorized traffic relate to each other. Table 14.1 provides an overview of the design and policy interventions that would increase the safety of walkers and bikers.

These proposed interventions target both auto drivers and pedestrians/cyclists. You will see that some of these suggestions match the ones we have proposed in this section (e.g., make automobile travel more expensive). In addition, however, are proposals to provide physical infrastructure for nonmotorized modes, reduce travel speeds, and improve education.

Walking and biking would be far more attractive to, and safe for, U.S. travelers if appropriate infrastructure, such as sidewalks and bikeways, were available. Where it is (as in Davis, California), people use it. Such infrastructure needs to be integrated and comprehensive; all too often in the United States a sidewalk or a bikeway suddenly ends in the middle of nowhere. It also needs to be maintained; glass on bikeways and uneven pavement on sidewalks are safety hazards. As Table 14.1 points out, appropriate infrastructure would lead to improved safety for walkers and bikers by reducing vehicular traffic as well as by separating nonmotorized from motorized modes.

Where walkers and cyclists must share road space, the speed of motorized vehicles needs to be reduced. Policies to slow traffic, known as "traffic calming," include building speed bumps, placing stop signs on every block, narrowing traffic lanes, and encouraging on-street parking. Traffic calming in residential neighborhoods and in shopping districts has enabled bikes and people to share the road more safely with autos. Slower traffic makes for a more pleasant pedestrian environment, while on-street parking provides a buffer between the traffic and the pedestrian way. Traffic-calming measures in Europe have been successful in improving pedestrian and cyclist safety (Pucher & Dijkstra, 2000).

Finally, education of cyclists, pedestrians, and automobile drivers enhances safety. The more automobile drivers can gain experience as cyclists and pedestrians (which is likely only when biking and walking become safer modes of travel), the more likely drivers should be, we believe, to drive safely when nonmotorized travelers are on the road. The suite of policies listed in the far right column of Table 14.1 (or combinations of subsets from this list) will increase pedestrian and cycling safety, which should significantly increase the share of U.S. travel that is nonmotorized, a goal of every sustainable transportation plan.

TABLE 14.1. Design and Policy Interventions to Mitigate the Effects of Traffic

Target	Objective	Physical planning and design actions	Policy actions

Drivers	• Manage/regulate vehicular traffic.	• Traffic control devices:
		  • Traffic signals
		  • Roadway signs
		  • Crosswalks
		  • Pavement markings
			• Enforce traffic regulations.
			• Impose fees and penalties on noncomplying drivers.

	• Reduce vehicular traffic volume.	• Infrastructure-accommodating alternative modes:
		  • Sidewalks, paths, trails
		  • Bike lanes
		  • Busways
		  • Carpool lanes
			• Make private automobile travel more expensive through gasoline and parking prices, license fees, and other taxes.
			• Use congestion pricing.

	• Reduce traffic speed.	• Traffic-calming devices:
		  • Vertical deflections
		  • Horizontal deflections
		  • Road narrowing
		  • Central island and medians
		  • Creation of cul-de-sacs
			• Designate slow speed areas:
			  • School safety zones
			  • Home zones

Pedestrians, cyclists	• Increase safety for pedestrians and cyclists.	• Provide physical infrastructure for pedestrians and cyclists:
		  • Sidewalks
		  • Bike paths
		  • Crosswalks
		  • Lighting
		• Maintain sidewalks and eliminate sidewalk obstructions.
			• Preferential treatment of nonmotorized modes when they intersect with motorized modes
			• Crossing aids for school children
			• Escort-to-school programs
			• Enforcement of helmets

	• Educate, inform about dangers of traffic.		• Training programs about safe driving, walking, and biking

Source: From Loukaitou-Sideris (2003). Adapted by permission of the author.

4. Take Advantage of New Technology

Vehicle emissions reductions, fuel efficiency savings, and safety improvements are primarily the result of technological advances. Through "technology forcing" regulation—imposing performance standards that cannot be met with current technologies—vehicle manufacturers have been "forced" to develop lighter weight and more aerodynamic vehicle designs, more efficient engines, and so on. Indeed, vehicle emissions have declined dramatically despite the almost doubling of VMT over the past 30 years. As of the 2004 model year, 13 different makes of "super ultra-low emissions vehicles" were offered in the retail market.[4] New regulations will soon lead to the "clean diesel" truck engine. And in California, the first law to limit vehicle carbon emissions was passed in 2002.

Technology-forcing regulation has been the policy of choice because it focuses on the vehicle manufacturer rather than the vehicle user, and therefore does not require a change in people's behavior. Efforts to change behavior (e.g., encourage more transit and ridesharing, or purchase of more fuel-efficient vehicles) have been quite un-successful, as noted earlier, because we as a society have been unwilling to change the relative price of private vehicle travel suffi-ciently to induce significant changes in travel behavior.

We can expect technology to continue to provide solutions. Although the zero emis-sions vehicle (ZEV) mandate did not result in a marketable electric vehicle, Greene (Chapter 11, this volume) argues that it contributed to the development of the hy-brid vehicle, the most likely candidate for a medium-term solution to both emissions problems and energy consumption. As this volume goes to press, both Toyota and Honda are selling hybrids in the U.S. retail marketplace. In the longer term, hydrogen fuel cells, solar energy, or some technology yet to be invented will give us a clean and renewable fuel source.

Technology also promises to greatly im-prove traveler safety. Automated braking systems, breathalyzers, sensors that notice when a driver is falling asleep, night vision enhancements, and lateral guidance systems will make all vehicles (and pedestrians) safer. Flashing cross-walks and smart signals that sense pedestrians in the roadway may fur-ther increase pedestrian safety. Information and telecommunications technologies will improve surveillance and emergency re-sponse for both private and public trans-port.

Technology has great potential for im-proving public transit through improve-ments to existing services and development of new services. Automated guidance sys-tems will smooth bus rides, provide for shorter headways, and allow vehicles to travel both on and off of exclusive guide-ways. The combination of cell phones and GPS makes it possible to provide real-time route guidance to transit users. With the right equipment onboard, travelers can use debit cards for fare payments as well as for parking and tolls. A new form of "on the fly" ridesharing could occur, as information technology makes possible confirmation that the persons involved are safe, insured drivers with no criminal record. Finally, technology should help to make public transit service "seamless," coordinating all modes and pro-viders to perform as a single system.

Technology also makes possible entirely new transportation services. Car sharing, a good example, is described in Box 14.2.

5. Remove Barriers to Flexible Use of the Transportation System

If people could use our existing transporta-tion system more flexibly, that is, if they could easily exercise more choices in terms of when and how they traveled, the effi-ciency of the system would be increased. Many constraints, including those of sched-uling, system design, and the nature of the workplace in the United States, prevent more flexible use.

Scheduling constraints include the week-day concentration of work and school start times in the morning and work end times in the afternoon, the major source of peak-period congestion. If workers could choose their work hours, for example, they might start earlier so as to have more time with family or friends at the end of the day, re-ducing peak congestion. They might take the bus instead of drive, if they could sched-ule work around the transit schedule and have some day-to-day flexibility in arrival time at work. Parents with young children have particularly restrictive schedules, com-plicated by childcare that may not be avail-able early enough or late enough to allow a change in work schedule. Moreover, few people enjoy the privilege of being able to change their work arrival and departure times from day to day, even by small incre-ments of time, especially on short notice; this constraint makes it difficult for people to deal with the inevitable vagaries of ev-eryday life, such as bus riders who encoun-ter congestion because of an accident or parents whose children become ill.

Scheduling constraints appear to be at the root of many truck-related congestion

---

BOX 14.2. Car Sharing

Car sharing is a good example of what technology makes possible. The car sharing concept emerged in Europe in the late 1980s as a highly localized, grassroots effort. Car sharing grew rapidly in Europe, with governments or auto manufacturers institutionalizing the concept and implementing it on a larger scale. Car sharing was introduced in the United States about a decade ago. Most U.S. programs are subsidized.

Car sharing is essentially the timeshare concept applied to cars: people purchase partial ownership of a vehicle and in return receive limited use of the vehicle. Just as with vacation housing timeshares, some form of central management is required to allocate time slots, collect fees, and maintain the vehicles. Technology makes these functions easier. The typical U.S. car share program requires a one-time membership fee plus a monthly fee. Car use is charged by the hour and by the mile. Zipcar is a company that offers car sharing in four U.S. cities. Their fee is $8.50–$10.50/hour plus $.18/mile after 125 free miles.[1] Members reserve cars in advance via the Internet and are notified where to pick up the car. A smart key-card is used to gain access to the assigned car. The car is returned to the site at the end of the trip and locked with the key-card. All card transactions are logged, so the central manager knows at all times the availability of each vehicle in the fleet.

The purpose of car sharing is to reduce overall car use. When someone owns a car, the additional costs of an extra trip are very small. Therefore car owners have an incentive to use their cars as much as possible. But for many people—particularly those living in large, dense cities—cars are not needed all the time. Car sharing makes it possible to use a car without paying the fixed costs of owning it. Of course, when an individual must pay a fee for each use that reflects full financial costs, she or he will choose the car only when it is worth the price. U.S. car-sharing programs emphasize environmental benefits and often use very clean, fuel-efficient cars. Several demonstration projects around the country use electric vehicles. One study of car sharing in the San Francisco Bay Area indicated that car-sharing participants traveled less by car than they would have had they not been in the car-sharing program.[2]

---

1. See *www.zipcar.com* for details.
2. See *www.stncar.com*.

---

problems. Deliveries to wholesale and retail establishments are determined by the receiver; many restrict delivery to daytime hours. Consequently, truck traffic adds to peak-hour congestion. Some restrictions are due to labor constraints (afternoon- or night-shift workers typically are paid a shift premium); others are due to local ordinances that limit delivery hours. Again, lifting these restrictions in places where local residents would not be adversely affected would smooth out the flow of truck traffic and make better use of existing capacity.

A second constraint on flexible use of our transportation system has to do with the design of the transportation system itself, in particular the difficulty, in many, if not most, places, of making trips that involve combining different travel modes. Making sure that travelers can readily combine modes (e.g., airports and public transit; bikeways and bus lines, together with bike racks on buses; sidewalks and transit; adequate parking at transit stations for cars and bikes) would enhance the flexible and therefore the efficient use of the transportation system.

Perhaps the ultimate form of flexibility is eliminating the need for travel entirely, via the use of information technology (IT), but the work process in most U.S. workplaces together with employees' enjoyment of the sociability of working with others preclude the easy and widespread substitution of IT for travel. Telecommuting and home-based work allow people to work at home at least part of the time. Telephones, video conferences, and web-based collaborative worksites make it possible for people to collaborate across the globe. Yet studies of telecommuting indicate that telecommuting is feasible for only a small share of all jobs, and almost all people who telecommute do so for only a minority of their workdays. Employees recognize the importance of "presentism" (i.e., being present in the workplace) to an upward career trajectory, and are therefore unwilling to become invisible by telecommuting extensively. Few are interested in working in isolation at home.

IT makes networked production possible, but management of these processes requires face-to-face contact. The dramatic increase in long-distance air transport over the last 20 years is consistent with these ideas. Studies of telecommuting indicate that whereas telecommuting does reduce the number of work trips, total travel may or may not be reduced as work trips may be replaced by other trips (Helling & Mokhtarian, 2001). And, of course, e-shopping allows you to shop from the comfort of your home, but a truck delivers your goods either to your home or, in Japan, to the nearest convenience store. Like Janelle (Chapter 4, this volume), we conclude that the likelihood for trip reduction via IT is limited, despite its promise for increased flexibility.

6. *Selectively* Increase and Improve Highway Capacity

It may be surprising to the reader that we suggest increasing highway capacity as an appropriate strategy for solving transportation problems. Most advocates of sustainable transportation would argue just the opposite: highways are blamed for urban sprawl, loss of open space, deterioration of urban cores, white flight, central-city fiscal problems, air pollution and other environmental damages, global warming, and most recently the obesity epidemic. While some of these claims may have merit, we must face facts. U.S. metropolitan areas are utterly dependent upon the highway system; 90% of all person trips and about two-thirds of all goods movements are made in private vehicles on roadways. The mobility provided by the car is superior to other alternatives under most circumstances. Car ownership is the key to job access, as well as to medical care, recreational opportunities, and reasonably priced groceries. Even if the mode share for transit and nonmotorized modes were to double or triple, about 80% of all person trips would still be in private vehicles. Thus, while we must reduce the damage cars generate, we must also recognize that cars are here to stay.

In metropolitan areas where population and employment are growing, it will be necessary to add highway capacity. But we can add capacity in ways that minimize both the effects of induced demand (see Johnston, Chapter 5, this volume) and environmental damage. New roads can and should be toll roads, as discussed in Strategy 1. Environmental damage can be reduced by context-sensitive design. There is no reason why every new highway must meet federal interstate standards. Some highways might be modern versions of the parkway, allowing for tighter curves and steeper grades and requiring less right-of-way. Such facilities would be more consistent with the fabric of local development and topography and perhaps less likely to be opposed by local residents.[5] Limited-access facilities have higher capacity per lane and therefore require less land to serve a given volume of traffic than an equivalent arterial street. Thus adding capacity in the form of limited-access highways is the most effective option for fast-growing areas.

Somewhat less controversial are strategies to increase the capacity of existing roadways. Traffic signalization technology is one of the big success stories of traffic engineering. It is estimated that optimization of signal timing can increase arterial capacity by 5–10% and greatly reduce delay. Reduced delay means reduced emissions, as running emissions are greatest at very low and very high speeds. Reduced delay at intersections means less exposure of pedestrians to carbon monoxide "hot spots," particulates, and other toxics. It also means greater reliability and shorter running times for public transit.

In the early 1990s, ITS (intelligent transportation systems) emerged as the "magic bullet" for increasing highway capacity. Proponents envisioned that new technology would allow development of automated highways, meaning highways where vehicles are controlled by a system manager, rather than by the driver. An automated system would require much shorter spaces between vehicles (because all vehicles are under central control and therefore "know" what all other vehicles are doing), and hence could double or triple existing highway capacity. The reality of developing such technology, not to speak of the practical problems of implementing such a system, has resulted in slow progress and far less optimistic forecasts of capacity increases. Nevertheless, progress is being made. Automated or "smart" cruise control is now available in some new cars. Smart cruise control tracks the car ahead and adjusts speed based on safe headways. Combining this with automated braking will indeed reduce headways and increase traffic flow.

7. Promote More Flexible Land Development and Redevelopment

It is clear that transportation and land use patterns are interdependent. An important part of reducing dependence on the automobile is therefore fostering land use patterns that make nonmotorized and transit travel more attractive. We must keep in mind, however, that land use change is a long-term strategy; most of the built environment of 2030 is already here. Moreover, individual and household circumstances

(e.g., gender, income, employment status, age, and number of children) have more influence over people's travel behavior than do land use patterns. These points highlight the limitations of land use strategies. Smart growth or transit-oriented development is not a magic bullet for solving urban transportation problems.

Substantial evidence shows that local controls on land use often make difficult or prevent high-density redevelopment near transit stations or moderate-density mixed-used development near suburban employment centers. Many urban planners believe that there is a significant market for in-town, transit-oriented, mixed-use development, yet local zoning restrictions inhibit such development. If we are to learn whether there is increasing demand for walkable, transit-oriented communities, it must be possible for real estate developers to build them. A first step is to remove the planning and zoning barriers that can prevent such development. A second step is to convince existing residents that such developments will benefit the entire community by providing more housing, activity, and mobility choices.

New development is equally important. Land use policies that restrict apartment houses, condominiums, apartments over stores, or mother-in-law apartments in suburban houses limit choice and often result in longer commutes or poorer employment opportunities. Policies that do not require sidewalks or bike paths as part of new development give a further advantage to car travel. At a minimum, local land use policy should be mode-neutral. Ideally, new development should be structured to allow for growth and densification over time, so that today's suburbs can gracefully mature into tomorrow's satellite cities.

If we provide choices, those who prefer to walk to the local coffee shop or bike to school will do so. As noted above, we have some examples of communities where the nonmotorized share is quite large. However, nonmotorized modes are substitutes for short, local trips, and therefore cannot be expected to reduce congestion or improve job access for the disadvantaged. Building communities with abundant walking and biking opportunities may be more about livability than solving transportation problems.

Other strategies are aimed at reducing the advantage of the private vehicle by slowing it down or prohibiting it from certain spaces. Traffic calming, discussed in Strategy 3 above, involves a range of techniques for slowing motorized traffic to enhance the residential environment and make walking and biking safer. Auto-free zones eliminate private vehicles altogether. In some places, a street is closed and transformed into a pedestrian mall; in other places, street closures may be temporary to provide a weekly farmer's market, a seasonal pedestrian shopping zone, or an ethnic street festival.

Adaptive reuse is another land use strategy. Proponents argue that instead of developing land at the fringe of urban areas, thereby contributing to sprawl, we should be making better use of the existing buildings, streets, and infrastructure at the core of U.S. cities. Deteriorated commercial zones often include warehouses or office buildings with sufficient architectural attractiveness to merit refurbishment and reuse. Factories and warehouses can be rebuilt to provide loft apartments; office buildings can be restructured for mixed use with street-level retail spaces and apartments above. These areas typically have high levels of public transit access and high accessibility to employment and services, making them ideal candidates for in-town transit-oriented developments.

Older downtown districts in cities throughout the United States are seeing adaptive reuse projects. Such redevelopment projects in former industrial inner-city areas, however, must first clean the site of toxic wastes deposited there by earlier industrial activity. A serious barrier to the redevelopment of such areas (known as "brownfields") is that the owner is responsible for cleaning the site if he or she knows that it contains contaminants. This regulation discourages owners from testing for toxic wastes and significantly reduces the attractiveness of brownfield sites to potential buyers, who will be responsible for cleanup if the land proves to be contaminated.

Finally, land use policy can and should take maximum advantage of the investments we have already made in rail transit systems. Higher-density, mixed-use development around transit stations can be encouraged by reducing parking requirements, allowing development density above current zoning limits, creating public–private partnerships to share infrastructure costs, and selectively offering tax reductions.

8. Promote Reinforcing Suites of Strategies That Are Appropriate for Local Conditions

Our last point has to do with successful outcomes. If we are to solve urban transportation problems, we must develop policies that work together synergistically rather than conflict with one another, and we must create policies that are tailored to the specific circumstances of a specific location. We have noted that many public policies work at cross-purposes, as in the provision of large subsidies to public transit in an effort to attract travelers out of their cars, while also providing free parking almost everywhere.

What would a suite of strategies be? Let's take a truly challenging and difficult problem: increasing the mobility of low-income people in older suburbs. Strategies might include (1) improved public transit service, with jitney-type service on major arterials; (2) a transportation "lifeline allowance" to qualifying households; (3) neighborhood-based car sharing; and (4) redevelopment of local commercial areas to expand availability of basic goods and services. This suite of programs would involve significant public costs. How might we finance it? A few suggestions include (1) considering using private contractors for the jitney service to hold down operating costs; (2) enticing an auto manufacturer to underwrite the carsharing program, since a successful program will result in additional car sales; (3) fostering economic redevelopment by establishing a business improvement district (which allows local tax revenues to be used for local improvements, including public safety); and (4) facilitating access to small business loans and permitting street vendors.

Our focus on place-specific solutions leads us to suggest strategies that limit the private vehicle in some places, while accommodating it in others. It makes sense to foster alternatives to the car, and the best incentive is to correct the current underpricing of the private vehicle. If we did so, all the other strategies we have discussed would be more effective. If Atlanta, Miami, and Los Angeles continue to grow at their historical rates, this growth cannot be accommodated through infill and densification; new development at the edges of these metropolises will continue. Growing population and employment will require more highways, as well as more public transit. Therefore, it also makes sense to develop better ways to build those highways—ways that consider the local context and ways that place the cost of these facilities on the people and firms who use them.

HOW DO WE FOSTER CHANGE?

Given the enormity of urban transportation problems and the array of forces that reinforce the status quo, it may seem quixotic to expect that we can achieve significant change. But change does happen. It happens as people's perceptions change, and perceptions often change as a result of the persuasion of key individuals or groups. We close this chapter with some examples of how a few major changes have been accomplished and how you might make a difference.

ISTEA and STPP

A landmark change in U.S. transportation policy took place with the passage of ISTEA in 1991 (see Chapters 1, 5, and 6, this volume). ISTEA changed the rules for regional transportation planning by linking transportation planning with air quality planning. Any region not meeting federal air quality standards is now required to submit a regional transportation plan that will help achieve those standards in a target year. In addition, ISTEA greatly increased the flexibility of transportation funding, allowing use of highway trust fund moneys for other purposes, with the discretion left to regions and states. ISTEA therefore provided the framework for fundamentally changing regional transportation planning and policy.

How did this major change come about? Part of the explanation lies with the actions of a small group of environmental activists who formed an organization, the Surface Transportation Policy Project (STPP), for the purpose of lobbying congressional leaders. Some group members had access to Senator Patrick Moynihan, then a powerful member of the Senate Environment and Public Works Committee, the key Senate committee for the federal transportation bill. STPP and Senator Moynihan were convinced that the Clean Air Act needed a means for enforcement if the United States was ever to solve urban air quality problems. In addition, Senator Moynihan was well aware that flexible funding would benefit his original home state of Massachusetts, where the Central Artery project (also known as the "Big Dig") was consuming every available transportation capital dollar. The activism of STPP and advocacy of Senator Moynihan were essential to the successful passage of ISTEA in 1991.

STPP has continued on as a major transportation lobbying organization. Housed in Washington, D.C., it issues reports, tracks legislation, and effectively communicates its agenda. Participation in advocacy organizations is one obvious way to help influence public policy.

The San Francisco Freeway Revolt

San Francisco was one of the cities where grassroots opposition to proposed urban freeways resulted in dramatic action. Plans developed in the 1940s called for a grid of freeways in San Francisco. Communities throughout the city strongly opposed the freeway plans as they became publicized in the 1950s. Citizen groups organized, held meetings, and signed petitions. In 1959 the San Francisco Board of Supervisors responded by voting to cancel seven of the 10 planned routes. Particularly hated was the Embarcadero Freeway, which was to link the Golden Gate and Bay Bridges via a route along the city's waterfront at Fisherman's Wharf. About 1 mile of the Embarcadero Freeway was built by 1959 before the full force of the revolt was felt. In the face of widespread public opposition, the project was stalled, and nearly a decade later the Board of Supervisors voted to stop construction.

For many years the unfinished freeway stood (or, more accurately, hung in midair) as a monument to the highway revolt that spread around the country and marked the beginning of the end for many of the urban segments of the interstate system. Decades later, the Embarcadero freeway suffered structural damage in the 1989 Loma Prieta earthquake. A 1985 citywide referendum had called for tearing down the freeway, and the earthquake damage provided the final impetus for its demolition. Today the area is redeveloped and home to a new sports stadium and mixed-use development.[6]

The Embarcadero Freeway revolt demonstrates the potential power of ordinary citizens. If, instead of opposing the freeway, the residents of San Francisco had accepted it as inevitable (given its strong support by the federal and state governments), the Embarcadero would probably have been completed and be standing to this day. Other examples of successful citizen-led opposition to proposed transportation projects are described in *At Road's End* (Carlson, Wormser, & Ulberg, 1995).

Individuals Who Made a Difference

Sometimes individuals provide the leadership for policy change. Here we briefly describe two who were instrumental in effecting improved automobile safety via federal regulation of the automobile industry. The first is Ralph Nader, who, in 1965, when he published his now-classic book, *Unsafe at Any Speed* (Nader, 1965), was a little-known consumer activist. In this book Nader took on the automobile industry by pointing out, inter alia, that the Chevrolet Corvair made by General Motors (GM) had a suspension design flaw that made it highly likely to roll over. When it was revealed that GM had hired a detective to discredit Nader, the resulting scandal led to Congressional hearings on automobile safety, the passage of two new auto safety laws, and the establishment in 1970 of a federal agency to oversee highway safety (this agency is now called the National Highway Traffic Safety Administration—NHTSA).[7]

Our second example, also related to auto safety, is that of exploding fuel tanks in the Ford Pinto, exposed in 1977 by Mark Dowie in an article ("Pinto Madness") in *Mother Jones* magazine.[8] Dowie described the Ford Pinto as a "firetrap" because the fuel tank, located in the rear of the car, would burst into flames when hit from behind at relatively low impact speeds (c. 25–30 miles per hour); he claimed that some 500 people had died from the fires resulting from such collisions. Dowie's article prompted a NHTSA investigation of the Pinto, and, under pressure from Nader's organization (Public Citizen, founded in 1971) and others, the Pinto was eventually recalled. In Indiana, Ford was charged with criminal homicide. The high-profile trial ended in an acquittal of Ford in 1980.

Thanks to these two individuals, Nader and Dowie, the United States now regulates automobile manufacturers to enforce auto safety standards.

Local Activists Seek to Limit Greenhouse Gas Emissions

One contemporary example of individuals promoting change in their local communities is the Cities for Climate Protection (CCP) campaign.[9] In 1991 the International Council for Local Environmental Initiatives (ICLEI; see *www.iclei.org*) launched the CCP campaign, in which the citizens and leaders of cities around the globe work to reduce their city's energy consumption in order to limit greenhouse gas emissions, which have been implicated in global climate change. The campaign now involves more than 500 local governments worldwide, including more than 140 cities and counties in the United States.

An international body of leading scientists, the Intergovernmental Panel on Climate Change (IPCC), has determined that human activity, particularly via the burning of fossil fuels and the resulting greenhouse gas emissions, is contributing to global climate change (International Panel on Climate Change, 1990). As Greene has noted (Chapter 10, this volume), the United States contributes substantially to greenhouse gas emissions worldwide, and nearly one-third of those emissions come from the transportation sector. The U.S. federal government has refused to join international efforts that would require the United States to reduce greenhouse gas emissions by specified amounts by a specific year in the future. In the absence of any federal policy to address global climate change, the CCP campaign, which involves a network of local groups to effect change at the global level, is significant.

Participating cities first pass a resolution that commits the city to an energy-reduction and waste-reduction plan. The city then conducts an emissions inventory and designs a local action plan to reduce carbon dioxide (CO_2) emissions. The target set by most participating U.S. cities is to reduce their 1990 levels of CO_2 emissions by 20% by the year 2010. The action plans are tailored to local conditions, but in general they emphasize specific ways to improve energy efficiency (e.g., via installing energy-efficient light bulbs and insulated windows; reducing travel in single-occupant vehicles) and the monetary savings that energy efficiency can yield for the city. ICLEI maintains a regional support network for participating cities throughout the United States.

The overall message of the CCP campaign is that global change starts in local places and that, moreover, global change (in this case, in reducing greenhouse gas emissions) *requires* coordinated change at the local level. The local focus of the CCP campaign fits well with the increasingly local focus of transportation planning and decision making. As more and more local governments in the United States join in this campaign, it will be interesting to see if this form of local activism will compel the federal government to adopt measures aimed at reducing greenhouse gas emissions. A federal agreement to specific target emissions reductions would add another example to those we have already described of individuals and activist groups bringing about significant urban transportation-related change at the federal level.

CONCLUSION

From the outset, the authors of this book have emphasized that transportation is essential to everyday life and that transportation investment and policy decisions have shaped the structure of today's urban areas. Ongoing processes, particularly globalization and information technology, are changing the supply of and demand for urban transportation. Because transportation planning is at the heart of planning for more livable cities, there is considerable interest among citizens and policymakers to find ways to make urban transportation more sustainable.

The transportation planning process is a direct descendent of highway planning in that the analytical tools of transportation planning were initially developed to guide regional highway investment. Fortunately, planning tools have improved so that they now enable consideration of land use and induced demand impacts, the incorporation of all modes, and representation and analysis via GIS. Although these improvements enhance the potential of planning, transportation investment decisions remain inherently political. Who stands to gain and who stands to lose from a proposed transportation system change is often an intensely geographic question.

Throughout this book, authors have stressed the many policy problems that urban transportation poses, including energy consumption, air pollution and other environmental impacts, and inequities in access to transportation. Authors have also discussed specific aspects of transportation, such as land use impacts and transportation finance, that should help you to understand the geography of urban transportation and think through potential solutions to policy problems. Although the United States has made considerable progress in some of these areas, such as environmental quality, we have not made much progress in other areas, such as energy consumption and transportation inequalities. Much remains to be done if we are to achieve the goals of sustainable transportation systems and livable cities.

In this chapter, we have explored a variety of approaches to solving urban transportation problems, specifically by managing the auto. We have outlined eight proposals for improving urban transportation and for creating more livable, and more sustainable, cities and communities. Clearly, urban transportation presents society, and especially U.S. society, with many difficult challenges. Our hope is that this book has provided you with the background and the analytical tools needed to take up these challenges and to help create badly needed change in our urban transportation systems. This change is likely to come slowly and incrementally and to be initiated by citizen activists. As the examples in this chapter have shown, you can make a difference!

NOTES

1. The exception is war-related legislation, as the recent establishment of the Department of Homeland Security demonstrates.
2. See Giuliano, Hwang, and Wachs (1993).
3. The rate is 29 per 100 million trips for pedestrians, 26 for cyclists, and only 9 for car occupants (Pucher & Dijkstra, 2000).
4. See *http://www.arb.ca.gov/msprog/ccbg/ccbg.htm*.
5. Citizen opposition to a proposed widening and straightening of Connecticut's Merritt Parkway in the early 1990s exemplifies the value that many people place on such non-Interstate Highway facilities (see Carlson, Wormser, & Ulberg, 1995, pp. 32–39).
6. Drawn from *www.cahighways.org, www.bikesummer. org, www.ocf.berkeley.edu*.
7. See *http://www.pbs.org/wgbh/pages/frontline/ shows/rollover/unsafe/cron.html*.
8. The article is available at *http://www.motherjones. com/news/feature/1977/09/dowie.html*.
9. See *www.iclei.org/co2/*.

REFERENCES

Brown, J., Hess, D., & Shoup, D. (2003). Fare-free public transit at universities: An evaluation. *Journal of Planning Education and Research, 23,* 69–82.

Carlson, D., Wormser, L., & Ulberg, C. (1995). *At road's end: Transportation and land use choices for communities.* Washington, DC: Island Press.

Giuliano, G., Hwang, K., & Wachs, M. (1993). Employee trip reduction in Southern California: First year results. *Transportation Research A, 27A*(2), 125–137.

Helling, A., & Mohktarian, P. (2001). Worker telecommunication and mobility in transition: Consequences for planning. *Journal of Planning Literature, 15*(4), 511–525.

International Panel for Climate Change. (1990). *Climate change: The IPCC scientific assessment.* Cambridge, UK: Cambridge University Press.

Kay, J. H. (1997). *Asphalt nation*. New York: Random House.

Loukaitou-Sideris, A. (2003, November). *Transportation, land use, and physical activity: Safety and security considerations.* Paper presented at workshop of Committee on Physical Activity, Transportation, and Land Use, Washington, DC.

Nader, R. (1965). *Unsafe at any speed: The designed-in dangers of the American automobile.* New York: Grossman.

Nadis, S., & MacKenzie, J. J. (1993). *Car trouble.* Boston: Beacon Press.

National Safety Council. (2001). *Fact sheet library.* Available online at http:www.nsc.org/library/pedstrns.htm

Nivola, P., & Crandall, R. (1995). *The extra mile.* Washington, DC: Brookings Institution Press.

Pucher, J., & Dijkstra, L. (2000). Making walking and cycling safer: Lessons from Europe. *Transportation Quarterly, 54*(3), 25–50.

Shoup, D. (1997). Evaluating the effect of cashing out employer-paid parking: Eight case studies. *Transport Policy, 4*(4), 201–216.

Shoup, D., & Willson, R. (1992). Employer-paid parking: The problem and proposed solutions. *Transportation Quarterly, 46*(2), 169–192.

Transportation Research Board. (2001). *Making transit work: Insight from Western Europe, Canada, and the United States* (Special Report No. 257). Washington, DC: Transportation Research Board, National Research Council.

Whitelegg, J. (1993). *Transport for a sustainable future: The case for Europe.* London: Bellhaven Press.

Index

About the Editors

Susan Hanson is the Jan and Larry Landry University Professor and a Professor and Director of the School of Geography at Clark University. She is an urban geographer with teaching and research interests in gender and economy, the geography of everyday life, and sustainability. Dr. Hanson has been editor of three geography journals—*Economic Geography*, the *Annals of the Association of American Geographers*, and *The Professional Geographer*—and serves on the editorial boards of several other journals. She is currently on the Executive Committee of the Transportation Research Board. Dr. Hanson also is a member of the National Academy of Sciences, a Fellow of the American Association for the Advancement of Science, and the American Academy of Arts and Sciences, and past president of the Association of American Geographers.

Genevieve Giuliano is Professor in the School of Policy, Planning, and Development at the University of Southern California and Director of the METRANS Transportation Center. Her research interests center on transportation policy and on relationships between land use and transportation. Dr. Giuliano is a former editor of *Urban Studies* and serves on the editorial boards of several journals. She is a former faculty fellow of the Lincoln Institute of Land Policy and a former member of the Executive Board of the Association of Collegiate Schools of Planning. Dr. Giuliano also is a member and past chair of the Executive Committee of the Transportation Research Board and is a National Associate of the National Academy of Sciences.

Contributors

Chang-Hee Christine Bae, PhD, Department of Urban Design and Planning, University of Washington, Seattle, Washington

Devajyoti Deka, PhD, New Jersey Transportation Planning Authority, Newark, New Jersey

David L. Greene, PhD, Oak Ridge National Laboratory, Oak Ridge, Tennessee

Genevieve Giuliano, PhD, School of Policy, Planning, and Development, University of Southern California, Los Angeles, California

Susan Hanson, PhD, School of Geography, Clark University, Worcester, Massachusetts

Donald G. Janelle, PhD, Department of Geography, University of California, Santa Barbara, California

Robert A. Johnston, MS, Department of Environmental Science and Policy, University of California, Davis, California

Thomas R. Leinbach, PhD, Department of Geography, University of Kentucky, Lexington, Kentucky

Peter O. Muller, PhD, Department of Geography and Regional Studies, University of Miami, Coral Gables, Florida

Timothy L. Nyerges, PhD, Department of Geography, University of Washington, Seattle, Washington

John Pucher, PhD, Department of Urban Planning, Rutgers University, New Brunswick, New Jersey

Brian D. Taylor, PhD, Department of Urban Planning, and Director, Institute of Transportation Studies, University of California, Los Angeles, California

Martin Wachs, PhD, FAICP, Institute of Transportation Studies, University of California, Berkeley, California